HIGH-RESOLUTION IMAGING

HIGH-RESOLUTION IMAGING

DETECTORS AND APPLICATIONS

Swapan K. Saha

PAN STANFORD PUBLISHING

Published by

Pan Stanford Publishing Pte. Ltd.
Penthouse Level, Suntec Tower 3
8 Temasek Boulevard
Singapore 038988

Email: editorial@panstanford.com
Web: www.panstanford.com

British Library Cataloguing-in-Publication Data
A catalogue record for this book is available from the British Library.

High-Resolution Imaging: Detectors and Applications

ISBN 978-981-4613-27-9 (Hardcover)
ISBN 978-981-4613-28-6 (eBook)

Printed in the USA

To my siblings
and
in memory of my parents

Contents

Preface

Existence of a quantum limit in light detection has led to a quest, on through the last century (and still going on), for the perfect detector that is asymptotically feasible. In 1901, Planck postulated that radiative transfers would occur by discrete packets of energy named "quanta," following which in 1905, Einstein explained the photoelectric effect. The existence of photons means that for a given collecting area, there exists a physical limit on the minimum light intensity for any observed phenomenon. The perfect detector would be the one that is capable of detecting individual photo-event in the image plane.

The technology of developing detectors has evolved rather fast. The main application of such detectors is in the field of astronomy. Optical telescopes collect the radiation from the faint stellar objects, and the photons received from these objects are feeble. The situation becomes worse in the presence of atmospheric turbulence, particularly in the case of speckle imaging (Labeyrie, 1970) where post-detection data-processing algorithms are required to decipher diffraction-limited spatial Fourier spectrum and image features of such objects. In the early 1970s, the problem was the data processing; computers were not powerful enough for real-time processing and video recorders were expensive. Another application for such detectors is the signal sensing, especially wavefront sensing for adaptive optics (AO) and fringe tracking, which is important for long-baseline optical interferometry.

Research in solid-state electronic imaging detector commenced at the end of the 1960s. Boyle and Smith (1970) introduced the concept of charge-coupled device (CCD). Soon CCDs replaced TV camera tubes for the photon-counting cameras using image intensifiers. However, the possibility to have a fully solid-state photon-counting camera is quite recent. Modern detectors with (i) high sensitivity, (ii) high selectivity and detectivity, (iii) fast response, (iv) high efficiency, (v) low noise, and (vi) high read-out speed play a key role in astronomical applications. A photon-counting system allows the accurate photon centroiding and provides the dynamic range needed for measurements of source characteristics. At present, the large-format CCD and complementary metal oxide semiconductor array mosaics, electron-multiplication CCDs, electron avalanche photodiode arrays, quantum-well infrared photon detectors, etc., are available. However, the requirements of artifact-free photon shot-noise limited images include higher sensitivity and quantum efficiency, reduced noise that includes dark current, read-out and amplifier noise, smaller point spread functions, and higher spectral bandwidth.

The interest in using the image information in the form of coordinate sets came from the high-angular-resolution techniques (Saha, 1999, 2002, 2007, 2010; Labeyrie et al., 2006, and references therein) for astronomy, which require to take images with exposure times shorter than the atmospheric coherence time (see Section 1.5.3), which is, in general, less than 20 ms depending on the high altitude wind. At this exposure time, the signal-to-noise ratio in each frame is so low that image intensification is usually required. In view of the reasons stated above, I felt the necessity to write a monograph on the modern detector, though a chapter is dedicated on this topic in my first book entitled, *Diffraction-Limited Imaging with Large and Moderate Telescopes*, 2007, World-Scientific. At the fall of 2009, Dr. Stanford Chong, Director and Publisher, Pan Stanford Publishing Co., invited me, for which I am indebted to, for expanding the talk on this topic, which I delivered at AOMD-2008, India, into a book, which I duly complied with to accept the offer. Studded with tens of figures and footnotes, this book aims to address modern detecting systems, technologies and design, evaluation and calibration, control electronics, scientific applications, and results.

The first chapter deals with the fundamentals of important aspects, such as electromagnetic radiations, optical wavelengths in particular, emanating from the celestial sources and their distortions due to the optical elements and Earth's atmosphere while passing through to the detector. Also, discussed in brief are semiconductor physics and laser and its related phenomena in the second and third chapters, respectively. Chapters 4 through 6 address the photon detection process, photodetectors, and charge transfer devices. The photon-counting devices in the visible wavelength are discussed in the penultimate Chapter 7. The last but not the least, Chapter 8 presents the radiation detectors in infrared wavelengths. My recent interaction with a noted quantum physicist on multi-photon extension of intensity interferometry (Hanbury Brown, 1974) gives me hope that the book would kindle spirits to pursue the fields yet to benefit from the knowledge gained in these chapters. Barring the radio detection technique and intensity interferometry, which were elucidated at length in my second book entitled, *Aperture Synthesis: Methods and Applications to Optical Astronomy*, 2010, Springer, I have also highlighted the detector systems for high energies in brief in Chapters 5 and 7, since their use in astronomy as well as in medical sciences is noteworthy.

I am indebted to my PhD supervisor, late Prof. A. K. Sen, who initiated me into experimental science at the Institute of Radio Physics and Electronics, University of Calcutta. I express my gratitude to A. Labeyrie, S. K. Sarkar, A. Satya Narayanan, Kavita S. Rao, B. P. Pal, R. Ghosh, S. P. Bagare, S. S. Negi, R. Ramesh, M. S. Sundararajan, R. Srinivasan, and U. S. Kamath for assistance as readers of the draft chapters. Thanks are also due to G. P. Weigelt, R. Osterbart, I. S. McLead, A. Popowicz, S. Pal, Ya-Lin Wu, F. Malbet, P. Nisenson (late), V. Chinnappan, N. K. Rao, L. Close, S. Morel, A. Blazit, E. Pedretti, D. Mourard, D. Ives, K. B. Jinesh, N. Bezawada, A. R. Rao, J. Chatterjee, S. Ghosh, S. Srivastav, G. C. Anupama, D. Ojha, and S. Bhadra for providing the images, spectra, figures, etc., and granting permission for their reproduction. The services rendered by Baba A. Varghese, A. Surya, R. K. Sharma, S. P. Tewari, V. Valsan, R. Mondal, P. Anbazhagan, and S. K. Dhara are greatly acknowledged.

References

Boyle W. S. Smith G. E., 1970, *Bell System Tech. J.*, **49**, 587.

Hanbury Brown R., 1974, *The Intensity Interferometry: Its Applications to Astronomy*, Taylor & Francis, London.

Labeyrie A., 1970, *Astron. Astrophys.*, **6**, 85.

Labeyrie A., Lipson S. G., Nisenson P., 2006, *An Introduction to Optical Stellar Interferometry*, Cambridge University Press, UK.

Saha S. K., 1999, *Bull. Astron. Soc. Ind.*, **27**, 443.

Saha S. K., 2002, *Rev. Mod. Phys.*, **74**, 551.

Saha S. K., 2007, *Diffraction-limited Imaging with Large and Moderate Telescopes*, World-Scientific, New Jersey.

Saha S. K., 2010, *Aperture Synthesis: Methods and Applications to Optical Astronomy*, Springer, New York.

Swapan K. Saha
November 2014

Principal Symbols

A_{21}	Einstein coefficient for spontaneous emission
$\mathtt{A}_{\mathrm{d}}$	Effective area of the detector
$\mathbf{B}(\mathbf{r}, t)$	Time dependent magnetic field
B_{21}	Einstein coefficient for stimulated emission
$B_{\nu}(T)$	Spectral radiancy
$\mathcal{B}_{\mathrm{n}}(\mathbf{r})$	Covariance function
C	Capacitance
C	Coulomb
$\mathcal{C}_{\mathrm{n}}^{2}$	Refractive index structure constant
D	Diameter of the aperture
D^{*}	Specific detectivity
D_{e}	Diffusion coefficient of electrons
D_{h}	Diffusion coefficient of holes
$\mathbf{D}$	Electric displacement vector
$\mathcal{D}_{\mathrm{n}}(\mathbf{r})$	Refractive index structure function
E_{C}	Conduction band energy
E_{D}	Donor level energy
$\mathcal{E}_{\mathrm{e}}$	Irradiance
$\mathcal{E}_{\mathrm{e}}(\lambda)$	Spectral irradiance
E_{F}	Fermi level energy
E_{g}	Bandgap energy
E_{i}	Intrinsic Fermi level energy
E_{k}	Kinetic energy
$\mathcal{E}_{\nu}$	Illuminance
$erfc$	Error function
$\mathbf{E}(\mathbf{r}, t)$	Time dependent electric field
E_{V}	Valence band energy

F	Flux
f	Focal length
f_{e}	Effective focal length
f_{G}	Greenwood frequency
F_{max}	Maximum frame rate
$F_{\#}$	Aperture ratio
G	Gain
$\mathcal{G}$	Heat conductance
$\mathbf{H}$	Magnetic field
$\mathcal{H}$	Heat capacity
$\hat{H}$	Hamiltonian operator
$\hbar(= h/2\pi)$	Reduced Planck constant
I	Intensity of light
i	Current
i_{B}	Base current
i_{C}	Collector current
i_{D}	Dark current
i_{d}	Detector current
I_{DC}	Direct current
i_{e}	Electron current
i_{E}	Emitter current
$\mathcal{I}_{\mathrm{e}}$	Radiant intensity
i_{FN}	Flicker noise current
i_{h}	Hole current
$\mathcal{I}_{\lambda}$	Spectral radiant intensity
$\mathcal{I}_{\nu}$	Luminous intensity
i_{n}	Dark current noise
i_{ph}	Photocurrent
i_{Q}	Shot-noise or quantum noise
i_{s}	Signal current
i_{sc}	Short-circuit signal current
$i(t)$	instantaneous photocurrent
i_0	Reverse saturated current
$\widehat{I}(\mathbf{u})$	Image spectrum
$\widehat{I}(\mathbf{x})$	Intensity distribution at the image plane
$\mathbf{J}$	Electric current density
$\mathbf{J}_{\mathrm{d}}$	Drift current

$\mathbf{J}_{\text{diff}}$	Diffusion current
j	= 1, 2, 3
L_e	Diffusion length of electrons
L_h	Diffusion length of holes
L_0	Outer scale length
l	Characteristic size of viscous fluid
l_c	Coherence length
l_0	Inner scale length
$\mathcal{L}_e$	Radiance
$\mathcal{L}_\nu$	Luminance (brightness)
$\mathcal{L}_\star$	Stellar luminosity
M	Multiplication factor
$\mathtt{M}$	Magnification
$\mathcal{M}_e$	Radiant emittance
$\mathcal{M}_e$	Spectral radiant emittance
$\mathcal{M}_\nu$	Luminous emittance
$\mathcal{M}_q$	Photon emittance
$\mathcal{M}_q(\lambda)$	Spectral photon emittance
m_v	Apparent visual magnitude
M_v	Absolute visual magnitude
m^*	Effective mass
$M_\star$	Stellar mass
n	Refractive index
N	Integer value
N_e	Magnitude of free electron concentration
N_h	Magnitude of hole concentration
N_{pe}	Number of photoelectrons
N_{ph}	Number of photons
n_{ro}	Read-out noise
$n(\mathbf{r}, t)$	Refractive index of the atmosphere
$O(\mathbf{x})$	Object illumination
$\widehat{O}(\mathbf{u})$	Object spectrum
P	Pressure
$\mathcal{P}$	Probability distribution
$\mathbf{p}$	Momentum
P_o	Output optical power
P_0	Incident optical power

q	Electron charge
$\mathcal{Q}_e$	Radiant energy
$\mathcal{Q}_\lambda$	Spectral radiant energy
$\mathcal{Q}_\nu$	Luminous energy
$\mathbf{r}(= x, y, z)$	Position vector of a point in space
R	Resistance
$\mathcal{R}$	Responsivity
$\mathcal{R}(\lambda)$	Spectral responsivity
$\mathcal{R}_\nu$	Luminous flux responsivity
R_e	Reynolds number
R_f	Entrance face reflectivity (Fresnel reflectivity)
R_H	Rydberg constant
r_0	Fried's parameter
$R_\star$	Stellar radius
$\hat{\mathbf{s}}$	Unit vector
$S(\mathbf{x})$	Point spread function
$\mathcal{S}_r$	Strehl's ratio
$\langle \widehat{S}(\mathbf{u}) \rangle$	Transfer function for long-exposure image
$\widehat{S}(\mathbf{u})$	Optical transfer function
$\lvert \widehat{S}(\mathbf{u}) \rvert^2$	Modulus transfer function
sr	Steradian
$s(\lambda)$	Spectral response of the detector
t	Time
$t(x, y)$	Transmittance
T	Temperature
$\mathbf{u} = (u, v)$	Spatial frequency vector
V	Electrostatic potential
$\mathcal{V}$	Visibility
v_a	Average velocity of a viscous fluid
V_{br}	Breakdown voltage
v_d	Drift velocity
v_g	Group velocity
V_{oc}	Open-circuit voltage
v_p	Phase velocity
w	Width of depletion region
$\mathcal{W}$	Radiation power
$\mathbf{x}(= x, y)$	Two-dimensional (2-D) position vector
$Z_j(\rho, \theta)$	Zernike polynomial of order j

Greek symbols

α	Extinction coefficient
α_e	Electron-ionization coefficient
α_h	Hole-ionization coefficient
β	Ratio of electron and hole mobilities
γ	Zenith angle
$\Gamma(\tau)$	FT of auto-correlation function
δ	Phase difference
$\delta(x)$	Dirac delta function
Δ	Relative core-cladding index difference
$\Delta\nu$	Spectral width
Δf	Electrical bandwidth
Δt	Integration time
ϵ_0	Permittivity in vacuum
ε	Energy dissipation
η_d	Quantum efficiency of the detector
Θ	Seebeck coefficient
θ	Angular diameter
θ_0	Isoplanatic angle
(θ, ϕ)	Polar coordinates
κ	Wave number
$\boldsymbol{\kappa}$	Wave vector
λ	Wavelength
λ_0	Wavelength in vacuum
μ_0	Permeability in vacuum
μ_e	Electron mobility
μ_h	Hole mobility
ν	Frequency
$\boldsymbol{\xi}(\xi, \eta)$	2-D position vector
$\langle\sigma\rangle$	Standard deviation
$\langle\sigma\rangle^2$	Variance
σ_I	Scintillation index
τ_0	Atmospheric coherence time
τ_c	Coherence time
τ_d	Response time
τ_e	Lifetime of electrons
τ_g	Group delay

τ_{h}	Lifetime of holes
τ_{r}	Total rise time
τ_{RC}	RC time-constant
τ_{w}	Delay time due to waveguide dispersion
ϕ_0	Work function
ϕ_{m}	Work function of metal
ϕ_{s}	Work function of semiconductor
Φ_{e}	Radiant power/flux
Φ_{λ}	Spectral radiant flux
Φ_{ν}	Luminous power/flux
$\Phi_{\mathrm{n}}(\boldsymbol{\kappa})$	Power spectral density
χ	Electron affinity
Ψ	Time-dependent wave-function
Psi_{B}	Potential difference between E_{F} and E_{i}
Ψ_{n}	Eigen functions
Ω	Solid angle
ω	Angular frequency
$*$	Complex operator
$\star$	Convolution operator
$\otimes$	Correlation
$\langle\ \rangle$	Ensemble average
$\widehat{\ }$	Fourier transform operator
$\boldsymbol{\nabla}$	Linear vector differential operator
∇^2	Laplacian operator

Some numerical values of physical and astronomical constants

c	Speed of light in free space	3×10^8 m.s^{-1}
c_1	First radiation constant	$2\pi hc^2 = 3.7418 \times 10^4$ Wμm^4.cm^{-2}
c_2	Second radiation constant	$hc/k_B = 1.4388 \times 10^4$ μm K
eV	Electron volt	1.60×10^{-19} J
h	Planck's constant	6.626096×10^{-34} J.s
Jy	Jansky	10^{-26} W m^{-2} Hz^{-1}
k_B	Boltzmann's constant	1.380662×10^{-23} J.K^{-1}
K_m	Luminous efficacy (photopic vision)	683 lm.W^{-1}
$\mathcal{L}_\odot$	Solar luminosity	3.839×10^{26} W
lx	Lux	lm.m^{-2}
$M_\odot$	Solar mass	1.9889×10^{30} kg
q	Elementary charge	1.6×10^{-19} C
R_L	Load resistance	
$R_\odot$	Solar radius	6.96×10^8 m
$T_\odot$	Solar effective temperature	5780 K
ϵ_0	Permittivity constant	8.8541×10^{-12} F.m^{-1}
μ_0	Permeability constant	1.26×10^{-6} H.m^{-1}
σ	Stefan–Boltzmann's constant	5.67×10^{-8} W m^{-2} K^{-4}

List of Acronyms

AC	Alternating current
ACT	Atmospheric Cerenkov Telescope
A/D	Analog-to-digital
AGN	Active galactic nuclei
AMBER	Astronomical Multiple BEam Recombiner
AO	Adaptive optics
APD	Avalanche photodiode
ASM	Adaptive secondary mirror
ASTROSAT	ASTROnomy SATellite
AT	Auxiliary Telescope
AU	Astronomical unit
BID	Blind iterative deconvolution
BJT	Bipolar junction transistor
BLIP	Background-limited performance
CCD	Charge coupled device
cd	Candela
CDS	Correlated double sampling
CHARA	Center for High Angular Resolution Astronomy
CIC	Clock-induced charge
CMOS	Complementary metal oxide semiconductor
CP40	Comptage de photon 40 mm
CPNG	Comptage de photon nouvelle gènèration
CPU	Central processing unit
COAST	Cambridge Optical Aperture Synthesis Telescope
CRT	Cathode ray tube
CS	Curvature sensor
CTE	Charge transfer efficiency
CTF	Contrast transfer function
CZT	Cadmium zinc telluride

CZTI	Cadmium zinc telluride imager
dB	Decibel
DC	Direct current
DM	Deformable mirror
DSP	Digital signal processor
EELT	European Extremely Large Telescope
EMCCD	Electron multiplying CCD
EMF	Electromotive force
ESA	European Space Agency
ESO	European Southern Observatory
FET	Field-effect transistor
FLUOR	Fiber-Linked Unit for Optical Recombination
FOV	Field-of-view
FPA	Focal plane array
FPGA	Field programmable gate array
FPS	Frames per second
FT	Fourier transform
FUSE	Far-Ultraviolet Spectroscopic Explorer
FWHM	Full width at half maximum
GALEX	Galaxy Evolution Explorer
G-APD	Geiger-mode APD
GI2T	Grand Interféromètre à deux Télescopes
GR	Generation-recombination
GRAVITY	General relativity analysis via VLT interferometry
HAGAR	High Altitude Gamma ray telescope ARray
HR	Hertzsprung-Russell
HAWAII	HgCdTe Astronomical Wide Area Infrared Imager
HCT	Himalayan Chandra Telescope
HST	Hubble Space Telescope
Hz	Hertz
IAO	Indian Astronomical Observatory
IC	Integrated circuit
ICCD	Intensified CCD
IF	Intermediate frequency
IO	Integrated optics
IONIC	Integrated optics near-IR combiner
IOTA	Infrared Optical Telescope Array

IPCS	Image Photon-Counting System
IR	Infrared
IRAF	Image Reduction and Analysis Facility
IRAS	InfraRed Astronomical Satellite
IRTS	InfraRed Telescope in Space
ISI	Infrared Spatial Interferometer
ISM	Interstellar medium
ISO	Infrared Space Observatory
I2T	Interféromètre à deux Télescopes
ITF	Intensity transfer function
kV	Kilovolt
Laser	Light amplification by stimulated emission of radiation
LBOI	Long-baseline optical interferometry
LBT	Large Binocular Telescope
LBV	Luminous blue variable
LCD	Liquid crystal display
LED	Light-emitting diode
LGS	Laser guide star
LHS	Left hand side
LIGO	Laser Interferometer Gravitational-Wave Observatory
LSF	Line spread function
LO	Local oscillator
L3CCD	Low light level CCD
LWIR	Long-wave infrared
ly	Light year
mA	Milli-ampere
MAMA	Multi Anode Micro-channel Array
mas	Milliarcseconds
Maser	Microwave amplification by stimulated emission of radiation
MBE	Molecular beam epitexy
MCP	Micro-channel plate
MCT	Mercury cadmium telluride
MHz	Megahertz
MIRC	Michigan Infra-Red Combiner
MMDM	Micro-machined DM
MMF	Magnetomotive force
MO	Microscope objective

MOS	Metal-oxide semiconductor
MOSFET	Metal-oxide semiconductor FET
MOSFIRE	Multi-Object Spectrometer for InfraRed Exploration
MMT	Multi Mirror Telescope
MRI	Magnetic resonance imaging
MSGIS	Mammary Gland Specific Gamma Imaging System
MTF	Modulation transfer function
MWIR	Mid-wave infrared
NA	Numerical aperture
NASA	National Aeronautics and Space Administration
NEA	Negative electron affinity
NEAT	Nearby Earth Astrometric Telescope
NEΔT	Minimum detectable temperature difference
NEP	Noise equivalent power
NETD	Noise equivalent temperature difference
NICMOS	Near Infrared Camera and Multi-Object Spectrometer
NIRSPEC	Near Infrared echelle Spectrometer
nm	Nanometer
OCT	Optical coherence tomography
Op-Amp	Operational amplifier
OPD	Optical path difference
OTF	Optical transfer function
PAPA	Precision Analog Photon Address
PAVO	Precision Astronomical Visible Observations
pc	Parsec
PC	Photoconductive
PDS	Photometric data system
PE	Photoemissive
PEM	Photoelectromagnetic
PHD	Pulse height distribution
pixel	PICture ELement
PMT	Photomultiplier tube
POSS	Palomar Observatory Sky Survey
PSD	Position-sensing device
PSF	Point spread function
PSPMT	Position sensitive photomultiplier technology
PTF	Phase transfer function
PV	Photovoltaic

PZT	Piezo-electric transducer
QE	Quantum efficiency
QDIP	Quantum dot IR photodetector
QWIP	Quantum well IR photodetector
RHS	Right hand side
RMS	Root-mean-square
ROIC	Read-out integrated circuit
SAO	Special Astrophysical Observatory
SH	Shack–Hartmann
SL	Super-lattice
SLS	Strained layer super-lattice
S/N	Signal-to-noise
sr	Steradian
STJ	Superconducting tunnel junction
SUSI	Sydney University Stellar Interferometer
SWIR	Short-wave infrared
SXT	Soft X-ray Telescope
TC	Triple-correlation
TDI	Time-delay and integration
TMT	Thirty Meter Telescope
TSM	Tomographic speckle imaging
TV	Television
UT	Unit Telescope
UV	Ultra-violet
UVIT	Ultra-Violet Imaging Telescope
VBO	Vainu Bappu Observatory
VBT	Vainu Bappu Telescope
VEGA	Visible spEctroGraph and polArimeter
VINCI	VLT INterferometer Commissioning Instrument
VLTI	Very Large Telescope Interferometer
WFS	Wavefront sensor

Chapter 1

Properties of Radiation

1.1 Preamble

Early studies of the properties of electric and magnetic fields, their interplay and the resulting emission of radiation, led to formation of several fundamental laws by Carl F. Gauss (1777–1855), Michael Faraday (1791–1867), André M. Ampère (1775–1836), and James C. Maxwell (1831–1879). These were presented in a compact differential formalism by Maxwell (1865), which are referred to as Maxwell's equations:

$$\nabla \cdot \mathbf{D}(\mathbf{r}, t) = \rho(\mathbf{r}, t), \tag{1.1}$$

$$\nabla \cdot \mathbf{B}(\mathbf{r}, t) = 0, \tag{1.2}$$

$$\nabla \times \mathbf{E}(\mathbf{r}, t) = -\frac{\partial \mathbf{B}(\mathbf{r}, t)}{\partial t}, \tag{1.3}$$

$$\nabla \times \mathbf{H}(\mathbf{r}, t) = \mathbf{J}(\mathbf{r}, t) + \frac{\partial \mathbf{D}(\mathbf{r}, t)}{\partial t}. \tag{1.4}$$

The electric $\mathbf{E}(\mathbf{r}, t)$ and the magnetic fields $\mathbf{B}(\mathbf{r}, t)$, are time dependent and can be specified at every point in space and time, in which $\mathbf{r}(= x, y, z)$ denotes the position vector, t is the time, $\mathbf{J} = \sigma\mathbf{E}$, the electric current density, $q(= 1.6 \times 10^{-19}$ C) the elementary charge, $\mathbf{E}$ and $\mathbf{B}$, the electric field strength and the magnetic induction, respectively, $\mathbf{H} = \mathbf{B}/\mu$, the magnetic field, $\mathbf{D} = \epsilon\mathbf{E}$, the

High-Resolution Imaging: Detectors and Applications
Swapan K. Saha

ISBN 978-981-4613-27-9 (Hardcover), 978-981-4613-28-6 (eBook)
www.panstanford.com

electric displacement, ρ, the volume density of free charge, ϵ, μ, and σ the respective dielectric permittivity, the magnetic permeability, and conductivity of the medium, "$\times$," the cross product, and

$$\nabla = \mathbf{i}\frac{\partial}{\partial x} + \mathbf{j}\frac{\partial}{\partial y} + \mathbf{k}\frac{\partial}{\partial z}. \tag{1.5}$$

represents a linear vector differential operator.

The unit of the electric field intensity, $\mathbf{E}$, is expressed as volt (V) m^{-1}, and that for the magnetic flux density $|\mathbf{B}|$ (in tesla; T = Wb m^{-2}). Equation (1.1) represents Gauss' theorem in electrostatics. Equation (1.2) tells that whatever might be the origin of $\mathbf{B}$, be it static magnetic poles or conduction current, the divergence is zero. Equation (1.3) is an outcome of Faraday's law of electromagnetic induction, which states that a changing magnetic field produces an electric field. Equation (1.4) originates from the circuital form of Ampère's law subject to satisfying the equation of continuity. What follows from circuital form of Ampère's law is

$$\nabla \times \mathbf{H} = \mathbf{J}. \tag{1.6}$$

From equation (1.6), one may obtain, $\nabla \cdot (\nabla \times \mathbf{H}) = 0$, which is an identity. That is

$$\nabla \cdot \mathbf{J} = 0, \tag{1.7}$$

which is the equation of continuity in a restricted sense of steady charge.

The equation of continuity actually is

$$\nabla \cdot \mathbf{J} + \frac{\partial \rho}{\partial t} = 0. \tag{1.8}$$

On using equation (1.1), one can rewrite equation (1.8)

$$\nabla \cdot \mathbf{J} + \frac{\partial}{\partial t}(\nabla \cdot \mathbf{D}) = 0. \tag{1.9}$$

Therefore,

$$\nabla \cdot \left(\mathbf{J} + \frac{\partial D}{\partial t}\right) = 0. \tag{1.10}$$

Using equation (1.10), Maxwell modified equation (1.6), by introducing the additional term, $\partial \mathbf{D}/\partial t$ on the right hand side (RHS), named it the displacement current density to obtain equation (1.4), which is also known as Ampère–Maxwell law.

By taking surface integral on both sides of equation (1.4) and converting the left hand side (LHS) into a line integral, one obtains the magnetomotive force (MMF) as $\oint_{\mathrm{l}} \mathbf{H} \cdot d\mathbf{l}$. That is

$$\oint_{\mathrm{l}} \mathbf{H} \cdot d\mathbf{l} = \int_{\mathrm{S}} \left(\mathbf{J} + \frac{\partial \mathbf{D}}{\partial t} \right) \cdot d\mathbf{S}. \tag{1.11}$$

The first term on the RHS indicates that MMF originates from current carrying conductor; this was postulated by Ampère and observed by Oersted. The second term indicates that time variation of the flux of electric displacement vector is a source of MMF. This reflects a symmetry of nature because it shows that as according to Faraday's law of electromagnetic induction, variation in the flux of magnetic induction gives rise to electromotive force (EMF), the variation of the flux of electric induction vector gives rise to MMF.

An exact description of electromagnetic radiation and its behavior requires a knowledge of Maxwell's electromagnetic field equations that describe the temporal and spatial dependence of electromagnetic fields and provide good agreement with observed phenomena over a wide range of frequencies in the spectrum. Equations (1.3) and (1.4) in free space (i.e., $\rho = 0$, $\mathbf{J} = 0$, $\epsilon = 0$), may be written as

$$\nabla \times \mathbf{E}(\mathbf{r}, t) = -\mu_0 \frac{\partial \mathbf{H}(\mathbf{r}, t)}{\partial t}, \tag{1.12}$$

$$\nabla \times \mathbf{H}(\mathbf{r}, t) = \epsilon_0 \frac{\partial \mathbf{E}(\mathbf{r}, t)}{\partial t}, \tag{1.13}$$

in which $\mu_0 (= 4\pi k = 4\pi \times 10^{-7}$ henrys (H).m^{-1}) is the permeability in free space (vacuum) and $\epsilon_0 (= 8.8541 \times 10^{-12}$ farads (F).m^{-1}) the permittivity in vacuum.

The curl of equation (1.12) is expressed as

$$\nabla \times \nabla \times \mathbf{E}(\mathbf{r}, t) = -\mu_0 \epsilon_0 \frac{\partial^2 \mathbf{E}(\mathbf{r}, t)}{\partial t^2}, \tag{1.14}$$

with $\nabla \times \nabla \times \mathbf{E} = \nabla(\nabla \cdot \mathbf{E}) - \nabla^2 \mathbf{E}$. Since $\nabla \cdot \mathbf{E} = 0$, in free space, one may derive the wave equation as

$$\nabla^2 \mathbf{E}(\mathbf{r}, t) - \mu_0 \epsilon_0 \frac{\partial^2 \mathbf{E}(\mathbf{r}, t)}{\partial t^2} = 0, \tag{1.15}$$

$$\nabla^2 \mathbf{E}(\mathbf{r}, t) - \frac{1}{c^2} \frac{\partial^2 \mathbf{E}(\mathbf{r}, t)}{\partial t^2} = 0. \tag{1.16}$$

Similarly, one may derive for **B**,

$$\nabla^2\mathbf{B}(\mathbf{r},t) - \frac{1}{c^2}\frac{\partial^2}{\partial t^2}[\mathbf{B}(\mathbf{r},t)] = 0. \tag{1.17}$$

The permittivity constant ϵ_0 and the permeability constant μ_0 in a vacuum are related to the speed of light, c

$$c = \frac{1}{\sqrt{\mu_0\epsilon_0}} \approx 3\times 10^8 \text{ meter (m)/second (s)}. \tag{1.18}$$

The above expressions (1.16–1.17) are known as the electromagnetic wave equations, which indicate that electromagnetic disturbances (waves) are propagated through free space. This result gives rise to Maxwell's electromagnetic theory of light. The propagation velocity, $v(= 1/\sqrt{\epsilon\mu})$, of the waves obey the wave equations; in free space $v = c = 1/\sqrt{\mu_0\epsilon_0}$.

The union of electromagnetic theory with optics began when these equations described waves traveling at the velocity of light. However, the development of quantum theory of light at the dawn of the last century, dealing with absorption and emission, cannot be described with just Maxwell's equations. While interpreting the spectrum of electromagnetic radiation emitted by a blackbody, Max Karl Ernst Ludwig Planck (1858–1947) postulated that the oscillating electric charges which gives rise to light emission can have discrete energies (Planck, 1900). The name, blackbody, comes from the assumption that the body absorbs at every frequency and hence would look black at low temperature. The radiation emitted by a body as a result of its temperature is called thermal radiation. All bodies emit thermal radiation to their surroundings and absorb such radiations from them where the rates depend upon the temperature difference.

The nature of the spectrum of thermal radiation is continuous for matter in a condensed state (i.e. solid and liquid) and is strongly dependent on temperature. Generally speaking, the detailed nature of the radiation spectrum depends somewhat on the composition of the body. However, experiments show that there is one class of hot bodies that emits thermal spectra of universal character. These are called black bodies, which have surfaces absorbing all the thermal radiation incident upon them (in the ideal case) and re-radiate energy, which is characteristic of these radiating systems.

However, one cannot explain the radiation emitted from such bodies on the basis of the laws of classical physics. Following Planck's suggestions, Albert Einstein (1879–1955) put forward that light is not distributed in continuous waves, but travels as a finite number of points of energy quanta, called *lichtquanten*, which are localized at points in space and move without dividing and can be absorbed or generated only in complete units (Einstein, 1905a). They travel at the speed of light and have properties like energy $\hbar\omega$, zero rest mass, momentum, spin $\hbar/2$, with $\hbar = h/(2\pi)$, $h(= 6.626096 \times 10^{-34}$ joules (J) seconds (s) $= 4.136 \times 10^{-15}$ eVs) as Planck's constant. This gave way to a term, photon, introduced by a chemist Gilbert Newton Lewis (1875–1946; Lewis, 1926), which eventually led to the development of quantum theory.

In fact, the birth of quantum mechanics is intimately linked with the theories and discoveries relating to the nature of light. Sir Isaac Newton (1642–1727) had described light as a stream of particles, what was termed as corpuscular (little particles) theory. This view had been superseded by the wave picture of light (Huygens' wavefront theory), a picture that culminated in the electromagnetic theory of Maxwell.

The energy of a photon depends on the radiation frequency, ν, of the light; the shorter the wavelength, the more energy each of these quanta has, the longer the wavelength the smaller the energy of the quanta. There are photons of all energies from high-energy gamma rays to low-energy infrared and radio waves. Photons are bosons,[a] having zero mass and rest energy and one unit of spin. Planck postulated that radiation of frequency, ν, can exchange energy with matter in units of $h\nu$. In other words, radiation of frequency, ν,

---

[a]Bosons, named after Satyendra Nath Bose (1894–1974), are particles (integer spin) that are associated with integral multiples of $\hbar$. These particles account for the phenomena of laser (acronym for light amplification by stimulated emission of radiation) (see Section 3.2), superconductivity, and superfluidity. Unlike in the case of fermions, which were named after Enrico Fermi (1901–1954), the particles with spin having odd multiples of $(1/2)\hbar$ obeying Pauli's exclusion principle, bosons can go into the ground state. A photon is a boson, whereas an electron is a fermion. Pauli's exclusion principle states that no two electrons in an atom can have identical quantum number; each energy level can accommodate a maximum of two electrons, one with spin-up $(+1/2)$ and the other with spin-down $(-1/2)$.

behaves like a particle, called a photon, of energy,

$$E = h\nu = \frac{h\omega}{2\pi} = \hbar\omega, \tag{1.19}$$

and momentum, defined by the product of the mass of a particle and its linear velocity, **p**, is given by,

$$\mathbf{p} = \hbar\boldsymbol{\kappa} = m\mathbf{c}, \tag{1.20}$$

with $\kappa = \omega/c = 2\pi\nu/c = 2\pi/\lambda$ as the wave number, $\boldsymbol{\kappa}(= \kappa\hat{\mathbf{s}})$ the wave vector in medium in the direction of propagation, $\hat{\mathbf{s}}$, ω the angular frequency, λ the wavelength, and c the velocity of light in a vacuum.

Equations (1.19) and (1.20) connect particle properties $(E, \mathbf{p})$ to the corresponding wave properties (ω, κ).

Angular momentum The angular momentum of a system of particles is due to both the spin of individual particles and their orbital momentum (Marcuse, 1980). The spin angular momentum is represented by electrons rotating about their own axes; it comes in units of $\hbar/2$, with $\hbar \equiv h/2\pi$ as the Planck constant divided by 2π. The orbital angular momentum for an atomic electron may be visualized in terms of a vector model where the angular momentum vector is seen as precessing about a direction in space. The total angular momentum is written as

$$\mathbf{J} = \mathbf{L} + \mathbf{S}, \tag{1.21}$$

where $\mathbf{J} = \sqrt{j(j+1)}\hbar$,

$$j = \begin{cases} 0,\ 1,\ 2,\ \cdots & \text{boson,} \\ 1/2,\ 3/2,\ 5/2, & \text{fermion,} \end{cases}$$

which is either an integer or a half integer, **L** and **S** the orbital and spin angular momentum respectively, which are quantized according to the following rules

$$\mathbf{L} = \sqrt{l(l+1)}\hbar \qquad \mathbf{S} = \sqrt{s(s+1)}\hbar,$$

l is an integer and s takes integer or half integer values depending on the particles. Spin angular momentum consists of two orthogonal states; on the other hand, the orbital angular momentum consists of an infinite number of orthogonal states.

Quantum theory A number of observations had been made towards the end of the nineteenth century establishing the velocity of light independent of the velocity of the source or of the observer. This result could not be explained by classical theory. Einstein (1905b) introduced the special theory of relativity. He extended and improved Maxwell's theory by modifying the foundations of classical mechanics to remove the apparent contradictions between mechanics and electrodynamics. This theory has completely changed the concepts of space and time. He showed that space and time are closely linked and affect each other. Space, time, simultaneity, mass are all relative, but speed of light is absolute being independent of the notion of the source and the observer. The mass-energy equivalence described by the equation, $E = mc^2$, arose from the theory of relativity, where he explained the varying kinetic energies of an object emitting radiation by considering the amount of the object's change in kinetic energy, as being equivalent to change in its mass.

However, in making the transition from Planck's quantized oscillators to quanta of radiation, Einstein introduced the idea of corpuscular behavior of radiation, the revival of which posed a severe conceptual problem, one of reconciling wave and particle like behavior of radiation. Quantum theory of atomic structure, proposed by Niels Henrik David Bohr (1885–1962), known as Bohr's hypothesis, in 1913 and the discovery of dual nature of electromagnetic radiation (wave-particle duality), predicted by Louis de Broglie (1892–1987), what is referred to as de Broglie's hypothesis (de Broglie, 1925), led to the development of quantum mechanics.

Wave-particle duality Wave and particle natures are dual. When one is exhibited, the other remains suppressed. If radiation can exhibit particle nature, a particle can also show wave nature. Following this argument, de Broglie stated that equations (1.19) and (1.20) would hold for a material particle, such as an electron. From wave nature of particle, de Broglie showed that the wavelength of a particle is directly related to the Planck constant and inversely proportional to the mass and velocity of the particle; the larger the object or faster it is moving, the shorter the wavelength. The de Broglie equations are

$$\lambda = \frac{h}{p} = \frac{h}{\gamma m_0 v} = \frac{h}{m_0 v}\sqrt{1 - \frac{v^2}{c^2}}, \tag{1.22}$$

$$\nu = \frac{E}{h} = \frac{\gamma m_0 c^2}{h} = \frac{m_0 c^2}{h}\frac{1}{\sqrt{1 - v^2/c^2}}, \tag{1.23}$$

in which m_0 designates the particle's rest mass, v the velocity of the particle, c the speed of the light in a vacuum, and $\gamma(\equiv 1/\sqrt{1 - v^2/c^2})$ the Lorentz factor.

These two equations (1.22 and 1.23) relate the wavelength, λ, of a particle to its momentum and the frequency of the wave associated with the particle to the total energy of the particle, respectively. For instance, an electron with an accurately measured value of mass and charge could undergo diffraction in a manner similar to that of light waves.

Poynting vector The energy density of a photon, w in a volume V is given by

$$\mathrm{w} = \frac{\hbar\omega}{\mathrm{V}}. \tag{1.24}$$

Equation (1.24) states that if a plane wave with an energy quantum is contained in a volume, V. The momentum density is expressed as

$$\frac{\mathbf{p}}{\mathrm{V}} = \frac{\mathbf{S}}{c^2} = \frac{\mathrm{w}}{c}\frac{\boldsymbol{\kappa}}{\kappa} = \frac{\hbar\omega/c}{\mathrm{V}}\frac{\boldsymbol{\kappa}}{\kappa} = \frac{\hbar\boldsymbol{\kappa}}{\mathrm{V}}, \tag{1.25}$$

where

$$\mathbf{S}(\mathbf{r}, t) = \mathbf{E}(\mathbf{r}, t) \times \mathbf{H}(\mathbf{r}, t), \tag{1.26}$$

is the Poynting vector, which represents the energy flux of an electromagnetic field having the units of energy per unit area per unit time (J m^{-2} s^{-1}); it is named after its inventor John Henry Poynting (1852–1914).

Uncertainty principle Planck's constant also occurs in statements of Werner Heisenberg's (1901–1976) uncertainty principle (Heisenberg, 1927) which is an important consequence of the wave-particle duality of matter and radiation and is inherent to the quantum description of nature (Beiser, 1973, and references therein). This provides a lower bound on the product of the standard deviations of two conjugate variables, such as position and momentum for a system or the angular momentum and angular displacement for a system or the energy and time for a moving particle. This principle applies to all conjugate variables and also to the notion of the wave packet. A pair of physical variables describing a quantum-mechanical system such that their commutator is a nonzero constant are also known as complementary variables. The precision in measurement of one of them destroys the possibility of measurement of the other. There is a minimum for the product of the uncertainties of these two measurements

$$\begin{aligned}\Delta x \Delta p &\geq \frac{\hbar}{2},\\ \Delta L_z \Delta \phi &\geq \frac{\hbar}{2},\\ \Delta E \Delta t &\geq \frac{\hbar}{2},\end{aligned} \tag{1.27}$$

where the uncertainty is given as the standard deviation of the measured value from its expected value. Uncertainty should not be construed as a limitation of measurement on the measuring device. It is an inadequacy inbuilt in the process of measurement. The product of two uncertainties is always greater than, or at best, the same as $\hbar/2$.

Schrödinger's equation If there is a wave, there must be a wave equation. Erwin Schrödinger (1887–1961) developed the wave-particle duality equation (Schrödinger, 1926)

$$i\hbar\frac{\partial \Psi}{\partial t} = \hat{H}\Psi, \tag{1.28}$$

in which $i = \sqrt{-1}$, Ψ is the time-dependent wave function, and in one dimension

$$\hat{H} = \frac{-\hbar^2}{2m}\frac{\partial^2}{\partial x^2} + V(x), \tag{1.29}$$

is the Hamiltonian operator, a quantum mechanical operator introduced by Schrödinger in quantum mechanics, representing the energy of the system, which is the sum of kinetic and potential energies.

Operating on the wave function, Ψ, with the Hamiltonian produces Schrödinger's equation. In the time independent Schrödinger's equation, the operation may produce specific values for the energy called energy eigen values, for example

$$\hat{H}\Psi_n = E_n\Psi_n, \tag{1.30}$$

where the functions, Ψ_n, are called eigen functions (or wave functions), possessing the properties that they are single valued, finite everywhere, and also their first derivatives with respect to their variables are continuous everywhere; E the multiplying factor, called eigen value.

For each value of n, there is a energy value and corresponding wave function. Equation (1.29) is a partial differential equation elucidating how the quantum state of some physical system changes with time. The first term on the RHS is the operator associated with kinetic energy and the second term the potential energy. The most precise computations are performed quantum mechanically by calculating the potential precisely and solving this equation. However, the least precise calculations are performed classically, neglecting quantization and using Newton's laws. A semi-classical computation, in which an atom's energy levels that are associated with attraction of the electrons to the nucleus, as well as with the effects of the angular momentum of the electrons, are computed quantum mechanically, but additional effects, such as light waves, are treated classically is an intermediate case. The observations associated with blackbody radiation, which is derived quantum mechanically, the photoelectric effect (see Section 2.1), and other related phenomena associated with the emission or transformation of light could be understood if the energy of light is discontinuously distributed in space. Advantages in technology have also led to matter wave interference experiments being demonstrated successfully with electrons, neutrons, atoms, and molecules.

Quantum statistics Light is quantized, and the statistics of the photons results in uncertainties of the amplitude and the phase of the light field. A quantum state can be occupied by several particles, what is referred to as quantum statistics. This applies to indistinguishable particles. Max Born (1882–1970) formulated the now-standard interpretation of the probability density function for

$\Psi\Psi^*$ in Schrdinger's equation of quantum mechanics (Born, 1926). A probability amplitude is, in general, a complex number whose modulus squared represents a probability or probability density; probability amplitudes may also correspond to probabilities of discrete outcomes. For example, the values taken by a normalized wave function, Ψ, are amplitudes, since $\Psi\Psi^* = |\Psi(x)|^2$ provides the probability density at position x. Probability amplitudes, defined as complex-number-valued function of position, may also correspond to probabilities of discrete outcomes. It is a quantity whose value is a definite complex number at any point in space. The probability of finding the particle described by the wave function (e.g., an electron in an atom) at that point is proportional to square of the absolute value of the probability amplitude. That the physical meaning of the wave function is probabilistic was proposed by Max Born.

1.1.1 *Planck's Law*

At the turn of the last century, Rayleigh and Jeans made a calculation of the energy density, $E_\nu(T)$, of cavity radiation using classical arguments. The radiated energy from a blackbody may be considered to be produced by standing wave or resonant modes of the cavity. By considering the radiation inside a cavity of absolute temperature T whose walls are perfect reflectors to be a series of standing electromagnetic waves. The number of independent standing waves, $G(\nu)d\nu$ in the frequency interval between ν and $d\nu$ per unit volume in the cavity is

$$G(\nu)d\nu = \frac{8\pi\nu^2 d\nu}{c^3}. \tag{1.31}$$

The average total energy, $\bar{\mathcal{E}}$, of these waves in thermal equilibrium at a temperature T can be found out using classical kinetic theory and equipartition law,[a] i.e., assigning an average energy, $\bar{\mathcal{E}}$ per standing wave

$$\bar{\mathcal{E}} = k_B T, \tag{1.32}$$

[a]The theorem of equipartition of energy states that the mean energy per degree of freedom of an entity, for example, a molecule of an ideal gas, which is a member of a system of such entities in thermal equilibrium at the temperature, T, is $1/2k_B T$.

in which $k_{\rm B}(= 1.380662 \times 10^{-23}$ J K$^{-1})$ is Boltzmann's constant, to each mode of vibration.

The amount of radiation emitted in a given frequency range should be proportional to the number of modes in that range. They showed that the number of modes was proportional to the square of the frequency. The energy density is related to the spectral distribution of blackbody radiation (spectral radiancy), $B_\nu(T)$, of the spectrum emitted by the hole on the cavity by the following equation

$$\begin{aligned} E_\nu(T)d\nu = \mathcal{E}G(\nu)d\nu &= \frac{4\pi}{c}B_\nu(T)d\nu \\ &= \frac{8\pi\nu^2 k_{\rm B}T}{c^3}d\nu, \end{aligned} \tag{1.33}$$

which has units of energy per unit volume per unit frequency (J m^{-3} Hz^{-1}).

The number of standing waves in the frequency interval times the average energy of the modes divided by the volume of the cavity gives the average energy per unit volume in the frequency interval ν to $\nu + d\nu$, i.e., the energy density $E_\nu(T)$. The spectral radiancy is defined in such a way that $B_\nu(T)d\nu$ equals the energy emitted per unit time in radiation of frequency in the interval of ν to $\nu + d\nu$ from a unit area of surface at absolute temperature, T. Rayleigh and Jean's result was not successful to justify the experimental results pointing out serious conflict between classical physics and experimental observations. They considered a cavity with metallic walls heated uniformly to temperature T. The walls emit electromagnetic radiation in the thermal range of frequencies. This happens due to the accelerated motions of the electrons in the metallic walls arising from the thermal agitation. In this treatment, the oscillations of the electrons were not considered, instead, attention was focused on the behavior of the standing electromagnetic waves inside the cavity. However, Rayleigh-Jeans' law laid the foundation in understanding of the equilibrium radiation from a blackbody (Rayleigh, 1900, 1905; Jeans, 1905).

Planck considered the energy as a discrete variable instead of a continuous variable. Like the speed of light, c, that sets an upper limit on any physical velocity; Planck's constant, h, turned out to be a universal constant which sets a lower limit on action (linear momentum multiplied by distance). This constant opened up new vistas of physics. From the assumption that the electromagnetic modes in a cavity were quantized in energy, Planck (1901) obtained a radiation formula. The average energy, $\bar{\mathcal{E}}$[a] per mode (or quantum) is the energy of the quantum times the probability that it would occupy, which is given by

$$\bar{\mathcal{E}}(\nu) = \frac{h\nu}{e^{(h\nu/k_B T)} - 1}, \tag{1.34}$$

where k_B is Boltzmann's constant, T the temperature in Kelvin (K), e the base of natural logarithms (2.718 ...), and ν the frequency.

The spectral energy density in the blackbody spectrum is expressed as

$$E_\nu(T)d\nu = \frac{8\pi\nu^2}{c^3}\frac{h\nu}{e^{(h\nu/k_B T)} - 1}d\nu. \tag{1.35}$$

At high frequencies, $h\nu \gg k_B T$ and $e^{h\nu/k_B T} \rightarrow \infty$, which means the spectral energy density, $E_\nu d\nu \rightarrow 0$. There is no more ultraviolet (UV) catastrophe. At low frequencies, $h\nu \ll K_B T$ and $h\nu/K_B T \ll 1$, Rayleigh-Jeans' formula holds good and is valid for radio wavelengths. Following

$$e^x = 1 + x + \frac{x^2}{2!} + \frac{x^3}{3!} + \cdots,$$

and if x is small, $e^x \approx 1 + x$, one gets

$$\frac{1}{e^{(h\nu/k_B T)} - 1} \approx \frac{1}{1 + \frac{h\nu}{k_B T} - 1} \approx \frac{k_B T}{h\nu} \qquad h\nu \ll k_B T. \tag{1.36}$$

Thus at low frequencies, equation (1.35) for Planck's law is modified to as

$$E_\nu(T)d\nu \approx \frac{8\pi k_B T \nu^2}{c^3}d\nu, \tag{1.37}$$

which is Rayleigh-Jeans' formula.

---

[a]The quantity, $\bar{\mathcal{E}}$, can be evaluated from the ratio of sums

$$\bar{\mathcal{E}} = \frac{\int_{n=0}^{\infty} \mathcal{E}\mathcal{P}(\mathcal{E})d\mathcal{E}}{\int_{n=0}^{\infty} \mathcal{P}(\mathcal{E})d\mathcal{E}},$$

in which $\mathcal{P}(\mathcal{E})d\mathcal{E}$ is the probability of finding a given entity of a system with energy in the interval between $\mathcal{E}$ and $\mathcal{E} + d\mathcal{E}$, when the density of states is independent of $\mathcal{E}$.

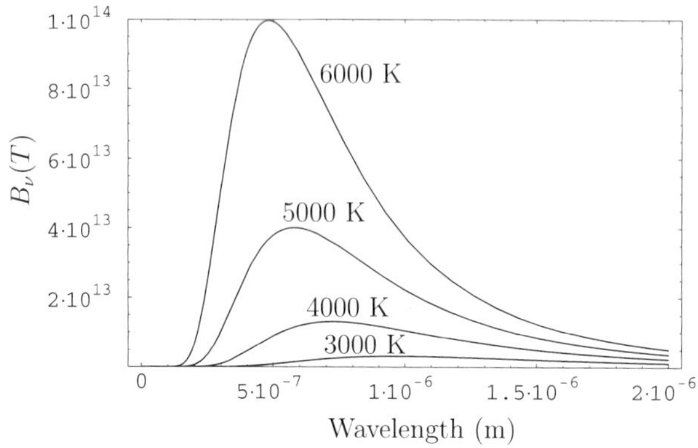

Figure 1.1 Intensity distributions of blackbody radiation at various temperatures, viz., 3000, 4000, 5000, and 6000 K. The abscissa contains the wavelengths and the ordinate depicts the intensity. The wavelength at which the curve is a maximum decreases with the increase of temperature.

A blackbody emits radiation of all wavelengths and the wavelength distribution of the radiation follows Planck's law. This pattern is characterized by

$$B_\nu(T) = \frac{2h}{c^2}\frac{\nu^3}{e^{(h\nu/k_B T)} - 1}, \tag{1.38}$$

which has units of energy per unit time per unit surface area per unit solid angle per unit frequency (J s^{-1} m^{-2} Sr^{-1} Hz^{-1}).

Figure 1.1 depicts the spectral radiancy, $B_\nu(T)$, plot as function of wavelength; each curve is for a given temperature, and the peak of the curve represents the maximum emission. Stars have an overall spectrum,[a] sources of opacity both lines as well as continuum process, which is governed largely by equation (1.38).

[a]There are three types of spectra, such as (i) continuous spectrum that is produced by an excited dense gas radiating at all wavelengths, (ii) emission spectrum, which is produced by an excited low-density gas radiating at a specific wavelength, and (iii) absorption spectrum that results from a light comprising a continuous spectrum while passing through a cool, low-density gas. The gas absorbs some of the wavelengths of light depending on the composition of the gas and the temperature of the light source. The emission spectra, used to study nebulae, are produced when a glowing gas emits energy at specific wavelengths, characteristic of the element composing the gas.

1.1.2 *Stefan–Boltzmann's Law*

Josef Stefan (1835–1893) reported an empirical relation (Stefan, 1879) expressing that the total energy emitted from a blackbody appeared to be proportional to the fourth power of its absolute temperature, T. Later, Ludig Boltzmann (1844–1906) derived the relation using thermodynamic arguments (Boltzmann, 1884), what is referred to as Stefan–Boltzmann's law. It explains the growth in the height of the curve as the temperature increases. Let Planck's law be expressed as a function of wavelength, $B_\nu d\nu = -B_\lambda d\lambda$. The negative sign appears since the wavelength decreases with increasing frequency, and $\nu = c/\lambda$, in which c is the velocity of light. Hence

$$B_\lambda = -B_\nu \frac{d\nu}{d\lambda} = B_\nu \frac{c}{\lambda^2}, \tag{1.39}$$

with $d\nu/d\lambda = -c/\lambda^2$.

Hence equation (1.38) can be recast as

$$B_\lambda(T) = \frac{2hc^2}{\lambda^5} \frac{1}{e^{(hc/\lambda k_B T)} - 1}. \tag{1.40}$$

The total brightness distribution is obtained as

$$B(T) = \int_0^\infty B_\nu d\nu = \frac{2h}{c^2} \int_0^\infty \frac{\nu^3 d\nu}{e^{(h\nu/k_B T)} - 1}. \tag{1.41}$$

Let the integration variable be changed to $x = h\nu/k_B T$, so that equation (1.41) is written as

$$B(T) = \frac{2h}{c^2} \frac{k_B^4 T^4}{h^4} \int_0^\infty \frac{x^3 dx}{e^x - 1} = \sigma T^4, \tag{1.42}$$

with $\sigma = \frac{2\pi^5 k_B^4}{15c^2h^3}$ and $d\nu = (k_B T/h)\, dx$; the definite integral here is a real number.

The relation between the luminosity and temperature of a star can be obtained from Stefan–Boltzmann's law. The stars emit energy, hence one needs to determine the luminosity that is defined as the amount of energy (total flux) a star radiates per unit time in the form of electromagnetic radiation in all directions. It is expressed in watts, or in terms of solar luminosity, $\mathcal{L}_\odot (= 3.845 \times 10^{26}$ W). $\mathcal{L}_\lambda$ or $\mathcal{L}_\nu$ is the corresponding monochromatic quantity. The surface flux of stars, F, equals the luminosity divided by the surface area, that is

$$F = \frac{\mathcal{L}_\star}{4\pi R_\star^2}, \tag{1.43}$$

where $R_\star$ is the radius of a spherical star and its surface area is $4\pi R_\star^2$.

Unlike the observed apparent brightness that is related to distance with an inverse square law relationship (see Section 1.2.1), luminosity is an intrinsic constant independent of distance. The brightness of a celestial object can be derived in terms of observed flux density,[a] $F_\nu (= \text{W m}^{-2})$, the energy received per unit time per unit telescope area per unit frequency. The flux, F is defined as the total amount of radiation crossing a unit area of surface in all directions and in unit frequency interval. In the case of isotropic radiation,[b] the amount of radiation leaving the surface, $F = \pi I$, with I as the total intensity, which includes all possible frequencies, ν, and is obtained by integrating I_ν, over all frequencies

$$I = \int_0^\infty I_\nu d\nu, \tag{1.44}$$

where I_ν denotes the specific intensity of radiation – the amount of energy passing a unit area of surface, placed perpendicular to the direction of propagation, per unit time, per unit solid angle, per unit frequency interval – over the entire solid angle element, $d\Omega$, which is equal to a surface element on a unit sphere; in spherical coordinates it is given by $d\Omega = \sin\theta \, d\theta \, d\phi$, in which ϕ is the azimuthal angle, and θ the polar angle between the solid angle $d\Omega$ and the normal to the surface.

The specific intensity, I_ν, has units of W.m^{-2}.s^{-1}.Hz^{-1} per steradian,[c] which is a function of position, direction, and time. The flux emitted by a star into a solid angle, Ω is related to its luminosity, $\mathcal{L}_\star$

$$\mathcal{L}_\star = \Omega r^2 F, \tag{1.45}$$

[a]In radio astronomy, flux densities are often expressed in janskys; one jansky (1 Jy) equals to 10^{-26} W m^{-2} Hz^{-1}.

[b]The isotropic sources are spherical, in which the total emitted flux is a function of the intensity.

[c]Steradian is a unit of solid angle and is used to define two-dimensional angular spans in three-dimensional space. It is the solid angle subtended at the center of a sphere of radius r by a portion of the surface of the sphere having an area r^2. There are 4π steradians (sr) about a point enclosed by a spherical surface.

where $F(= \mathcal{L}_\star/4\pi r^2)$ designates the flux density observed at a distance, r at the detector (energy/sec/m^2); the energy is diluted over the surface area of the sphere of radius, r.

The brightness of the star in different colors is associated with the temperature. This parameter and the luminosity lead to derive the stellar radius. The integrated intensity of a blackbody is proportional to T^4, and therefore the flux density F for isotropic radiation of intensity, B is

$$F = \pi B = \sigma T^4, \tag{1.46}$$

in which $\sigma(= 5.67 \times 10^{-8}$ W m^{-2} K$^{-4})$ is Stefan–Boltzmann's constant.

1.1.3 *Effective Temperature of a Star*

The light from a star is usually concentrated in a rather narrow range of wavelengths. The spectrum of star's light is approximately a thermal spectrum (blackbody spectrum). Assuming the star is spherical, and following equations (1.43), equation (1.46) is recast as

$$\mathcal{L}_\star = 4\pi\sigma R_\star^2 T^4. \tag{1.47}$$

The effective temperature of a star is the same as the temperature of a blackbody that emits energy at the same rate per unit area, which can be derived from equation (1.47) if the distance to the star is known. This relationship helps account for the large range of stellar luminosities. A small increase in effective temperature can significantly increase the energy[a] emitted per second from

[a]While observing a radiation source, one measures the energy collected by the detector during a period of time, which equals the flux density integrated over the radiation collecting area of the instrument and the time interval. The energy, E, striking an optical imaging system, say a telescope or a lens, per second, is given by

$$E = \mathtt{A} \times \text{flux at Earth} = \mathtt{A} \int I \cos\alpha d\Omega,$$

with

$$d\Omega = \frac{\cos\theta d\Sigma}{r^2},$$

$\mathtt{A}$ as the area of the lens, I the intensity of starlight, α the angle between the direction of light and the normal of the lens, which is considered to be very small, i.e., $\cos\alpha = 1$, $d\Sigma(= R^2 \sin\theta d\theta d\phi)$ the element of star's surface area, θ the angle between the each square meter of a star's surface. It is evident that the quantities, namely, luminosity, radius, and temperature of a star are independent and are also related to the absolute bolometric magnitude of the star.

The magnitude scale for the measurement of brightness of a star, defined by N. R. Pogson (1857), is a logarithmic scale with a base 2.512. It is an inverse scale with brighter stars having smaller numerical values for their magnitudes, i.e., a second magnitude star is brighter than a third magnitude star with a brightness ratio of 2.512 ($\sqrt[5]{100}$). The faintest star that can be seen with a normal human visual system (human eye; see Section 4.2.1) on a clear, dark night has a magnitude of six. These stars are defined so that the brighter stars has the smaller magnitude. The present definition for the magnitude scale is

$$m_1 - m_2 = -2.5 \log \frac{F_1}{F_2}, \tag{1.48}$$

in which the subscripts 1 and 2 refer to two different spectral regions, and F_1 and F_2 are the fluxes of the said stars at the Earth; m_1 and m_2 are their respective magnitudes.

Equation (1.48) is the apparent magnitude of a star, which depends on its flux as observed from the Earth. The brightness of any celestial object is often derived in terms of observed flux density, $F_\nu (= \mathrm{W\ m^{-2}})$. However, this quantity is a convolution of the true brightness and the effect of distance on the observed brightness, as well as any other absorption in the light path. Measuring such a quantity depends on the instrument since the sensitivity (capability of absorbing photons and generating a corresponding amount of electrons) of the detector is different at different wavelengths. Here, the flux received from the light is proportional to its intrinsic brightness or luminosity and is inversely proportional to the square of the distance.

---

normal to $d\Sigma$ and the direction to Earth, ϕ the azimuthal angle, r the distance of the star, and R the radius of the lens.

Therefore, the flux, F, is derived as

$$F = \frac{R^2}{r^2} \int_0^{2\pi} d\phi \int_0^{\pi/2} I \cos\theta \sin\theta d\theta.$$

Astronomers often characterize a star's intrinsic brightness in terms of absolute magnitude, M_V, which is defined as the apparent magnitude if the star is at a distance of 10 parsec[a] away from the Earth. If the star's absolute magnitude is known, one can infer its distance by comparing absolute and apparent magnitudes

$$m - M = -2.5 \log \frac{F(r)}{F(10)} = 5 \log \frac{r}{10\ \text{pc}}. \tag{1.49}$$

If the distance is expressed in parsec, the (1.52) may be expressed as

$$m - M = 5 \log r - 5. \tag{1.50}$$

To note, if the flux is monochromatic at some wavelength, the magnitude, m is a monochromatic magnitude at the wavelength. However, in case of flux, F, of an object being integrated over all frequencies, so that it takes into account the total amount of energy radiated, the magnitude, m, becomes the bolometric magnitude, m_{bol}. Let $\phi(\lambda)$ be the efficiency of the equipment for responding to radiation of wavelength, λ. The measured flux is related to the flux incident on the telescope by

$$F_{\text{measured}} = \int_0^\infty \phi(\lambda) F_\lambda(r) d\lambda. \tag{1.51}$$

The flux reaching the Earth depends both on the distance of the star, as well as on the intrinsic brightness or luminosity of the star. The intensity of light emitted by any celestial object varies strongly with wavelength. The absolute bolometric magnitudes can be expressed in terms of the luminosity, $\mathcal{L}_\star = 4\pi r^2 F$, with F as the total flux at a distance $r = 10$ pc. Let $F_\odot$ be the flux for the Sun, hence

$$\begin{aligned} M_{\text{bol}} - M_{\text{bol}_\odot} &= -2.5 \log \frac{F}{F_\odot} \\ &= -2.5 \log \frac{\mathcal{L}_\star / 4\pi r^2}{\mathcal{L}_\odot / 4\pi r_\odot^2} = -2.5 \log \frac{\mathcal{L}_\star}{\mathcal{L}_\odot}. \end{aligned} \tag{1.52}$$

[a] Parsec (parallax second; pc) is defined by the distance to a star that exhibits a parallax angle of one arcsecond employing a baseline of 1 AU (the average distance between the Earth and the Sun). For instance, at a distance of 100 pc, the parallax is 10 milliarcsec (mas); 1 pc = 3.26 ly (light-year).

In terms of radii and temperatures, equation (1.52) is written as

$$M_{\text{bol}} - M_{\text{bol}_\odot} = -2.5 \log \frac{R_\star^2 T^4}{R_\odot^2 T_\odot^4}$$
$$= -5 \log \frac{R_\star}{R_\odot} - 10 \log \frac{T}{T_\odot}, \tag{1.53}$$

where $R_\odot (= 6.96 \times 10^8$ m) is the radius of the Sun.

Combining the bolometric flux with the measured angular diameter of a star, although bolometric corrections do depend on models to an extent, one gets the effective temperature. It is known that the star's total luminosity is given by its radiant emittance times its surface area (see equation 1.47). If the star does have the spherical symmetry, at a distance r, the incident bolometric flux (total radiant flux per unit collecting area) is expressed as

$$F = \frac{\mathcal{L}_\star}{4\pi r^2} = \left(\frac{R_\star}{r}\right)^2 \sigma T_e^4 = \left(\frac{\theta_{LD}}{2}\right)^2 \sigma T_e^4 \quad \text{erg/cm}^2\text{/sec}, \tag{1.54}$$

in which θ_{LD} denotes the limb-darkened diameter[a] of the source.

Observations of limb-darkening measurements may provide the atmospheric temperature structure of a source. Long-baseline interferometry (Labeyrie et al., 2006, Saha 2010 and references therein) is required to collect data in the vicinity of and beyond the first zero or minimum of the visibility function. So, the effective temperature, T_e is recast as

$$T_e = \left(\frac{4F}{\theta_{LD}^2 \sigma}\right)^{1/4}. \tag{1.55}$$

1.1.4 *Wien's Law*

Wien's law, named after Wilhelm Carl Wien (1864–1928), also called Wien's displacement law, states that the frequency at which the radiated power is most intense increases with increasing

[a]The limb-darkening occurs since the temperature of the star decreases for an increasing distance from the center to the periphery of the star. Since surface brightness scales as T^4 for thermal radiation, the limb of stars appear fainter than the central regions of their disks. However, the temperature for the Sun does not uniformly drop as the radius increases, and for certain spectral lines, the optical depth is unity in a region of increasing temperature. In such a situation, one observes the phenomenon of limb brightening.

temperature. From the curve (see Figure 1.1), it is observed that the wavelength of the maximum intensity decreases with increasing total intensity. In the 1890s, Wien (1898) studied the frequency distribution of blackbody radiation, where he found that the product $\lambda_{\max} T (= 0.0028978$ K.m) is an absolute constant. This law allows one to derive the temperature of an object from the color of the emitted light. For example, the effective temperature of the Sun is 5760 K, and using this law, this temperature corresponds to a peak emission at a wavelength of 0.0028978 K.m/5760 K $=$ 503 nm (1 nm $= 10^{-9}$ m). Wien's Law provides the wavelength of the peak of the radiation distribution and explains the shift of the peak to shorter wavelengths as the temperature increases. When the wavelength, $\lambda \approx \lambda_{\max}$ or $hc/(\lambda k_B T) \gg 1$, one obtains, $e^{hc/\lambda k_B T} \gg 1$, so that Equation (1.40) is approximated (Wien's approximation) to

$$B_\lambda(T) \approx \frac{2hc^2}{\lambda^5} \frac{1}{1 + (hc/\lambda k_B T) + \cdots - 1} \approx \frac{2ck_B T}{\lambda^4}. \tag{1.56}$$

This approximation to Planck's spectrum is called the Rayleigh-Jeans approximation and is valid at the radio wavelengths. Equation (1.56) is useful in radio astronomy; it gives the brightness temperature of a radio source, which is a measure of the intensity of radiation thermally emitted by an object, which is an equivalent blackbody, providing the same power output per Å as the source at a wavelength, λ. In the isotropic case, the observed flux density, F'_λ is given by

$$F'_\lambda = \frac{R_\star^2}{r^2} F_\lambda = \left(\frac{\theta_\star}{2}\right)^2 \pi B_\lambda(T_b), \tag{1.57}$$

where $F_\lambda = \pi B_\lambda(T_b)$, T_b is the brightness temperature.

1.1.5 *Color Temperature*

In the visible part of the electromagnetic spectrum (see Section 1.1.6), the shape of the emission spectrum determines the color of the light as perceived by the human eye. The color of a star is defined as the ratio of brightness in two different wavelengths. As determined by the measurements at two different spectral regions give information about the star's temperature. Let the bolometric luminosity, $\mathcal{L}_{\rm bol}$, be

$$\mathcal{L}_{\rm bol} = \int_0^\infty \mathcal{L}_\lambda d\lambda. \quad (1.58)$$

In reality, one measures the flux in a certain bandpass, hence for a filter with bandpass, f_λ, one writes

$$\mathcal{L}_f = \int_0^\infty \mathcal{L}_\lambda f_\lambda d\lambda. \quad (1.59)$$

The energy distribution could be traced if one can measure the flux at many wavelengths. In order to achieve this, a system of filters is required. The peaks for transmission for these filters are in the ultraviolet (U), blue (B), visible (V), red (R), and infrared (I), respectively. The UBV system,[a] also referred to as Johnson system, is a photometric system for classifying stars according to their colors (Johnson and Morgan, 1953). The categories of these filters are given in Table 1.1.

The color of a star is measured by comparing its brightness in two different wavelength bands. One observes magnitude of an object successively through two different filters. For example, the (B-V) provides the color index, which is the difference between the magnitudes in the blue and visual regions of the spectrum and the (U-B) color index is the analogous difference between the ultraviolet and blue regions of the spectrum and so on. The (B-V) is a function of the effective temperature T_e. The hotter a star is, the smaller its color index (B-V).

[a]Strömgren photometric system ($uvby$) is four color photometric system with H_β for stellar classification (Strömgren, 1956, Crawford, 1958). The filters whose respective mean wavelength for u, v, b, y are 350 nm, 410 nm, 470 nm, 550 nm and the respective half-widths of response function are 30 nm, 20 nm, 20 nm, and 20 nm. The mean wavelength for β_{narrow} is 485.8 nm and for the β_{wide} is 485 nm, while the respective half-width for these wavelengths are 2.9 nm and 12.9 nm. The color difference $c_1 = (u - v) - (v - b)$ is a measure of the continuous hydrogen absorption; $(v - b)$ defines a color gradient and c_1 measures how much u differs from that gradient. c_1 is a rather clean measure of the Balmer discontinuity which is at maximum at spectral type A 2. For hotter stars, c_1 measures the temperature and absolute magnitude (gravity) for cooler stars in the main sequence. In order to measure the blanketing, the color difference is given by $m_1 = (v - b) - (b - y)$, in which $(b - y)$ defines the color gradient and m_1 measures how much v differs from that gradient, and hence is a measure of the blanketing, which can be associated to the metallicity. H_β is free of reddening and measures temperature for cool stars and gravity for hot stars.

Table 1.1 Categories of filters

Filters	Effective wavelength λ_e	Filter width $\Delta\lambda$
U	365 nm	68 nm
B	440 nm	98 nm
V	550 nm	89 nm
R	700 nm	220 nm
I	900 nm	240 nm

Because of the absorption and scattering of starlight by matter between the emitting object and the observer, the extinction of starlight occurs. Considering $(B\text{-}V)_0$ is the intrinsic color of the star, and the color excess is given by $(B\text{-}V)\text{–}(B\text{-}V)_0$. The *UBV* system provides only the V magnitude and the color indices $(U\text{-}B)$ and $(B\text{-}V)$. The energy flux above the Earth's atmosphere due to a star of $B–V = 0 = m_B–m_V$ emits flux, $\mathrm{F}_0 = 10^8$ photons $\mathrm{s^{-1}\,m^{-2}\,Hz^{-1}}$ at 550 nm. The magnitude differences form a color and it is correlated with the effective temperature.

Stars appear in different colors from blue to red. These colors indicate about the temperature of the star. The color temperature, T_c, of a star is determined by the comparison of the spectral distribution of the star's radiation at two different wavelengths, λ_1 and λ_2. Assuming the intensity distribution follows Planck's law, the ratio of these flux densities are the same as the ratio obtained from this law. Hence

$$\frac{F'_{\lambda_1}}{F'_{\lambda_2}} = \frac{B_{\lambda_1}(T)}{B_{\lambda_2}(T)} = \left(\frac{\lambda_2}{\lambda_1}\right)^5 \frac{e^{(hc/\lambda_2 k_B T)} - 1}{e^{(hc/\lambda_1 k_B T)} - 1}, \tag{1.60}$$

with h as Planck's constant, c the velocity of light, and k_B Boltzmann's constant.

The observed flux densities corresponds to certain magnitudes, m_{λ_1} and m_{λ_2}, that is

$$m_{\lambda_1} - m_{\lambda_2} = -2.5 \log\left(\frac{F'_{\lambda_1}}{F'_{\lambda_2}}\right) + C, \tag{1.61}$$

with C as the constant which is a consequence of the different zero points of the magnitude scales.

From the Wien's approximation in the visible spectrum of light, the magnitude difference is given by

$$\begin{aligned} m_{\lambda_1} - m_{\lambda_2} &= -2.5 \log \left(\frac{B_{\lambda_1}}{B_{\lambda_2}} \right) + C \\ &= -2.5 \log \left(\frac{\lambda_2}{\lambda_1} \right)^5 + 2.5 \frac{hc}{k_B T} \left(\frac{1}{\lambda_1} - \frac{1}{\lambda_2} \right) \log e + C. \end{aligned} \tag{1.62}$$

This can be expressed as

$$m_{\lambda_1} - m_{\lambda_2} = a + b/T_c. \tag{1.63}$$

The Color-Temperature correlation needs to be calibrated empirically for specific groups of stars, for example, dwarfs, supergiants, spectral type range, etc. For main sequence stars, the relationship is given by $B - V = (7000\ K/T_e) - 0.56$.

Extinction The extinction is defined as the attenuation in the amount of radiation passing through the atmosphere. It arises from both the interstellar medium and the Earth's atmosphere and is related to a coefficient, which depends on the medium and wavelength of transmission. The relative loss of flux of the starlight on traveling the distance $dz(= dr \cos\theta)$, in which dr is the optical path and θ the angle between the z-axis and the direction of the propagation of the radiation, is $\alpha_a dz$; therefore, $dF = -F\alpha_a dz$. The flux observed on the ground, F is given by

$$F = F_0 e^{-\int_0^\infty \alpha_a dz}, \tag{1.64}$$

where F_0 denotes the incident flux on the top of the atmosphere and α_a the absorption coefficient.

Optical depth The optical depth, τ, a dimensionless quantity, is a measure of how much radiation is absorbed in passing through a medium. This quantity is measured along the vertical optical path dz, and, in the planetary science convention, is measured downwards from the top of the atmosphere, i.e., $\tau_{top} = 0$ and increases downward as z decreases. In an empty space, where the opacity is zero, the optical depth does not increase and the flux remains same. When the optical thickness is very large, the corresponding region is nearly opaque to the radiation. In terms of optical depth, τ, the measured flux on the Earth is

$$F = F_0 e^{-\tau}. \tag{1.65}$$

If the zenith distance (90°-altitude), γ, of the star is not too large, one may approximate the atmosphere by plane parallel stratified layers. With the increase of zenith distance ($\gamma \geq 60^\circ$), the refraction effects, curvature of the atmosphere, and variations of air density with height may become significant. The path length through the atmosphere, called air-mass, $\chi (= \sec \gamma)$.

1.1.6 *Electromagnetic Spectrum*

Light is an electromagnetic radiation, whose spectrum (see Figure 1.2) is a display of radiation from an object as a function of wavelength spanning from gamma rays (frequencies range between 10^{19} and 10^{21} Hz) to radio waves (10^4 and 10^{11} Hz). The visible light is a tiny fraction of this spectrum. The Earth's atmosphere blocks off the incoming radiation from space beyond the violet edge of the visible spectrum, while it attenuates the infrared (IR). The detection can be made from the ground through the atmospheric windows, which allow the radiations in the visible, IR, and radio wavelengths to penetrate the atmosphere. It is noted that the human eye (see

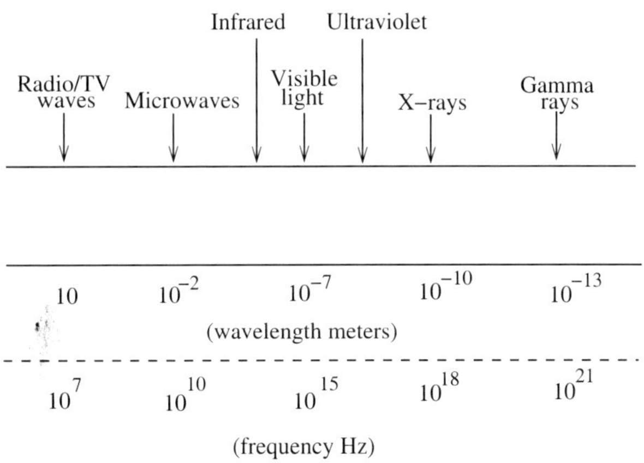

Figure 1.2 Electromagnetic spectrum.

Section 4.2.1) responds to visible light, while detecting the rest of the spectrum requires various types of instruments; in ultraviolet (UV) and shorter wavelengths, observations are carried out from space.

While explaining the spectrum of hydrogen, Bohr (1913) developed a theory based on the following concepts:

(1) The negatively charged electrons confined to atomic shells encircle a small positively charged nucleus; electrostatic forces provide attraction. The electrons can jump from one energy level to another; however, they cannot have orbits with energies other than the allowed energy levels.
(2) An electron jump between orbits must be accompanied by an emitted or absorbed amount of electromagnetic energy, $h\nu$, in which h is Planck's constant that is related to the angular momentum of the electron in orbit around the nucleus, as well as with the frequency, ν, of the emitted radiation.
(3) The energy of the particles in the Bohr atom is restricted to certain discrete values. The energy is quantized, which means that only certain orbits with certain radii are allowed. The electromagnetic field interacts with its surroundings in discrete energy, $E = h\nu$, or multiples thereof and that energy differences in atomic jumps are emitted in these discrete amounts of energy (quantum atomic levels) as electromagnetic radiation:

$$h\nu = E_{n_2} - E_{n_1}. \tag{1.66}$$

Following Coulomb's law,[a] which provides a derivation of the force between two charged objects and Newton's second law pertaining to the net force acting on a body, which is equated to the product of its mass times the acceleration, the total energy of

[a]Coulomb's law states that the force of attraction (or repulsion) between two charged bodies is directly proportional to the product of the charges and inversely proportional to the square of the distance between them. A similar law holds good for particles with mass, i.e., the force, F, pulling the electron towards the proton is

$$F = \frac{q^2}{4\pi\epsilon_0 r_n^2}.$$

an electron in the orbit n is written as

$$E_n = \frac{1}{2}m_e v_n^2 - \frac{q^2}{4\pi\epsilon_0 r_n}$$

$$= -\frac{m_e q^4}{32\pi^2\epsilon_0^2\hbar^2}\frac{1}{n^2} = -C\frac{1}{n^2}, \tag{1.67}$$

with $\hbar = h/(2\pi)$, v as the velocity of the electron, r_n the distance between the electron and the proton, ϵ_0 the permittivity in vacuum, q the electron charge, and m_e the electron mass[a] C a constant, and n an integer.

For each value of n, known as quantum number, a horizontal line can be drawn. These are, in general, arranged vertically in accordance with the numerical values derived from Equation (1.67). Such a pictorial representation is referred to as an energy level diagram. In the quantized energy levels for a hydrogen atom, labeled by such a quantum number, the lowest energy state is generally termed as the normal (or ground) state ($n = 1$), corresponds to the lowest value of the angular momentum. In this orbit, the energy of electron of hydrogen atom is -13.595 eV (electron volts); it is customary to express the energy value of the stationary value of the stationary states in eV. The other orbits correspond to higher energies and are called excited, radiating, critical, or resonance, levels. As the electron gains more energy, it moves into stationary states, which are further away from the nucleus. Beyond an energy, called the ionization potential, the single electron of the hydrogen atom is no longer bound to the atom. The other orbits correspond to higher energies. The successive states possessing more energy than the ground state are known as the first excited state, the second excited state, and so on. Beyond an energy called the ionization potential the single electron of the hydrogen atom is no longer bound to the atom.

---

[a]The reduced mass, μ, of the combined electron and proton system is expressed as

$$\mu = \frac{m_e M_P}{M_P + m_e},$$

in which M_P is the proton mass.

The transition quantum energy from $E_{n_2} \rightarrow E_{n_1}$ is:

$$h\nu = C\left(\frac{1}{n_1^2} - \frac{1}{n_2^2}\right), \tag{1.68}$$

In terms of wavelength, λ, this may be written as

$$\frac{1}{\lambda} = \frac{C}{hc}\left(\frac{1}{n_1^2} - \frac{1}{n_2^2}\right) = R_H\left(\frac{1}{n_1^2} - \frac{1}{n_2^2}\right), \tag{1.69}$$

in which $R_H (= 1.097 \times 10^7\ \text{m}^{-1})$ designates the Rydberg constant and C the constant.

The spectrum of a star is the spectrum of a blackbody with absorption lines due to transitions near the surface of the star, what is referred to as photosphere, as well as some emission lines due to transitions of ions above the photosphere. Hydrogen is the most abundant element, transitions between different states of hydrogen are very important. The emission or absorption lines in the visible part of the hydrogen spectrum, known as Balmer series, are produced by transitions ($E_n \rightarrow E_2$) between the second (or first excited) state and higher energy states of the hydrogen atom. For example, the transition from E_3 to E_2 is called Hα (656.3 nm) and is the first line of Balmer series. The other lines are Hβ (486.1 nm), Hγ (434.1 nm), and Hδ (410.2 nm). To note, the stars are made mostly of hydrogen and helium; about 90% atoms in the Universe are hydrogen atoms. Most prominent lines in many astronomical objects are Balmer lines of hydrogen. Very weak Balmer lines may be envisaged either in very cool stars or in too hot stars, while strong hydrogen lines can be witnessed in the stars having intermediate temperature, which is of the order of 10,000 K. On the contrary, helium absorption only occurs in the very hottest stars, and calcium and other metal lines are strong at cooler temperatures; the absorption by molecules happens in the very coolest stars.

The series of lines in the hydrogen spectrum associated with transitions to or from the first energy level or ground state ($E_n \rightarrow E_1$) gives the Lyman series, which are in the ultraviolet region (UV; Lα at 121.6 nm; Lβ at 102.6 nm) with a series limit at 91.2 nm. The other series are the Paschen series ($E_n \rightarrow E_3$) in the short wave infrared, Brackett series ($E_n \rightarrow E_4$), and Pfund series ($E_n \rightarrow E_5$) in the long wave infrared.

1.2 Light Intensity

The light waves are characterized by a wavelength, λ and a frequency, ν. These quantities are related through $\nu = c/\lambda$, in which c is the velocity of the light. The intensity I of light has been defined as the time average of the amount of energy which crosses in unit time, a unit area perpendicular to the direction of the energy flow. Maxwell's theory provides good agreement with observed phenomena over a wide range of frequencies. Also, it leads to quadratic terms associated with the flow of energy. Describing the optical field in terms of energy, known as intensity, led to results in agreement with the fringes with respect to their intensity and spacing. The frequency of oscillations of visible light is too high for the changing electric and magnetic fields to be directly measured. By taking time averages of the quadratic field components over an interval, which is much greater than the time period, $T = 2\pi/\omega$, the intensity is derived as

$$I \propto \langle \mathbf{E}^2 \rangle = \lim_{T\to\infty} \frac{1}{2T} \int_{-T}^{T} \mathbf{E}^2 dt, \tag{1.70}$$

where the angle brackets, $\langle .. \rangle$, stand for the time average of the quantity; the unit of intensity is W.sr^{-1}.

The quantity within the angle brackets is due to the assumed ergodicity of the field; therefore, the intensity, I, is

$$I = \langle V^2 \rangle = \frac{1}{2}\mathbf{A}(\mathbf{r})\mathbf{A}^*(\mathbf{r}). \tag{1.71}$$

Here $\mathbf{A}(\mathbf{r})$ is the amplitude of the wave and * the complex conjugate. The unit of intensity is expressed as the joule per square meter per second, (J m^{-2}s^{-1}), or watt per square meter, (W m^{-2}).

The photocurrent at the detector, that transforms the light intensity into electrical signal providing the number of photoevents collected during the time of measurement and an additive random noise, is proportional to the light intensity, and the detector circuitry averages over a time, which is longer than a cycle of oscillation. A detector receives an average of the effects produced by the different values of the amplitude, which is sensitive to the square of $\mathbf{E}(\mathbf{r}, t)$; therefore, the time average of the intensity tends to a finite value as the averaging interval is increased indefinitely. The intensity of the wave averaged over the time interval $2T$ is needed to make an observation. Assuming a stationary wave field, taking the time average of the energy over an interval much greater than the period, $T = 2\pi/\omega$, the intensity is deduced as

$$I(\mathbf{r}, t) = \lim_{T\to\infty} \frac{1}{2T} \int_{-T}^{T} |E(\mathbf{r}, t)|^2 \, dt = \langle E E^* \rangle. \tag{1.72}$$

1.2.1 *Radiometry*

Radiometry describes and quantifies the light energy or power transfer from a source transmitted through a medium or through an optical system (Wolfe, 1998). It measures energy mostly in the regions of the electromagnetic radiation, namely, ultraviolet (UV), visible, and infrared (IR). The optical power is, in general, expressed in radiometric unit (Siegel and Howell, 1981; Wyszecki and Stiles, 1982). In what follows, the fundamental quantities and the related parameters (see Table AI given in Appendix A) are elucidated in brief.

Radiant energy Measured in joules, this term is defined as the energy that is emitted or propagated in the form of particles or electromagnetic radiation. It is derived by integrating radiant flux with respect to time. The radiant energy is used for radiant heating, for example, this energy of the microwaves can be converted into thermal energy (heat). Also, it can be generated electrically by IR lamps or can be absorbed sunlight to heat water. The spectral radiant energy, which is the amount of radiant energy per unit wavelength interval at wavelength, λ, is given by

$$\mathcal{Q}_\lambda = \frac{d\mathcal{Q}_e}{d\lambda} \qquad \mathrm{J.nm}^{-1}, \tag{1.73}$$

where $\mathcal{Q}_e$ is the radiant energy.

Radiant flux Also known as the radiant power, the radiant flux is defined to be the energy per unit time, t

$$\Phi_e = \frac{d\mathcal{Q}_e}{dt} \qquad \mathrm{W}. \tag{1.74}$$

The spectral radiant flux is given by

$$\Phi_\lambda = \frac{d\Phi_e}{d\lambda} \qquad \text{W.nm}^{-1}. \tag{1.75}$$

Radiant flux density Measured in watts per square meter, the radiant flux density is defined as the radiant flux per unit area at a point on a surface. The flux may arrive at the surface, in which case it is called irradiance, $\mathcal{E}_e$; the flux can arrive from any direction above the surface (see Figure 1.3a). It is defined as

$$\mathcal{E}_e = \frac{d\Phi_e}{dA} \qquad \text{W.m}^{-2}, \tag{1.76}$$

where Φ designates the radiant flux arriving at the point and the area of intersection with the surface has a differential cross-sectional area dA.

On the contrary, the flux leaving the surface due to emission and/or reflection (see Figure 1.3b), the radiant flux density is referred to as radiant emittance, $\mathcal{M}_e$; the flux can leave in any direction above the surface. It is given by

$$\mathcal{M}_e = \frac{d\Phi_e}{dA} \qquad \text{W.m}^{-2}, \tag{1.77}$$

in which Φ_e is the radiant flux leaving the point.

The radiant flux density can be measured anywhere in three-dimensional space. The spectral irradiance, $\mathcal{E}_e(\lambda)$, and the spectral radiant emittance, $\mathcal{M}_e(\lambda)$, are given by the respective following equations

$$\mathcal{E}_e(\lambda) = \frac{d\mathcal{E}_e}{d\lambda} \qquad \text{W.m}^{-2}.\text{Hz}^{-1};$$
$$\mathcal{M}_e(\lambda) = \frac{d\mathcal{M}_e}{d\lambda} \qquad \text{W.m}^{-2}.\text{Hz}^{-1}. \tag{1.78}$$

Both $\mathcal{E}_e(\lambda)$ and $\mathcal{M}_e(\lambda)$ are functions of wavelength.

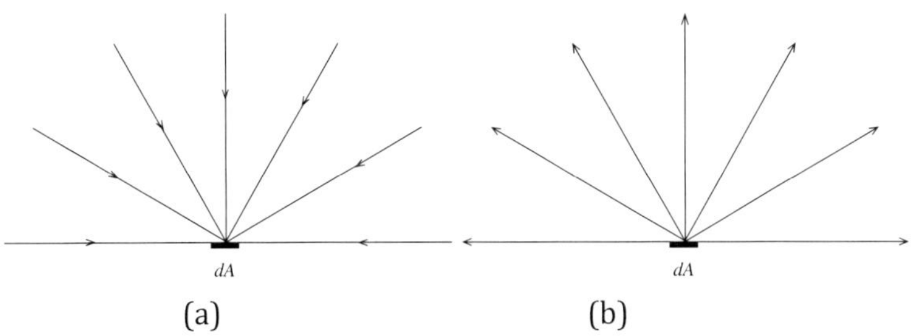

Figure 1.3 (a) The flux arriving at the surface and (b) radiant emittance in three dimensions.

Radiance The radiance, $\mathcal{L}_e$, is the amount of energy traveling at some point in a specified direction, per unit time, per unit area, perpendicular to the direction of travel, per unit solid angle $d\Omega$ (measured in steradian; sr). It is a fundamental quantity, which is invariant for an optical system that has no absorption or reflections. It remains constant as the radiation traverses through the optical system. It is pertinent to note that when radiation is incident on an object, a portion of it is transmitted, some absorbed, and some reflected. The reflectivity[a] and absorptivity,[b] which can be derived from Maxwell's equations (see Section 1.1), represent the values for an ideal material. However, the real materials deviate from the ideal properties and possess transmittance[c] reflectance, and absorptance. The total of these three terms is unity.

The radiance contains both areal and solid angle concept that is required to derive the radiant flux incident onto a system. The response of a sensor is proportional to the radiance. The radiance characterizes both emission from diffuse sources and reflection from diffuse surfaces; the diffusing materials illuminated by a light source may behave as a secondary source. The response of a camera or human eye is proportional to the radiance. If the angle between the surface normal and the ray is given by θ and dA the differential area surrounding the point, the differential form is $dA_{proj} = \cos\theta dA$. If an elemental cone $d\Omega$ containing a ray of light

---

[a]Reflectivity, ρ, is defined as the ratio of the power reflected (per unit area) to the power incident, while the emissivity, ϵ, is a measure of the ability of a body to radiate energy and is given by the ratio of the power radiated by the body (per unit area) to the power radiated by a blackbody at the same temperature. The relationship between these two terms, according to Kirchhoff's Law is

$$\rho = 1 - \epsilon.$$

[b]Absorptivity is the property of a body that determines the fraction of incident radiation absorbed by the body and the absorptance of a medium refers to its ability to absorb the wave energy per unit time as the wave propagates in the medium.

[c]Transmittance is defined as the ratio of the radiant flux transmitted through a material to that incident upon it. The probability of transmission of the photon that has propagated a distance l is described by Beer's law

$$T = e^{-\alpha(\lambda)l},$$

in which $\alpha(\lambda)(= \alpha_a + \alpha_s)$ denotes the extinction coefficient, α_a and α_s the absorption and scattering coefficients, respectively.

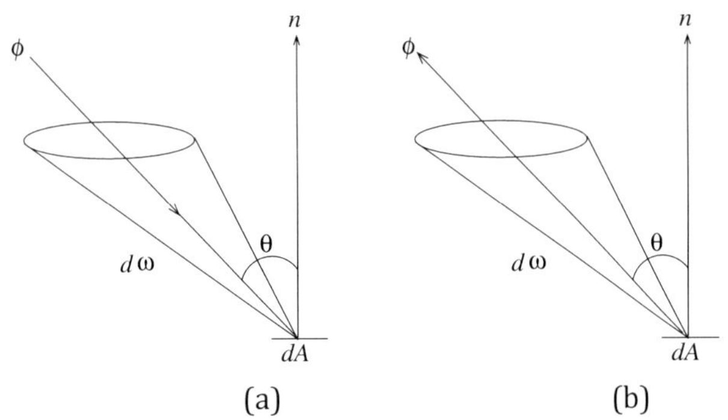

Figure 1.4 (a) Radiance arriving at the surface and (b) radiance leaving at the surface.

that is arriving at or leaving a surface (see Figure 1.4), the observed radiance in the direction, θ, is defined by

$$\mathcal{L}_e = \frac{d^2\Phi}{(\cos\theta d\Omega)dA} \approx \frac{\Phi}{\Omega A\cos\theta} \qquad \text{W.m}^{-2}.\text{sr}^{-1}, \tag{1.79}$$

with Φ as the total radiant flux and Ω the solid angle subtended by the observation; the approximation holds for the small area of the surface A and Ω, in which $\cos\theta$ is approximately constant.

Unlike the radiant flux density, the radiance does not distinguish between flux arriving at or leaving a surface. The spectral radiance is the radiance per unit wavelength interval at wavelength, λ, which is given by

$$\mathcal{L}_\lambda = \frac{d^3\Phi}{(\cos\theta d\Omega)dAd\lambda} \qquad \text{W.m}^{-2}.\text{sr}^{-1}.\text{nm}^{-1}. \tag{1.80}$$

Similarly, $\mathcal{L}_q$ is the spectral photon sterance expressed in photons per m^2.sr.nm. $\mathcal{L}_e$ and $\mathcal{L}_q$ are invariant for an optical system that lacks reflections or absorptions, which means that these parameters remains constant as the radiation traverses through such a system. For Lambertian sources (see Section 1.2.3), the radiance is emitted into a hemisphere having a solid angle π. The spectral emittance is related to the spectral radiance sterance by:

$$\mathcal{L}_e(\lambda, T) = \frac{\mathcal{M}_e(\lambda, T)}{\pi} \qquad \text{W.m}^{-2}.\text{sr}^{-1}.\text{nm}^{-1};$$
$$\mathcal{L}_q(\lambda, T) = \frac{\mathcal{M}_q(\lambda, T)}{\pi} \qquad \text{ph.m}^{-2}.\text{sr}^{-1}.\text{nm}^{-1}, \tag{1.81}$$

where $\mathcal{M}_q(\lambda, T)$ denotes the spectral photon emittance, which is defined as the spectral radiant emittance, $\mathcal{M}_e(\lambda, T)$, divided by the energy of one photon, $h\nu$, h Planck's constant, ν the frequency, λ the wavelength, and T the absolute temperature.

Radiant intensity It is a measure of the amount of radiation emitted from a point source, which radiates equally in every direction in three-dimensional space, into a unit solid angle in a particular direction. It is pertinent to note that according to the radiative heat transfer theory (for example, infrared light) does not use the concept of point source, but uses the radiant intensity to describe radiance (watts per unit area per unit solid angle). This quantity is expressed as the radiant flux per unit solid angle leaving this source. From the definition of a differential solid angle, $d\Omega (= dA/r^2)$, the radiant intensity, $\mathcal{I}_e$, is given by

$$\mathcal{I}_e = \frac{d\Phi_e}{d\Omega} \qquad \text{W.sr}^{-1}. \tag{1.82}$$

Since the radiant flux may intercept dA at an angle θ, using the inverse square law for a point source (see Figure 1.5), the radiant flux density (see equation 1.76) is written as

$$\mathcal{E}_e = \frac{\mathcal{I}_e \cos\theta}{d^2}, \tag{1.83}$$

where d is the distance from the source to the surface element, dA.

Inverse square law The inverse square law defines the relationship between illumination from a constant intensity light source and its distance from a surface. The intensity per unit area on the surface is inversely proportional to the square of the distance between the

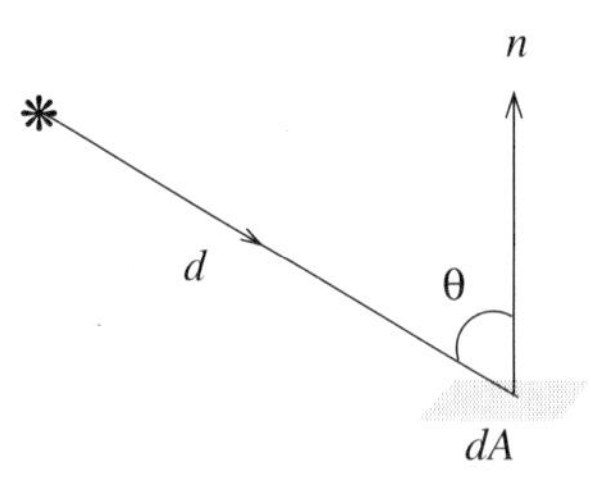

Figure 1.5 Apparent brightness follows inverse square law.

source and surface. The spectral radiant intensity at wavelength, λ, is defined as

$$\mathcal{I}_\lambda = \frac{d\mathcal{I}_e}{d\lambda} \qquad \text{W. sr}^{-1}.\text{nm}^{-1}. \tag{1.84}$$

1.2.2 *Photometry*

Photometry is measurement of light in units that are quantified according to the sensitivity of the human eye (see Section 4.2.1), the foundations of which were laid down by Bouguer (1729). In astronomy, photometry is the measurement of flux or intensity of stars and other celestial objects, such as nebulae, galaxies, planets, etc., at several wavelengths. Measurement of the spectral distributions of these objects is known as spectro-photometry. If the distance of the measured object is known, photometry may provide information about the total energy emitted by the object, its size, temperature, and other physical properties. The difference between radiometric and photometric theory is in their units of measurement. The unit of these quantities are differentiated by the subscripts, for instance, the subscript *e* is used for the former, while ν is used for the latter. The photometric units are obtained by integrating the corresponding radiometric unit multiplied with the wavelength sensitivity of the eye, over all visible wavelengths, and then multiplying by a suitable factor. Photometric lights are used when a physically-based lighting simulation is desired to achieve illumination effects (Walsh, 1958). In what follows, the photometric quantities and the related parameters (see Table AII given in Appendix A). are enumerated in brief.

Luminous flux Luminous flux, Φ_ν, is the amount of light energy arriving, leaving or going through a surface per unit time (dQ/dt). The measurement of this flux requires all of a source's power to be concentrated on a detector. Its unit of measurement is the lumen (lm; 1 lm ≈ *l*.46 × 10 3 W), which is defined as the luminous flux of a light source emitting one candela of luminous intensity over a solid angle of one steradian (sr), i.e., 1 lm = 1 cd. sr. The candela (one lumen per steradian), a Latin word for "candle," is the luminous intensity in a specified direction, of a light source emitting monochromatic radiation of frequency 540×10^{12} Hz, and it has a radiant intensity in that direction of $1/683$ W.sr^{-1}. An infinitesimally narrow-band source of light possessing 1 W at a peak wavelength of 555 nm of the relative luminous efficiency curve yields in luminous flux of 683 lm.

A source of radiative energy may be characterized by its spectral energy distribution, $\mathcal{E}_\lambda$, which specifies the time rate of energy the source emits per unit wavelength interval. The total power (radiant flux) emitted by a source is given by the integral of the spectral energy distribution

$$\Phi_e = \int_{0.36\ \mu\text{m}}^{0.8\ \mu\text{m}} \mathcal{E}_\lambda d\lambda \qquad \text{W.}\mu\text{m}^{-1},$$

where $\mathcal{E}_\lambda$ stands for the power in units of W.μm^{-1}.

The brightness sensation evoked by a light source with spectral energy distribution, $\mathcal{E}_\lambda$, is specified by its luminous flux, Φ_ν

$$\Phi_\nu = K_\text{m} \int_{0.36\ \mu\text{m}}^{0.8\ \mu\text{m}} \mathcal{E}_\lambda V(\lambda) d\lambda \qquad \text{lm}, \tag{1.85}$$

in which $K_\text{m}(= 683$ lm.W$^{-1})$ is the luminous efficacy for photopic vision (see Section 4.2.1) and $V(\lambda)$ the spectral luminous efficiency for photopic vision.

In order to determine the luminous flux from a source radiating over a spectrum, it is necessary to determine the spectral power distribution (SPD) for the particular source. Such a distribution, measured by a spectro-photometer, is a plot of the energy levels of a light source through a range of wavelengths of light. It contains all the basic physical parameters about the light and serves as the starting point for quantitative analyzes of color. After doing so, one may calculate the luminous flux at each wavelength, or at regular intervals for continuous spectra. Adding up the flux at each wavelength provides the total flux produced by a source in the visible spectrum.

Illuminance It is a measure of the amount of visible light incident on a surface of unit area. The SI derived quantity of which denotes luminous flux density and has a name called lux (lx), which is generated by a flux of one lumen (lm) incident on an area of 1 m^2, i.e., 1 lx = 1 lm.m^{-2}. The illuminance, $\mathcal{E}_\nu$, is used to describe illumination levels on a surface without being size-dependent. The measurement of illuminance is, in general, susceptible to errors introduced by off-axis light.

Luminous intensity The luminous intensity is defined as the quantity of luminous flux emitted uniformly from a point source within a solid angle of one steradian (sr). It measures light energy per unit time in a given direction; this measurement is not applicable to a collimated light source. The unit of luminous intensity is one lumen per steradian. This unit of luminous intensity is also called the standard candle, or candela (cd). The luminous intensity, $\mathcal{I}_\nu$, for a monochromatic source of a particular wavelength, λ is expressed as

$$\mathcal{I}_\nu = 683\overline{y}(\lambda)\mathcal{I}_e \qquad \text{cd},$$

in which $\mathcal{I}_e$ denotes the radiant intensity in W.sr^{-1} and $\overline{y}(\lambda)$ the standard luminosity function.

In the presence of more than one wavelength, $\mathcal{I}_\nu$ is given by

$$\mathcal{I}_\nu = 683\int_0^\infty \overline{y}(\lambda)\mathcal{I}_e(\lambda)d\lambda. \tag{1.86}$$

Luminance Also known as photometric brightness, it is a measure of the flux reflected off a surface in a given direction. The technique takes into account the area of the surface measured, and the angle subtended by an observer. This quantity is converted to colors to create a realistic rendering. Analogous to radiance, differentiating the lumen with respect to both area and direction, its unit is given by cd. m^{-2}.

Luminous energy This term is a measure of the rate of flow of flux. The measurement of any photometric quantity on a time-dependent basis can also be perceived. Measured in lumen seconds ($lm.s$), it is a photometrically weighted radiant energy.

Luminous emittance Measured in lux, it is defined as the luminous flux per unit area emitted from a surface.

1.2.3 *Lambert's Cosine Law*

Lambert's cosine law states that the radiant (or luminous) intensity observed from an ideal diffuse radiator is directly proportional to the cosine of the angle, θ, between the observer's line of sight and the normal to the surface. It may be applied to all Lambertian emitters in photometry. This law is dependent upon the cosine of the angle between the viewing direction and the normal to the surface and is given by the relation

$$I(\theta) = I_0 \cos\theta, \tag{1.87}$$

where $I(\theta)$ represents intensity in a direction θ with respect to a line normal to the emitting area, I_0 the brightness as seen from a direction normal to the surface, and θ the angle of the direction of view from the normal to the surface.

This indicates that the brightness of light on a surface of a fixed area varies with incident angle, which happens since the projected surface area in the plane perpendicular to the incidence is proportionally reduced. Thus, a sensor requires cosine correction in order to account for off-axis light. The intensity, $I(\theta)$, of a flat Lambertian light source obeys Lambert's law (for example, an ideal diffuser) from any direction less than 90° from the normal to the surface. A surface that obeys Lambert's law is referred to as Lambertian, and it exhibits Lambertian, reflectance. It is a flat radiating surface that reflects all incoming light without any losses and appears equally bright from all directions. It can be an elemental area of a curved surface, which may be an active (or a passive) reflective surface. Here, the intensity falls off as the cosine of the observation angle with respect to the surface normal. The radiance is independent of direction. The surface brightness can be expressed in candelas per projected square meter.

1.3 Opto-Electronic System Theory

The field of optical communication system has envisaged a remarkable success following the development of opto-electronic components, such as semiconductor lasers (see Section 3.3.8), low loss optical fibers (see Section 3.4), and photoelectric devices (see Chapter 5). The subject deals with the fiber-optic communication system, wave optics, Fourier optics, photon optics, coherent and non-linear light wave communications.

1.3.1 *Linear System*

A system is defined to be a mapping of a set of input functions into a set of output functions. In an electrical network, the inputs and outputs are real functions of 1-D independent variable (time). The response to a unit-impulse must be independent of the time at which that input is applied, known as time-invariant system, since their characteristics do not change with time. Its impulse response, $h(t, \tau)$ (its response at time t to a unit impulse applied at time τ) depends on the time difference $(t - \tau)$. In an optical imaging system, the input and output can be real-valued function (intensity) or the complex-valued function (field amplitude) of 2-D independent variable (space). In this, the point spread function (PSF), defined as the intensity distribution in diffraction pattern for apertures (pupil) caused by a point source, with or without aberrations, in the field-of-view (FOV; see Section 4.4.1.4), must be the same over the entire field, what is known as the space-invariant (isoplanatic) system. It is proportional to the modulus square of the Fourier transform of its pupil transmission function. The aberration in the beam is due to imperfect, dirty, or damaged optics, which influences causing different rays to converge to different points. Aberrations generate from subsets of a larger centered bundle of spherically aberrated rays. Typically, an optical system suffers from spherical aberration, coma, astigmatism, and chromatic aberration.

A linear system is characterized by its response to a delta function, known as impulse response, for example, the term, $\mathcal{L}[\delta(\mathbf{x})]$, is the impulse response of an optical system. Here $\mathcal{L}$ is the mathematical operator and $\mathbf{x}(= x, y)$ the 2-D position vector. In a time- and/ or shift-invariant linear system, the eigen functions are exponential e^{iux}. Such a system responds to an harmonic input by an harmonic output at the same frequency u and these responses specify the properties of the system. Many physical processes may be approximated as being linear shift-invariant systems. If the function $g_o(\mathbf{x}_0)$ represents the input to a system and $g_i(\mathbf{x}_1)$ represents the corresponding output in an optical system, by the definition of $\mathcal{L}$, the two functions are related by the following equation

$$g_i(\mathbf{x}_1) = \mathcal{L}\left[g_o(\mathbf{x}_0)\right]. \tag{1.88}$$

The Fourier transform[a] of the convolution of a single aperture function with the Dirac delta function (unit impulse function, δ, array is equal to the product of the individual transform; the Dirac delta function is defined by the property

$$\delta(x) = \begin{cases} 0 & x \neq 0, \\ \infty & x = 0, \end{cases}$$

with the integral of the Dirac delta from any negative limit to any positive limit as 1, i.e., $\int_{-\infty}^{\infty} \delta(x)dx = 1$. Pulsed light sources are sometimes referred to as delta light sources when the length of the pulse is short with respect to the response time of a detecting instrument, say a photomultiplier (see Section 5.2.2).

The input is decomposed by the shifting property of the Dirac delta function, δ as

$$g_o(\mathbf{x}_0) = \int_{-\infty}^{\infty} g_o(\boldsymbol{\xi})\delta(\mathbf{x}_0 - \boldsymbol{\xi})d\boldsymbol{\xi}, \tag{1.89}$$

where the coefficient, $g_o(\boldsymbol{\xi})$, is the transmittance pattern for an optical system and $\boldsymbol{\xi}(= \xi, \eta)$ the 2-D position vector.

A transfer function is a complex function representing the relation between the input and output of a linear time-invariant system. Its magnitude is the ratio of the intensity modulation in the image to that in the object, what is referred to as the modulation transfer function (MTF). Also, referred to as the spatial frequency response, it decreases as the frequency rises. The argument of the transfer function is called the phase transfer function (PTF).

The sharpness of a photographic imaging system or of a component of the system, for example, film, lens, image sensor, etc., is characterized by this parameter. To note, the image quality

[a]At a certain time, the wave group can be represented by the Fourier integral where the function representing the amplitudes of the wave vary with the frequency. This function is called the Fourier transform. In the space domain, the Fourier transform pair can be expressed as

$$\widehat{f}(u) = \int_{-\infty}^{\infty} f(x)e^{-i2\pi ux}dx,$$

$$f(x) = \int_{-\infty}^{\infty} \widehat{f}(u)e^{i2\pi ux}du.$$

Since there is considerable symmetry within each of these pairs of equations, $\widehat{f}(u)$ and $f(x)$ are each described as the Fourier transform of each other.

measures the perceived degradation with respect to an ideal image. A method called sharpness factor, employed by British Broadcasting Company, determines the amount of detail an image can convey. It is defined as the ratio of the area contained under the MTF curve up to the cut-off frequency of the system, divided by the area under an idealized MTF (a rectangle of unit amplitude) up to the cut-off frequency. Equation (1.89) is regarded as the expression of g_o as a linear combination of weighted and displaced function. Suppose for a rectangular slit

$$g_o(\boldsymbol{\xi}) = \begin{cases} 1 \text{ for,} & |\xi| \leq a/2; |\eta| \leq b/2, \\ 0 & \text{otherwise;} \end{cases}$$

and therefore, on substituting equation (1.89) into equation (1.88), one gets

$$\begin{aligned} g_i(\mathbf{x}_1) &= \mathcal{L}\left[\int_{-\infty}^{\infty} g_o(\boldsymbol{\xi})\delta(\mathbf{x}_0 - \boldsymbol{\xi})d\boldsymbol{\xi}\right] \\ &= \int_{-\infty}^{\infty} g_o(\boldsymbol{\xi})\mathcal{L}\left[\delta(\mathbf{x}_0 - \boldsymbol{\xi})\right] d\boldsymbol{\xi}. \end{aligned} \tag{1.90}$$

A system is linear if the principle of superposition[a] applies. This principle states that the system response to the sum of two inputs is equal to the responses to the each of the two inputs as expressed in equation (1.90). In order to ensure the shift-invariance, the isoplanatism (space invariance) condition imposes that the supports of the functions g_o and g_i is restricted to the isoplanatic (space-invariant) patch.

A linear imaging system is said to be a space-invariant if its impulse response $h(\mathbf{x}_1, \boldsymbol{\xi})$, depends on the distance, $(\mathbf{x}_1 - \boldsymbol{\xi})$. For such a system, it is written as

$$h(\mathbf{x}_1 - \boldsymbol{\xi}) = h(\mathbf{x}_1, \boldsymbol{\xi}),$$

which denotes the response of the system at point $\mathbf{x}_1$ of the output space to a function input at coordinates $\boldsymbol{\xi}$ of the input space. In reality, imaging systems are seldom isoplanatic over their object

[a] A system is said to be linear if the following superposition property is obeyed for all input functions, $g_{01}(\mathbf{x}_0)$ and $g_{02}(\mathbf{x}_0)$ and all complex constants, a and b

$$\mathcal{L}[ag_{01}(\mathbf{x}_0) + bg_{02}(\mathbf{x}_0)] = a\mathcal{L}[g_{01}(\mathbf{x}_0)] + b\mathcal{L}[g_{02}(\mathbf{x}_0)].$$

field; however, it is possible to divide the object field into small regions within which the system is approximately invariant. In the presence of atmospheric turbulence, the time-invariance condition for the impulse response is needed. If these conditions are met, the impulse response is constant, and

$$\mathcal{L}\left[\delta(\mathbf{x}_0 - \boldsymbol{\xi})\right] = h(\mathbf{x}_1 - \boldsymbol{\xi}). \tag{1.91}$$

The system input-output is related by the following expression:

$$g_i(\mathbf{x}_1) = \int_{-\infty}^{\infty} g_o(\boldsymbol{\xi}) h(\mathbf{x}_1 - \boldsymbol{\xi}) d\boldsymbol{\xi} \tag{1.92}$$

$$= g_o(\mathbf{x}) \star h(\mathbf{x}), \tag{1.93}$$

where $\star$ denotes 2-D convolution.[a]

Equation (1.92) demonstrates that if the image formation system is space-invariant, the PSF depends on the differences of the corresponding coordinates. The effects of imaging elements of an imaging system can be described by specifying the images of point sources located throughout the object field. The output of a shift-invariant system is the convolution of the input and the impulse response of the system. In this case, a displacement of the object would result in a displacement of the image but not a change of image configuration. The convolution of two functions results in a function which is broader than both, the implication of which is that all optical systems decrease the resolution of an image. In a perfect image formation system, the image of a point source is not blurry. A linear and space-invariant image formation system is represented in the spatial frequency domain. After taking the Fourier transform

[a]The convolution of two functions simulates phenomena, such as a blurring of a photograph, in which each point of object is replaced by a spread function. The mathematical description of convolution of two functions is of the form

$$h(x) = \int_{-\infty}^{\infty} g(x_0) f[(x - x_0)] dx_0 = f(x) \star g(x),$$

where $h(x)$ is the output value at a particular point x and $g(x)$ the blurring function, and geometrically, the convolution operation $\star$ is common area between two functions.

In the Fourier plane the effect turns out to be a multiplication, point by point, of the transform of $\widehat{f}(u)$ with the transfer function, $\widehat{g}(u)$, that is

$$\widehat{h}(u) = \widehat{f}(u)\widehat{g}(u).$$

(FT) of both sides of equation (1.93), the spectra, $\widehat{g_i}(\mathbf{u})$ and $\widehat{g_o}(\mathbf{u})$ of the system output and input are seen to be related by the following expression:

$$\widehat{g}_i(\mathbf{u}) = \widehat{g}_o(\mathbf{u})\widehat{h}(\mathbf{u}), \tag{1.94}$$

in which $\widehat{h}(\mathbf{u})$ is the transfer function of the system, $\mathbf{u}(= u, v)$ the 2-D spatial frequency vector, $u = x/\lambda$, $v = y/\lambda$, and $\hat{}$ the Fourier transform

$$\widehat{h}(\mathbf{u}) = \int_{-\infty}^{\infty} h(\boldsymbol{\xi})e^{-i2\pi\mathbf{u}\cdot\boldsymbol{\xi}}d\boldsymbol{\xi}. \tag{1.95}$$

This function, $\widehat{h}(\mathbf{u})$ indicates the effect of the system in the frequency domain. To note, the transformation in optics is from the spatial domain to a spatial frequency spectrum (cycles per mm), while in electronics the transformation is from time to time-frequency spectrum (cycles per sec). A cosinusoidal input to a real time invariant linear system produces a cosinusoidal output.

1.3.2 *Transfer Functions of Optical and Opto-Electronics Systems*

Any mechanism operating on either electrical or optical information modifies the input signal. The difference between input and output is related to a describable property of the device that makes the MTF useful. With optical imaging devices, the transfer function is the Fourier transform of the PSF, thus the resolution at the image plane is determined by the width of this PSF. Once the transfer function of a particular device is known, a manipulation of that function allows to describe the differences between the input and output information without prior knowledge of the nature of the information. The advantage of such an approach to system description is that the transfer functions of each individual components may be multiplied together for providing the overall system transfer function.

The simplest optical system is the section of free space length, s, in the z-direction of the Cartesian co-ordinate system. A plane wave propagates as $e^{-i\boldsymbol{\kappa}\cdot\mathbf{r}}$ in which $\boldsymbol{\kappa}(= \kappa_x\mathbf{i} + \kappa_y\mathbf{j} + \kappa_z\mathbf{k})$, $\mathbf{r}$ is the position vector of a point (x, y, z), with a uniform velocity; the wave vector, in general, has components along the three axes of the Cartesian coordinate system, $\boldsymbol{\kappa} \cdot \mathbf{r}(= \kappa_x x + \kappa_y y + \kappa_z z)$ represents planes in space of constant phase. Thus, the transfer function of the system is given by

$$e^{-i\kappa_z d} = e^{-id\sqrt{\kappa_0^2-\kappa_x^2-\kappa_y^2}}, \tag{1.96}$$

with $\kappa_z = \sqrt{\kappa_0^2 - \kappa_x^2 - \kappa_y^2}$, and $\kappa_0 = 2\pi/\lambda_0$.

The wave attenuates if $\kappa_x^2 + \kappa_y^2 > \kappa_0^2$, hence, a section of the free space works as a low pass filter. In cylindrical coordinate system, on writing, $\kappa_x^2 + \kappa_y^2 = \kappa_\rho^2$, one obtains

$$\widehat{h}(\kappa_\rho) = e^{-id\sqrt{\kappa_0^2-\kappa_\rho^2}}, \tag{1.97}$$

and the system attenuates all spatial frequencies, $\kappa_\rho > \kappa_0$.

In an optical system where the object is illuminated from behind (trans-illuminated), the image is formed from the transmitted light. No true object can be perfectly thin; hence, an incident ray exits at slightly different transverse coordinates. A transmitting object, such as a lens may be considered thin lens (whose thickness is negligible compared to its aperture) if a ray entering at co-ordinates (x, y) on one face emerges at approximately the same on the opposite face. Such a lens delays an incident wavefront by an amount proportional to the thickness of the lens at each point. Ordinary lenses are pieces of glass with spherical surfaces. The line passing through the centers of curvature of the surfaces is called the axis of the lens. A lens is described based on its focal length (mm) defined by the distance at which a lens focuses parallel rays of light. The focal length specifies the view angle through the lens, which determines the field-of-view (FOV; see Section 4.4.1.4) based on the distance to the object. A quantity, called aperture speed, $F_\#$, is a number, defined as the ratio of the lens focal length, f, divided by the effective diameter, D, of the aperture, i.e., $F_\# = f/D$ (Smith, 2007). It refers to the speed in which a lens can transmit light. As the $F_\#$ gets larger, the light speed decreases, resulting in a lower image quality. The chief varieties of lenses are (i) biconvex, (ii) biconcave, (iii) plano-convex, (iv) plano-concave, and (v) convexo-concave or meniscus.

For a non-uniform thickness of the object, the refractive index varies from point to point, refraction within the object modifies the position at which a given ray exits. The transmittance of a thin lens of focal length f at a wavelength, λ is given by

$$t(x, y) = e^{i\kappa n\Delta_0} e^{-i\kappa(x^2+y^2)/(2f)}, \tag{1.98}$$

where Δ_0 is the maximum thickness, n the refractive index of the material.

On dropping out the constant phase term, $e^{i\kappa n\Delta_0}$, equation (1.98) is recast as

$$t(x, y) = e^{-i\kappa(x^2+y^2)/(2f)}. \tag{1.99}$$

1.4 Image Formation

An image of an extended object is essentially a convolution of the brightness distribution in the object with the diffraction pattern of a point source (impulse response) produced by the imaging system. A perfect image of a small object lying at a large distance from an imaging system is an exact replica of the object except for its magnification. The magnification is a number expressing change in object to image size. Such an image is known to be the Gaussian image.

Let an object function $f(x, y)$ be placed at a distance s_o in front of a thin lens in order to determine the transfer function at a distance s_i behind the lens (see Figure 1.6). It is considered that the object and the image lie in mutually parallel planes that are perpendicular to the optical axis of the system; the entrance and the exit pupils, respectively, lie in planes that are parallel to the object and the image planes. The propagation of light inside the system is limited by one aperture. The transfer of light from the object to the image plane, barring the effects of diffraction at the pupil, is treated by the laws

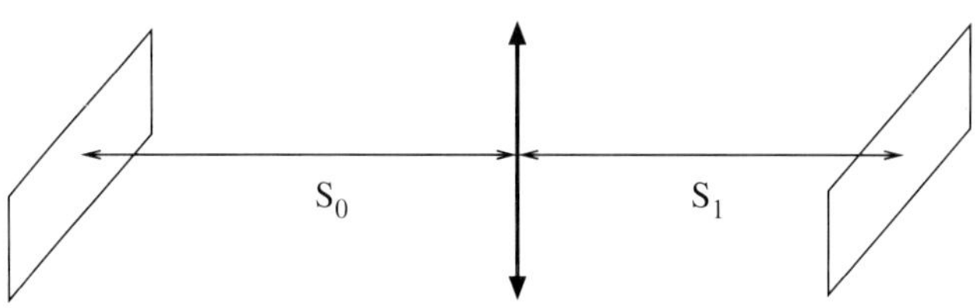

Figure 1.6 Geometry for the image formation in which s_o is the distance between the source and the entrance aperture of an imaging system, for instance, a lens in this case and s_i the distance between the lens and the image of the source.

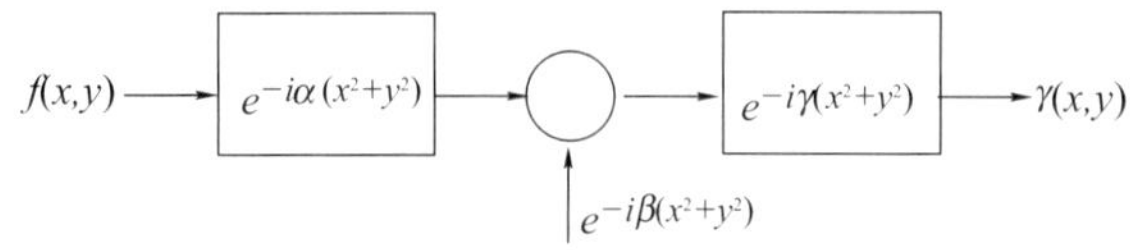

Figure 1.7 Cascade of subsystems representing the image formation system given in Figure 1.6.

of geometrical optics. Since a section of free space of length, s is represented in terms of PSF, $h(x, y)$

$$h(x, y) = e^{-i\kappa(x^2+y^2)/(2s)}, \tag{1.100}$$

the coherent transfer function, $\widehat{h}(\kappa_x, \kappa_y)$ is given by

$$\widehat{h}(\kappa_x, \kappa_y) = e^{is\kappa_\rho^2/(2\kappa)}. \tag{1.101}$$

It is assumed that the PSF is independent of the image-plane coordinates. The PSF is shift-invariant and the Fourier transform of which is the coherent transfer function. The image formation property of the composite system may be explained by a cascade of sub-networks representing the components of building blocks with the known PSF and transfer function as given in Figure 1.7. The optical transfer function (OTF) of an optical system describes the imaging quality representing the complex factor applied by an optical imaging system to the frequency components of object intensity distribution relative to the factor applied to the zero frequency component. It is the normalized (scaling so that the value at the zero frequency turns out to be one) Fourier transform of its PSF, which can be used to express OTF in terms of the autocorrelation[a] of its pupil function. The output of the network is

$$\gamma(x, y) = \left\{\left[f(x, y) * e^{-i\alpha(x^2+y^2)}\right] e^{-i\beta(x^2+y^2)}\right\} * e^{-i\gamma(x^2+y^2)}, \tag{1.102}$$

[a]The autocorrelation of $f(x)$ in the plane x, is the correlation of $f(x)$ and $f(x)$ multiplied by the complex exponential factor with zero spatial frequency; therefore, the resulting representation is

$$\mathcal{F}\left[\int_{-\infty}^{\infty} f(x')f^*(x'-x)dx'\right] = \mathcal{F}\left[f(x) \otimes f(x)\right] = \left|\widehat{f}(u)\right|^2.$$

where $\otimes$ stands for correlation.

with

$$\alpha = \frac{\kappa}{2s_o}, \ \beta = \frac{\kappa}{2f}, \ \text{and} \ \gamma = \frac{\kappa}{2s_i}. \tag{1.103}$$

In the case of $s_o = f = s_i$, the output field distribution becomes,

$$\gamma(x_i, y_i) = \iint_{-\infty}^{\infty} f(x, y) e^{-i2\pi(xx_i + yy_i)/(\lambda f)} dxdy. \tag{1.104}$$

If the object is placed in front focal plane of the lens, the phase curvature disappears, leaving an exact Fourier transform relation at the back focal plane.

In the case of an incoherent source, which is considered to be a mosaic of point sources, the intensity distribution of the image of an incoherent object is obtained by adding the intensity distributions of the images of its elements; the impulse response in the image plane vary statistically independent fashions. According to van Cittert–Zernike theorem, the degree of coherence as a function of spatial separation is the same as the diffraction pattern due to an aperture of the same size and shape as source. They must, therefore, be added on a received power or intensity basis rather than on actual electric field amplitude and phase. The intensity, $I(\mathbf{x}_1)$ is of the form

$$I(\mathbf{x}_1) = \int_{-\infty}^{\infty} O(\mathbf{x}_0) |h(\mathbf{x}_1 - \mathbf{x}_0)|^2 d\mathbf{x}_0, \tag{1.105}$$

in which $O(\mathbf{x}_0)$ represents the spatial intensity distribution in the object plane, where the light is assumed to be incoherent (as in an actual source).

Equation (1.105) is a convolution of the intensity distribution in the object with the squared modulus of the transmission function

$$I(\mathbf{x}) = O(\mathbf{x}) \star |h(\mathbf{x})|^2. \tag{1.106}$$

On defining the normalized frequency spectra

$$\mathcal{G}_o(\kappa_x, \kappa_y) = \frac{\iint_{-\infty}^{\infty} O(x_0, y_0) e^{i(\kappa_x x_0 + \kappa_y y_0)} dx_0 dy_0}{\iint_{-\infty}^{\infty} O(x_0, y_0) dx_0 dy_0} \tag{1.107}$$

$$\mathcal{G}_i(\kappa_x, \kappa_y) = \frac{\iint_{-\infty}^{\infty} I(x_1, y_1) e^{i(\kappa_x x_1 + \kappa_y y_1)} dx_1 dy_1}{\iint_{-\infty}^{\infty} I(x_1, y_1) dx_1 dy_1}. \tag{1.108}$$

On employing convolution theorem

$$\mathcal{G}_i(\kappa_x, \kappa_y) = \mathcal{G}_o(\kappa_x, \kappa_y) \widehat{h}(\kappa_x, \kappa_y), \tag{1.109}$$

with

$$\widehat{h}(\kappa_x, \kappa_y) = \frac{\iint_{-\infty}^{\infty} |\widehat{h}(x, y)|^2 e^{i(\kappa_x x + \kappa_y y)} dx dy}{\iint_{-\infty}^{\infty} |\widehat{h}(x, y)|^2 dx dy}. \quad (1.110)$$

The function $\widehat{h}(\kappa_x, \kappa_y)$ is referred to as optical transfer function (OTF) of the system. The spatial frequency spectrum of the diffracted image of an isoplanatic incoherent object is equal to the product of the spectrum of its Gaussian image and the OTF of the system. The spatial frequency spectrum of the diffracted image of an isoplanatic incoherent object is equal to the product of the spectrum of its Gaussian image and the OTF of the system. The OTF obeys the following properties (Goodman, 1968)

$$\widehat{h}(0, 0) = 1,$$

$$\widehat{h}(-\kappa_x, -\kappa_y) = \widehat{h}^*(\kappa_x, \kappa_y), \quad (1.111)$$

$$|\widehat{h}(\kappa_x, \kappa_y)| \leq |\widehat{h}(0, 0)|.$$

The measuring arrangement, in the case of point spread function (PSF) or line spread function (LSF) analysis, is simple; however, it loses accuracy owing to the lower photometric value and the signal-to-noise (S/N) ratio. This difficulty may be avoided in the case of sine wave analysis, in which a slit is used as object and the image is scanned by a sinusoidal object pattern or vice versa. The transfer functions of certain important elements are also necessary to be determined, which are enumerated below.

1.4.1 *Electro-Optical System*

An electro-optic system is capable of producing and controlling a beam of electrons in order to produce an image of an object; it uses magnetic and/or electrostatic fields. In order to evaluate the imaging properties of such a system, it is essential to know how various sources of aberration combine to blur the size of the final image or spot. The transfer function of the electro-optical lens incorporated in the device needs to be defined. The degradation of the image quality is due to

- the existence of non-linear effects (aberrations) in the image transformation by electron optical tract as a consequence of the difference of optical length for the trajectories commenced from the different points of the object to the screen cross section by these trajectories,

- image transport from a flat surface to curvilinear one,
- parasitic screen illumination caused by a portion of the secondary electrons that comes out owing to the excitation of the molecules of special coating on the screen by the electrons, and
- the technology deviations from the ideal device, which may be in the form of the deviations of the shift or tilt of the axes, elliptic and other deformations of the electrodes.

An optical or electro-optical system may be characterized by analyzing the image produced by a narrow slit object or pulse in space or to a sinusoidally varying test object; the result may be interchanged by the Fourier transform. The frequency region in which the system operates is, in general, well corrected for PTF; therefore, it may not contribute significant effect to the determination of the transfer function. Figure 1.8 shows an electro-optical system, a sub-system in proximity focused image intensifier. The proximity focus intensifiers of new generation are of compact mechanical construction with their length being smaller than their diameter. If the emission of photoelectron in the input plane is

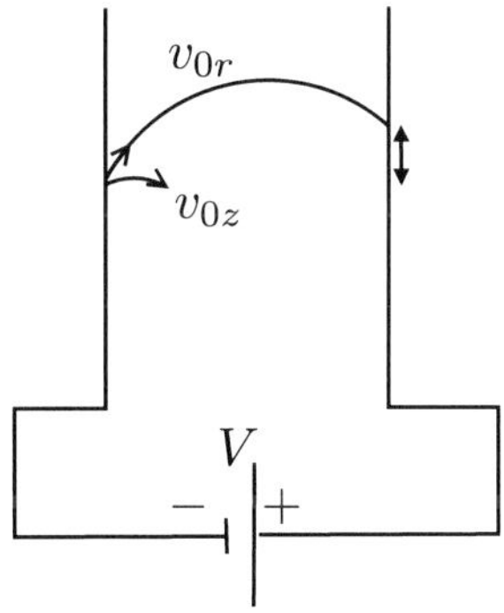

Figure 1.8 Geometry of a simple electro-optical transmission system that has no focusing mechanism; it is a sub-system in proximity focused image intensifier.

accompanied with a radial velocity,[a] v_r, along with the desirable axial velocity, v_z, a point object on the input plane may appear smeared at the output image surface with a radius of confusion,[b] r. Therefore, the MTF of the system can be found as

$$M(\kappa_\rho) = 2\pi \int_0^\infty \mathcal{P}(r) I_0(\kappa_\rho r) dr, \tag{1.112}$$

where $\mathcal{P}(r)$ is the probability distribution of electrons, $I_0(\kappa_\rho r)$ the current density distribution, and $\kappa_\rho (= 2\pi f_\rho)$ the spatial frequency.

1.4.2 *Photographic Film*

L. Daguerre (1787–1851) developed a photochemical process, known as the Daguerreotype, in 1839, which became the basis of photography. The photosensitive emulsion of silver (Ag) absorbs light. The crystalline grains of silver bromide (AgBr), which were used from around 1880, are suspended in a gelatin emulsion. When an incident photon excites an electron of the emulsion into the conduction band, and leaves a positively charged hole (see Section 2.2.4.1). The released electron is recaptured by a defect in the crystal lattice (see Section 2.2.2). The chemical reaction is given by $\text{Ag}^+\text{Br}^- + h\nu \rightarrow \text{Ag}^+ + \text{Br} + e^-$. The silver ion can combine with the electron to form a neutral silver atom, i.e., $\text{Ag}^+ + e^- \rightarrow \text{Ag}$. The free silver produced in the exposed silver halide makes up the latent image. The latent image is later amplified in the developing process. The deposit of silver produces a dark area in the film. The noise, or grain, of the plate is due to the formation of silver grains at the

---

[a]The radial velocity, v_r, defined as the velocity of a celestial object in the direction of the line of sight

$$v_r = \frac{c(\lambda_o - \lambda)}{\lambda} = \frac{c(\nu - \nu_o)}{\nu_o}$$

is detected by looking for Doppler shifts in the star's spectral lines. An observer can measure the radial velocity accurately by taking a high-resolution spectrum and comparing the measured wavelengths of known spectral lines to wavelengths from laboratory measurements. This can be used to estimate the masses of the stars, as well as some orbital elements, such as eccentricity and semi-major axis. Such a method has also been used to detect exo-solar planets.

[b]The circle of confusion is an optical spot caused by a cone of light rays from a lens not coming to a perfect focus while imaging a point source; it is also known as blur spot.

lattice defects without the action of an incident photon. The step for developing process are

- developing the plate over mercury heated to 75°C, which causes the mercury to amalgamate with the silver, which is followed by
- fixing the image in a warm solution of common salt or sodium sulfate, and
- rinsing the plate in hot distilled water.

This yields a negative from which copies may be made. However, this process had a few weaknesses, such as (i) the surfaces were extremely delicate, (ii) the image was reversed laterally, and (iii) the chemicals that were used, for example, bromine and chlorine fumes and hot mercury, were toxic.

After several decades or so, the quality of photographs improved significantly, particularly with the development of rolled paper film and dry emulsion plates. These plates were extensively used by astronomers. However, the analytical derivation of the MTF of photographic material is complicated due to high degree of non-linearities and adjacency effects. Therefore, some designers rely on the empirically derived expression for the MTF. The following equation for the MTF, $|\widehat{h}(\kappa_\rho)|$, provides satisfactorily result:

$$|\widehat{h}(\kappa_\rho)| = \frac{1}{[1 + (K'\kappa_\rho)^2]^{3/2}}, \tag{1.113}$$

in which K' denotes the film constant.

1.5 Effects of Atmosphere on the Star Images

In ground-based telescopes, the diffraction-limited resolution is never achieved as the telescopes have to look through the Earth's atmosphere. Three factors related to atmospheric processes that affect optical wave propagation are absorption, scattering, and turbulence. The absorption and scattering by the constituent gases and particulates of the atmosphere give rise to the attenuation of an optical wave. The atmosphere is a highly turbulent medium. Due to the turbulence and the concomitant development of thermal convection in the atmosphere, the density of air fluctuates in space and time, which leads to irradiance fluctuations, beam broadening, and loss of spatial coherence of the optical wave, among other effects. These effects have consequences on the transmission of optical waves through the atmosphere, astronomical imaging in particular.

The Earth's atmosphere is a mixture of gases and is primarily composed of nitrogen (N_2; ~78%), oxygen (O_2; ~21%), and argon (Ar; ~1%). The notable other components are water (H_2O; ~0–7%), ozone (O_3; ~0–0.01%), and carbon dioxide (CO_2; ~0.01–0.1%). The atmosphere is thickest near the surface of the Earth and thins out with height until it merges with interplanetary space. Depending on the thermal characteristics, chemical compositions, movement, and density that decays exponentially with altitude, it is divided into four primarily layers, such as (i) troposphere (extends to ~8–14 km from the surface of the Earth), (ii) stratosphere (begins above troposphere and extends to ~50 km), (iii) mesosphere (commences above the stratosphere and extends to ~85 km), and thermosphere (starts above the mesosphere and extends to as high as ~600 km). These layers are separated by three isothermal boundaries called tropopause, stratopause, and mesopause, respectively. The atmosphere extends to great heights, with density declining by a factor of $e(2.718)$ over an altitude interval given by a variable quantity, called the scale height $H[= k_B T/(\mu g)]$, with k_B as Boltzmann's constant, T the mean surface temperature, $\mu(= 1.3\ \mathrm{kgm^{-3}})$ is related to specific mass of air, and g the acceleration due to gravity on the Earth's surface; the scale height has the dimension of length.

1.5.1 *Absorption*

Absorption is a quantum process where atmospheric molecules absorb energy from incident photons. For instance, as the electromagnetic waves propagate in the Earth's atmosphere, some or all of the energy contained in the waves are absorbed by the atmosphere and get converted into heat, resulting in attenuation of the wave energy (Weik, 1997). When a photon is absorbed by a system, it increases the internal energy of the system. This energy increase or decrease with changes in (i) translation kinetic energy of molecules, (ii) both vibrational and rotational kinetic energies of polyatomic molecules, and (iii) distributions of electric charge within a molecule. Collisions between molecules tend to equalize the distribution of total energy in a gas. The molecular energy states arise from the following aspects:

Rotational absorption spectrum The incident electromagnetic waves can excite the rotational levels of molecules if they possess an electric dipole moment. The electromagnetic field exerts a torque on the molecule. The rotational states are separated by a small energy intervals (typically 10^{-3} eV; Beiser, 1973). The spectra for rotational transitions of molecules is typically in the microwave region with wavelengths of 0.1 mm to 1 cm. For a diatomic molecule, the rotational energy is obtained from Schrödinger's equation (see equation 1.28) with the Hamiltonian expressed in terms of the angular momentum operator. The rotational energy level is

$$E_j = \frac{j(j+1)\hbar^2}{2I},$$

in which j is the rotational quantum number (see Section 1.1) and I the moment of inertia that is defined as the summation of all the mass points, m_i times the square of the distances r_i from those points to the center of mass of the system, i.e., $I = \sum_i m_i r_i^2$. The rotational absorption spectra, which arise from transitions between rotational energy states, is $\hbar(j+1)/(2\pi I)$. In the case of rigid molecule, the frequency of the absorbed photon is

$$\nu_{j\rightarrow j+1} = \frac{\Delta E}{h} = \frac{\hbar(j+1)}{2\pi I}.$$

When sufficiently excited, a molecule may vibrate, as well as rotate.

Vibrational translations The vibrational transitions tend to be associated with larger energy intervals (typically 0.1 eV); hence, absorptions or emissions occur at shorter wavelengths of 1 μm to 0.1 mm. A diatomic molecule behaves like a simple harmonic oscillator with resonant frequency

$$\nu' = \frac{1}{2\pi}\sqrt{\frac{k}{m'}},$$

which is quantized to actual vibrational frequency, k the constant, and m' the reduced mass.

On solving the harmonic oscillator problem quantum mechanically, the energy of the oscillator turns out to be

$$E_v = \left(v + \frac{1}{2}\right) h\nu',$$

with $v(= 0, 1, 2, 3, \cdots)$ as the vibrational quantum number.

The vibrational energy levels of a diatomic molecule are specified by (Beiser, 1973)

$$E_v = \left(v + \frac{1}{2}\right) \hbar \sqrt{\frac{k}{m'}}.$$

The selection rule for transitions between vibrational states is

$$\Delta v = \pm 1.$$

Electronic transitions This transition involves the energy levels of electrons orbiting an atomic nucleus. The molecular electron states have the highest energies, with typical separations between the energy levels of outer electrons of several eV. The corresponding spectra are in the visible and UV regions.

Absorption helps the Earth by limiting exposure to harmful radiation, such as X-ray and UV. Also, it acts as a heat source for the Earth. The molecules in the atmosphere contributing to absorption are O_2, N_2, CO_2, Ar, methane (CH_4), nitrous oxide (N_2O), carbon monoxide (CO), water vapor (H_2O), and O_3. The ultraviolet (UV) light below 0.2 μm is absorbed by O_2 and O_3, but little absorption occurs at visible wavelengths except for H_2O absorption between 0.65 and 0.85 μm. Both CO_2O and water vapor are the major absorbents of infrared (IR) radiation.

The H_2O-vapor and O_3 concentrations have significant spatial and temporal variation. The water vapor content determines the current synoptic situation, where advection of moist or dry air would dictate the local moisture profile. The O_3 concentration peaks in the stratosphere, which lies between 15 and 30 km, and is produced by photochemical processes. It varies with the synoptic situation as well, where stratospheric air is often ingested into the troposphere during a cold frontal passage. The O_3 often produced from combustion engine exhaust in the lower atmosphere by photochemical reactions.

Air-borne aerosols also contribute to the radiative transfer mechanism; the large particles can extinguish light through scattering and absorption. They are tiny particles, such as dust particles, water droplets in the form of haze or fog, smoke, pollutants, snow, rain, etc., which can be of different sizes, shapes, distributions, constituents, and concentrations and are found at all altitudes and at different geographic locations. The large particles can extinguish light through scattering and absorption. Aerosols are, in general, created at the Earth's surface and their concentration drops at higher altitude. These particles absorb in the entire optical region, and also add to the background emission in the infrared. Information on ambient temperature and relative humidity can be used to obtain the surface water vapor pressure. This value, in conjunction with measured water vapor column, provides an estimate of water vapor scale height. These statistics are of interest towards an astronomical site characterization.

1.5.2 *Scattering*

Scattering results from photons colliding with atmospheric particles. In this process, all (or some) of the optical power in a mode is transferred into another mode, for example, the light waves are reflected or scattered by tiny particles in the atmosphere (Hulst, 1981). The scattering medium introduces random path fluctuations on reflection or transmission. The large and rapid variations of the phases are the product of these fluctuations with wave vector, $\boldsymbol{\kappa}$. A change in frequency of the light changes the scale of the phase fluctuations. The scattering occurs when atoms or other particles within the fiber spread the light. It is a general physical process that occurs whenever a light wave travels through a medium having scattering objects. Based on a dimensionless size parameter, $\beta = \pi D_p/\lambda$, in which πD_p is the circumference of a particle and λ the wavelength of incident radiation, the models of light scattering is divided into following domains.

Rayleigh scattering Named after Lord Rayleigh, Rayleigh scattering is caused by tiny dielectric (non-absorbing) and spherical particles, such as bubble, droplet, and air molecules (mainly nitrogen molecules, N_2) in the atmosphere between 10 and 20 km altitude whose dimensions are smaller than its wavelength, i.e., $\beta \ll 1$. The degree of scattering varies as a function of the ratio of the particle diameter to the wavelength of the radiation, in addition to other factors including polarization, angle, and coherence. The scattering coefficient is proportional to λ^{-1}, known as Rayleigh's law, which means that light of longer wavelength would experience less scattering as compared to light with shorter wavelength. For air molecules, scattering is negligible at $\lambda > 3$ μm and at $\lambda < 1$ μm, Rayleigh scattering produces the blue color of the sky because blue light is more scattered than red light; hence, the star looks redder when seen through the cloud.

Mie scattering Named after Gustav Mie, Mie scattering arises from dust, predominantly for particles larger than the Rayleigh range, i.e., $\beta \approx 1$. Scattering losses decrease rapidly with increasing wavelength. Mie scattering has a spectral dependence of approximately λ^{-1}, whose presence is felt in particle-rich air at visible and near-IR wavelengths (Sigrist, 1994). The presence of aerosols in the atmosphere have pronounced scattering effects on light. Mie scattering produces the white glare around the Sun in presence of particulate material in the air. It produces a pattern like lobe, with a sharper and intense forward lobe for larger particles. If the ratio of particle diameter to wavelength more than $\beta \gg 1$, the laws of geometric optics are applied to describe the interaction of light with the particle; at this point the interaction is not generally described as scattering.

Both Rayleigh scattering and Mie scattering are considered as elastic scattering that involves no loss or gain of energy by the radiation. Brillouin scattering where the shift in frequency, $\Delta\bar{\nu} \leq 0.1$ cm^{-1}, Raman scattering in which $\Delta\bar{\nu} \leq 10^4$ cm^{-1}, and Compton scattering (see Section 5.2.5) are considered to be inelastic scattering. In Raman scattering, the scattered photons have either a higher or lower photon energy; the incident photons interacting with the molecules in a fashion that energy is gained or lost so that the scattered photons are shifted in frequency. Both Rayleigh and Raman scattering depend on polarizability of the molecules.

1.5.3 *Atmospheric Turbulence*

Turbulence is a non-linear process. On heating of the Earth's atmosphere by solar radiation, the large warm packets of air closer to the ground move up due to buoyancy and initiate convection causing the turbulence near the ground. They dissipate their kinetic energy continuously and randomly into smaller and smaller packets of air, each having a unique temperature. These packets are called turbulent eddies. Convection changes with insolation and disappears during night time, but horizontal circulation of air commences. The random fluctuations in the atmospheric motions occur predominantly due to

- the friction encountered by the air flow at the Earth's surface and consequent formation of a wind-velocity profile with large vertical gradients,
- differential heating of different portions of the Earth's surface and the concomitant development of thermal convection,
- processes associated with formation of clouds involving release of heat of condensation and crystallization, and subsequent changes in the nature of temperature and wind velocity fields,
- convergence and interaction of air-masses with various atmospheric fronts, and
- obstruction of air-flows by mountain barriers that generate wave-like disturbances and rotor motions on their lee-side.

Formation of eddies When the average velocity, v_0 (m/s), of a viscous fluid of characteristic size, l (m), is gradually increased, two distinct states of fluid motion are observed (Tatarski, 1961), namely, (i) laminar (regular and smooth in space and time), at very low v, and (ii) unstable and random fluid motion at v greater than some critical value. Between these two extreme conditions, the flow passes through a series of unstable states. The energy in the form of large scale disturbances breaks down into smaller and smaller structures. This process is characteristic of turbulent flow, which occurs when the Reynolds number, Re, exceeds a critical value. The dimensionless quantity Reynolds number is defined as

$$Re = \frac{\mathrm{v}_0 l}{\nu_v}, \tag{1.114}$$

where ν_v denotes the kinematic viscosity of the fluid.

For a pipe flow, if the value of Re is below 2100, it indicates laminar flow, while values nearing 4000 suggest turbulence. For $Re \sim 10^6$, the medium becomes chaotic in both space and time, exhibiting considerable spatial structure. Swirling water, puffs of smoke, and changing dust motion of sunlight exemplify such a chaotic condition as well. They are all unpredictable, random phenomena of patterns (eddies or vortices) that emerge inside them. Such patterns dissolve as fast as they are created. An important property of atmospheric eddies is that they exist in a variety of length scales and their distribution is random. There exists an upper limit, L_0, decided by the process that generates turbulence, what is referred to as outer-scale and a lower limit, l_0, decided by the size at which viscous dissipation overtakes the breakup process. The spectral energy cascades proceed at a constant rate governed by the eddy turn over time. The energy enters the flow at scale length, L_0, called outer-scale, and spatial frequency, $\kappa_{L_0} = 2\pi/L_0$, as a direct result of the non-linearity of Navier–Stokes' equation governing fluid motion.

Kolmogorov turbulence model The optically important property of the Kolmogorov law is that the refractive index fluctuations are largest for the largest turbulent elements up to the outer scale of the turbulence. Kolmogorov (1941a) proposed a model for the velocity motion in a fluid medium by investigating the mechanical structure of turbulence. For one-dimension, the Kolmogorov power spectrum is proportional to wave number as

$$\mathcal{E}(\kappa) \propto \kappa^{-5/3}, \tag{1.115}$$

and for the three-dimensional case it is found to be $\Phi(\boldsymbol{\kappa})$ and is proportional to the wave number, $\kappa (= 2\pi/\lambda)$

$$\Phi(\boldsymbol{\kappa}) = 0.033 C_n^2 \kappa^{-11/3}, \tag{1.116}$$

in which C_n^2 designates the refractive index structure constant describing the strength of the turbulence.

The severity of wavefront aberration, beam broadening, beam wander, and scintillation is related to C_n^2, which reflects the strength of the atmospheric turbulence; the larger C_n^2 is, the stronger is the turbulence. The C_n^2 is a function of the path length through the atmosphere and varies with time and geographic location.

It is, in general, lower at night and highest around noon. Also, C_n^2 is dependent on the variation in air-density. When the air-density variation decreases, for instance, at high altitudes, it decreases. Kolmogorov's spectrum of atmospheric turbulence can be represented by Zernike polynomials. Fried (1966) used a form of such polynomials to describe the statistical strengths of aberrations produced by atmospheric turbulence.

Zernike polynomials Zernike polynomials were developed as a convenient set for representing wavefront aberrations over a circular pupil (Born and Wolf, 1984). These are an orthogonal expansion over unit circle and are defined as

$$Z_j(\rho, \theta) = \sqrt{n+1} R_n^m(\rho) \begin{cases} \sqrt{2}\cos(m\theta), & m \neq 0, \quad \text{for j even}, \\ \sqrt{2}\sin(m\theta), & m \neq 0, \quad \text{for j odd}, \\ 1, & m = 0, \end{cases}$$

with

$$R_n^m(\rho) = \sum_{s=0}^{(n-m)/2} \frac{(-1)^s (n-s)!}{s!\,[(n+m)/2 - s]!\,[(n-m)/2 - s]!} \rho^{n-2s},$$

$Z_j(\rho, \theta)$ the Zernike polynomial of order j, the radial functions, $R_n^m(\rho)$, are polynomials in ρ containing the powers $\rho^n, \rho^{n-2}, \cdots \rho^m$, n the radial degree of the Zernike polynomials (see Appendix A; Table AIII), and m the azimuthal frequency.

The index j is used to order the mode and s is a mode-ordering number. Each mode has a coefficient, which is a number with positive or negative sign. The Zernike coefficient (in μm) specifies how much of that mode (aberration) is present. Zernike modes are grouped into Zernike orders. A wavefront can be decomposed using a series of orthogonal functions. The wavefront distortions are expanded in terms of series of Zernike basis $Z_J(r)$. The phase is thus computed by

$$\psi_J(\boldsymbol{\rho}) = \sum_{j=1}^{J} a_j Z_j(\boldsymbol{\rho}). \tag{1.117}$$

The Zernike coefficients are given by

$$\sqrt{\langle a_j^2 \rangle} = \sqrt{N_j} \left(\frac{D}{r_0}\right)^{5/6}.$$

Refractive index structure function Turbulent air motions cause variations in the density, pressure, temperature and humidity of the air from one point to another. While the local temperature fluctuations of the air could be of the order of a few hundredth of a degree throughout the atmosphere, fluctuations of a few tenths of a degree or more are typical in the lowest layer of the atmosphere. The temperature fluctuations in small patches of air cause changes in index of refraction. The fluctuations in the refractive index are found at the interface of different air layers. The refractive index of the atmosphere may be expressed as

$$n(\mathbf{r}, t) = n_0 + n_1(\mathbf{r}, t), \tag{1.118}$$

where, $n_0 \approx 1$ is the mean refractive index of air, $n_1(\mathbf{r}, t)$ the randomly fluctuating term, $\mathbf{r}$ the three-dimensional (3-D) position vector and t the time. The refractive index varies due to the temperature inhomogeneities. The dependence of the refractive index of air upon pressure, P (millibar) and temperature, T (Kelvin), at optical wavelengths is given by (Ishimaru, 1978)

$$n_1 \cong n - 1 = 77.6 \times 10^{-6} \frac{P}{T}. \tag{1.119}$$

As the refractive index of the air is highly sensitive to the temperature, it varies randomly from one point to another. Fluctuations in the refractive index induce random optical path lengths to the ray that are normal to the wavefront arriving at the top of the atmosphere from a distant star. Consequently, the wavefront reaches the external pupil of a ground-based telescope and gets corrupted in the sense that the surface of constant phase is no longer planar; it has an overall tilt and small-scale corrugations on top of it. The RMS value of the phase perturbations increases with the size of the wavefront.

Since the atmospheric effects are random, they can only be described statistically using strength estimates. Following Tatarski (1961), the variance of the difference between the two values of the refractive index at any two points within a given layer is described by their second order structure function:

$$\mathcal{D}_n(\mathbf{r}) = \langle |n(\mathbf{r}) - n(\boldsymbol{\rho} + \mathbf{r})|^2 \rangle, \tag{1.120}$$

where $\langle\,\rangle$ denotes an ensemble average over the repeated parameter ρ, $n(r)$ the refractive index at position $\mathbf{r}$ and $\mathcal{D}_n(\mathbf{r})$ the statistical variance in refractive index between two parts of the wavefront separated by a distance **r** in an atmospheric layer.

Equation (1.120) statistically describes the spatial distribution of the inhomogeneities at a given time, *t*. For the case of an isotropic layer following Kolmogorov's model (1941a), this structure function, $\mathcal{D}_n(\mathbf{r})$ depends on the strength of the turbulence

$$\mathcal{D}_n(\mathbf{r}) = \mathcal{C}_n^2 r^{2/3}, \qquad l_0 \ll r \ll L_0. \tag{1.121}$$

Similarly, the velocity structure function, $\mathcal{D}_v(\mathbf{r}) = \mathcal{C}_v^2 r^{2/3}$, and temperature structure function, $\mathcal{D}_T(\mathbf{r}) = \mathcal{C}_T^2 r^{2/3}$, can also be derived (Roddier, 1981).

For the case of an atmosphere stratified into a series of horizontal layers, $\mathcal{C}_n^2$ may be taken as a function of the *h* above the ground level. Under these conditions, equation (1.121) is valid within a layer of constant $\mathcal{C}_n^2$.

Atmospheric coherence diameter Variations in refractive index in the atmosphere above the telescopes lead to severe distortion of the wavefronts coming from stars. The resultant images are shifted and broadened to sizes many times larger than the diffraction limit. This broadening process is known as the atmospheric seeing. The image degradation produced by atmospheric turbulence is characterized by the size of the atmospheric coherence diameter (Fried, 1966), also referred to as Fried's parameter, r_0. This parameter is the cophasal cell size in the atmosphere over which the wavefront distortion has a root-mean-square (RMS) value not exceeding 1 radian. It is an important parameter for testing the atmospheric condition at the existing, as well as at the prospective astronomical sites. Systematic measurements of r_0 (Saha and Yeswanth, 2004) would help in understanding the various causes of the local seeing, viz., thermal inhomogeneities associated with the building, aberrations in the design, manufacture and alignment of the optical train, etc. The phase perturbations introduced into wavefronts by the layered atmosphere is described by the second order structure function for the phase perturbations:

$$\mathcal{D}_\psi(\boldsymbol{\xi}) = \langle |\psi(\mathbf{x}) - \psi(\mathbf{x}+\boldsymbol{\xi})|^2 \rangle \tag{1.122}$$

$$\simeq 6.88 \left(\frac{\boldsymbol{\xi}}{r_0}\right)^{5/3}. \tag{1.123}$$

The structure function (equation 1.122) is dependent on the integral of $C_n^2(h)$ taken over the path traversed by the light and the wavenumber, κ, as follows:

$$\mathcal{D}_\psi(\boldsymbol{\xi}) = 2.914\kappa^2\xi^{5/3}\int_0^L C_n^2(z)dz, \tag{1.124}$$

where L is the mean effective height of the turbulence and z the transmission altitude.

Equation (1.124) can be described in terms of angular distance of the source from the zenith, γ as

$$\mathcal{D}_\psi(\boldsymbol{\xi}) = 2.914\kappa^2\xi^{5/3}\sec\gamma\int_0^L C_n^2(z)dz. \tag{1.125}$$

Comparison of equation (1.123) with equation (1.125) yields the value of r_0 in terms of C_n^2

$$r_0 \simeq \left(\frac{6.88}{\mathcal{D}_\psi}\right)^{3/5},$$

which is derived as

$$r_0 = \begin{cases} \left[0.423\kappa^2\sec\gamma\int_0^L C_n^2(z)dz\right]^{-3/5} & \text{for plane waves,} \\ \left[0.423\kappa^2\sec\gamma\int_0^L C_n^2\left(\frac{z}{L}\right)dz\right]^{-3/5} & \text{for spherical waves.} \end{cases} \tag{1.126}$$

For electromagnetic wave transmission along a horizontal path, the Fried parameter, r_0 reduces to

$$r_0 = \begin{cases} 1.68\left(C_n^2 L\kappa^2\right)^{-3/5} & \text{for plane waves,} \\ 3\left(C_n^2 L\kappa^2\right)^{-3/5} & \text{for spherical waves.} \end{cases} \tag{1.127}$$

The amplitude of the refractive index fluctuations, described by $C_n^2(z)$ varies with wavelength λ, so the variation of r_0 with wavelength can be approximated as

$$r_0 \propto \lambda^{6/5}. \tag{1.128}$$

Atmospheric time constant Another important parameter is the so called atmospheric time constant or atmospheric coherence time, τ_0, which is determined by the stability of the atmosphere along the line of sight to the source. This parameter is a function of r_0 and the transverse component of the wind velocity and is given by

$$\tau_0 = 0.314\frac{r_0}{v_w}, \tag{1.129}$$

with v_w as the wind velocity averaged over the altitude; the typical value of w_v is about 20 meter per second.

Isoplanatic angle The instantaneous atmospheric phase aberrations depends on the isoplanatic angle, θ_0, which is defined to be the cone within which the beam does not possess different phase variance due to turbulence and is expressed as

$$\theta_0 = 0.314 \frac{r_0}{L \sec \gamma}. \tag{1.130}$$

This parameter, θ_0, limits the distance between guide star and the scientific objects.

Scintillation Scintillation refers to the temporal or spatial fluctuations in the signal irradiance of an optical wave caused by small random refractive-index fluctuations. The scintillation index, σ_I^2, defined as the normalized variance of the irradiance fluctuations scaled by the square of the mean irradiance, is used to describe this intensity fluctuation effects. For weak fluctuation regime (scintillation index less than 1), this parameter is proportional to the Rytov variance of a plane wave, which is given by

$$\sigma_I^2 = 1.23 C_n^2 \kappa^{7/6} L^{11/6}. \tag{1.131}$$

1.5.4 *Resolving Power of a Telescope*

All telescopes have an inherent limitation of their angular resolution; the limit is set by the diffraction at the aperture of the telescope. When a point source, such as a star, is observed through a telescope with a circular aperture, the image appears to be a disk surrounded by a series of concentric rings with a bright central spot. These rings are produced by Fraunhofer diffraction, also known as far-field approximation of the light by the aperture. In this case, the intensity, $I(\theta)$, can be approximately expressed in terms of a Bessel function of the first order

$$I(\theta) = I(0) \left[\frac{2J_1(\kappa a \sin \theta)}{\kappa a \sin \theta} \right]^2, \tag{1.132}$$

where $I(0)$ is the peak intensity at the center of the diffraction pattern, $D = 2a$ the diameter of the aperture, $J_1(u)$ the first order Bessel function, and κ the wave number.

Fraunhofer diffraction occurs if the distances of the source of light and observation screen are effectively large compared to the dimension of the diffraction aperture so that wavefronts arriving at the aperture and the observation screen may be considered as plane, i.e., $D^2/\lambda z \ll 1$, in which D is the size of the aperture, λ the wavelength, and z the distance from the aperture. The near-field or the Fresnel approximation occurs when $D^2/\lambda z \geq 1$. Using the small-angle approximation, $\sin\theta \simeq \theta$, where θ is measured in radians. The central region of the profile, from the peak to the first minimum, is called Airy's disk. It has an angular radius given by

$$\theta \simeq \frac{1.22\lambda}{D}, \tag{1.133}$$

with D as the diameter of the telescope.

If two point sources are very close together, it is possible to resolve them if the central peaks of the two diffraction patterns are no closer than the radius of the Airy disk. When the first diffraction minimum of the image of one point source coincides with the maximum of another, it is known as Rayleigh's criterion.

In the ideal condition, the resolution that can be achieved in an imaging experiment, $\mathcal{R}$, is limited only by the imperfections in the optical system and according to Strehl's criterion, is given by the integral of its transfer function

$$\mathcal{R} = \int_{-\infty}^{\infty} \widehat{S}(\mathbf{u})d\mathbf{u}, \tag{1.134}$$

where $\widehat{S}(\mathbf{u})$ is the optical transfer function (OTF) of the ground-based telescope, whose PSF is $S(\mathbf{x})$.

For the 2-D incoherent source, the complex amplitude in the image-plane is the convolution of complex disturbances of the wave, $U(\mathbf{x})$, and the pupil transmission function, $P(\mathbf{x})$

$$S(\mathbf{x}) = P(\mathbf{x}) \star U(\mathbf{x}). \tag{1.135}$$

If $\mathcal{R} = S(\mathbf{0})$, the central intensity in the image is known as Strehl's intensity. The image quality can be characterized by a number called Strehl's ratio, $\mathcal{S}_r$, which is defined as the ratio of the maximum intensity at a point of the image due to an aberrated optical system to the on axis intensity in the image due to the same but unaberrated system (Strehl, 1902). The on-axis $\mathcal{S}_r$ obtained in the image plane of an imaging system is determined by integrating the probability distribution for the wavefront phase variance across the aperture plane. For an error less than $\lambda/2\pi$, Strehl's ratio is expressed by

$$\mathcal{S}_r \simeq e^{-(2\pi/\lambda)^2 \langle\sigma\rangle^2}, \tag{1.136}$$

where $\langle\sigma\rangle^2$ designates the variance.[a]

The perturbations in the wavefront produce effects similar to optical aberrations in the telescope and thus degrade the image quality. When a very small aperture is used, a small portion of the wavefront is intercepted and the phase of the wavefront is uniform over the aperture. If the amplitudes of the small-scale corrugations of the wavefront are much smaller than the wavelength of the light, the instantaneous image of a star is sharp resembling to the classical diffraction pattern taken through an ideal telescope, in which the point spread function (PSF) is invariant to spatial shifts. The resolution is limited either by the telescope aperture or by the atmosphere

$$\mathcal{R} = \frac{\pi}{4}\left(\frac{D}{\lambda}\right)^2 \qquad D \ll r_0. \tag{1.137}$$

In the absence of the atmospheric turbulence, the full-width half maximum (FWHM) is the diffraction-limit of telescope (see equation 1.137). Incorporating both Strehl's ratio (equation 1.136) and the diffraction-limited resolution (equation 1.137), one gets Strehl's resolution, that is

$$\mathcal{R} = \frac{\pi \mathcal{S}_r}{4}\left(\frac{D}{\lambda}\right)^2. \tag{1.138}$$

The discrete layers of turbulence are blown by wind across the telescope aperture; hence, a change in tilt occurs, which in turn, causes random motion of the star's image at the focal plane. As the aperture size increases, there is a decrease in the sharpness and amplitude of the motion; the amplitude of the random variation of phase across the intercepted wavefront tends to become larger. This

---

[a]The variance measures of statistical dispersion; the higher the variance, the larger the spread of values. It is computed as the average squared deviation of each number from its mean. Mathematically, the variance in a population is given by

$$\langle\sigma_{st}\rangle^2 = \frac{\sum(X-m)^2}{N},$$

with m as the arithmetic mean, X the variable, and N the number of scores, while the positive square root of the variance is known as the standard deviation that has the same units as the original variable.

leads to the blurring of the image, which is larger than the Airy's disk of the telescope (see equation 1.138). The blurring suffered by such an image is modeled by a convolution with a point spread function (PSF)

$$\mathcal{R} = \frac{\pi}{4}\left(\frac{r_0}{\lambda}\right)^2 \qquad D \gg r_0. \tag{1.139}$$

Seeing The image motion and blurring together are referred to as atmospheric seeing or simply seeing. It is the total effect of distortion in the path of the light via different contributing layers of the atmosphere to the detector placed at the focus of the telescope. The seeing disk is defined as the FWHM of a Gaussian distribution function fitted to a histogram of the image in arcseconds. The quality of seeing is characterized by $\theta_s = 0.976\lambda/r_0$.

The seeing fluctuates on all time scales down to minutes and seconds. Figure 1.9 depicts the micro-fluctuations of r_0 at a step of ~150 ms observed at the 2.34 meter VBT, Kavalur, India, on February 28, 1997. The major sources of image degradation

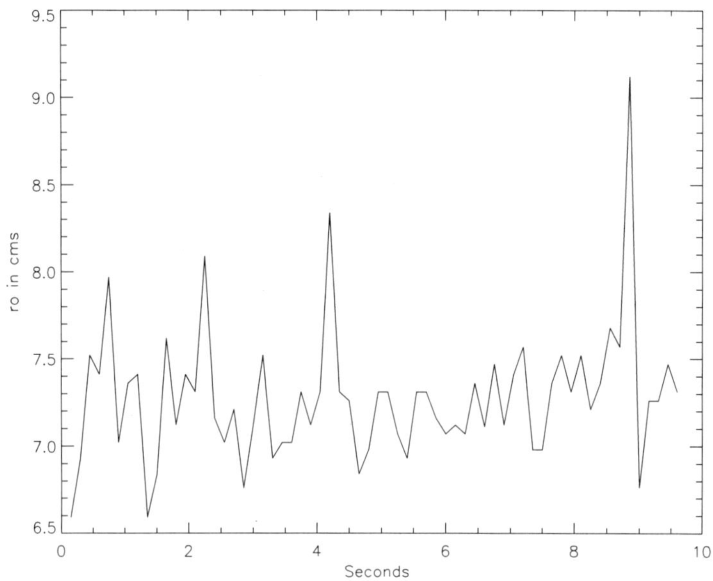

Figure 1.9 Micro-fluctuations of r_0 as observed at the 2.34 meter VBT, Kavalur, India on 28 February, 1997 (Saha and Yeswanth, 2004).

predominantly comes from the surface layer, as well as from the aero-dynamical disturbances in the atmosphere surrounding the telescope and its enclosure, namely,

- thermal distortion of primary and secondary mirrors when they get heated up,
- dissipation of heat by the latter mirror,
- rise in temperature at the primary cell,
- at the focal point causing temperature gradient close to the detector, etc.

Degradation in image quality can occur due to opto-mechanical aberrations, such as aberrations in the design, manufacture and alignment of the optical train, mechanical vibrations of optical system as well.

1.5.5 *Passive Approach to Get over Turbulence Effect*

As stated in Section 1.5.3, the atmospheric turbulence distorts the characteristics of downward propagating light; the longer the path, the more it suffers deflection. The power spectral density of refractive index fluctuations follows a power law with large eddies having greater power (Tatarski, 1961). When a wavefront passes down through the atmosphere, it suffers phase fluctuations and reaches the entrance pupil of a telescope with patches of random excursions in phase (Fried, 1966). Light reaching the entrance pupil of an imaging system is coherent only within Fried's parameter, $\approx r_0$; therefore, it causes blurring of the image limiting the theoretical performance of any terrestrial optical telescope larger than r_0.

Images recorded at the telescope are convolved with the combined point spread functions (PSF) of the atmosphere and the telescope. If the exposure time is shorter ($<$20 millisec) than the evolution time of the phase inhomogeneities, then each patch of the wavefront with diameter r_0 would act independently of the rest of the wavefront resulting in many bright spots—speckles—spread over the area defined by the long-exposure image. Figure 1.10 depicts the simulated short-exposure and long-exposure images of an unresolved star. The long-exposure time is equivalent to the exposure time in a standard film camera, $\Delta t \gg \tau_0$, in which Δt is

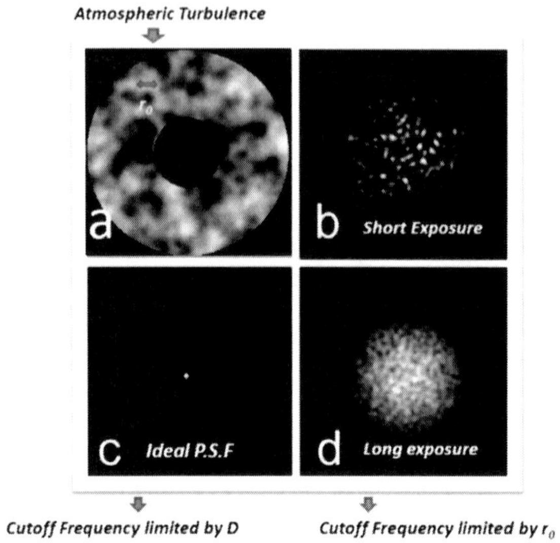

Figure 1.10 (a) Atmospheric turbulence simulated over a primary mirror having a Cassegrain hole, (b) simulated short-exposure images of an unresolved star, (c) ideal point spread function of the telescope, and (d) conventional long-exposure image of the same star.

the integration time and τ_0 the atmospheric coherence constant. It is, in general, more than 20 msec in visible wavelengths, and for the IR wavebands, it is of the order of more than 100 msec.

These speckles can occur randomly along any direction within an angular patch of diameter $1.22\lambda/r_0$. The speckle life time describes the longest possible exposure time for recording speckle interferograms (specklegram), which is of the order of $\sim$0.05 to $\sim$0.001 sec in optical wavelengths. The average size of the speckles is of the same order of magnitude as the Airy's disk of the telescope. The number of correlation cells is proportional to the square of D/r_0 and the number of photons, but is independent of its diameter. The normalized optical transfer function (OTF) of a long-exposure image is the Fourier transform of its point spread function (PSF) and can be expressed as

$$OTF(\nu) = e^{-3.44(\lambda\nu/r_0)^{5/3}}. \qquad (1.140)$$

Following equation (1.93), the intensity, $I(\mathbf{x})$, of the image of a quasi-monochromatic incoherent source is defined as

$$I(\mathbf{x}) = O(\mathbf{x}) \star S(\mathbf{x}), \tag{1.141}$$

where $\mathbf{x}(= \ x, y)$ is the 2-D position vector, $O(\mathbf{x})$ the intensity of the object, which is convolved with the function representing the combined effects of the turbulent atmosphere and the image forming optical system, $S(\mathbf{x})$.

In the Fourier plane, it is written as

$$\widehat{I}(\mathbf{u}) = \widehat{O}(\mathbf{u})\widehat{S}(\mathbf{u}). \tag{1.142}$$

with $\mathbf{u}(= \ u, v)$ as the 2-D spatial frequency vector, $\widehat{O}(\mathbf{u})$ the transform of the object, and $\widehat{S}(\mathbf{u})$ the transfer function of the telescope-atmospheric combination.

Conventional imaging in astronomy is associated with long-exposure integration time. The effect of both telescope and atmosphere is usually considered as a random linear filtering. The relation between the average intensity, $\langle I(\mathbf{x})\rangle$, and the irradiance $O(\mathbf{x})$ of the resolved object is given by

$$\langle I(\mathbf{x})\rangle = O(\mathbf{x}) \star \langle S(\mathbf{x})\rangle, \tag{1.143}$$

and its Fourier transform is

$$\langle \widehat{I}(\mathbf{u})\rangle = \widehat{O}(\mathbf{u})\langle \widehat{S}(\mathbf{u})\rangle, \tag{1.144}$$

where, $\langle \widehat{I}(\mathbf{u})\rangle$ is the long-exposure image spectrum, $\widehat{O}(\mathbf{u})$ the object spectrum, and $\langle \widehat{S}(\mathbf{u})\rangle$ the transfer function for long-exposure images.

The diffraction-limited resolution of a ground-based telescope can be achieved by employing post-detection processing of a large data set of short-exposure images using Fourier domain methods. Certain specialized moments of the Fourier transform of a short-exposure image contain diffraction-limited information about the object of interest. This method, referred to as speckle interferometry, was described by Labeyrie (1970). The specklegram represents the resultant of diffraction-limited incoherent imaging of the object intensity, $O(\mathbf{x})$ as described by equation (1.141). This is averaged over the duration of the short, narrow bandpass exposure. An ensemble of such specklegrams, $I_k(\mathbf{x})$, $k = t_1, t_2, t_3, \ldots, t_M$, constitute an astronomical speckle observation. The size of the

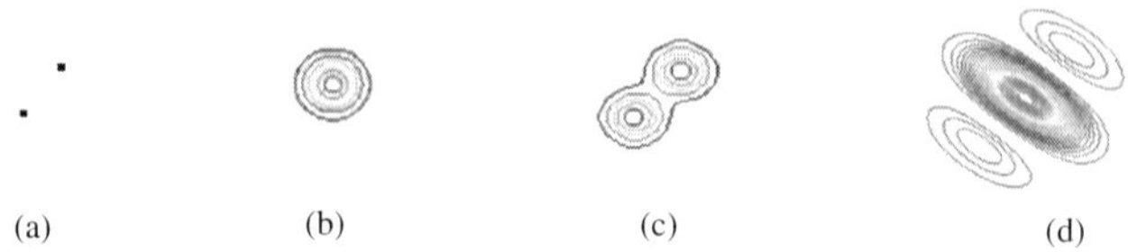

Figure 1.11 (a) Simulated binary system, (b) Gaussian PSF, (c) convolved binary with the PSF, and (d) power spectrum of (c).

data sets is constrained by the consideration of the S/N ratio. Integration time of each exposure varies from a few milliseconds to 20 millisec, depending on the condition of seeing, to freeze the single realization of the turbulence. Usually specklegrams of the brightest possible reference star are recorded to ensure that the S/N ratio of reference star is much higher than the S/N ratio of the program star. Following equation (1.142), the ensemble averaged power spectrum (see Figure 1.11) is obtained for a large set of short-exposure images:

$$\langle|\widehat{I}(\mathbf{u})|^2\rangle = |\widehat{O}(\mathbf{u})|^2\langle|\widehat{S}(\mathbf{u})|^2\rangle. \qquad (1.145)$$

The form of transfer function $\langle|\widehat{S}(\mathbf{u})|^2\rangle$ can be obtained by deriving Wiener spectrum of the instantaneous intensity distribution from the reference star (unresolved star). The object is reconstructed from the power spectrum of the image knowing the PSF power spectrum and the true object autocorrelation is then obtained by Fourier inversion. This method in case of components in a group of stars retrieves the separation, position angle with a 180° ambiguity, and the relative magnitude difference (Labeyrie, 1970; Saha 1999, 2002, 2007, and references therein).

In order to reconstruct Fourier phase of such objects, several post-detection image processing techniques have been developed. Among them, the bispectrum[a] method (Lohmann et al., 1983) provides a way of finding the phase of the object image. Its

[a]The Fourier transform of the triple correlation is called bispectrum. The bispectrum of an image, $I(\mathbf{x})$ is given by the triple product

$$\widehat{I}^3(\mathbf{u}_1, \mathbf{u}_2) = \widehat{I}(\mathbf{u}_1)\widehat{I}^*(\mathbf{u}_1 + \mathbf{u}_2)\widehat{I}(\mathbf{u}_2),$$

where $\widehat{I}(\mathbf{u}_1)$ denotes the Fourier transform of image, $I(\mathbf{x}, \mathbf{u}_1, \mathbf{u}_2$, and $\mathbf{u}_1 + \mathbf{u}_2$ the 2-D coordinates in Fourier space. The Fourier phase of object is retrieved using recursive techniques from phase of $\widehat{I}(\mathbf{u}_1, \mathbf{u}_2)$.

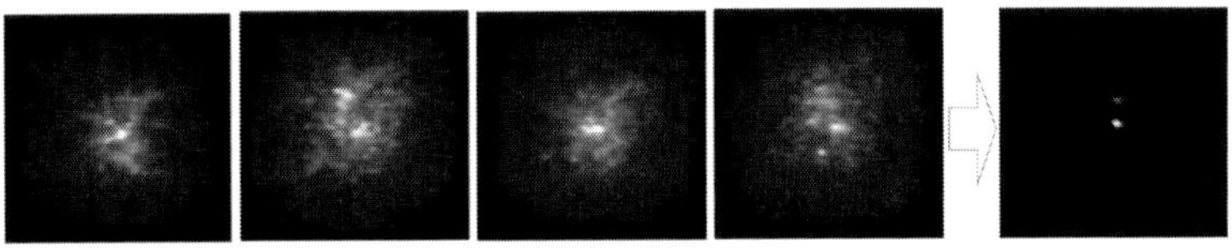

(a) (b)

Figure 1.12 (a) A set of specklegrams of a close binary star, HR 5747 (β Coronae Borealis); these frames were taken by means of the speckle camera system developed by Saha et al. (1987) in March 1990 at the Cassegrain focus of the 2.34 m Vainu Bappu telescope (VBT), Vainu Bappu Observatory (VBO), India and (b) reconstructed image using Radon transform based tomographic speckle imaging (TSM) algorithm. The separation found to be 0.21 arcsec (Surya and Saha, 2014) which is consistent with results from blind iterative deconvolution (BID) technique (Ayers and Dainty, 1988) obtained by Saha and Venkatakrishnan (1997).

applications are also of great importance in the context of upcoming multi-aperture interferometer, for example, the hypertelescope (see Section 3.2.1.2). In the absence of proper adaptive optics system, the future large telescopes like European Extremely Large Telescope (EELT) or Thirty Meter Telescope (TMT) may completely depend on this technique to fully utilize the large aperture (Surya et al., 2014).

Surya and Saha (2014) have developed both direct bispectrum and Radon transform[a] based tomographic speckle masking (TSM) algorithms to process speckle frames (see Figure 1.12). The latter is a much computationally efficient technique because it works with one-dimensional image projections and provides good image recovery. In order to simulate aperture masking interferometry in Fizeau mode combination with 5–20 subapertures, a laboratory setup was developed. The results from TSM algorithm with the images obtained with this aperture masking setup (see Figure 1.13; top). Figure 1.13, bottom, illustrates the results obtained from the simulated data.

[a]Radon transform transforms a 2-D image in to set of projections and allows an inverse transform to get back the image; the inverse radon transform can be carried out using filtered back projection algorithm.

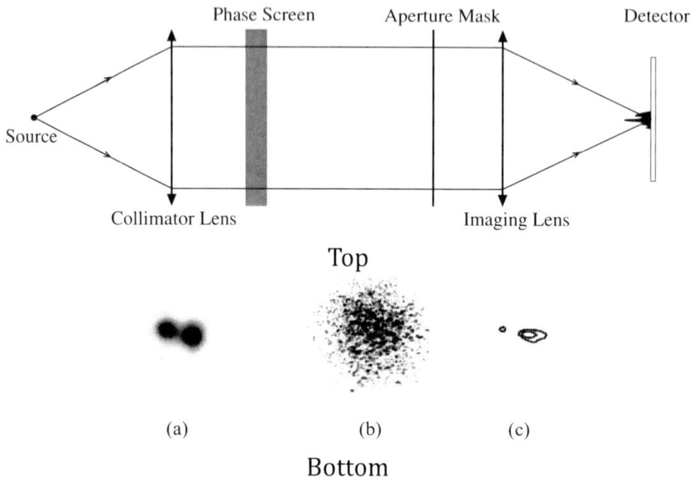

Figure 1.13 Top panel: Optical layout of the laboratory setup to simulate aperture masking interferometry. Bottom: (a) simulated binary system, (b) turbulence distorted image of the same, and (c) the result obtained by the TSM algorithm (Surya and Saha, 2014).

1.5.6 *Real-Time Correction Technique*

Another development in the field of high angular resolution imaging is to mitigate the effects of the turbulence in real time, which was suggested by Babcock (1953). The purpose of this technique, what is referred to as adaptive optics (AO) system, is to (i) sense the wavefront perturbations, and (ii) compensate for them in real time. Such an approach has the advantages over post-detection image restoration techniques that are limited by noise and restores the image quality to a useful fraction of the diffraction-limited capability of a telescope. With the use of such a system, the quality of the image changes to a greater extent in observational astronomy; the structure of the stars or the galaxies can be seen in depth. This technique has been employed in other branches of physics as well (Liang et al., 1997). AO systems are useful for spectroscopic observations, as well as for photon-starved imaging with very large telescopes, and ground-based long-baseline optical interferometers (Labeyrie, 1975; Mourard et al., 1989; Labeyrie et al., 2006; Saha, 2010, and references therein). However, most of these results that are obtained so far are from the near-infrared wavebands (Saha, 2007, and references therein). It is easier to achieve diffraction-limited information using AO systems at longer wavelengths since $r_\circ$ is proportional to the six-fifths power of the wavelength (see equation 1.128), following which one may find that the width of seeing-limited images

$$\frac{1.22\lambda}{r_0} \propto \lambda^{-1/5}, \tag{1.146}$$

varies with λ. In optical band, the most difficult problem arises from the number of actuators on the deformable mirror, since the atmospheric coherence length decreases with $\lambda^{5/6}$. The number of degrees of freedom and the number of sub-aperture in the wavefront sensor in an AO system should be determined by

$$\left(\frac{D}{r_0}\right)^2 \propto \lambda^{-12/5}, \tag{1.147}$$

in which D designates the telescope aperture.

1.5.6.1 Greenwood frequency

As the wind moves the eddies past the telescope aperture, the tilt of the intercepted wavefront changes. Therefore, the wind velocity also dictates the speed of measurement and correction. In order to take a corrective measure, Greenwood (1977) derives the mean square residual wavefront error as a function of servo-loop bandwidth[a] for a first order controller

$$\sigma_{cl}^2 = \left(\frac{\mathrm{f}_G}{\mathrm{f}_{3\ dB}}\right)^{5/3} \quad \mathrm{rad}^2, \tag{1.148}$$

where $f_G = 0.426v/r_0$ designates Greenwood frequency, which is higher-order wavefront variance due to temporal constraints, v the

[a]The bandwidth, Δf, is defined to be the range of frequencies to which a detector and its electric circuitry are capable of responding, which is expressed in hertz (cycles per second), that is

$$\Delta f = \frac{1}{A^2}\int_0^\infty |H(j\omega)|^2 df,$$

with $H(j\omega)$ as the transfer function of the circuit, A the maximum absolute value of $H(j\omega)$, f and ω the frequency and angular frequency, respectively; the bandwidth is generally taken to be the lowest frequency at which the magnitude electrical response is less than −3 dB from the DC (zero frequency) response.

wind velocity in the turbulent layer of air, $f_{3\,dB}$ is the closed loop bandwidth of the wavefront compensator, $dB(= 10 \log P_o/P_i)$ the decibel that is unit of the ratio of output to input power, P_o the output power, and P_i the input power.

Greenwood frequency is a measure of the characteristic frequency of the tilt of atmospheric turbulence. It may represent the entire spectrum for adaptive optics. The required frequency bandwidth of the control system is dictated by such frequency; an adaptive optics system with a closed-loop servo response should reject most of the phase fluctuations. If the turbulent layer moves at a speed of 10 m per second, the closed-loop bandwidth for $r_0 \approx$ 11 cm, in the optical band (550 nm) is around 39 Hz; for strong winds and UV wavelengths, it may reach 600 Hz.

1.5.6.2 Adaptive optics system

An adaptive optics (AO) system, a scientific and engineering approach, mainly consists of three main components: (i) a wavefront corrector, (ii) a wavefront sensor, and (iii) a control system. Figure 1.14 shows a generic block diagram of an adaptive optics imaging system. The lowest-order component of the wavefront corruption consists of a simple tilt, which manifests itself as x-y deflection of the star images at the focal plane of a telescope. Hence, the image of a point source dances randomly. In order to remove this component of distortion is carried out with a fast steering mirror, called a tip-tilt mirror, in the path of the optical beam. After traveling further, it reflects off of a deformable mirror (DM) that eliminates high frequency wavefront errors. A beam-splitter divides the beam into two parts: one is directed to the wavefront sensor to measure the residual error in the wavefront and to provide information to the actuator control computer to compute the DM actuator voltages and the other is focused to form an image (Roddier, 1999; Saha, 2007, and references therein).

Wavefront correctors The wavefront correctors can be in the form of a tip-tilt mirror and a deformable mirror. These mirrors are flexible, whose surface can be electronically controlled in real time such that perturbed incident wavefront phase-shifts are canceled

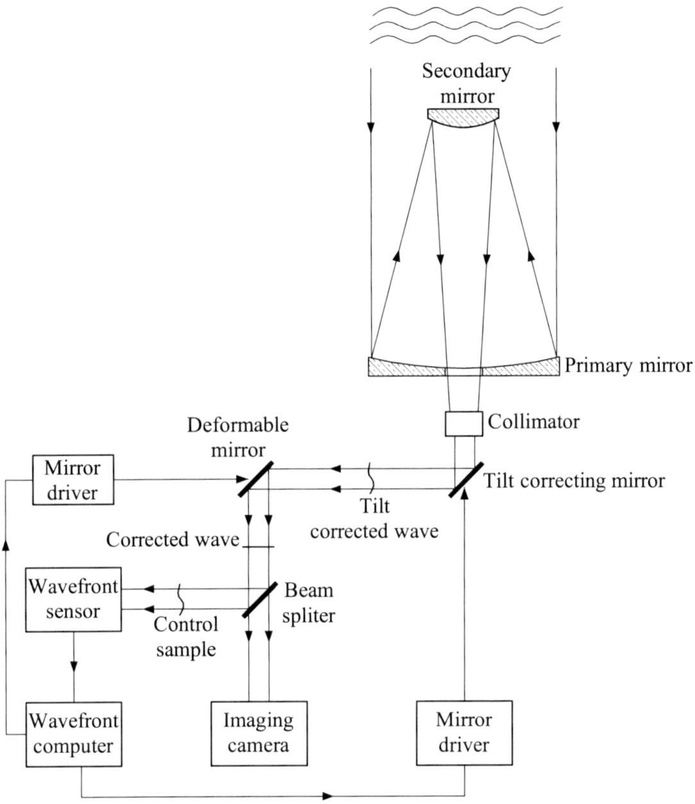

Figure 1.14 Schematic of the adaptive optics imaging system.

as the optical field bounces from their surfaces. In conjunction with closed-loop control electronics, the implementation of the dynamically controlled active optical components consisting of a tip-tilt mirror system has several advantages: (i) conceptually the system is simple, (ii) fainter guide stars increase the sky coverage, and (iii) field-of-view (FOV) is wider (Glindemann, 1997). These systems are limited to two Zernike modes (x and y-tilt), while a higher order system compensating many Zernike mode is required to remove high frequency errors. A variety of DMs have been developed for the applications of (i) high energy laser focusing, (ii) laser cavity control (see Section 3.3.6), (iii) compensated imagery through atmospheric turbulence, etc.

Wavefront sensor A wavefront sensor is a device for measuring the imperfections of optical wavefront. Several wavefront sensors, such as (i) lateral shearing interferometer, (ii) Shack–Hartmann (SH) sensor, and (iii) the curvature sensor are in use as well. The most commonly employed method of wavefront sensing is SH wavefront sensing. It consists of an array of micro lenslets, which samples the incoming distorted wavefront. Due to non-planarity of the sampled wavefront at each lenslet, the image displaces its position from the reference position. By measuring the displacement from the reference point, the slope of the wavefront can be calculated. By using the slope of the wavefront, its phase is estimated. Since the turbulence variations are taking place in the order of milliseconds, the necessary algorithm has to be faster than the atmospheric variations. For a diffraction-limited image, $\theta_b = \lambda/d$, in which d denotes the size of the lenslet, while under atmospheric turbulent conditions, $\theta_d \approx \lambda/r_0$. Figure 1.15a depicts the distribution of intensity at the focal plane of a lenslet array illuminated by the test beam, while Figure 1.15b shows the aberrated intensity wavefront taken through a Cassegrain telescope. Careful observation of the lenslet spots would reveal deviations in the spot position. It is to be noted that four missing spots in the middle is due to the central hole in the primary mirror of a Cassegrain telescope.

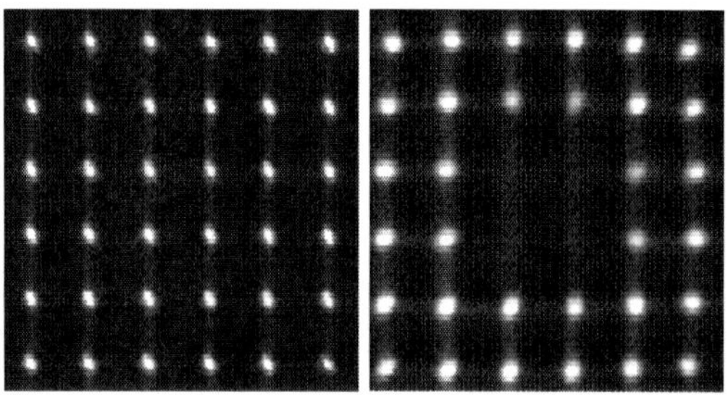

(a) (b)

Figure 1.15 Intensity distribution at the focal plane of the 6×6 lenslet array captured by a Peltier cooled electron multiplying CCD (EMCCD; see Section 7.7.1) camera (a) for an ideal case at the laboratory and (b) an aberrated wavefront taken through a Cassegrain telescope. (Courtesy: V. Chinnappan).

Wavefront controller The control unit computes and transmits signals proportional to the signals that are supposed to be applied to the deformable mirror from the computer to the proper electrodes of the deformable mirror. Such a system acts as feedback loop; the next cycle corrects the small errors of the last cycle. This process should be at speeds commensurate with the rate of change of the corrugated wavefront phase errors. The phase reconstruction method finds the relationship between the measurements and the unknown values of the wavefronts and can be categorized as being either zonal or modal, depending on whether the estimate is either a phase value in a local zone or a coefficient of an aperture function (Rousset, 1999). In the case of curvature sensing, the computed sensor signals are multiplied by a control matrix to convert wavefront slopes to actuator control signals, the output of which are the increments to be applied to the control voltages on the DM. A conjugate shape is created using these data by controlling a DM.

Real-time computations The real-time computation of the wavefront error, as well as correction of wavefront distortion involves digital manipulation of wavefront sensor data in the wavefront sensor processor, the re-constructor and the low-pass filter, and converting to analog drive signals for the DM actuators. The functions are (i) to compute sub-aperture gradients, phases at the corners of each sub-aperture, low-pass filter phases, and (ii) to provide actuator offsets to compensate the fixed optical system errors and real time actuator commands for wavefront corrections.

Guide star In addition, a laser guide star (LGS; see Section 3.5.4) is also required to improve the S/N ratio for the wavefront signal since the bright natural guide stars are not always available within isoplanatic patch.

Chapter 2

Photoelectric Concept

2.1 Photoelectric Effect

In 1887, while receiving the electromagnetic waves in an induction coil with a spark gap generator, Heinrich Hertz (1857–1894) observed the reduction of maximum spark length when the apparatus was put in a black box (Hertz, 1887). He noticed that the spark occurred in the air gap between the two small metal spheres of the transmitter when ultraviolet radiation was directed at one of these spheres. The box absorbed such radiation, which helped the electrons in jumping across the gap. Hertz reported the observations but did not pursue the experiment further, and also did not make any attempt to explain them, but others did. In 1897, J. J. Thomson (1856–1940) showed that electrons are emitted when light falls on a metal surface (Thomson, 1897). These are now referred to as photoelectrons (the prefix "photo" refers to the incident light).

Around 1900, Philipp Lenard (1862–1947), an assistant of Hertz, had followed up the afore-mentioned experiment. The metal sample was housed in an evacuated glass tube with a second metal plate mounted at the opposite end. The advantage of the vacuum tube is to avoid any surface contaminants or oxidation on the metal, which would help to study the effect on the metal alone. Following this, the

High-Resolution Imaging: Detectors and Applications
Swapan K. Saha

ISBN 978-981-4613-27-9 (Hardcover), 978-981-4613-28-6 (eBook)
www.panstanford.com

tube was positioned in a manner so as to enable the light to shine on the first metal plate. This plate was made out of photoemissive (PE) material (see Section 5.2.1) under investigation. This tube, known as photocell, was connected to a circuit with a variable power supply, a voltmeter, and a micro-ammeter. He had observed that when ultraviolet radiation falls on the first metal surface, it got positively charged.

A micro-ammeter[a] measured this small amount of current (a flow of charge), which also serves as a measure of the rate at which photoelectrons are leaving the surface of the PE material. It is pertinent to note that with the negative end being connected to the non-illuminated metal surface, which sets up a potential difference, pushes the photoelectrons back into the PE surface. When the power supply is set to a low voltage, it traps the least energetic electrons, reducing the current through the micro-ammeter. With the increase of the voltage, it drives increasingly more energetic electrons back until none of them is able to leave the metal surface and the micro-ammeter reads zero. The potential at which this occurs is called the cut-off or stopping potential, V_c. It is a measure of the maximum kinetic energy, $E_k (= eV_c)$, of the electrons emitted as a result of the photoelectric effect. When light at 480 nm strikes a photosensitive surface, the emitted electrons have a stopping potential of 0.45 V. If the charge (q) is given in coulombs, the energy is calculated in joules, while if charge is given in elementary charges, the energy is calculated in electron-volts (eV), where $1 \text{ eV} = 1.60 \times 10^{-19}$ J. The photon energy, $h\nu$, with h as Planck's constant and ν the frequency of the incident photon (see Section 1.1), at different wavelengths, viz., 400 nm, 550 nm, and 700 nm is, respectively, derived as 3.11, 2.26, and 1.61 eV. Lenard (1902) assumed that the intensity of the light varied inversely as the square of the distance between the source and the detector. His observations said that

- the kinetic energy of the emitted electrons was independent of the intensity of the incident light and
- the energy of the emitted electron increased when the frequency of the incident light was increased.

[a] A micro-ammeter is a sensitive galvanometer with a maximum deflection of only a few micro-amperes.

The photon concept (see Section 1.1.1) helped Einstein (1905a) to obtain his photoelectric equation, for which he was awarded the Nobel Prize in 1921, and led to the conclusion that the number of electrons released would be proportional to the intensity of the incident light. In the particle (quantum) picture, one can postulate the light quantum to transfer its entire energy to a single electron. Einstein argued that the emission of a photoelectron was the result of the interaction of a single quantum with an electron. According to him, the monochromatic radiation behaves as if it consists of mutually independent energy quanta of magnitude, $h\nu$. This theory is the explanation of the photoelectric effect. A photoelectric light detection occurs when a photon interacts with a sensor to create an electrical signal. The interaction of light of appropriate frequency with materials results in the absorption of photons and the creation of electron–hole pairs (see Section 2.2.4.1). The photon absorption is accomplished by a change in the energy state of an electron in the sensor material. If the energy is sufficient to release the electrons from the substance, the charge carriers under the influence of an external electric field drift to the electrodes of opposite polarity produce an electric current, called photocurrent.

For the light with frequencies below a certain cut-off value, ν_c, the photoemission does not occur. For $\nu > \nu_c$, where ν_c is the cut-off frequency of the photon required to produce photoelectric emission and is a constant for the material, a major fraction of the excess energy $[= h(\nu - \nu_c)]$ appears as kinetic energy of the emitted electron. The non-measurable time lag between the incidence of the radiation and the ejection of the electron follows from the corpuscular nature of the radiation. When an oscillator emits a photon, it drops from energy, $nh\nu$ to a level $(n - 1)h\nu$. The energy of the photon is expended in liberating the electron from the metal and imparting a velocity to it. The maximum kinetic energy, E_k, of the ejected electrons is linearly related to the energy of the absorbed photons, $E(= h\nu)$, and the work function, ϕ_0, of the PE surface (see Table 5.1)

$$\begin{aligned} E_k &= h\nu - \phi_0 \\ &= h\nu - h\nu_c = h(\nu - \nu_c), \end{aligned} \tag{2.1}$$

with $E_k(= 1/2mv^2)$ as the kinetic energy of the ejected photoelectron from the metal surface, m the mass of the electron, v the velocity of the electron, and ν_c the cut-off frequency, which is a characteristic of the metal.

The work function of a metal is defined as the minimum energy, measured in electron volts, needed to release an electron from atomic binding. To note, when two or more atoms are located in proximity of each other (within distances of the order of atomic dimensions), certain forces act between them, which bind the atoms together to make molecules. If the two atoms are similar, such as H_2 and O_2, the binding takes the form of covalent binding, while in the case of dissimilar atoms, for example, sodium chloride (Nacl) or calcium fluoride (CaF_2), it is called ionic binding. In a semiconductor crystal, two electrons are shared by each pair of ionic neighbors. Such a configuration is called a covalent bond.

The validity of equation (2.1) was established in a series of experiments by Robert Millikan (1868–1953), best known for the oil-drop experiment, in 1914, who also made the first direct photoelectric determination of Planck's constant, h (Millikan, 1914, 1916). He observed that the photoelectrons constitute a current between two plates that were kept in a vacuum, if light of a certain frequency is allowed to fall on one of the plates (cathode). A cathode is an emissive material that emits electrons when light of wavelength shorter than the cut-off wavelength falls on the surface. By applying an electric field in the right direction, the electrons can be swept away from the emitter, which creates an electric current through the tube and around the circuit. The emitted electrons are collected by other plate (anode). An anode is a positively charged element of device to which electrons are attracted. When the voltage across the plates is varied, the current also varied. In short, one may summarize as follows:

(1) At zero voltage, there is a finite value of the current implying that some of the emitted photoelectrons reach the anode.
(2) As the accelerating voltage is increased, the current increases until it reaches a saturation (a condition, in which there is no further increase in detector response as the input light intensity

is increased) value, where all the photoelectrons are collected by the anode.

(3) If the anode is kept at a slightly negative potential, one can detect a weak current, which implies that some of the photoelectrons do manage to reach the anode. However, beyond a certain voltage, the current turns out to be zero, which is termed as the cut-off voltage, V_c, for example, for sodium, $V_c \approx$ 2.3 V and for copper, $V_c \approx 4.7$ V.

(4) If the wavelength of the incident radiation is not changed but made it more intense, the magnitude of the current becomes larger, implying a greater emission of photoelectrons, however the value of the cut-off potential remains the same. The maximum kinetic energy of the emitted photoelectrons does not depend on the intensity of the incident radiation. A greater intensity leads to a larger number of electrons constituting a larger current; a faint violet light would eject electrons of greater kinetic energy than an intense yellow light although the latter would produce a large number of electrons.

(5) The energy, $h\nu$, of the light quanta must depend on the frequency of the light, ν, which in turn, should be reflected in the energy of the electrons that have been ejected. With the proper frequency, photoemission commences immediately after the radiation is incident on the metal surface. The time lag between the incidence and emission of a photoelectron is less than a nanosecond (ns). The frequency of the incident radiation and the material on the surface affect the maximum kinetic energy of the photoelectrons.

(6) With the increase of frequency of the incident radiation, the cut-off potential, and hence the maximum kinetic energy of the electrons varies linearly with the frequency; for frequencies less than a critical value, ν_c, there is no emission of photoelectrons.

For $\nu > \nu_c$, the emitted electrons have some extra energy, characterized by velocity, v, and is dictated by the energy equation

$$h\nu = \frac{1}{2}mv^2 + \phi_0, \tag{2.2}$$

in which m represents the mass of an electron.

Equation (2.2) states that the incident photon energy is equal to the sum of work function and kinetic energy of the electron just after leaving the photocathode. With the increase of temperature, some of the electrons acquire extra energy and eventually they may come out of the metal as thermionic emission, which obeys Richardson's law. This law states that the emitted current density, **J** is related to temperature, T by the equation

$$\mathbf{J} = A_{\mathrm{R}} T^2 e^{-\phi_{\mathrm{m}}/(k_{\mathrm{B}} T)}, \tag{2.3}$$

where k_{B} $(= 1.38 \times 10^{-23}\mathrm{JK}^{-1})$ is Boltzmann's constant, and $A_{\mathrm{R}}(= 4\pi\, m q\, k_{\mathrm{B}}^2/h^3)$ the proportionality constant, known as Richardson's constant independent of T, T the temperature of the device, q the absolute value of electron charge, and ϕ_{m} the work function of the metal,

2.2 A Brief Account of Semiconductors

A semiconductor is a material that has intermediate electrical conductivity between a conductor and an insulator; depending on the condition it may act as either a conductor or an insulator. An insulator is a material with a very low electrical conductivity or thermal conductivity (for instance, glass,[a] air, cloth, and rubber are electrical insulators; feathers and wool make good thermal insulators). Semiconductors differ from metals in their characteristic property of decreasing electrical resistivity with increasing temperature (Yacobi, 2003). These materials are made up of individual atoms bonded together in a regular, periodic structure to form an arrangement. An atom consists of electrons orbiting around a nucleus, and according to Neils Bohr, the electrons have fixed orbits around the positively charged nucleus of an atom. They are restricted to orbits with certain energies. Electrons can jump from one energy level to another, but would not have orbits with energies other than the allowed energy levels. The potential energy of the electrons is proportional to their distance from the nucleus. The electrons are found at discrete stable energy levels. According to Pauli's exclusion principle the maximum number of electrons that can occupy a specific energy level is dictated by $2n^2$, with n as the

[a]Glass is an optically transparent fusion product of inorganic materials that has cooled to a rigid state without crystallizing.

number of energy level (principal quantum number) in question. The properties of any solid material, including semiconductors, depend upon the nature of the constituent atoms, as well as upon the way the atoms are grouped together.

2.2.1 *Crystal Structure*

Elements and their chemical compounds are found to occur in three states, namely, solids, in which the atoms are tightly bound, liquids, where the atoms are loosely bound, and gases, with atoms being freely or randomly bound. Solids can be in the following form:

Amorphous A solid substance with its atoms held apart at equilibrium spacing, but with no long-range periodicity in atom location in its structure is an amorphous solid. It has no definite geometrical shape, exhibits isotropic properties (same in all directions), such as glass, rubber, and wood.

Crystalline When the atoms come together to form a solid crystal, the outer or valence electrons interact as in covalent bonding to bind the atoms together. A crystal (ideal) is made up of regular and periodic three-dimensional repetition of a certain unit pattern of atom or molecules in space called the crystal structure (Dekker, 1981). It has a definite geometrical shape, exhibits anisotropic properties (different in crystallographic orientations), viz., zinc (Zn), copper (Cu), NaCl, zinc sulfide (ZnS), etc.

2.2.1.1 Crystal lattice

The patterns of a crystal are located upon the points of a lattice, an arrangement in a particular array of points repeating periodically in three dimensions. A lattice obeys the minimum symmetry conditions of the particular crystal systems it represents. The environment about any point is identical to the environment about any other points. The crystal lattice may be regarded as made up of an aggregate of a set of parallel equidistant planes passing through a lattice point, which are known as crystal planes. Each lattice point with one or more assembly of atoms or molecules identical in composition represents a crystal structure and is called basis or a pattern. The actual crystal structure is obtained when the basis is repeated with correct periodicity in all possible directions. The crystal structure is real, while the lattice is imaginary. The representation of the position of the lattice points in terms of vectors in space lattice is called basis vectors, while a geometrical representation of crystal structure in terms of lattice points is known as space lattice.

Unit cell The crystal structure is specified by describing the arrangement within the solid of a small representative group of atoms or molecules, called unit cell. By multiplying identical cells in three directions, the location of all the particles in the crystal is determined. When a unit cell of a lattice containing lattice points is repeated along the direction of the basis vectors, the space lattice is generated. In fact, a space lattice is considered as comprising of a large number of identical unit cells. There exist two types of unit cells, such as

(1) **Primitive cell**: The primitive cells are the simplest imaginary building blocks of a lattice that have lattice points at the corners; they have the smallest volume.
(2) **Non-primitive cell**: The non-primitive cells are the basic building blocks of a lattice that have lattice points at the corner, as well as at the faces or in the interior. Typically, these unit cells have a few times the volume of the primitive cell. They can fill space without overlaps and gaps through translational vectors, which are sums of multiples of lattice constants.

Bravais lattice If each lattice point in a space lattice is a substitution of identical set of one or more atoms, the lattice points are equivalent to each other and is termed as Bravais lattice (Bravais, 1850). A Bravais lattice is an infinite array of discrete points generated by a set of discrete translation operations described by a vector, $\mathbf{r}$

$$\mathbf{r} = k\mathbf{a}_1 + l\mathbf{a}_2 + m\mathbf{a}_3,$$

where k, l, and m represent a set of integers whose values depend on a particular lattice site and the primitive translation vectors, $\mathbf{a}_{j=1,2,3}$, define a Bravais lattice, which lie in different directions and span the lattice.

This discrete set of vectors must be closed under vector addition and subtraction. For any choice of **r**, the lattice is identical (in every aspect) when viewed from any of its lattice points. Both unit cell and Bravais lattice together form a crystal lattice. In the event the lattice points are non-equivalent, it is called non-Bravais lattice.

Brillouin zone A Brillouin zone is defined as a Wigner-Seitz primitive cell in the reciprocal lattice (Brillouin, 1953). It is constructed as the set of points enclosed by Bragg's planes, the planes perpendicular to a connection line from the origin to each lattice point and passing through the midpoint. The region in $\boldsymbol{\kappa}$-space that low-κ electrons can occupy without being diffracted is called first Brillouin zone, with $\boldsymbol{\kappa} = \kappa_x, \kappa_y$. It is the smallest volume enclosed by the planes that are perpendicular bisectors of the reciprocal lattice vectors drawn from the origin. The zone boundaries are $\boldsymbol{\kappa} = \pm\pi/a$, in which a is the lattice parameter for the length of a cube. The second Brillouin zone contains electrons with κ values from π/a to $2\pi/a$ for electrons moving in the $\pm x$ and $\pm y$ directions. These electrons do not fit into the first zone yet which have sufficiently small wave numbers to avoid diffraction by the diagonal sets of atomic planes. Further Brillouin zones can be constructed in the same manner.

Miller indices Miller indices are symbolic vectors that are used as a system of notation for planes within a crystal or a lattice. The number of indices should match with the dimension of the lattice or the crystal. Miller indices, h, k, and l, are based on the intercepts of a plane with the three crystal axes, each intercept with an axis being measured in terms of unit cell dimensions (a, b, and c) along that axis. There are several related notations, in which the negative numbers or directions are denoted with a bar on top of the number:

- $[hkl]$: The square brackets [ ], designate a direction in the basis of the direct lattice vectors from the origin to a point.
- (hkl): The parentheses, ( ), denote a crystal face or a family of planes throughout a crystal lattice.
- $\langle hkl \rangle$: The angle brackets, $\langle\,\rangle$, indicate a set of equivalent directions that are equivalent to $[hkl]$ by symmetry.

- $\{hkl\}$: The braces, $\{\}$, represent a set of face planes that are equivalent to (hkl) by the symmetry of the crystal lattice.

2.2.1.2 Arrangements of space lattice

There are fourteen ways of arranging points in space lattice, such that all the lattice points have the same surrounding. Each of these lattices belongs to one of the seven crystal systems (Kittel, 1996):

Cubic In a cubic structure, the axial lengths, as well as inter-axial angles are equal, i.e., $a = b = c$ and $\alpha = \beta = \gamma = 90^\circ$, where a, b, and c are the respective axial lengths and α, β, and γ the respective inter-axial angles. The three axes have repetition intervals of equal length, but all three axes are perpendicular to each other. For cubic crystals with the spacing, d, between adjacent (hkl) lattice planes is given by

$$d_{hkl} = \frac{a}{\sqrt{h^2 + k^2 + l^2}}. \tag{2.4}$$

where a the lattice parameter for the length of a cube. There are three types of cubic structures:

(1) **Simple cubic**: In this type, the atoms are at the corners of a cube and form a unit cell. A corner atom has four neighbors in the same plane and two nearest neighbors in a vertical plane; the coordination number is six.

(2) **Body centered cubic (BCC)**: In this system, an atom is at the center of each cube. The distance between two adjacent planes characterized by the Miller indices is $a/\sqrt{2}$, in which a represents lattice constant. A center atom has eight neighbors at the corner of the cube; the coordination number is eight.

(3) **Face centered cubic (FCC)**: In this, an atom is at the center of each of the six faces of the cube, which has four atoms in the same plane, four atoms below this plane, and four atoms above the plane; the coordination number is twelve. The BCC and FCC lattices are Fourier transforms of one another.

Tetragonal Two of the axes have repetition interval of equal length, but all three axes are perpendicular to each other. The geometrical configuration is given by $a = b \neq c$ and $\alpha = \beta = \gamma = 90^\circ$, and the types of Bravais lattices are in the form of (i) simple cubic and (ii) body centered cubic.

Orthorhombic In this three axes have repetition interval of unequal lengths, but they are perpendicular to each other. The geometrical configuration is given by $a \neq b \neq c$ and $\alpha = \beta = \gamma = 90^\circ$, and the types of Bravais lattices are in the form of (i) simple cubic and (ii) body centered cubic, (iii) face centered cubic, and (iv) base-centered.

Monoclinic All the three axes have repetition intervals of unequal length; two of them are not perpendicular to each other, but are perpendicular to the third one. The geometrical configuration is given by $a \neq b \neq c$ and $\alpha = \gamma = 90^\circ \neq \beta$, and the types of Bravais lattices are in the form of simple cubic and base-centered.

Triclinic None of the axes are equal to or perpendicular to any of the other, whose geometrical configuration is given by $a \neq b \neq c$ and $\alpha \neq \beta \neq \gamma \neq 90^\circ$, and the types of Bravais lattices are in the form of simple cubic.

Trigonal or rhombohedral All the three axes have repetition intervals of equal length and they are oriented to each other at equal angle, but other than 90°. The geometrical configuration is given by $a = b = c$ and $\alpha = \beta = \gamma \neq 90^\circ$, and the types of Bravais lattices are in the form of simple cubic.

Hexagonal Two of the axes have repetition interval of equal length, between which there is angle of 120°. The third axis is oriented at 90° to the former two axes. The geometrical configuration is given by $a = b \neq c$ and $\alpha = \beta = 90^\circ$; $\gamma = 120^\circ$ the types of Bravais lattices are in the form of simple cubic.

2.2.2 *Crystal Defects*

It is improbable to get a perfect crystal with every atom of the same type in the correct position. All crystals have some defects that contribute to the mechanical properties of metals. A state of the lattice in which the atomic arrangement has departed from regularity is referred to as defect. The influence of defects on the physical properties of the crystal depends on the defect dimensionality. The defects may scatter conduction electrons in a metal, increasing its electrical resistance in many pure metals, and decrease the strength of the crystal. Many important defects are electrically active. Defects that can contribute free electrons to the host crystal are known as donors, while defects contributing holes, i.e., removing free electrons, are known as acceptors. In what follows, the basic classes of defects are enumerated briefly.

2.2.2.1 Point defects

A point (or zero-dimensional) defect involves a single atom change to the normal crystal array. The point defects are irregularities and are found at the site of an atom or at a point within the crystal. The major types of point defect are as follows:

Interstitial impurities An impurity is the substitution of a regular lattice atom with an atom that does not normally occupy that site. An interstitial is an atom that occupies a place outside the normal lattice position. The interstitial impurity occurs when an extra atom is present at the regular or interstitial position (a point intermediate between regular lattice points) in the lattice, which may be an impurity atom or an atom of the same type as the regular lattice atoms. When an atom is trapped in the interstitial point, it is called an interstitialcy. The interstitial impurities introduce less distortion to the lattice and are more common in real materials.

Vacancy A vacancy is the absence of an atom from a regular lattice position, which is present invariably in all materials. In a simple crystal lattice, it leads to the displacement of the nearest atoms in the direction of the vacancy position. At high temperatures, the atoms randomly change their positions leaving behind empty lattice sites. In most cases, diffusion (mass transport by atomic motion; see Section 2.2.4.3) occurs due to the vacancies. The number of vacancies, N_v, increases exponentially with the absolute temperature, T (= 273 K), and is estimated using Maxwell–

Boltzmann's distribution[a]

$$N_v = N_s e^{-E_v/(k_B T)}, \tag{2.5}$$

where k_B represents Boltzmann's constant, N_s the number of regular lattice sites, and E_v the energy needed to form a vacant lattice site in a perfect crystal.

In order that an atom can jump into a vacancy site, it is necessary for it to possess enough energy (for instance, thermal energy) to be squeezed from its neighbors. The energy necessary for the motion, E_m, is known as the activation energy for vacancy motion. The average thermal energy of an atom is small in comparison to the activation energy, E_m with a large deficit in energy, which is to be pooled together from neighboring atoms in a small volume, for a jump. The probability of frequency of jumps, R_j, depends exponentially on temperature and is described by

$$R_j = R_0 e^{-E_m/(k_B/T)}, \tag{2.6}$$

in which R_0 is the attempt frequency proportional to the atomic vibration.

(1) **Schottky defect**: It is created by transferring an atom or an ion from the original correct position to the surface of the sample.
(2) **Frenkel defect**: If the atom is transferred to an interstitial position instead of surface, the defect is known as Frenkel defect. In this case, two imperfections, such as interstitial atom and a vacancy, are created. The electrical conductivity in some ionic salts, like many alkali halides, is caused by Frenkel defects (i.e., ions, not by electrons). The conductivity in an alkali halide can be modeled by the diffusion of Frenkel defects through the crystal.
(3) **F-center**: The term F-center came from the German word for color "farbe," as crystals containing these point defects are highly colored. It results from the absence of a negatively

[a]Under thermodynamic equilibrium the population of the atomic energy levels is determined by Maxwell–Boltzmann's distribution

$$\frac{N_2}{N_1} = \frac{g_2}{g_1} e^{-(E_2-E_1)/(k_B T)},$$

where k_B is Boltzmann's constant, $g_{1,2}$ the degeneracies of state 1 and 2, and T the temperature.

charged ion from a particular point in an ionic solid. When a vacancy is left behind by an ion and an electron is trapped within this vacancy, the energy levels of the electron are quantized. The term color center came later to include all point lattice defects that absorb light in a spectral region where the crystal itself does not absorb. Such defects may constitute intrinsic centers or impurity centers. The former encompass both positive-ion and negative-ion vacancies, as well as interstitial ions: the latter, impurity atoms and ions. The color defects in a crystal may be produced in a number of ways, for example,

- by introducing chemical impurities,
- by bombarding crystals with X-rays, γ-rays, electrons, and neutrons,
- by heating the crystal in the alkali metal vapor and quenched to room temperature, which would show an absorption band (known as F-band) in the ultraviolet (UV) or visible, whereas the original crystals were transparent in that region; for instance, the excess of potassium (K) in potassium chloride (KCl) colors it blue, while sodium (Na) in NaCl turns the crystal yellow, and
- by electrolysis.

Substitutional impurity If the foreign atom replaces or substitutes a matrix atom, it is referred to as a substitutional impurity. An impurity atom in the lattice may be present at any interstitial position or at any substitutional position, i.e., in place of regular lattice sites.

2.2.2.2 Line defects

If a regular atomic arrangement is broken in the small vicinity of a certain line, the corresponding defect is called linear (or 1-D) where the typical examples are dislocations. A dislocation is a region of strain in the regular crystal structure in which a part of the lattice has undergone a shearing strain equal to one lattice vector, called Burger's vector. Burger's vector represents the magnitude of the structural defect. There are two basic types of dislocations:

Edge dislocation An edge dislocation in a metal occurs when the periodicity of the atomic lattice array is interrupted along certain directions in a crystal. The edge dislocations occur along the rows of the crystal structure, what is referred to as line defects. An edge location is found by the insertion (or removal) of an extra half plane of atoms in the crystal structure. Burger's vector is perpendicular to the dislocation line. The dislocations move in steps. When the half-plane reaches a free surface, it produces a slip step.

Screw dislocation A screw dislocation, which derives its name from the spiral or helical path (like a screw), changes the character of the atom planes. The atom planes no longer exist separately from each other. It is traced around the dislocation line by the atomic planes of atoms, and may be thought of as being formed by a shear stress that is applied to produce the distortion. In this, Burger's vector is parallel to the dislocation line and it has a row of atoms about which a normal crystallographic plane appears to be spiral.

2.2.2.3 Planar (surface) defects

A planar (two-dimensional) defect in the crystal may occur when a regular atomic arrangement is violated along the part of some surface with a thickness of the order of inter-atomic distances.

Grain boundaries Each crystal is known as a grain. A grain boundary is a general planar defect that separates regions of different crystalline orientation (i.e., grains) within a polycrystalline solid. To note, most crystalline solids are an aggregate of several crystals, which are referred to as polycrystalline. The orientation of the crystals changes across the grain boundary. The grain boundaries are, in general, the result of uneven growth when the solid is crystallizing and act as obstacles to dislocation motion; hence, the presence of more grain boundaries (finer grain size of the order of a few μm) would increase the strength of the crystal. The size of the grains can be controlled by the cooling rate when the sample is produced from the melt. The rapid cooling produces smaller grains, whereas slow cooling results in larger grains. At room temperature, the large grains result in low strength, hardness, and ductility. There exists some atomic mismatch in a transition from the crystal orientation of one grain to that of an adjacent one within the grain boundary. Various degrees of crystallographic misalignment between adjacent grains are possible.

Tilt boundaries A tilt boundary between two slightly misaligned grains appears as an array of edge dislocations. When the orientation mismatch is on the order of a few degrees, it is known as low angle grain boundary. Twist low angle grain boundaries result from array of screw dislocations. The regions of material separated by small-angle grain boundaries are called subgrains.

Twin boundaries Twins are either grown during crystallization or the result of thermal work. A twin boundary occurs when the crystals on either side of a plane are mirror images of each other. The boundary between the twinned crystals will be a single plane of atoms.

2.2.2.4 Bulk (or volume) defects

Bulk defects include the following:

Voids These are regions, where a large number of atoms missing from the lattice. They can occur due to

- pores that form due to the gas entrapment during melting and pouring; it is commonly known as porosity and
- the shrinkage of a material as it solidifies, which is referred to as cavitation.

Cracks A crack occurs where internal broken bonds create new surfaces. Cracks are due to the non-uniformity of heating or cooling, thermal mismatch, constrained expansion or contraction leading to stress development, foreign (non-metallic) inclusions, such as slag, oxide particles or sand entrapment, and other phases. These are generally introduced during processing and fabrication processes for various reasons and are harmful to the material. There is a tendency to form micro-cracks ($\sim$10 μm in size), which are formed when there is abrasion with dust particles, on the surface of a solid. They are formed at the grain boundaries as well.

Precipitates These regions occur when impurity atoms cluster together to form small regions of a different phase; the term "phase" refers to that region of space occupied by a physically homogeneous material.

2.2.3 *Energy Bands in Semiconductor*

Electrons interact with the nucleus of an atom to form a range of bound energy levels (see Section 1.1.6). Because the electrons are shared by more than one atom in the crystal, the energy levels allowed, take on many more closely spaced states. This large number of discrete but closely spaced energy levels is called an energy band. The two highest energy bands are the valence band and the conduction band. These bands are separated by a region that designates energies that the electrons in the solid cannot possess. This region is called the forbidden gap or bandgap, E_g, which is the energy difference between the maximum valence band (valence-band edge) energy, E_V and the minimum conduction band (conduction-band edge) energy, E_C. In order to obtain an expression for the energy bands, one considers Schrödinger's wave equation for electrons

$$\nabla^2\Psi + \frac{8\pi^2 m_e}{h^2}(E - V)\Psi = 0, \tag{2.7}$$

where the Laplacian operator, ∇^2, in Cartesian coordinates reads

$$\nabla^2 = \frac{\partial^2}{\partial x^2} + \frac{\partial^2}{\partial y^2} + \frac{\partial^2}{\partial z^2}, \tag{2.8}$$

E the energy, $V = V(x, y, z)$, the potential energy in the medium in which the particle propagates.

The energy, E includes both kinetic and potential energies, that is

$$E = \frac{1}{2}m_e v^2 + V = \frac{p^2}{2m_e} + V, \tag{2.9}$$

with $p(= m_e v)$ as the momentum of the electron.

Assuming that no forces act on the electron, the potential energy is constant; the electron moves in a field-free space described by a wave function

$$\Psi = e^{i\kappa\cdot\mathbf{r}}, \tag{2.10}$$

that represents a plane wave propagating in a specific direction of the wave vector, $\boldsymbol{\kappa}$.

2.2.3.1 Bandgap energy

The energy of an electron is a function of its quantum mechanical wave vector $\boldsymbol{\kappa}$. In a semiconductor, this dependence of electron energy on its wave vector forms the band structure of the semiconductor. The upper wavelength cut-off, λ_c, for any semiconductor can be determined by its energy gap. Equation (2.1) yields the following relationship by inserting the numerical values for Planck's constant, h and the velocity of light, c

$$\lambda_c(\mu m) = \frac{hc}{E_2 - E_1} = \frac{1.24}{E_g(eV)}, \tag{2.11}$$

E_g is the bandgap energy; the bandgap of a semiconductor is typically less than 4 eV.

In order for an electron to jump from a valence band to a conduction band, it needs a specific amount of energy for the transition. For example, silicon (Si) has a bandgap of 1.12 eV and germanium (Ge) has a bandgap of 0.72 eV. Energies of this magnitude, in general, cannot be acquired from an applied field; hence, the valance band remains full, the conduction band empty. The width of the bandgap determines the type of material. Based on the general band structures, the semiconductor materials are classified as the following types.

Direct bandgap compounds In semiconductor, the minimum energy state in the conduction band, and the maximum energy state in the valence band are characterized by the wave vector, $\boldsymbol{\kappa}$, in the Brillouin zone (see Section 2.2.1.1). If the wave vectors are the same, it is called a direct bandgap material, i.e., the maximum energy of the valence band occurs at the same momentum value as the minimum energy of the conduction band. The conduction band minimum and valence band maximum occurs at $\kappa = \pm\pi/a$, in which a is the lattice parameter for the length of a cube. The III–V materials, such as gallium arsenide (GaAs)/AlGaAs, and indium phosphide (InP), fall under this category, where a large, sharply rising absorption constant is found immediately above the bandgap energy, with

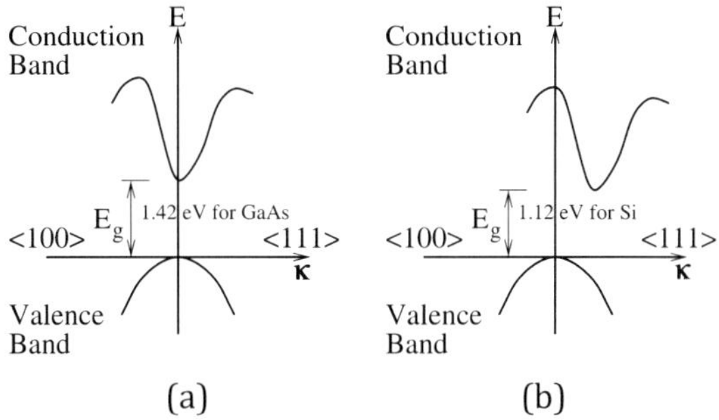

Figure 2.1 Energy band structure diagrams of (a) GaAs and (b) Si depicting energy versus wave vector, κ (Ahmed, 2007). The number in the angle brackets (<100> and <111>) are the Miller indices (see Section 2.2.1.1).

values of absorption coefficient, α, of the order of 10^4 to 10^5 cm^{-1}. They have the minimum of the conduction bands and the maximum of the valence bands occurring at the same value of the wave vector, κ. In GaAs, each gallium atom has three electrons in its outermost shell of electrons and each arsenic atom has five. When a trace of an impurity element with two outer electrons, such as zinc, is added to the crystal, the result is the shortage of one electron from one of the pairs, causing an imbalance in which there is a hole for an electron but there is no electron available. This forms a P-type semiconductor. When a trace of an impurity element with six outer electrons, such as selenium, is added to a crystal of GaAs, it provides an additional electron, which is not needed for the bonding. This electron can be free to move through the crystal. Thus, it provides a mechanism for electrical conductivity (N-type semiconductor). Figure 2.1a,b depicts direct and indirect electron transitions in semiconductors, respectively.

Indirect bandgap compounds If the κ-vectors are different, it is called an indirect bandgap material, such as Si and Ge. That is the minimum of conduction bands and the maximum of valence bands do not occur at the same κ value. The absorption of threshold is a more slowly increasing function of energy and the absorption coefficient, α is generally much lower; values $\sim 10^2$ to 10^3 cm^{-1} are common for photon energies several tenths of an eV above the bandgap. However, the electron escape depths are much longer in these compounds than the direct materials, making net efficiency (proportional to α times escape depth) comparable. Both silicon (Si) and germanium (Ge) are the most useful intrinsic semiconductors. Both these elements have four electrons in the outer orbit allowing them to form crystals. The electrons in the outermost shell of an atom pair form perfect covalent bonds with the electrons of the neighboring atoms, which makes Si to act as nearly an insulator, where very little electricity would flow through it. Although germanium at a given temperature would have more free electrons and a higher conductivity and was used for developing first transistors, Si is more widely used semiconductor material for electronics. It is the basic material used in the development of integrated circuit. Also, it can be used at much higher temperatures than germanium.

While the Si-based solid-state devices were evolving through various phases of the integrated technology, advances were also made in developing photoelectronics, such as the photosensitive diodes and transistors, exploring the Si properties to the electromagnetic radiation. The covalent bonds can be broken by the incident photons and a charge pattern accumulates in the Si lattice, which reflects that of the incident light.

2.2.3.2 Periodic potentials

For atomic states, V is negative and leads to the development of discrete negative bound energy states or levels as indicated in equation (2.7) and their associated eigen functions. The probability distribution function for the electron described by this wave function has an infinite extent and uniform value in any plane perpendicular to the direction of propagation. Applying equation (2.9) as a solution to Schrödinger's equation leads to

$$E = \frac{\hbar^2 \kappa^2}{2m_e} + V. \tag{2.12}$$

A semi-classical description is obtained by taking into account de Broglie's relationship, $p = m_e v = \hbar\kappa$. The electron kinetic energy, E_k, is written as

$$E_k = E - V = \frac{1}{2}m_e v^2$$

$$= \frac{p^2}{2m_e} = \frac{\hbar^2 \kappa^2}{2m_e}. \tag{2.13}$$

Equation (2.13) is a semi-classical representation of the propagation of the particle describing the wave-like character of the particle by the wave vector, κ, while it preserves the deterministic nature of the classical particle.

Consider that an electron is propagating along the direction of a vector, $\mathbf{r}$, in the lattice, which is then exposed to periodic variations of the potential caused by the charged nuclei of the atoms forming the lattice and by the electrons of these atoms. The periodicity of the potential energy in the lattice is given by

$$U(\mathbf{r}) = U(\mathbf{r} + \mathbf{R}),$$

where $\mathbf{R}$ the vector of translational symmetry of the lattice.

The potential energy of the lattice is periodic with a period, $\mathbf{R}$; the potential is due to the effect of the ion cores and all other electrons. The band structure, in principle, can be calculated from the periodic arrangements of the atoms and their atomic order number.

Bloch's theorem According to Bloch's theorem (Bloch, 1928), the wave function of an electron in a periodic lattice is written as

$$\Psi_{n\kappa}(\mathbf{r}) = u_{n\kappa}(\mathbf{r})e^{i\kappa \cdot \mathbf{r}}, \tag{2.14}$$

with $u_{n\kappa}(\mathbf{r}) = u_{n\kappa}(\mathbf{r} + \mathbf{R}) = u_{n\kappa}(\mathbf{r} + 2\mathbf{R}) = \cdots = u_{n\kappa}(\mathbf{r} + n\mathbf{R})$ as a lattice periodic part, and has the same periodicity as the periodic potential, $e^{i\kappa \cdot \mathbf{r}}$ the wave function of a free particle, the integer $n(= 1, 2, 3, \cdots)$ the band index, and κ the electron wave vector.

Functions of this type (equation 2.14) are referred to as Bloch's function, which, in general, depend on κ and n. The function, $u_{n\kappa}(\mathbf{r})$, is periodic with period, $R = |\mathbf{R}|$ and has a unique shape for the different energy bands of the semiconductor. This function does not vary significantly, for $|\mathbf{r}| \ll |\mathbf{R}|$; it would be dominated by the factor, $e^{i\kappa \cdot \mathbf{r}}$, if $|\kappa| \ll 2\pi/|\mathbf{R}|$, i.e., in the region of small κ, the electron wave function is described by the free particle wave function. Therefore, the dispersion relation of a particle in a periodic potential, $E(\kappa)$, would become parabolic in this region, similar to the parabolic dispersion relation of free particle given in equation (2.13). For larger values of κ, significant deviations from the parabolic dispersion are expected in this region. The relation between the energy, E and the wave number, κ, is referred to as energy band structure of the lattice.

Kronig–Penney model The Schrödinger equation (see Section 1.1) for a one-dimensional periodic potential was solved by Kronig and Penney (1930). When V is a periodic potential with regular, repeated alignment of the atoms, the potential is expressed as

$$V(x) = V(x + a) = V(x + 2a) = \cdots = V(x + na), \qquad (2.15)$$

where a stands for the periodic spacing between identical atoms or repeated groups of different atoms and n the integer.

The solution of Schrödinger's equation describing the probability distribution associated with the location of the electron is

$$\Psi = u_\kappa(x + na)e^{i\kappa\cdot\mathbf{r}}, \qquad (2.16)$$

$u_k(x + na)$ is a periodic function of x with period a, the subscript, κ is a quantum number that is used to label the specific state of the wave function given in equation (2.14).

The relationship between the kinetic energy and κ is approximated in the small region of κ by employing the energy of a free electron (see equation 2.12)

$$E_\mathrm{k} = \frac{\hbar^2\kappa^2}{2m_\mathrm{C}}, \qquad (2.17)$$

where m_C is the effective electron mass[a] in the conduction band; an electron near the conduction band edge may be approximately described as a particle of effective mass confined to a 3-D box with perfectly reflecting walls.

Equation (2.17) is the energy above the band edge associated with a given state k. It is given in accordance to parabolic band approximation where E_k is measured from the bottom of the band. The energy band structure for the valence band, E_V, takes the shape of parabola curving downward (see Figure 2.1; the curvature of this

[a]The effective mass of an electron is not the same as the free electron mass (electron mass in free space) because of the wave-particle dual character of the electron and its interaction with the lattice (scattering).

band is determined by $d^2E/d\kappa^2$. The valence band is filled with electrons for which there is one electron for each of the available energy levels. The conduction band, E_C, is devoid of electrons since it corresponds to the unoccupied higher energy levels of the individual atoms. Unlike in the case of valence band, the energy band structure becomes a parabola curving upward for the conduction band (see Figure 2.1). The effective mass, m^*, of an electron in a periodic potential is given by

$$m^* = \frac{\hbar^2}{d^2E/d\kappa^2}. \tag{2.18}$$

Equation (2.18) states that the effective mass is inversely proportional to the second derivative of E with respect to κ, i.e., the effective mass is inversely proportional to the curvature of the dispersion relation. Depending on the nature of the periodic potential, the effective mass may be lighter or heavier than the free electron mass. However, the effective masses for the conduction and valence bands, m_C and m_V may not be the same, for instance, for the semiconductor, gallium-arsenide (GaAs), $m_C = 0.067\ m_e$ and $m_V = 0.48\ m_e$.

The energy spectrum of the electrons consists of a number of allowed energy bands separated by forbidden region (Dekker, 1981). The minimum energy required to move an electron from its bound state to a free state where it can participate in conduction is called the forbidden energy gap (bandgap energy), E_g. The width of the bandgap, which increases with increasing energy, determines the type of material. The width of a particular allowed band decreases with increasing binding energy (required amount of energy to free electrons from their atomic orbits) of the electrons.

2.2.4 *Electrical Properties of Semiconductors*

Every solid has its own characteristic energy band structure (see Figure 2.2). In order for a material to be conductive, both free electrons and empty states must be available. The energy band in a metal is partially filled with electrons. In the absence of an applied electric field, the electrons move in random directions colliding with random impurities and/or lattice imperfections in the crystal arising from thermal motion of ions about their equilibrium positions.

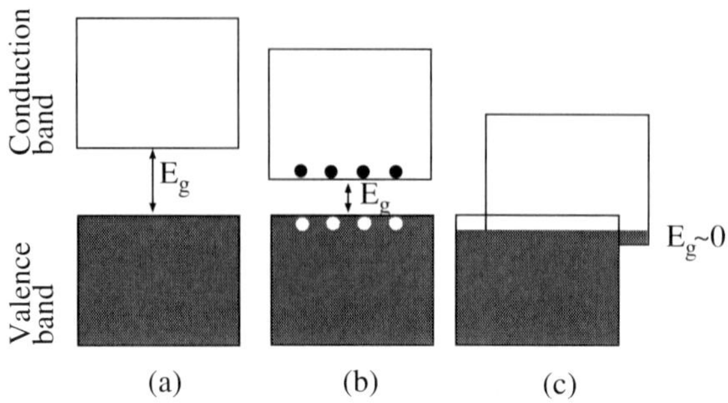

Figure 2.2 Energy band structures for (a) insulator, (b) semiconductor, and (c) conductor.

The frequency of electron-lattice imperfection collisions may be described by a mean free path that is defined as the average distance an electron travels between collisions. Under the influence of an applied electric field, the electrons acquire energy band move into higher states. Since these mobile electrons constitute a current, this substance is a conductor and the partly filled region is the conduction band. There is no bandgap in metals in which the valence band overlaps the conduction band. Semi-metals have their highest band filled. This filled band, however, overlaps with the next higher band, and therefore they are conductive but with slightly higher resistivity than normal metals, for instance, arsenic, bismuth, and antimony.

A solid for which a certain number of energy bands are completely filled, the other bands being completely empty, is an insulator. The valence electrons form strong bonds neighboring atoms, in which there are no free electrons to participate in current conduction. The forbidden energy gap is large (> 4 eV), where no electrons can reach the conduction band at ordinary temperature. An example is a diamond crystal, for which the forbidden gap is about 6 eV. The energy that can be supplied to an electron from an applied electric field is too small to carry the electron from the valence band up through the forbidden gap into a vacant band. This makes conduction improbable.

A semiconductor has a similar band structure as an insulator but with a much smaller bandgap, say ~1 eV. The bonds between the neighboring atoms are moderately strong. It needs a specific amount of energy for the transition by which the number of thermally excited electrons becomes appreciable. Some electrons can jump to the empty conduction band by thermal or optical excitation; hence, conduction becomes probable.

2.2.4.1 Electron–hole pairs

At any temperature above absolute zero temperature, there is a finite probability that some of the electrons residing in the valence band acquire thermal energy greater than the bandgap energy, E_g, and hence move to the conduction band across the forbidden energy gap. This produces a bound electron into the conduction band. The remaining net positive charge behaves as a positively charged particle, known as a hole (electron deficiency), in the valence band that represents the energy state of bound electrons. Thus, an electron–hole pair is created. The continual movement of the space for an electron can be stated as the movement of a positively charged particle through the crystal structure. At thermal equilibrium, the creation of electron–hole pairs is balanced by their recombination. The recombination is a process in which a conduction electron is captured by an ionized atom, releasing energy either as a photon or as thermal lattice energy, so that a mobile electron–hole pair disappear. If the energy of the photons, say between 1.1 eV and 4 eV, exceeds the energy gap of the semiconductor, E_g, a photon can be absorbed, which creates a single electron–hole pair; photons of higher energy generate multiple pairs. However, these generated electrons, if left untrapped, would recombine into the valence band within 100 μsec. The rate of recombination, R, is proportional to the product of the concentrations of electrons and holes, i.e., $R = \beta N_e N_h$, with β is the proportionality constant, N_e, N_h as the electron and hole densities, respectively. The recombination coefficient depends on the characteristics of the material, including its composition and defect density, and on temperature. In the steady state, the rate of recombination must equal the rate of generation. This process is responsible for the operation of semiconductor light sources. There are three types of recombination (Grundmann, 2010), for instance:

Radiative recombination The radiative recombination occurs when the energy is released by the electron, which may take the form of an emitted photon.

Non-radiative recombination The non-radiative recombination can occur due to the transfer of energy to lattice vibrations or to another free electron (Auger, 1923).

Shockley–Read–Hall recombination Recombination may occur indirectly through defect levels (Shockley and Read, 1952, Hall, 1952). These are energy levels associated with impurities or defects due to grain boundaries, dislocations, or other lattice imperfections (see Section 2.2.2), that lie within the energy bandgap. The impurity-assisted recombination can be either radiative or non-radiative.

In direct bandgap compounds (see Section 2.2.3.1), the recombination of electrons and holes takes place in a single step (the probability of recombination is more). On the contrary, in indirect compounds, the recombination process requires several intermediate stages; the generated photons tend to recombine resulting in reduction of the output intensity. Although indirect bandgap materials can emit some photons, the number of photons is of orders of smaller magnitude than for the direct bandgap materials.

Both the electrons and holes can participate in conduction and are called carriers. The more abundant charge carriers are known as majority carriers, while the less abundant charge carriers are called minority carriers. In N-type semiconductor, the majority carriers are negatively charged electrons and the minority carriers are holes. In P-type semiconductor, the majority carriers are positively charged holes and the minority carriers are electrons. The spontaneous carrier recombination rate is denoted by the parameter, $\gamma_s (= 1/\tau_s)$, in which τ_s is the total spontaneous carrier recombination lifetime, that is

$$\frac{1}{\tau_s} = \frac{1}{\tau_{\text{rad}}} + \frac{1}{\tau_{\text{nonrad}}},$$

τ_{rad} the lifetime of an excess electron–hole pair associated with the radiative recombination and τ_{nonrad} that associated with non-radiative carrier lifetime.

Unlike in a metal, where the current is due to the flow of electrons, the current in a semiconductor results from the movement of both electrons and holes. The electrical properties of semiconductors are determined by (i) the number of electrons and holes per unit volume and (ii) the mobility of the electrons and holes. Upon applying an electric field in a semiconductor, the transport of charges commences producing a current. This current is due to two distinct phenomena, such as (i) carriers drift in an electric field and (ii) carriers diffuse if a concentration gradient exists (Sze and Ng, 2007). The total majority carrier flow is the sum of a drift current and a diffusion current, which are enumerated below briefly.

2.2.4.2 Drift current

The term drift is the motion of the charged particle in response to an applied electric field. When such a field is applied across a semiconductor, the carriers start moving, producing a current. The positively charged holes move with the electric field, whereas the negatively charged electrons move against the electric field. The motion of each carrier is described by a constant drift velocity, v_d, which is in the direction opposite to that of the electric field. When a free electron collides with the lattice, its direction of motion would be continuously changing. Under the influence of an electric field, a free electron moves with a constant velocity in the steady state. The drift velocity is defined as

$$v_\mathrm{d} = \frac{e\mathbf{E}}{m}\tau, \tag{2.19}$$

where m is the mass of electron, $\mathbf{E}$ the electric field, $\tau(= l/v)$ the mean collision time defined as the average time interval between two successive collisions of an electron with the lattice corners, l the mean free path, which is the average distance traveled by the conduction electron between successive collisions with the lattice ions, and v the effective instantaneous speed of the random motion, which is much higher than the drift velocity.

The average drift velocity, $\langle v \rangle$, of the electrons under the influence of an electric field gives rise to an electric current. In the absence of this field, the free electrons in a metallic substance would move in random directions, so that the average velocity of free electrons in any direction is equal to zero, i.e., $\langle v \rangle = 0$. However, by applying an external field on the specimen, there would be a net positive value of average velocity, i.e., $\langle v \rangle = \langle v \rangle'$, in which $\langle v \rangle'$ is the average velocity of conduction electron in the presence of the field. In the case of sudden disappearance of the field across the metal, the average velocity, $\langle v \rangle$, of its conduction electrons decays exponential to zero from a value, $\langle v \rangle'$. The decay process is represented by

$$\langle v \rangle = \langle v \rangle' e^{-t/\tau_r}, \tag{2.20}$$

with t as the time counted from the instant the field is turned off, τ_r the relaxation time that is defined as the time taken by an electron to reduce its velocity to $1/e$ of its initial value; if $t = \tau_r$, the average velocity, $\langle v \rangle$ becomes equal to $\langle v \rangle'$.

The average velocity is proportional to the electric field, E, that is

$$\langle v \rangle = \mu \mathbf{E}, \tag{2.21}$$

in which $\mu (= q\tau/m)$ is called the mobility of the electrons (square meters per volt.second), m the mass of the carrier, and q the electron charge.

The resultant of carrier drift is known as drift current, which depends on the parameters, such as (i) the electron- or hole-concentration, (ii) the electric field, (iii) the mobility constant, and (iv) the charge.

2.2.4.3 Diffusion current

Diffusion is the process of particles distributing themselves from the regions of high concentration to regions of low concentration. In this process, the external forces are not required; the particles move due to thermal motion. The movement of charge results in a current, called diffusion current, which is driven by maintaining a concentration gradient. The diffusion current density follows Fick's law stating that the diffusion flux is proportional to the negative of the concentration gradient, that is

$$\mathbf{J} = -D\nabla N, \tag{2.22}$$

with $\nabla = \left(\frac{\partial}{\partial x}, \frac{\partial}{\partial y}, \frac{\partial}{\partial z} \right)$,

$$\mathbf{J} = \sigma \mathbf{E}, \tag{2.23}$$

as the current density (ions per area per time), D the diffusion coefficient in area per unit time ($\mathrm{m^2.s^{-1}}$), $\sigma(= Nq\mu)$ the conductivity of the metal, and N the density of the current carrying ions per unit volume; the direction of such a current depends on (i) the change in the carrier concentrations and (ii) the gradient of the concentration of holes and electrons.

2.2.4.4 Mobile charge carrier

Unlike a metal, which is unipolar, a semiconductor is bipolar. Two types of mobile charge carriers are available, for example, the negative electrons of mobility, μ_e and the positive holes of mobility, μ_h. These particles move in opposite directions in the electric field, $\mathbf{E}$, since they are of opposite sign, the current of each is in the same direction. The current density, $\mathbf{J}$ is given by

$$\mathbf{J} = (N_e\mu_e + N_h\mu_h)q\mathbf{E} = \sigma\mathbf{E}, \tag{2.24}$$

with $\sigma = (N_e\mu_e + N_h\mu_h)q$, N_e as the magnitude of free electron concentration, and N_h the magnitude of hole concentration.

2.2.5 *Electron Distribution in Semiconductors*

All electrons are identical to one another in mass, charge, spin, and magnetic moment. As stated earlier, at absolute zero temperature, the electrons would occupy the lowest state, where the kinetic energy is zero. All the possible energy levels would be filled up with two electrons each up to a maximum level, called the Fermi level. The number of electrons per unit volume occupying states in the energy range between E and $E + dE$ in any electronic system in thermal equilibrium (Sze and Ng, 2007) is given by

$$N(E)dE = f(E)\rho(E)dE, \tag{2.25}$$

with

$$f(E) = \frac{1}{1 + e^{(E-E_F)/(k_B T)}}, \tag{2.26}$$

as Fermi–Dirac's distribution (Fermi function for electrons in thermodynamical equilibrium (Fermi, 1926, Dirac, 1926), k_B Boltzmann's constant, E_F the Fermi level energy, T the temperature of the device, E the energy of the electron, and $\rho(E)$ the density of states that is defined as the number of available energy states per unit volume of a metal in an energy interval, E and $E + dE$; the density of states is zero at the band edge, and increases away from it at a rate that depends on the effective masses of the electrons and holes.

The Fermi energy is defined as the energy of the highest occupied quantum state in a system of fermions and is denoted by E_F, which is a characteristic of the metal (Kittel, 1996). The fermions are subatomic particles with half integer spin, which possess an intrinsic angular momentum of $\hbar/2$, with $\hbar = h/(2\pi)$, and obey Pauli's exclusion principle at absolute zero temperature. The energy level is referred to as the Fermi level. The Fermi function, a mathematical function having no units, i.e., $0 < f(E) < 1$, provides the probability of occupation of a given energy state for a material in thermal equilibrium in terms of the Fermi energy, Boltzmann's constant, and the temperature. The behavior of Fermi factor at different temperatures is enumerated briefly:

(1) At absolute zero temperature, ($T = 0$ K; there is no thermal energy), the kinetic energy of the electrons cannot exceed the Fermi energy, E_F. If $E < E_F$, the Fermi function, $f(E) = 1$, which means that all the quantum states are occupied for the levels having energy less than the Fermi energy at temperature 0 K, while if $E > E_F$, the Fermi function $f(E) = 0$, meaning that all the quantum states are empty for the levels having energy greater than the Fermi energy at $T = 0$ K.

(2) When $k_B T \ll E_F$ and for an electronic state with energy the same as E_F, i.e., $E = E_F$ for $T > 0$ K, $f(E_F) = 1/2$, some electrons are excited to the levels above the Fermi level, consequently a few levels below the Fermi level are partially empty and those above are partially filled. In such a situation, the Fermi level represents the energy at which the levels are half filled and half empty. The Fermi level is thus referred to as the half-occupancy level. For intrinsic semiconductors, which have number of positive and negative charge carriers, the Fermi level lies in the middle of the bandgap. In this level, the probability of electron occupancy is 1/2. The electrons inside the metal need certain amount of energy to come out of the metal and this

deficiency is called surface potential barrier. The $k_B T$ is small as compared with E_F for all metals below their melting point ($k_B T = 0.03$ eV at room temperature).

(3) At very high temperature, $E - E_F \ll k_B T$, the entire distribution turns out to be Maxwell–Boltzmann.

The Fermi distribution law is employed in the free electron theory of metals, where $\rho(E)$ is proportional to $E^{1/2}$ when E is measured from the bottom of potential well representing the metal. In this case, the Fermi energy, E_F at $T = 0$, represents the highest occupied state. In the case of insulators and semiconductors, E_F is usually located somewhere between these bands.

In thermal equilibrium, a single Fermi energy applies both to the valence and conduction bands. Under condition in which thermal equilibrium is disturbed, such as P-N junction with a current flow or a bulk semiconductor in which a large population of conduction electrons and holes is created by photoexcitation process, the semiconductor can reach a quasi-equilibrium state. In this case, electrons and holes are characterized by two separate quasi-Fermi levels, E_{FC} and E_{FV}. The concept of quasi-Fermi levels in excited system is valid if the carrier scattering time within a band is much shorter than the equilibrium time between the bands. Typically, the time constants vary from the order of 100 ps to the order of 1 millisecond ms, depending on the properties of the specific semiconductor and the carrier concentration. The total number of electrons per unit volume, N_C in the conduction band can be determined from the following expression:

$$N_C = \int_{E_C}^{\infty} N(E)dE = \int_{E_C}^{\infty} f(E)\rho(E)dE, \tag{2.27}$$

where n is the total number of electrons per unit volume.

At temperature different from zero, the density of electrons in the conduction band is

$$N_C = \frac{\rho}{1 + e^{(E_C - E_F)/(k_B T)}}.$$

The density of electrons in the conduction band, N_C is written as

$$N_C = 2\left[\frac{2\pi m_C^* k_B T}{h^2}\right]^{3/2} e^{(E_F - E_C)/(k_B T)}, \tag{2.28}$$

in which m_C^* is the effective mass of an electron near the conduction band.

A similar expression can be obtained for the number of holes in the valence band in the P-doped material. Equation (2.28) implies that as the Fermi energy, E_F, moves closer to the conduction band, the density of electrons in the band increases significantly.

2.2.6 *Semiconductor Materials*

The semiconductor materials, such as silicon (Si) and germanium (Ge), or a compound crystal, namely, gallium-arsenide (GaAs), indium phosphide (InP), and cadmium telluride (CdTe), is being used for developing transistors, various kinds of diodes, for example, light-emitting diodes (LEDs), photodiodes, solar cells, and digital and analog integrated circuits. The word, digital, is based on a binary weighted proportion; the term binary refers to the base 2 number system used. The digital input (or output) is arranged in bits. Each bit is worth half of the bit to its left and twice the bit to its right. The higher the number of bits, the finer the resolution; these bits are typically arranged in groups of four, called bytes.

Among the parameters of a semiconductor material for its operation in electronic devices or in photodetectors, (i) the bandgap energy, E_g (see Section 2.2.3.1), (ii) the number of free carriers available for conduction, and (iii) the generation and recombination of free carriers in response to the electric field, temperature and light shining on the material are important. Semiconductors are mainly classified into two categories, which are elucidated below.

2.2.6.1 Intrinsic semiconductor

An intrinsic (pure) semiconductor is made up of one pure element or pure compound and possesses poor conductivity. It is free of impurities and defects, and does not require dopants for bandgap modification. An intrinsic semiconductor has equal concentrations of both negative carriers (electrons) and positive carriers (holes; see Figure 2.3a). The intrinsic carrier concentration, defined as the number of electrons in the conduction band (or the number of holes in the valence band) per unit volume in an intrinsic material,

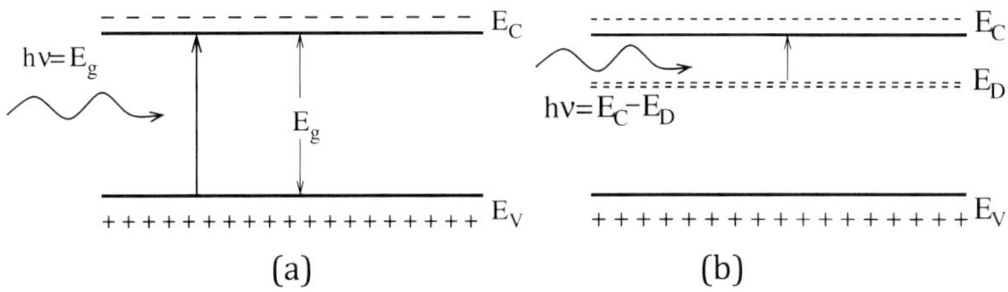

Figure 2.3 Energy band diagrams of (a) an intrinsic semiconductor and (b) the discrete donor energy state; E_C stands for the conduction band, E_V the valance band, and E_D the donor level.

depends on the bandgap of the material and on the temperature of the material, thus affecting the conductivity of the material. The intrinsic concentration, N_i is given by

$$N_i^2 = A_0 T^3 e^{-E_{g0}/(k_B T)}, \tag{2.29}$$

where k_B is Boltzmann's constant, E_{g0} the energy gap at 0 K in eV, A_0 the constant independent of T.

With increasing temperature, the electron–hole pairs increases and correspondingly the conductivity increases. The current flow in an intrinsic semiconductor is influenced by the density of energy states, which in turn, influences the electron density in the conduction band. Table (2.1) presents some of the important semiconductor materials, their forbidden energy bandgaps and corresponding optical cut-off wavelengths (Streetman and Banerjee, 2000; Seeger, 2004).

Table 2.1 Some intrinsic semiconductor materials and their bandgap energies

Semiconductor	Symbol	Bandgap (eV)	λ_c (μm)
Aluminum arsenide	AlAs	2.16	0.495
Cadmium telluride	CdTe	1.56	0.79
Gallium antimonide	GaSb	0.72	1.72
Gallium arsenide	GaAs	1.42	0.87
Germanium	Ge	0.66	1.88
Indium antimonide	InSb	0.17	7.29
Indium arsenide	InAs	0.36	3.44
Indium gallium arsenide phosphide	InGaAsP	0.38–2.25	0.55–3.260
Indium phosphide	InP	1.35	0.92
Silicon	Si	1.12	1.11

2.2.6.2 Extrinsic semiconductor

In an extrinsic semiconductor, a small amount of impurities is added into its crystal lattice (see Section 2.2.1.1). This process, called doping, alters the electrical properties of the semiconductor and improves its conductivity; the dopants make alloys or affect changes in properties. Doping with an element whose valence is different from the atom it replaces. The formation of alloy takes place when dopant anions and cations are close in size to original material. Addition of a donor impurity in a semiconductor creates an intermediate energy level, called intrinsic Fermi level energy, $E_{\mathrm{i}}(= E_{\mathrm{C}} - E_{\mathrm{g}}/2 = E_{\mathrm{V}} + E_{\mathrm{g}}/2)$, close to the conduction band in the energy diagram of an extrinsic semiconductor, which effectively extends the valence band to lower energy (see Figure 2.3b). For a P-type semiconductor, there are more holes in the valence band than there are electrons in the conduction band, while for an N-type semiconductor, there are more electrons in the conduction band than there are holes in the valence band. Two types of elements, such as (i) trivalent (aluminum [Al], gallium [Ga], boron [B], indium [In]) that has three valence electrons and (ii) pentavalent (phosphorus [P], arsenic [As], antimony [Sb], bismuth [Bi]) having five valence electrons, are used to dope silicon and germanium.

Doping process helps in producing two groups of semiconductors, such as the negative charge conductor (N-type) and the positive charge conductor (P-type) by changing the electron and hole-carrier concentrations at thermal equilibrium. If the dopant is a donor (pentavalent), in which it gives off its excess unpaired electron, the dopant energy levels can form below the conduction band. In the case of an acceptor, it is in need of an unpaired electron, the dopant energy levels can form above the valence band. The dopant produces excitation at an impurity center in the form of either a free electron-bound hole or a free hole-bound electron. The former type is for N-type semiconductor, which contains electrons that behave as the current carriers in its conduction band, while the latter type is for P-type semiconductor that has holes for current carriers in its valence band. The various dopants for Si, their respective energy levels and whether they become as electron donors or acceptors are given in Table 2.2.

Table 2.2 Energy gap and cut-off wavelengths of some extrinsic silicon dopants

Silicon dopants	Symbol	Energy gap (eV)	Electron acceptor (or donor)	λ_c (μm)
Aluminum	Al	0.057	acceptor	21.8
Antimony	Sb	0.039	donor	31.8
Arsenic	As	0.049	donor	25.3
Boron	B	0.045	acceptor	27.6
Bismuth	Bi	0.069	donor	18.0
Gallium	Ga	0.065	acceptor	19.1
Indium	In	0.16	acceptor	7.8
Thallium	Tl	0.26	acceptor	4.8
Zinc	Zn	0.31	acceptor	4.0

Adding N-type impurities decreases the number of holes. Similarly, doping with P-type impurities decreases the concentration of free electrons below that in the intrinsic semiconductor. Under thermal equilibrium, the product of the negative and positive concentrations is a constant independent of the amount of donor and acceptor impurity doping (Millman and Halkias, 1972). The density of carriers in intrinsic material, N_i, is a function of temperature. In an N-type material, the free-electron concentration is approximately equal to the density of donor atoms. The density of holes, N_h, in the N-type semiconductor is, $N_\mathrm{h} = N_\mathrm{i}^2/N_\mathrm{e}$, and the density of electrons, N_e, in the P-type semiconductor is $N_\mathrm{e} = N_\mathrm{i}^2/N_\mathrm{h}$.

2.2.7 *Diode*

A diode is a two terminal electronic device that restricts current flow to one direction. It is made up of two doped semiconductor materials forming a junction, a boundary between a region where the conductivity is due to holes and a region where the conductivity is due to electrons. This is termed P-N junction, which is attached to two electrical terminals; a vacuum tube diode is a vacuum tube with two electrodes, an anode and a cathode. At the junction of a diode, an electric field is present internally since there is a change in the level of the conduction and valence bands. This change leads to the electrical rectification effect produced by this junctions.

An ideal P-N diode has zero ohmic voltage drop across the body of the crystal, behaves as a short-circuit; in reality, it has a low resistance (which resists current flow) of the order of a few ohms. For such a diode, the potential barrier at the junction can be lowered by the applied forward voltage (bias). In this case, there is a significant increase in the concentration of electrons in the conduction band near the junction on the N-side and also the concentration of holes in the valence band near the junction on the P-side. This process under forward bias is known as minority carrier injection. The holes from the P-region diffuse to the N-region, where they recombine with free electrons, and the electrons from the N- to the P-region due to concentration gradients. To note, the depletion layer for the forward bias is smaller than the depletion layer for the reverse bias. However, the diffusion current (see Section 2.2.4.3) is absent in reverse current, while the diffusion current for the forward bias overwhelmingly built voltage from the depletion layer, which results in a forward bias current flowing due to the diffusion current. In the opposite direction, it has a high resistance (ideally infinite, behaves as an open circuit) of the order of several megaohms (MΩ) blocking the current. This behavior is known as rectification (see Figure 2.4) and is used to convert alternating current (AC) to direct current (DC) that is defined as the unidirectional flow of electric charge; it extracts modulation from radio signals from radio receivers.

During the concentration gradients, a layer of positive and negative charges are developed on the N-side and P-side of the

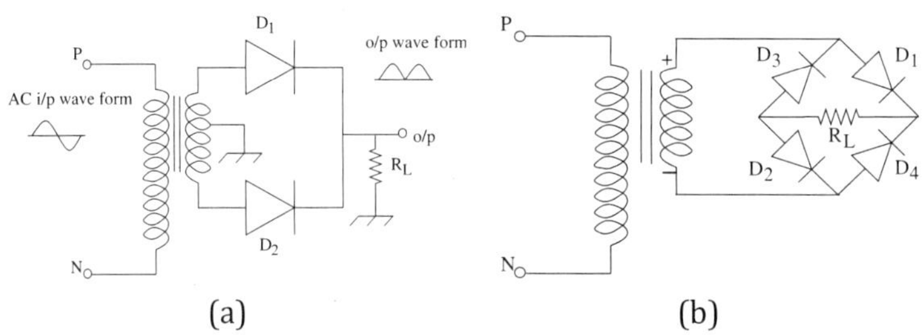

Figure 2.4 Circuit diagrams of (a) full-wave rectifier; i/p stands for the input, o/p the output, P, and N, the positive and negative, respectively, and (b) bridge rectifier using diodes.

junction, respectively. This space-charge region on either side of the junction is known as depletion layer (region), in which no charge carriers are present. The loss of electrons from the N-region and the gain of electrons by the P-region cause a potential difference across the junction. As they leave, they leave behind the ionized impurities (dopants) that created them. The migration stops when the electric field created by such impurities causes carriers to drift back at the same rate as they diffuse away. The depletion region is characterized by a large electric field, changes in hole and electron concentrations, large amounts of fixed charge, and a varying potential, which facilitates the use of Poisson's equation

$$\nabla \cdot \mathbf{E} = \frac{\rho}{\epsilon_0} = -\nabla^2 V, \tag{2.30}$$

where **E** is the electric field, ρ the electric charge, ϵ_0 the dielectric permittivity, and V the electrostatic potential.

The width of depletion region, w, is the total region comprising of the width of the depletion region on the N-side, x_{n} and the width of the depletion region on the P-side, x_{p}, i.e., $w = x_{\mathrm{n}} + x_{\mathrm{p}}$. Since both the drift current and diffusion current are present in equilibrium, the total current density $\mathbf{J} = 0$. The electrical charged atoms, called ions, produce electric field representing a potential difference across the junction, which is termed as barrier potential. After several collisions, the electric field repels the rest of the majority carriers away from the junction. The depletion region is produced immediately surrounding the junction that has no majority carriers. The width of depletion region depends on the doping level. With heavy doping, the depletion region becomes thin. This is because the diffusion charge need not travel for across the junction before recombination takes place.

Consider that a negative voltage is applied to the N-region of a diode. If the voltage drop is across the space-charge region, the current from N to P is i_0. The holes going from P- to N-region have to climb a small potential hill (Dekker, 1981) and would give rise to a current, $i_0 e^{V/V_{\mathrm{T}}}$. The hole current is made up of two terms: drift and diffusion currents (see Section 2.2.4). The net hole current is given by

$$i_{\mathrm{h}} = \frac{q D_{\mathrm{h}}}{L_{\mathrm{h}}} N_{\mathrm{h}_0} \left(e^{V/V_{\mathrm{T}}} - 1\right), \tag{2.31}$$

with V_T $(= k_B T/q = T/11,600 = 26$ mV) as the thermal voltage, for instance, at room temperature (300 K), T the absolute temperature, k_B Boltzmann's constant, V the applied voltage across the terminal of the diode in volts, which is positive for a forward voltage and negative for a reverse bias, D_h the diffusion coefficient of the holes, $L_h (= \sqrt{D_h \tau_h})$, the diffusion length of holes, an average path length over which a hole diffuses before recombination takes place, τ_h the lifetime of holes, and q the electron charge.

Hence, the net current, i_d, flowing through an ideal diode in amperes (Shockley[a] ideal diode equation) is defined by

$$i_d = i_0 \left(e^{V/(\eta V_T)} - 1 \right), \tag{2.32}$$

in which i_0 is the reverse saturation current, the diode leakage current density in the absence of light, and the symbol, $\eta = 1$ for germanium and approximately 2 for silicon at rated current.

The saturation current is defined by the maximum current, which can be obtained under certain conditions. For example, in a two element vacuum tube, it refers to either the space-charge limited current or the temperature-limited current. In the former case, further increase in filament temperature produces no significant increase in anode current, whereas in the latter case, a further increase in voltage produces a relatively small increase in current. In a gaseous-discharge device, it is the maximum current that can be obtained for a given mode of discharge. Attempts to increase the current result in a different type of discharge. In the case of a semiconductor, it is the maximum current that precedes a change in conduction mode, for instance, it is the maximum current a diode can allow to flow before overheating depending on the type of junction. The magnitude of the saturated current depends on the intensity of the incident light.

On applying a potential to the diode affects the current flowing through the diode. Figure 2.5 depicts the characteristic curve of a typical diode. If there is no bias, the applied voltage, V turns out to be zero, the diode is in equilibrium and for reverse bias, V is negative,

[a]William Shockley, John Bardeen, and Walter Brattain were jointly awarded the Nobel Prize in 1956 for the research on semiconductors and their discovery of the transistor effect in 1947, which was considered to be one of the greatest inventions of the 20th century.

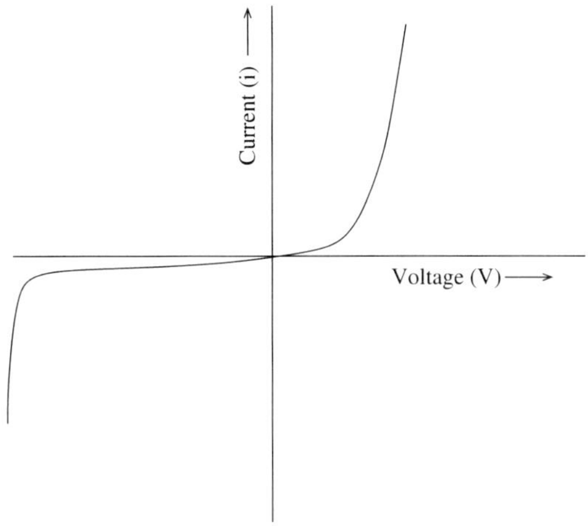

Figure 2.5 Current–voltage characteristic of a typical diode.

i.e., $e^{-V/\eta V_{\mathrm{T}}} \ll 1$; hence, $i_{\mathrm{d}} = -i_0$. A positive value of i_{d} means that the current flows from the P- to the N-side. The diode is forward biased if the voltage, V is positive, $e^{V/\eta V_{\mathrm{T}}} \gg 1$, and therefore

$$i_{\mathrm{d}} = i_0 e^{V_{\mathrm{d}}/(\eta V_T)}. \tag{2.33}$$

The diode acts as a capacitor (which stores charge) with the boundaries of the depletion region as the two plates of a parallel plate capacitor. To note, a capacitor, characterized by capacitance, contains at least two electrical conductors (plates) separated by a dielectric (insulator). If there is a potential difference (voltage) across these plates, a static electric field develops across the dielectric, causing positive charge to collect on one plate and negative charge on the other. The energy is stored in the electrostatic field. The capacitors are, in general, used in electronic circuits for blocking the direct current (DC), while allowing the alternating current (AC). The capacitance offers a low reactance path to AC components of the current. The capacitive reactance χ_{C} is equal to $1/(2\pi f_{\mathrm{c}})$, in which f_{c} is the cut-off frequency. The capacitive reactance would increase with, decrease in frequency and it would decrease with increase in frequency. For DC components, it behaves as an open circuit and all the DC current passes through the load, while a small part of the AC component passes through the load producing a small ripple voltage. The ripple factor is the ratio of AC voltage at the output to that of DC voltage at the output.

When the rectifier voltage is increasing, the capacitor (connected in shunt with the load resistor, R_L), changes to peak voltage. Just after the positive peak, the rectifier output voltage tries to fall. At this point, the capacitor has a charge V_m across it. Since the source voltage become slightly less than V_m, the capacitor tries to send current back through the diode. The diode disconnects the source from the load and the capacitor starts to discharge through the load preventing the load voltage from falling to zero. The capacitor continues to discharge until the source voltage becomes more than the capacitor voltage. The diode again starts conducting and the capacitor is charged to V_m. The rate at which capacitor discharge depends on the time constant, τ, i.e., $C R_L = \tau$. The ripple increases with increasing load current, which reduces DC output voltage.

As stated earlier in this section, a reverse bias causes major carriers to move away from the junction, and thereby uncovering immobile charges. The thickness of the space-charge layer at the junction increases with reverse bias potential. This increase in uncovered charge with applied voltage is considered a capacitive effect. The transition or depletion region capacitance, C_T, is defined as

$$C_T = \frac{dq}{dV},$$

in which dq is increase in charge due to increase in voltage.

The transition capacitance, C_T, is not constant but depends upon the magnitude of the reverse voltage; with the increase of depletion layer, the resulting C_T decreases. In the reverse bias, the depletion region acts as an insulator or dielectric material, while the P- and N-type regions on either side have low resistance and act as a plates. The current, i is given by

$$i = \frac{dq}{dt} = C_T \frac{dV}{dt}. \tag{2.34}$$

Another capacitive effect, called diffusion capacitance, C_D, is considered in the forward bias. When forward biased, the diffusion capacitance is evident since the depletion region reduces; the density of charge carriers is higher near the junction. Thus charge is stored on both sides of the junction and decays exponentially with distance. The diffusion capacitance, C_D, which is observed that the amount of stored charge varies with the applied voltage as for a capacitor; it is directly proportional to the forward current through the diode.

Zener diode With the increase of reverse bias, a large amount of current flows in the reverse direction. The applied voltage at this point is referred to as the breakdown voltage, V_{br}. The diode is said to be in the region of avalanche breakdown that occurs when high voltage causes free electrons to travel at high speeds, colliding with other electrons and knocking them out of their orbits, resulting in a enhanced amount of free electrons. The diodes of this kind are known as Zener diodes, which are designed with adequate power-dissipation capabilities to operate in the breakdown region may be employed as voltage reference devices. Such diodes are used to regulate the voltage across the load resistance against changes due to the variations in the load current and supply voltage.

Avalanche multiplication A current multiplying phenomenon, called avalanche multiplication (see Section 5.3.5) occurs in a diode that is reverse-biased close to its breakdown voltage, V_{br}. On applying such a voltage to the P-N junction, the electron–hole pairs are generated in the depletion layer of a photodiode. The electrons drift toward the N^+ side due to the electric field developed across this junction. The drift speed of these electron–hole pairs (or carriers) depends on the electric field strength. On increasing the electric field strength to a certain level, the carriers may collide with the crystal lattice; therefore, the drift speed turns out to be saturated at an average speed. On further increasing the reverse voltage, the carriers that escaped collision, collide with the crystal lattice; hence, the newly generated electron–hole pairs produce ionization. With each collision produces additional electron–hole pairs in a process like a chain reaction. The newly generated carriers in the presence of high electric field result in more ionization called avalanche effect. Avalanches generate the pulses of electric current that are registered by Geiger counters. The ionization ratio, k, a parameter to indicate device noise, is defined as

$$k = \frac{\beta}{\alpha},$$

with α as the ionization rate of electrons and β the ionization rate of holes.

2.2.8 *Transistor*

A bipolar junction transistor (BJT), which conducts by using both majority and minority carriers, is an active electronic device. It has three terminals, labeled base (in the middle), collector, and emitter. The base is lightly doped and very thin compared to the diffusion length of minority carriers. It passes most of the injected charges carriers from the emitter to the collector. The emitter is heavily doped so that it can inject a large number of charge carriers into base. The collector is moderately doped. Such a transistor works as a voltage-controlled current source and whose basic action is to control of current at one terminal by controlling voltage applied at other two terminals. Following its invention, it has become the fundamental building block of modern electronic devices. The main advantages of a transistor are (i) smaller in size and light weight allowing the development of miniaturized electronic devices, (ii) non-requirement of heating element, (iii) operates at lower voltages as compared to vacuum tubes, which makes it suitable for small battery powered applications, (iv) longer life with no aging effect, (v) insensitivity to mechanical shock and vibration, (vi) non-requirement of warm up period, (vii) instant availability for use, and (viii) rugged in construction. However, unlike the vacuum tubes, the silicon transistors typically do not operate at voltages higher than about 1000 V. They are sensitive to radiation and cosmic rays (extremely energetic subatomic charged particles) as well.

There are two types of transistors, such as P-N-P-type transistor and N-P-N-type transistor. In both cases, the layer in the center is the base, and the two outer layers are the emitter (the corresponding N-type region emits electrons into the P-region, what is called the base) and collector. In the former case, it consists of a Si (or Ge) crystal in which a thin layer of N-type Si (or Ge) is sandwiched between the two layers of P-type Si (or Ge), while in the latter case, a layer of P-type material is sandwiched between two layers of N-type material. The arrow mark on the emitter specifies the direction of current flow when emitter-base junction is forward biased. The direct current

(DC) biasing is necessary for proper operation of a transistor. If a voltage (or current) is applied to one pair of the terminals of a transistor, for example, N-P-N-type transistor, changes the current flowing through another pair of terminals. These electrons are collected at the junction with the reverse bias (collector). If the emitter-bias junction is forward biased, the emitter current flows through the base into collector, while if the collector-base junction is reverse biased, almost the entire emitter current flows through the collector. The forward bias applied to the emitter-base junction of a P-N-P transistor causes a lot of holes from the emitter region to cross over to the base region as the base is lightly doped with N-type impurity. Since the number of electrons in the base region is very small, the number of holes combined with electrons in N-type base region is also very small and small base current $i_{\rm B}$ (μA) would flow. The remaining holes cross over to the collector region to constitute the collector current, $i_{\rm C}$. The emitter current, $i_{\rm E}$ is given by

$$i_{\rm E} = i_{\rm B} + i_{\rm C}. \tag{2.35}$$

The collector current is being controlled by the base-emitter voltage. Since the base-emitter junction is forward biased, holes from the base are injected into the emitter. The ratio of the collector current, $i_{\rm C}$ to the base current is a dimensionless parameter, β, of the individual transistor. Mathematically, the gain of a transistor is expressed as

$$\begin{aligned}\beta &= \frac{i_{\rm C}}{i_{\rm B}} = \frac{i_{\rm C}}{i_{\rm E} - i_{\rm C}} \\ &= \left(\frac{1}{\alpha} - 1\right)^{-1} = \frac{\alpha}{1-\alpha},\end{aligned} \tag{2.36}$$

where α is the ratio of the collector current ($i_{\rm C}$) to the emitter current ($i_{\rm E}$), whose typical values are $0.99 < \alpha < 1.0$ for the state-of-the-art transistor; typically $\beta > 100$ for the state-of-the-art transistors, which is also designated as an equivalent hybrid (h)-parameter, $\beta = h_{\rm fe}$, that stands for forward current gain in the common-emitter (CE) configuration.

In the case of emitter being forward biased

$$i_{\rm C} = i_{\rm C0} - \alpha i_{\rm E}, \tag{2.37}$$

in which $i_{\rm C0}$ is the reverse saturation collector current.

The reverse saturation current flowing through reverse biased collector-base, i.e., collector to base leakage current with emitter open is expressed as

$$i_{\mathrm{C}} = i_{\mathrm{C}}(\text{majority}) + i_{\mathrm{C}}(\text{minority}). \tag{2.38}$$

On combining equations (2.35) and (2.37), one finds

$$i_{\mathrm{C}} = \frac{i_{\mathrm{C0}}}{1-\alpha} + \frac{\alpha i_{\mathrm{B}}}{1-\alpha}. \tag{2.39}$$

Using equation (2.36), this equation translates into

$$i_{\mathrm{C}} = (1+\beta) i_{\mathrm{C0}} + \beta i_{\mathrm{B}}, \tag{2.40}$$

in general, $i_{\mathrm{B}} \gg i_{\mathrm{C0}}$; hence, $i_{\mathrm{C}} = \beta i_{\mathrm{B}}$ in the active region.

The transistor amplifiers are circuits that provide signal gain, defined by the output divided by the input, having no units. The performance of an amplifier depends on its frequency response. A typical circuit diagram of a transistor amplifier in CE-configuration is given in Figure 2.6. Its voltage gain, A_{V} is defined as signal voltage out divided by signal voltage in, i.e., $V_{\mathrm{out}}/V_{\mathrm{in}}$. The input signal voltage, V_{in}, is equal to the terminal voltage of the signal source connected to the circuit. The signal voltage is equal to the signal emitter current times the signal impedance in the emitter. The output signal voltage, V_{out}, is equal to the output signal current multiplied by the signal impedance in the collector. The expression for i_{C} for any V_{C} and i_{E} is

$$i_{\mathrm{C}} = -\alpha i_{\mathrm{E}} + i_{\mathrm{C0}}\left(1 - e^{V_{\mathrm{C}}/V_{\mathrm{T}}}\right), \tag{2.41}$$

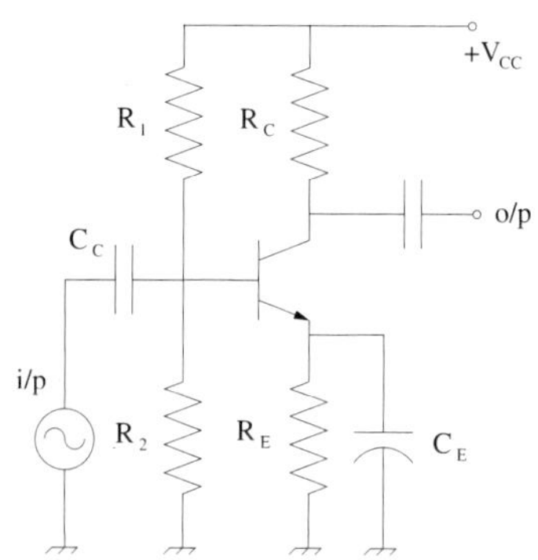

Figure 2.6 Circuit diagram of an amplifier using an N-P-N transistor.

where V_C designates the voltage drop across the collector junction from the P- to the N-side, $V_T = k_B T/q$, k_B is Boltzmann's constant, T the absolute temperature, and q the electron charge.

Let w be the width of the P-type base region and D_e the diffusion coefficient of electrons in the base. The time, t required for an electron to cross the base is equal to $t = w^2/2D$. The probability for an electron to recombine with a hole during the crossing is

$$\frac{t}{\tau_e} = \frac{w^2}{2D_e\tau_e} = \frac{w^2}{2L_e^2},$$

in which τ_e is the lifetime of electrons and L_e their diffusion length in the base material; in most cases, $t/\tau_e \ll 1$ since $w < L_e$.

All the electrons emitted by the emitter are collected by the collector. The base transport factor is the ratio of the number of electrons arriving at the collector and the emitted electrons. Assume that the current is flowing between the emitter and the base. The total current is made up of two parts, such as (i) a hole current, i_h, from the base into the emitter, and (ii) an electron current, i_e, from the emitter into the base. The ratio of these current is given by

$$\frac{i_e}{i_h} = \frac{L_h\sigma_e}{w\sigma_h},$$

with σ_e and σ_h as the conductivities of the emitter and base regions, respectively and L_h the diffusion length of holes in the emitter region.

The current flowing through the emitter junction is given by the sum of the hole current injected from the base into the emitter and the electron current injected from the emitter into the base. Following equation (2.31), the respective hole and electron currents (Dekker, 1981) are

$$i_h = \frac{qD_h}{L_h} N_{h_0} \left(e^{V/V_T} - 1\right), \tag{2.42}$$

$$i_e = \frac{qD_h}{w} N_{e_0} \left(e^{V/V_T} - 1\right), \tag{2.43}$$

where V is the applied voltage between base and emitter, N_{h_0} and N_{e_0} the equilibrium concentration of holes and electrons, respectively, and D_h the diffusion coefficient of the holes.

Although, like the hole current, the electron current from emitter to base is determined by the diffusion length of electrons in the base region, the width of the base is $w \ll L_e$ for electron current; hence, L_e is replaced by w. From equations (2.42) and (2.43), it follows that

$$\frac{i_e}{i_h} = \frac{D_e N_{e_0} L_h}{D_h N_{h_0} w}. \tag{2.44}$$

Using $N_{e_0} = N_i^2/N_h$ and $N_{h_0} = N_i^2/N_e$, in which N_i is the intrinsic concentration, N_h the concentration of holes in the base, and N_e the concentration of electrons in the emitter, one obtains

$$\frac{i_e}{i_h} = \frac{D_e L_h N_e}{D_h w N_h} = \frac{L_h \sigma_e}{w \sigma_h}, \tag{2.45}$$

with $\sigma(= Ne\mu)$ the conductivity of the metal and $\mu(= De/k_B T)$ as the mobility (see Section 2.2.4.4).

As the base-emitter voltage (V_{BE}) is increased, the base-emitter current, i_{BE} and hence the collector-emitter current (i_{CE}) increase exponentially. Because of such exponential relationship, the BJT has a higher transconductance than the field effect transistor (see Section 2.2.9). The collector impedance is very high because of the reverse bias; for voltages larger than a few times $k_B T/e$, the collector current is independent of the bias, i.e., the impedance approaches to infinity. The collector impedances are of the order of 10^6 Ω (ohms) or higher. Further, the resistance of the emitter is very low, for example, for 1 ma at room temperature, the resistance is 25 Ω.

All transistors have capacitances between their terminals. The voltage gain of an amplifier varies with signal frequency due to effect of variation in circuit capacitive reactance with signal frequency on the output voltage. The fall off in amplifier gain is due to the effect of the coupling capacitor and the bypass capacitors. The former passes an AC signal from one stage to the next, while blocking DC signal and the latter bypasses all the AC current from the emitter to the ground. Due to the coupling capacitor, the output across the load resistance, R_L, is free from the collector. It offers a very low reactance in high frequency and do not have any effect at medium and high frequencies. However, at low frequencies, the reactance of coupling capacitors is large, due to which the voltage drop across them increases with the reduction in frequency. The increased voltage drop reduces both the output voltage and gain of the amplifier in the low frequency region of the frequency response. If the bypass capacitor is not connected in the circuit, the AC voltage across the emitter resistance affects the input AC voltage. Such a feedback of AC signal is reduced by connecting such a capacitor. The bypass capacitor is good enough to provide an effective bypass to the lowest frequency of the signal. Furthermore, there are stray capacitances that are capacitances between wiring and ground, although they are very small.

2.2.9 *Field-Effect Transistor*

The field effect transistor (FET) is a high-input impedance ($\approx 10^{11}$ Ω) device, whose operation depends upon the flow of the majority carriers. It is a unipolar device and is less noisy than a bipolar transistor (Brennan, 2003). There exist two types of FETs: the junction field effect transistor (JFET) and the metal-oxide semiconductor FET (MOSFET; see Section 2.3.3). Both these categories can be fabricated as either an N-channel device or a P-channel device. The JFET consists of a slab of N-type silicon with P-type regions diffused on its two sides. The N-channel device is made by forming a channel of N-type material in a substrate of P-type material and the P-type regions are electrically connected together and form the gate. The channel is a region of N-type material between the two gate regions through which the majority carriers move from the source to the drain. The three terminals are connected to the device, which are labeled as source, drain, and gate (see Figure 2.7; top left), the functions of which are described below:

Source S It is a terminal through which the majority carriers (electrons) enter and move along the channel. On applying a voltage between the two end-wires, a current starts flowing along the channel and maintains this by continually putting electrons in the source.

Drain D This is a terminal thhough which the electrons leave. The effective resistance between the two ends depends upon the size and shape of the channel and the properties of the N-type material. The drain-to-source voltage, V_{DS}, is positive if the drain is more positive than the source.

Gate G The gate permits electrons to flow through or blocks their passage by creating or eliminating a channel between the source and drain. The electrons flow from the source terminal toward the drain terminal if influenced by an applied voltage. The gate voltage controls a current between source and drain.

Body The body refers to the bulk of the semiconductor in which the gate, source and drain lie. Usually the body terminal is connected to the highest or lowest voltage within the circuit, depending on the type of the FET. The body terminal and the source terminal are

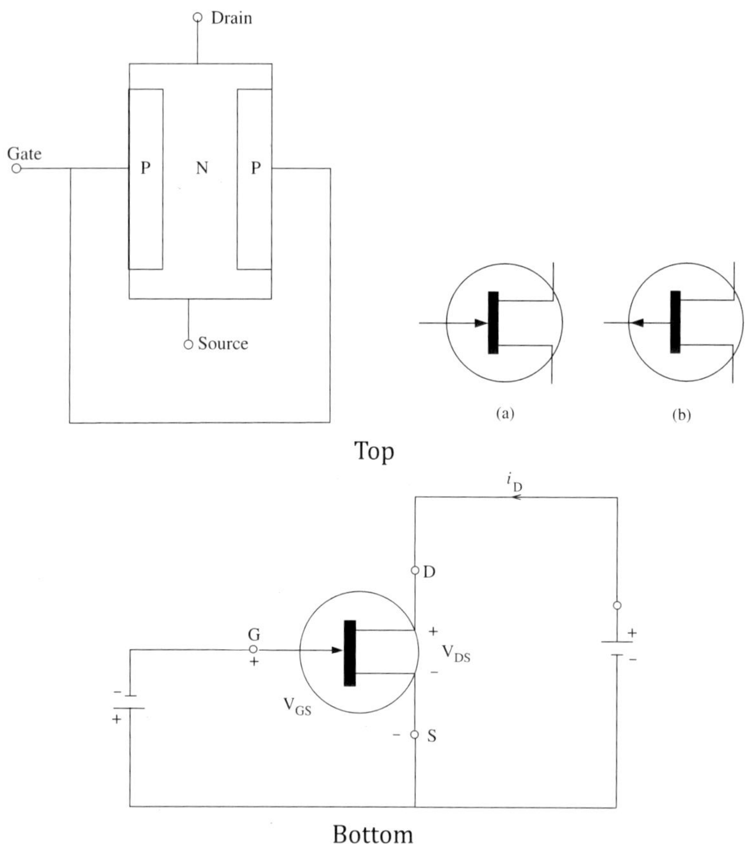

Figure 2.7 Top (Left): Basic structure of N-channel JFET, Top (Right): Circuit symbol of FET; (a) N-type FET and (b) P-type FET, and bottom: Bias circuit of FET.

sometimes connected together since the source is often connected to the highest or lowest voltage within the circuit, although there are several uses of FETs, which do not have such a configuration, such as transmission gates and cascade circuits.

The operation of such a device is based on reverse-biasing the P-N junction between the gate and the channel. The reverse bias across the junction controls the channel width, and hence the current flow from drain to source; as the reverse bias increases, so does the thickness of the region of immobile uncovered charges. The conductivity of this region is nominally zero due to the nonavailability of current carriers. The effective width of the channel becomes progressively decreased with increasing reverse bias. For a fixed drain-to-source voltage, V_{DS}, the drain current is a function of the reverse biasing voltage across the gate junction. Figure 2.7 (bottom) depicts the bias circuit of a JFET. In amplifier applications, the JFET is used in the region beyond current-saturation region (pinch-off). Let the saturation drain current be designated as i_{DS}, and its value with the gate shorted to the source, i.e., $V_{GS} = 0$, by i_{DSS}. The transfer characteristic provides the following relationship:

$$i_{DS} = i_{DSS}\left(1 - \frac{V_{GS}}{V_P}\right)^2, \tag{2.46}$$

which can be approximated by the parabola; here V_P is the pinch-off voltage.

The gate reverse current, also called the gate cut-off current, i_{GSS}, provides the gate-to-source current, with the drain shorted to the source for $|V_{GS}| > |V_P|$. Typically, i_{GSS} is of the order of a few nanoampheres for a silicon device. The current–voltage characteristics of the P-channel JFET are described by the same equations as the N-channel JFET; here V_P is positive, $0 \leq V_{GS} \leq V_P$. The drain current, i_D is a function of the gate voltage, V_{GS} and drain voltage, V_{DS}. The small scale mutual conductance (transconductance), g_m is given by

$$\begin{aligned} g_m &= g_{m0}\left(1 - \frac{V_{GS}}{V_P}\right) \\ &= \frac{2}{|V_P|}\sqrt{i_{DSS} i_{DS}}, \end{aligned} \tag{2.47}$$

in which $g_{m0}(= -2i_{DSS}/V_P)$ is the value of g_m for $V_{GS} = 0$.

An amplification factor, M, for an FET is defined as

$$M = R_\mathrm{d} g_\mathrm{m}, \tag{2.48}$$

with R_d is the drain resistance, the reciprocal of which is known as the drain conductance, g_d.

2.3 Integrated Circuits Technology

An integrated circuit (IC) is made out of a semiconductor material containing both active and passive components, such as transistors, diodes, capacitors, and resistors, and their interconnections. The ICs are used for a variety of devices, including microprocessors, audio and video equipment, etc. Depending on the manufacturing technology, ICs are divided into two groups, namely, hybrid and monolithic (single piece). They are further divided into analog and digital. The output voltage of an analog circuits is continuous, and follows changes in the input, while the output voltages for digital ICs is either low and high and it changes from one state to the other very quickly. The advantages of ICs over the discrete circuits are (i) low cost and small size, (ii) high reliability, (iii) high performance, because the components switch quickly and consume little power, as a result of the close proximity of the components, and (iv) high speed.

2.3.1 *Fabrication of ICs*

The fabrication of ICs depends on materials, processes, and design principles. The basic structure of an N-P-N transistor has four layers of material. The bottom layer is a semiconductor wafer, whose conductivity changes with either doping or temperature, made out of mono-crystal silicon (P-type silicon in this case), which serves as substrate upon which the integrated circuit is built. The second layer is a thin (a few μm) N-type material that is grown as a single crystal extension of the substrate. All active and passive components, which are made by implementing and diffusing P-type and N-type impurities, are built within this thin layer; capacitors and resistors are sometimes made in layers above this layer as well. In the fabrication of these components, it is desired to distribute impurities in certain precisely defined regions within this second layer. The selective implementation and diffusion of impurities is accomplished by employing silicon dioxide (SiO_2; commonly known as glass) that serves as an insulating layer at the middle, which protects portions of the wafer against the penetration of the impurities. The third layer of material is this SiO_2, which provides protection of the semiconductor surface against contamination. In the region where the implementation is to take place, the SiO_2 is etched away, leaving the rest of the wafer protected. In order to perform selective etching, the SiO_2 is subjected to a photolithographic process. The fourth metallic layer is added to supply the required interconnections among the components. In fact, there are many such layers, separated by the layers of insulators that do not permit charge exchange between metal and semiconductor, because the present day ICs requires multiple layers of interconnects.

2.3.1.1 Processing steps

The fabrication of such ICs follows a sequence of processes (Millman and Halkias, 1972) that include

Wafer processing The pure silicon is melted in a pot and a small seed containing the desired crystal orientation is inserted into molten silicon and slowly pulled out. The silicon crystal is manufactured as a cylinder (ingot; 25 mm to 300 mm in diameter) with a predetermined number of impurities. This crystal ingot is subsequently sliced into thin disks, what is referred to as wafers. One side of each wafer is marked for crystal orientation and polished for eliminating surface imperfections. Depending on the diameter of the wafer, its thickness is determined; the larger the diameter, the larger the thickness. A square wafer of 200 mm^2 can be divided into 100 chips with a surface area of surface area of 400 mm^2 each; a modern transistor can occupy a space of ≈ 0.4 μm. The cleanliness of wafers during processing is critical; complex sequence of acid and alkali solution are utilized to remove particles, organic films, metals, etc.

Buried layer The metallic or insulating buried layers are produced several microns within a semiconductor substrate. For a bipolar IC, an SiO_2 is grown on the P-type substrate. The oxide is removed from the selected areas of the surface by means of photolithography. The oxide left behind on the surface serves as a mask for the implementation step. Following this, the arsenic (As) is implemented into the exposed surface areas by an ion implementation process. Under diffusion process, the N-type impurities penetrate into the P-type substrate, which, in turn, form heavily doped N-type regions. These regions, called buried layers, serve to reduce the collector series resistance in the finally constructed bipolar transistors and also to electrically isolate the active elements at the surface of the wafer from the substrate.

Epitaxial growth The term epitaxy refers to the growth of crystals on a crystalline substrate, which determines their orientation. The epitaxial process produces a layer of single-crystal silicon on an existing crystal wafer of the same or different material. The epitaxial growth is performed in a reactor. Silicon tetrachloride ($SiCl_4$) is used as source for the epitaxial growth of pure silicon, that is

$$\mathrm{SiCl_4} + 2\mathrm{H_2} \rightleftarrows \mathrm{Si} + 4\mathrm{HCl}. \tag{2.49}$$

Impurity diffusion The diffusion of impurities into the silicon chip is an important process in the fabrication of ICs. The doping materials can be added to change the electrical characteristics of silicon locally through the diffusion dopants deposited on silicon move through the lattice by thermal diffusion. The thermal annealing allows doping impurities to diffuse further into the bulk band repairs lattice damage caused by the collisions with doping ions. The solution to the diffusion equation provides the effect of temperature and time on the diffusion distribution. The equation governing the diffusion of neutral atoms is given by

$$\frac{\partial N}{\partial t} = D\frac{\partial^2 N}{\partial x^2}, \tag{2.50}$$

in which N is the particle density in atoms per unit volume, as a function of x from the surface and time, t, and D the diffusion coefficient in dimension of $\mathrm{m^2.s^{-1}}$.

If an intrinsic silicon wafer is exposed to a volume of gas having uniform density, N_0 atoms per unit volume of N-type impurities, for example, phosphorus, these atoms diffuse into the silicon crystal. If the diffusion is continued for a much longer times, the silicon becomes uniformly doped with N_0 phosphorus atoms per unit volume. It is assumed that the surface density of impurity atoms remains at N_0 for all diffusion times and that $N(x) = 0$ at $t = 0$ for $x > 0$. If equation (2.50) is solved and the above boundary conditions are applied, one finds

$$\begin{aligned} N(x,t) &= N_0\left(1 - erf\,\frac{x}{2\sqrt{Dt}}\right) \\ &= N_0 erfc\,\frac{x}{2\sqrt{Dt}}, \end{aligned} \tag{2.51}$$

where $erfc\ y$ means the error-function[a] complement of y.

If a specific number, Q of impurity atoms per unit area are deposited on one face of the wafer and upon heating the material, the impurity atoms would diffuse again into the silicon. When the boundary conditions, $\int_0^\infty N(x)dx = Q$, for all times and $N(x) = 0$ at $t = 0$ for $x > 0$, one obtains a Gaussian distribution of this kind

$$N(x,t) = \frac{Q}{\sqrt{\pi Dt}} e^{-x^2/(4Dt)}. \tag{2.52}$$

Ion implementation This process permits more precise and reproducible control over the dopant profiles in the substrate. It is employed for introducing impurities into silicon wafers. Impurities with an electric charge are accelerated to high energy and highly energized donor (or acceptor) atoms impinge on the surface and travel below it. The side-effect of this process is that due to the bombardment with high energy ions, it causes significant damage to the single-crystal substrate. The amount of damage is a function of the incident ion energy and mass, and the substrate temperature. This process is made selective by using a photoresist (a light sensitive organic polymer) pattern to prevent impurity ions from reaching the silicon surface where no impurities are needed.

[a]The error-function of y is defined as

$$erf\ y = \frac{2}{\pi}\int_0^y e^{-\lambda^2}d\lambda.$$

Oxidation If silicon (Si) is exposed to oxygen or water vapor at high temperatures ($\sim$1000°C), silicon dioxide (SiO_2) is formed, that is

$$Si + O_2 \rightarrow SiO_2. \tag{2.53}$$

SiO_2 is deposited on materials other than Si through reaction between gaseous silicon compounds and oxidizers. It acts as insulation between different layers of metallization.

Etching Once the desired shape is patterned with photoresist, the etching process allows unprotected materials to be removed. In the early days all etches were based on liquid chemicals (wet etching), primarily acids. The wet etching is non-directional. It etches in all directions (isotropic) at the same rate, the result of which is that while etching down through a film, the etchant is etching underneath the edge of the photoresist. In the recent days, the dry or plasma etching that uses ionized gases are preferred. It can achieve faster etching in one direction than the other directions (anisotropic).

Photolithography The photolithography, a process used to transfer patterns to each layer of the IC, is used to mark the pattern of the substrate to be doped. The surface to be patterned is spin-coated with the photoresist. The mask pattern is developed on the photoresist, with UV light exposure; the soluble photoresist can be removed chemically. The developed photoresist acts as a mask for patterning of underlying layers and then is removed. Depending on the type of photoresist, the exposed or unexposed parts become resistant to certain types of solvents. The negative photoresist hardens in the areas exposed to the light, while the positive photoresist hardens in the area unexposed to the light.

Encapsulation During encapsulation, lead frames are placed onto mold plates and heated. Molten plastic material is pressed around each die to form its individual package. The mold is opened, and the lead frames are pressed out and cleaned.

2.3.1.2 Types of junction

There exist two types of metal semiconductor junctions:

Ohmic contact In two metal semiconductor junctions, one at each end of the diode, one expects a contact potential to develop across these additional junctions. Assuming that the metal semiconductor contacts have been manufactured in a way that they are non-rectifying. The contact potential across these junctions is constant, independent of the direction and magnitude of the current. A contact of this type is referred to as an ohmic contact; the contact potential across the junctions is desired when a lead is to be attached to a semiconductor.

Schottky barrier A Schottky barrier diode, also referred to as metal semiconductor diode, is formed by a metal-semiconductor contact, which has a low forward voltage drop and a very fast switching characteristics; a highly doped substrate provides a poor ohmic contact and a high series resistance. The semiconductor is separated from the metal by a layer of 10^{-4} to 10^{-5} cm thick of the same material but without donor levels (Mott, 1939). Further, it is assumed that any potential difference between the metal and semiconductor exists essentially across the insulating layer, the field strength being constant (Schottky, 1942). The thickness of such a layer is large compared with the mean free path for scattering of the electrons by lattice vibrations; therefore, the electron current through the layer is due to (i) the electric field and (ii) diffusion.

A Schottky diode is usually fabricated from a metal (see Section 5.3.4), such as aluminum (Al), which acts as a P-type impurity when in contact with Si. However, if it is attached as a lead to N-type Si, an ohmic contact is desired, which needs to be prevented. In order to do so, N^+ diffusions are made in the N-regions near the surface where the Al is deposited. If the N^+ is omitted and the Al is deposited directly upon the N-type Si, an equivalent P-N structure is formed, which results in a metal-semiconductor diode. If the metal and the semiconductor have work functions related in the proper way, this can be a rectifying barrier. The external volt-ampere characteristics of a metal-semiconductor diode are essentially the same as that of a P-N junction.

2.3.2 *Metal-Oxide Semiconductor Structure*

A basic metal-oxide semiconductor (MOS) is a semiconductor device (as a diode or a capacitor), which comprises of three layers, such as (i) a conductive metal electrode at the top, (ii) a metallic oxide (for example, SiO_2), and (iii) a conductive electrode at the bottom layer (see Figure 2.8).

The insulated metal gate electrode can be thought of as one of the plates of a capacitor. The insulator may be formed by direct oxidation of the semiconductor, or it may be deposited by vaporization, sputtering, or condensation in the vapor phase. An applied potential difference at the upper electrode, which creates an electric field, acts on mobile charges in the N-type semiconductor. If a positive potential is applied to the gate, the generated electrons could be collected under the electrode forming a region of holes. The holes would diffuse into the substrate and would be lost. Thus, the electrons generated by the incoming photons could be collected in the respective pixels. These electrons should be counted to reproduce the pattern of the incident light, which is termed as the image. For $V > 0$, the majority carriers (electrons) accumulate at the surface under the insulator, while for $V < 0$, the electrons are repelled far from the surface, creating a depletion zone. For $V \ll 0$, a positively charged inversion layer is formed at the surface, which is a potential well for the minority carriers (holes). The lower electrode

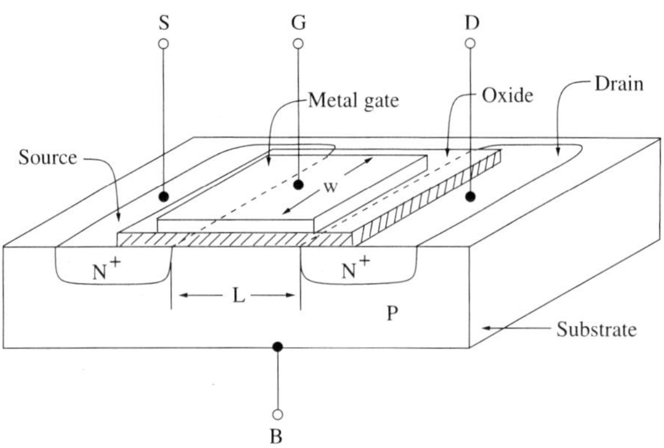

Figure 2.8 N-channel MOSFET structure showing channel length, *L*, and width *w*.

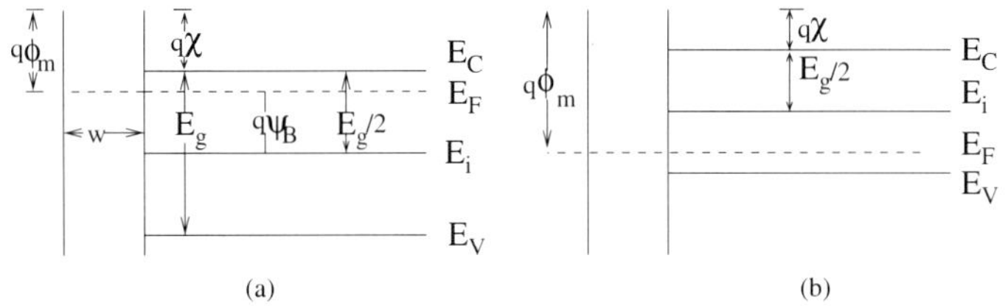

Figure 2.9 Band diagram of ideal metal-oxide semiconductor (MOS) diode: (a) N-type semiconductor and (b) P-type semiconductor; w is the thickness of the insulator.

may be made transparent to radiation. Figure 2.9 depicts the energy band diagram at no bias for both N-type and P-type semiconductors. An ideal MOS diode is defined as

(1) At zero applied bias, the energy difference between the metal work function, ϕ_{m}, and the semiconductor work function or the work function difference, ϕ_{ms}, is zero, that is

$$\phi_{\mathrm{ms}} = \phi_{\mathrm{m}} - \left(\chi + \frac{E_{\mathrm{g}}}{2q} - \Psi_B\right) = 0 \qquad \text{for N type,} \quad (2.54)$$

$$\phi_{\mathrm{ms}} = \phi_{\mathrm{m}} - \left(\chi + \frac{E_{\mathrm{g}}}{2q} + \Psi_B\right) = 0 \qquad \text{for P type,} \quad (2.55)$$

in other words, the bias is flat when there is no applied voltage, where χ is the semiconductor electron affinity, E_{g} the bandgap energy, q the electric charge, Ψ_B the potential difference between the Fermi level energy, E_{F} and the intrinsic Fermi level energy, E_{i}. For a semiconductor vacuum interface, the electron affinity is defined as the difference between near-surface vacuum energy, E_0 and near-surface conduction band edge, E_{C}, i.e., $\chi = E_0 - E_{\mathrm{C}}$.

(2) The charges exist in the structure under any biasing conditions are those in the semiconductor, as well as those on the metal surface adjacent to the insulator.

(3) There is no carrier transport through the insulator under DC biasing conditions, or the resistivity of the insulator is infinity.

The MOS structure is obtained by applying a positive voltage on a metal electrode (which could be a highly doped polycrystalline silicon conducting layer), also referred to as gate, on top of a

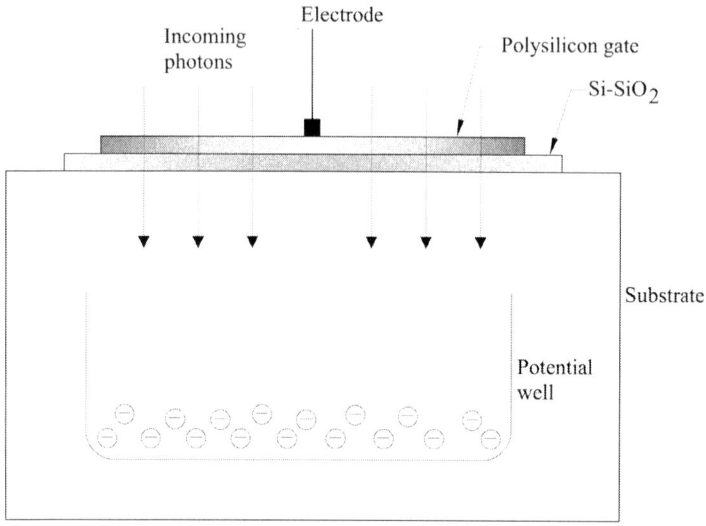

Figure 2.10 Typical potential well.

epitaxial P-type silicon substrate separated by a thin insulating layer made of SiO_2. In such a situation, the majority carriers (holes) are repelled from the region beneath the silicon–silicon dioxide ($Si-SiO_2$) interface and a depletion layer is formed under the electrode. This depletion layer is an electrostatic potential well (see Figure 2.10), whose depth is proportional to the applied voltage on the electrode, and can be used as photodetector of an optical input. This well is capable of storing any photocharges released in the semiconductor by an incident photon, as well as any thermally excited mobile charges, whose capacity can be altered by changing the voltage. The minority carriers (electrons), which are generated by this input, as well as by thermal energy, is attracted toward the electrode, and can thus, be shifted in sequence to the output, through on-chip amplifier. The holes, that are generated, are repelled out of the depleted layer, and are subsequently lost in the substrate. The carriers generated outside the depletion region diffuse slowly and recombine before they are attracted toward the depleted region. The electrons and the holes that are generated outside the depletion layer, recombine before they could be attracted toward the depletion layer.

The charge generation and collection can be understood in terms of a parallel plate MOS-capacitor holding the electrical charge. When the MOS structure is biased at three different voltages, V, three different situations at the semiconductor surface (see Figure 2.11) can, in principle, arise (considering P-type semiconductor), such as:

Accumulation The accumulation occurs when a negative voltage ($V < 0$) is applied to the metal plate (see Figure 2.11a), the top of the valence band bends upward and is closer to the Fermi level. For an ideal MOS, there is no current flow in the structure; hence, the Fermi level remains constant in the semiconductor. Since the carrier density depends exponentially on the energy difference ($E_F - E_V$), this band bending causes an accumulation of the majority carriers (holes) near the semiconductor surface. The negative charge on the gate attracts holes from the substrate to the oxide-semiconductor interface.

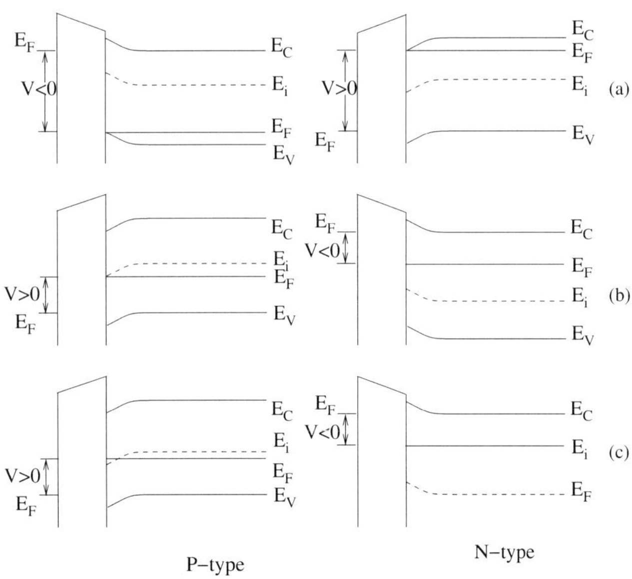

Figure 2.11 Energy band diagram of biased metal-oxide semiconductor (MOS) structure: (a) accumulation, (b) depletion, and (c) inversion. The left panel is for the P-type semiconductor, while the right panel is for the N-type semiconductor.

Depletion In the case of a small positive voltage ($V > 0$) being applied (see Figure 2.11b), the bands bend downward and the positive charge on the semiconductor surface region pushes the mobile holes into the substrate. Therefore, the semiconductor is depleted of majority carriers at the interface and the negative charge, due to the ionized acceptor ions, appears in the semiconductor surface region.

Inversion As the potential across the semiconductor increases (large positive voltage, $V > 0$), another type of negative charge emerges at the oxide-semiconductor interface (see Figure 2.11c). The bands bend even more downward such that the intrinsic Fermi level, E_{Fi} at the surface crosses over the Fermi level energy, E_{F}. At this point, high negative charge is induced in the semiconductor surface region. This charge is due to the minority carriers, which form a inversion layer. With further increase of such voltage, the depletion layer width barely increases further since the charge in the inversion layer increases exponentially with the surface potential.

However, in real case, it differs with the work function difference and interface traps and oxide charges. Similar results can be obtained for the N-type semiconductors as well. The interface trapped charge is due to the SiO_2-Si interface properties and dependent on the chemical composition of this interface; the interface trap density is orientation dependent as well. The fixed-oxide charge is located within $\approx$3 mm of the SiO_2-Si interface; this charge is fixed, which cannot be charged or discharged over a wide variation of surface potential. The oxide-trapped charges are associated with the defect in the SiO_2.

2.3.3 *Metal-Oxide Semiconductor FET*

A different type of field effect transistor (FET), called metal-oxide semiconductor FET (MOSFET), is a voltage-controlled device, in which the current flowing through the main channel between the drain and source is proportional to the input voltage (Brennan, 2003). The MOSFET is further classified under the following categories:

Enhancement type MOSFET An N-channel enhancement type MOSFET is fabricated on a P-type substrate. Two heavily doped N-type regions, such as source and drain) are created in the substrate. A thin layer of submicron size of SiO_2 is deposited between source and drain region. Metal is deposited on the insulator to form the gate of the device (thus, metal-oxide semiconductor). Metal contacts are also made to the source, drain, and body region. The applied voltage, V_{GS}, between the gate and the source, repels the holes in the P-type substrate near the gate region, lowering the concentration of the holes. With the increase of this voltage, hole concentration decreases, and the region near gate behaves like an intrinsic semiconductor material and then, finally, like an N-type material as electrons from the source and drain enter this region. As a result, when V_{GS} become larger than a threshold voltage, V_T, that is defined as the voltage at which a MOS transistor begins to conduct, a narrow layer between source and drain regions is created, which is populated with N-type charges; for voltages less than V_T, the channel is cut off. The thickness of this channel is controlled by the applied V_{GS}.

The physical structure of a P-type MOS (PMOS) is identical to a N-type MOS (NMOS) except that the semiconductor types are interchanged, i.e., body and gate are made of N-type material and source and drain are made of P-type material and a P-type channel is formed. As the sign of the charge carriers are reversed, all voltages and currents in a P-type MOS are reversed.

Depletion type MOSFET The depletion-type MOSFET has similar structure as the enhancement-type MOSFET, however it has a physically implanted channel. Such an N-type device has an N-type channel between the drain and the source. When a voltage between the drain and the source, V_{DS}, is applied, a current, i_D flows even for $V_{GS} = 0$. As V_{GS} increases, the channel turns out to be wider and the current increases. In a N-type depletion-type MOSFET, a negative V_{GS}, can also be applied to the device, which makes the channel smaller and reduces the current, i_D. The negative voltage, V_{GS}, depletes the channels from N-type carriers. A depletion-type MOSFET can be operated either in enhancement mode or in depletion mode (see Figure 2.12). A P-type depletion MOSFET operates to P-type enhancement MOSFET; the threshold voltage is

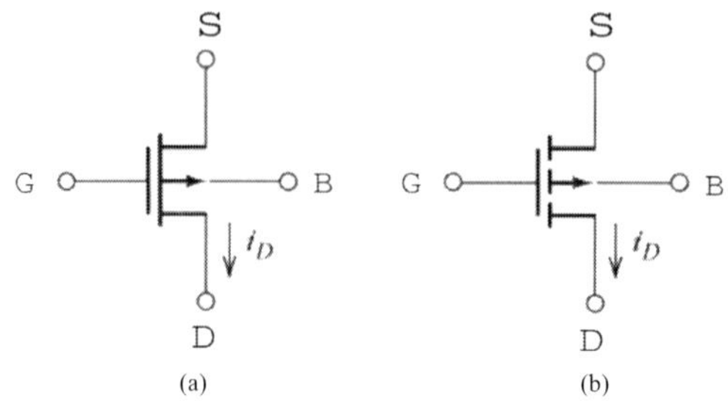

Figure 2.12 Circuit symbols for P-channel (a) Depletion type MOSFET and (b) Enhancement type MOSFET; for such an N-channel device, the direction of the arrow is reversed.

more than zero for depletion type, while it is less than zero for the enhancement type.

Three parameters such as V_{GS}, i_D, and V_{DS} describe behavior of an NMOS transistor:

$$i_D = 0;\ V_{GS} < V_T;\quad \text{Cutoff region: for any } V_{DS}, \tag{2.56}$$

$$i_D = K[2V_{DS}(V_{GS} - V_T) - V_{DS}^2];\ V_{DS} < V_{GS} - V_T;\ \text{Ohmic} : V_{GS} > V_T, \tag{2.57}$$

$$i_D = K(V_{GS} - V_T)^2;\ V_{DS} > V_{GS} - V_T;\quad \text{Active} : V_{GS} > V_T. \tag{2.58}$$

Here, K is a constant, which depends on manufacturing of the NMOS. For a small values of V_{DS}, a NMOS behaves as a resistor, R_{DS}, which is given by

$$R_{DS} = \frac{V_{DS}}{i_D} \approx \frac{1}{2K(V_{GS} - V_T)}. \tag{2.59}$$

For a PMOS transistor, the equations remain the same, except that the drain current is defined to be leaving the terminal, and all the equations are reversed, that is

$$i_D = 0;\ V_{GS} > V_T;\quad \text{Cutoff region: for any } V_{DS}, \tag{2.60}$$

$$i_D = K[2V_{DS}(V_{GS} - V_T) - V_{DS}^2];\ V_{DS} > V_{GS} - V_T;\ \text{Ohmic} : V_{GS} < V_T, \tag{2.61}$$

$$i_D = K(V_{GS} - V_T)^2;\ V_{DS} < V_{GS} - V_T;\quad \text{Active} : V_{GS} < V_T, \tag{2.62}$$

where V_T is negative for a PMOS transistor.

Complementary MOS (CMOS) A CMOS (see Section 6.3) employs a combination of PMOS and NMOS transistors (see Section 2.3.3) to implement logic gates to allow a path always to exist from the output to either the power source or ground. It requires more logic functions, for instance, those involving AND and OR gates to manipulate the paths between the gates to represent the logic.

2.4 Operational Amplifier

One of the most useful and important components of analog electronics is the operational amplifier (Op-Amp), which can amplify AC/DC signals and perform mathematical operations. It is referred to as the basic analog integrated circuit (IC). Its primary use is in amplifier, such as audio and video frequency pre-amplifiers and buffers (see Figure 2.13a), inverting amplifier (see Figure 2.13b), and non-inverting amplifier (see Figure 2.13c), summing amplifier, differential amplifiers. The Op-Amp is also employed in other related circuits, for example, differentiators and integrators, filters, rectifiers, voltage and current regulators, analog calculators, analog-to-digital (A/D) converters, digital-to-analog converters, and waveform generators. An analog computer is closely connected to the concept of negative feedback The advantages of such a feedback to an amplifier are (i) the gain of the circuit can be controlled, (ii) the non-linear distortions, as well as the effects of noise can be reduced substantially, (iii) both the input and output impedance of the amplifier can be modified, and (iv) the bandwidth of the amplifier can be enlarged.

An Op-Amp is composed of many transistors and resistors; the resulting circuit follows a certain set of rules. These devices arc classified by their construction, for instance: (i) discrete (built from individual transistors or valves), (ii) fabricated in an IC, and (iii) hybrid. Their internal structure consists of three stages, namely, (i) differential amplifier, (ii) voltage amplifier, and (iii) output amplifier. An Op-Amp is a five-terminal device comprising of two high impedance inputs (inverting input marked with a negative sign and non-inverting input, marked with a positive sign), V_+ the voltage source, which powers the amplifier and sets the maximum voltage

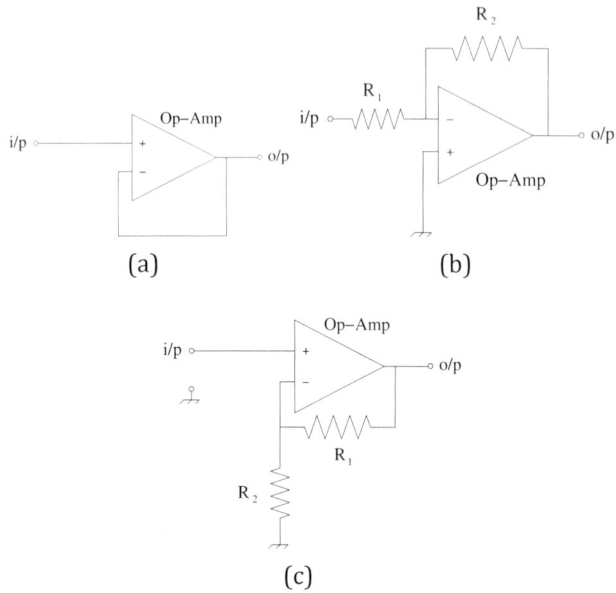

Figure 2.13 Typical circuit diagrams of (a) non-inverting buffer using Op-Amp, (b) inverting Op-Amp circuit, and (c) non-inverting Op-Amp circuit; i/p stands for the input, o/p the output.

output, V_- thought to be as ground, and an output port, where the amplified signal appears. The output voltage is the difference between the two input terminals, multiplied by the gain, i.e., $V_{\text{out}} = A(V_+ - V_-)$.

An ideal Op-Amp has characteristics, such as

- infinite open loop gain (practically $\approx 10^4$ to 10^5); the voltage gain of the amplifier is $V_{\text{out}}/V_{\text{in}}$ or $20 \log V_{\text{out}}/V_{\text{in}}$ in dB,
- infinite input impedance (practically $\approx$2–3 MΩ), i.e., no current flows into the input pins,
- zero output impedance (practically $\approx 75\Omega$), i.e., the Op-Amp can drive any load impedance to any voltage,
- infinite common mode rejection ratio (CMRR; practically 90 to 100 dB),
- infinite bandwidth (practically in MHz), and
- zero output voltage when the input voltage difference is zero.

The characteristics of Op-Amp do not drift with temperature. Also, it has zero noise and infinite power supply rejection ratio (PSRR). The input capacitance of a typical Op-Amp is about 1.4 pf (picofarad) and the total power dissipation is 85 mW (milliwatt). With negative feedback, and the voltage difference between the inputs of the Op-Amps is zero.

A large number of operational amplifier ICs are available to suit application from standard bipolar, precision, etc., in either standard configuration or with internal JFET transistors. Three types of Op-Amps, for example, (i) single (low power), (ii) dual (low noise), and (iii) quad (high-speed, high power, high-voltage, low offset) are available within one single device.

Chapter 3

Concept of Laser

3.1 Introduction

The theory of stimulated emission was put forwarded by A. Einstein (1917), which predicted that as light passed through a substance it could stimulate the emission of more light. In 1954, this phenomenon was used by Charles Townes and Arthur Schawlow in the construction of a microwave amplification by stimulated emission of radiation (maser) using ammonia gas and microwave radiation. Later, this operating principle of maser was put forth to the optical frequencies by them from the Bell Telephone laboratories in 1958, which led to the realization of the device, known as light amplification by stimulated emission of radiation (laser; Townes and Schawlow, 1958). A laser is a quantum optical device that produces a nearly monochromatic (single wavelength), and coherent beam of light by exciting atoms to a higher energy level and causing them to radiate their energy in phase.

The discovery of lasers has led to a wide range of applications in many diverse areas, for instance, medicine, communications, defense, measurement, etc. As a precise light source, they are employed in many scientific investigations as well. In medicine, laser surgery provides more invasive surgical techniques, which

High-Resolution Imaging: Detectors and Applications
Swapan K. Saha

ISBN 978-981-4613-27-9 (Hardcover), 978-981-4613-28-6 (eBook)
www.panstanford.com

yields in more rapid healing and reduction in complications. Medical lasers have the ability to produce thermal, physical, mechanical, and welding effects when exposed to tissues. Some of the applications of lasers include stone removal (laser lithotripsy), activation of specific drugs or molecules and denaturizing of tissues and cells in body. They are also used for guidance purposes in missiles, aircrafts and satellites and make up for a potential replacement of ballistic missiles. Lasers find applications in material processing, compact disks, laser fusion, surveying, remote sensing, lidar, welding, cutting, and drilling. Also, they are used for the measurement of pollutants in the atmosphere.

3.2 Main Characteristics

Lasers, which generate stimulated radiation in strongly non-equilibrium conditions, differ from the thermal radiation sources that emit spontaneous radiation in a thermal equilibrium state. The laser light possesses the following properties:

3.2.1 *Coherent Beam*

An important property of a laser is its coherence that is attributed to the ability of radiation to produce interference. The notion of coherence is defined by the correlation properties between the various quantities of an optical field. The optical coherence is related to the various forms of the correlations of the random processes (Born and Wolf, 1984; Mandel and Wolf, 1995). The interference phenomena stems from the principle of superposition, which reveals the correlations between light waves. Two sources of light are coherent, when there is a fixed or constant phase relationships between their wave-trains. Similarly, two photons are said to be in phase, when the crest of wave emitted by one photon falls on the crest of wave emitted by second photon. Hence, all of the photons that make up the laser beam have a fixed phase relationship (coherence) with respect to one another. There exist two types of coherence, such as temporal coherence and spatial coherence.

3.2.1.1 Temporal coherence

The temporal coherence is determined by its spectral bandwidth (see Section 4.1.1.1) and describes the interference of the emitted wave-trains; a laser beam can have extremely small spectral bandwidth, $\Delta\lambda \sim 10^{-6}$ Å. The narrower the frequency bandwidth, higher the temporal coherence. Michelson's classical interferometer (see Figure 3.1), which is directly related to the monochromaticity of radiation, is based on this principle; the higher the monochromaticity, the longer the coherence time, τ_c. This interferometer produces interference fringes by splitting a beam of monochromatic light so that one beam strikes a fixed mirror and the other a movable mirror. When the reflected beams are brought back together, an interference pattern results (Michelson and Morley, 1887). Results from the Michelson interferometer were used to verify special relativity. They are also being used in possible gravity-wave detection. Its other applications can be seen in the form of (i) measurement of length, (ii) optical coherence tomography (non-invasive technique for imaging subsurface tissue structure with micrometer-scale resolution), and (iii) studies of the upper atmosphere to get information about the winds and temperature the temperature deformation. It has also been used in aerodynamics,

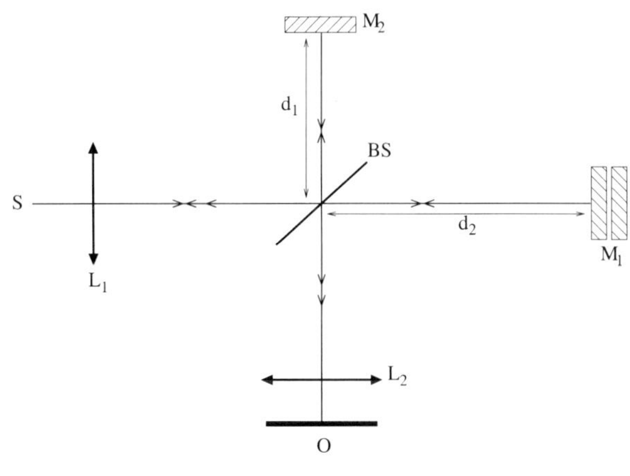

Figure 3.1 Schematic diagram of a classical Michelson interferometer; BS stands for the beam-splitter, M_1 and M_2 the two mirrors, and L_1 and L_2 the two lenses.

combustion, fluid mechanics, and in areas like diffusion, etc. The variations of Michelson's interferometer, such as Twyman–Green interferometer, Mach–Zehnder interferometer were also developed.

3.2.1.2 Spatial coherence

The spatial coherence determines the surface area that a light source with circular area can illuminate coherently with the light cone of its zeroth diffraction order. This can be improved by increasing the distance to the source and by reducing its size; the point sources are more coherent than those with extended source. The interferometry makes use of principle of superposition of waves to combine at least two separate waves to get information (see Figure 3.2a). These waves interfere, resulting in fringes (see Figure 3.2b). The pattern of such fringes is determined by the phase difference by the two waves. The waves that are in phase would undergo constructive interference (white bands), while the waves that are out of phase undergo destructive interference (dark bands). For constructive interferogram, the distance between the mirrors should be equal to integral multiple of $\lambda/2$, where, λ is the wavelength of light. The interference term enables the positions of the fringe intensity maxima, I_{max}, and minima, I_{min}, to be calculated. A method is widely to measure the contrast or visibility, $\mathcal{V}$, of the fringe is given by

$$\mathcal{V} = \frac{I_{\text{max}} - I_{\text{min}}}{I_{\text{max}} + I_{\text{min}}}. \tag{3.1}$$

The visibility contains information about both the spatial and spectral nature of the source; it turns out to be one when the

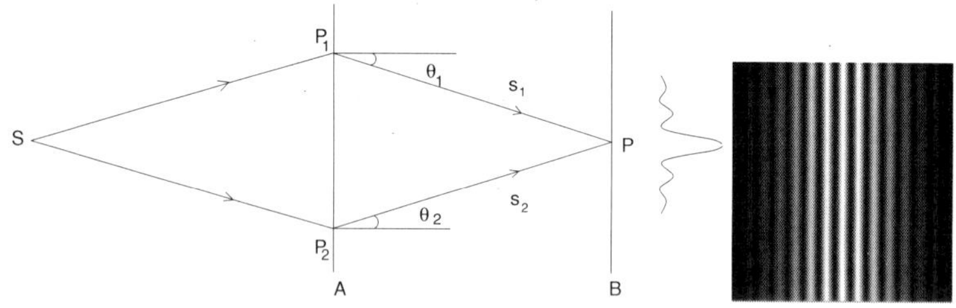

Figure 3.2 (a) Spatial coherence of two apertures, P_1 and P_2, illuminated by a source, σ and (b) double slit Young's fringe pattern.

intensity of both the waves are equal. In the case of Young's experiment (Young, 1802), there is no time-shift when two rays that pass through two slits and fall on a screen at a particular point, but they arrive from these points in space. The phase difference between the rays at this meeting point is constant (Saha, 2007).

Long-Baseline Optical Interferometry (LBOI) can obtain interference fringes by coherently combining the light of two separate telescopes (Labeyrie et al., 1975, 1986; Shao et al., 1988), in which only amplitude information can be obtained. When more than two telescopes are used, a quantity, called closure-phase, can be measured (Jennison, 1958). Such a method measures the amplitude and phase of the complex Fourier transforms of spatial brightness distribution of the source corresponding to the interferometer baselines. This technique provides the phase information needed to reconstruct images and has been applied in the field of radio astronomy. Such a method is immune to the atmospherically induced random phase errors, as well as to the permanent phase errors introduced by the imaging systems. The first image ever produced by an optical interferometer using separate telescopes was the binary star Capella (α Aurigae), that was obtained at the Cambridge Optical Aperture Synthesis Telescope (COAST; Baldwin et al., 1996). Of course, by applying the concept of hypertelescope imaging (Labeyrie, 1996; Labeyrie et al., 2013) with many diluted apertures, one may get direct images in a more efficient and sensitive way. Nevertheless, the LBOI has become a powerful tool, which enabled scientists to obtain results from the area of stellar angular diameters with implications for emergent fluxes, effective temperatures, luminosities and structure of the stellar atmosphere, dust and gas envelopes, binary star orbits with impact on cluster distances and stellar masses, relative sizes of emission-line stars and emission region, stellar rotation, limb-darkening, and astrometry. The disks around several young stellar objects (YSO), a few debris disks, core of an luminous blue variable (LBV) object, a nova, several active galactic nuclei (AGN) were resolved (Saha, 2010, and references therein).

COAST The Cambridge Optical Aperture-Synthesis Telescope (COAST) is based on closure-phase technique having a maximum baseline up to ~100 m providing images with a resolution down to 1 mas. Light from the telescopes is combined co-axially using a four-way combiner, where a beam splitter is used to combine two beams pairwise, of which one or both beams was already the combination of two others. The combined beam is focused onto one or more detectors for intensity read-out. Four telescopes, each comprising a 50 cm siderostat mirror that feed a fixed horizontal 40 cm Cassegrain telescope (f/5.5) with a magnification of 16 times, are arranged in a Y-layout with one telescope on each arm, movable to a number of fixed stations and one telescope at the center of the Y. The COAST is no longer in operation.

Hypertelescope Hypertelescope is an optical interferometric array with a set of mirrors arranged in a large mirror-shape, working together to resolve astronomical images at much higher angular resolutions. It uses the aperture synthesis technique, in which many telescopes are employed in an interferometric array, where most of the phase information can be retrieved. Aperture synthesis technique became highly successful at radio wavelengths where the heterodyne detection (see Section 4.3.1.4), instead allows a coherent reconstruction, with a phased combination of fringe signals recorded from different baselines at different times.

However, the hypertelescope differs from the Fizeau interferometers, which become inefficient when the subaperture spacing is large compared to their size. The reason is that most energy goes in a broad diffractive halo rather than in a narrow interference peak. Densifying the exit pupil, i.e., distorting it to increase the relative size of the subpupils, in such a way that the pattern of subaperture centers is preserved, concentrates the halo and intensifies the image (Labeyrie, 1996). Direct images may be obtained with such a system combining their focused beam on a camera. However, in the lack of proper adaptive optics (AO) system, the hypertelescope may completely depend on the bispectrum technique to fully utilize the aperture extend (Saha, 2010; Surya and Saha, 2014; Surya et al. 2014). The natural aperture rotation, relative to celestial North, during hours of observing (possibly intermittent) may improve the reconstruction. Also, with a Carlina geometry, the pupil drift across a

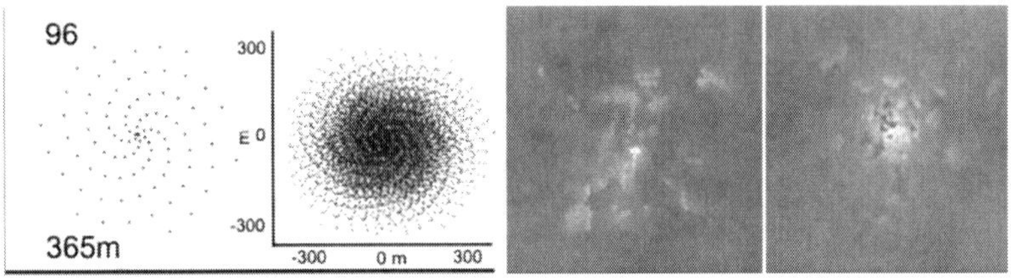

(a) (b) (c) (d)

Figure 3.3 Simulation of Imaging of Dusty Torus around AGN by Hypertelescopes; (a) apertures used for imaging, (b) corresponding u-v coverage, (c) Edge-on geometry model and (d) pole-on geometry of an active galactic nuclei (AGN).

fixed and non-redundant array of mirrors tends to vary the aperture pattern, which improves the reconstructed image.

The reconstructed and deconvolved images of the imaging of dusty torus around the active galactic nuclei (AGN) by hypertelescope are shown in Figure 3.3. For the AGN performance simulations, the model of Schartmann et al. (2012) is used. In this scenario, the purported, dust torus, which in the standard model for AGNs is assumed to shadow the nucleus proper, is assumed to comprise an assemblage of discrete clouds. Here an AGN model in pole-on and edge-on torus geometry is applied. The images from 3-D radiative transfer code MC3D (Wolf, 2003) are used, which provides the temperature and surface brightness distributions of the resulting heated dust distribution as inputs to the hypertelescope simulator code.

3.2.2 *Spatial Filter*

A spatial filter is an optical device that uses the principles of Fourier optics to alter the structure of a beam of coherent light or other electromagnetic radiation. It is used to remove the unwanted multiple-order energy peaks and pass only the central maximum of the diffraction pattern. The spatial filter in a Fourier transform spectrometer modifies the output by adjusting the amplitude and phase of the measured longitudinal correlation function. Furthermore, when a light beam passes through a system, dust in the air or on optical components and other non-uniformities can distort the beam and create scattered light, which can leave unwanted ring patterns in the beam profile.

Often a laser system does not produce a beam with a smooth intensity profile. In addition to the optical aberrations (see Section 1.3.2) caused by the optical elements, such as mirrors or laser cavity that acts as an optical resonator (see Section 3.3.6), the variations in the laser gain medium itself blurs the image. The resonator can be characterized by eigen values and eigen functions (Menn, 2004). These eigen functions are oscillations that may be self-generated in a sporadic manner by the resonator. The intensity distribution across the beam profile is determined by the transverse modes developed in the cavity. These modes constitute a two-dimensional (2-D) array of functions and are described by the symbol, $\mathbf{E}_{ij}^{t}$, in which i, j are the small integers describing the number of horizontal and vertical lines with minimal intensity in the beam profile. The first mode, $\mathbf{E}_{00}^{t}$, has a radial distribution described by the Gaussian distribution function, with the maximum energy density on the optical axis. A clean Gaussian laser beam can be produced by employing a spatial filter (see Figure 3.4), in which the transverse profile of the intensity of the beam is defined by

$$I(r) = I(0)e^{-2r^2/w^2}, \tag{3.2}$$

with I as the intensity at a point, r, from the maximum intensity, $I(0)$, and w the beam radius.[a]

The first mode is the preferred one since the beam profile has minimum diffraction losses and the wavefronts are almost spherical (Haus, 2010). The other modes, viz., $\mathbf{E}_{10}^{t}$, $\mathbf{E}_{01}^{t}$, $\mathbf{E}_{11}^{t}$, etc., have an intensity distribution differing from equation (3.2); they may have several points of maximum intensity or points of maximum intensity arranged as a ring shape, etc. Also, there can be longitudinal modes available in the resonator (Yariv, 1982). The generation of higher-order modes causes random fluctuations in the laser beam intensity, what is referred to as spatial noise, which needs to be cleaned out.

---

[a]The light beam has a radial profile of a Gaussian function at any cross section perpendicular to the optical axis. The Gaussian is overspread to infinity in the radial direction. The beam radius, w, is the distance from the beam axis to a point where the intensity drops to $1/e^2$ ($\approx$0.135) of the on-axis value, $I(0)$. As the beam propagates, the beam begins to diverge, causing the width the beam to increase, which gives rise to a wider intensity profile and a smaller value for $I(0)$.

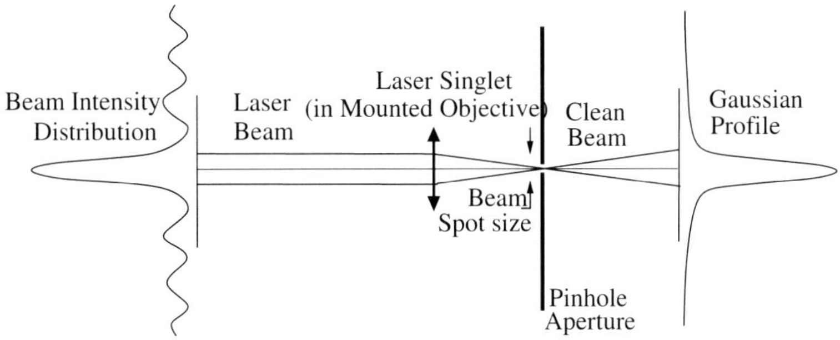

Figure 3.4 Concept of a spatial filter.

The first application of spatial filtering was used in a microscope, in Abbe's theory of image formation. The setup for a spatial filter, depicted in Figure 3.4, comprises a tiny high-power lens, typically a microscope objective, whose focal length, f, is specified in terms of its magnification when used in a standard imaging configuration with an image location of 160 mm from the back focal plane, i.e., $f = 160/\mathtt{M}$, in which M stands for magnification, and a micron-sized aperture (pinhole) or a fiber-optic system (see Figure 3.5). This arrangement essentially filters out the high spectral frequencies of the beam. In order to spatially filter the beam correctly, the pinhole should match the diameter of the point-spread function (PSF) of the converging lens. If the lens is limited by the diffraction, the diameter of the PSF is given by twice the Rayleigh limit being, $1.22\ \lambda f/D$, in which f is the focal length of the lens and D the diameter (width of the laser, which is typically much smaller than the actual objective). The lens is bringing the Fourier pattern of the light field to its back focal plane. With appropriate size of the pinhole and the objective combination, the central DC portion of the light field at the focal plane of the lens can be filtered.

In a long-baseline optical interferometer (see Saha, 2010, and references therein) that requires detection of very faint signals and reproduction of interferometric visibility's to a high precision, this approach helps to spatially filter the light being delivered to the beam-combiners, thus smoothening the turbulence-induced corrugated wavefronts and one finds that the contribution of variations are much reduced. However, the drawbacks of spatial filtering are a

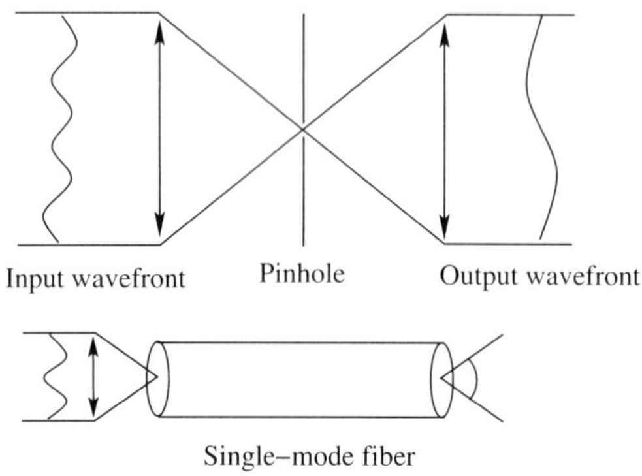

Figure 3.5 Principle of wavefront smoothing by spatial filtering with a pinhole (top panel) and with a single mode fiber (bottom panel; Saha and Morel, 2000).

loss of optical coupling efficiency and larger photometric variations due to the turbulence. Both pinholes and single-mode optical fiber can act as suitable spatial filters though many interferometric implementations use fiber components. However, it is wise to use spatial filter, if the size of a telescope exceeds a few times than Fried's parameter, r_0 (see Section 1.5.3).

3.2.3 *Beam Expansion*

The divergence of the laser light is very small, and hence in general, is limited by the diffraction (see Section 1.5.4). The beam size of a laser is typically 0.8 to 1.2 mm in diameter, and hence it becomes almost essential to expand the beam in most situation so that the entire object area is illuminated. The beam expansion becomes essential if it is required to be transmitted over a large distance. Also, because of high directionality of the laser beams, they can be focused to a very small spot of a few μm^2, can produce temperature of $10^{4\circ}$(C) at the focused point. In the first case, the divergence angle of the laser beam is reduced considerably in order to collimate (parallel rays) the beam, which can be performed by incorporating a beam expander. In the second case, the beam diameter should be increased in order to reduce the diffraction limit of the lens

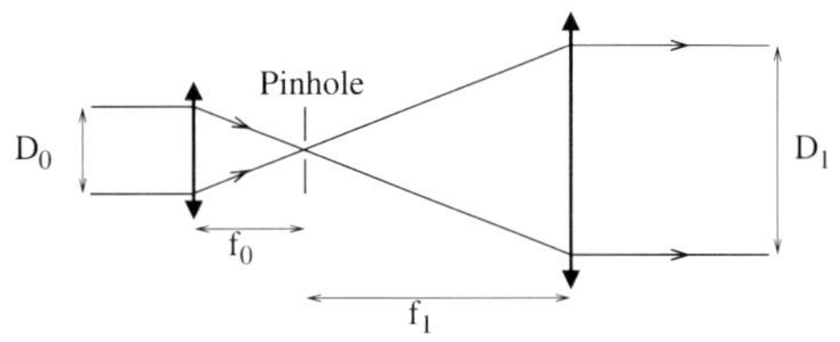

Figure 3.6 Schematic diagram of the laser beam expansion and collimation.

concentrator. An inverted Galilean telescope may be inserted after the laser to achieve both goals.

Figure 3.6 depicts such an arrangement in which the object rays, located at infinity, enter parallel to the optical axis of the internal optics and exit parallel to them as well. A second lens collimates the beam with pinhole located at its front focal plane. The system is an afocal system with an additional pinhole and the distance between the two lenses is the summation of their focal lengths. As for a standard telescope, the magnification is given by the ratio of the focal lengths; therefore, the expanded beam diameter is expressed as

$$D_1 = \frac{f_1}{f_0} D_0,$$

in which D_0 represents the diameter of the laser beam, f_0 and f_1 correspond to the focal lengths of the microscope objective (MO) and collimating lens, respectively.

The beam from a small He–Ne laser is typically, 0.8 to 1.2 mm in diameter with a Gaussian profile. The beam can be expanded with a lens, typically a microscope objective, having short focal length and the beam width, D_0, is much less than the focal length of the lens f_0, so the beam divergence would be, $\theta = D_0/f_0$. The beam divergence is dependent upon the diameters of the input laser beam and output collimated beam. Both the beam from the laser and the expanded beam should have a Gaussian intensity profile (see Section 1.4).

3.2.4 *Output Power*

The output of a laser, defined as the power that is emitted per unit solid angle, Ω, may either emit light in a continuous manner or emit light in short bursts (pulsed). If during the process of stimulated emission, the population inversion (see Section 3.3.4) is maintained by continuous pumping of energy, the laser action continues indefinitely, which results in a continuous wave laser, while if the pumping cannot be maintained continuously the output is a pulsed laser. Two parameters, such as (i) the average output power and (ii) the peak power that is approximated as $P_{peak} = E_{pulse}/\Delta t$, in which E_{pulse} represents the energy per pulse and $\Delta t(= 1/\Delta\nu)$ the full width at half maximum (FWHM) of the pulse; the duration of each pulse varies from microseconds to femtoseconds. They can emit optical power from several micro watts to megawatts. The power output from a laser ranges from a milliwatt to millions of watts; continuous wave lasers have power levels $\sim 10^5$ W and pulsed lasers have a total energy $\sim$50,000 J.

3.3 Principles of Emission

As stated in Section 1.1.6, the atoms are characterized by discrete energy states. According to Einstein (1917), there are three different ways through which an atom can interact with electromagnetic radiation (see Figure 3.7), which are described below.

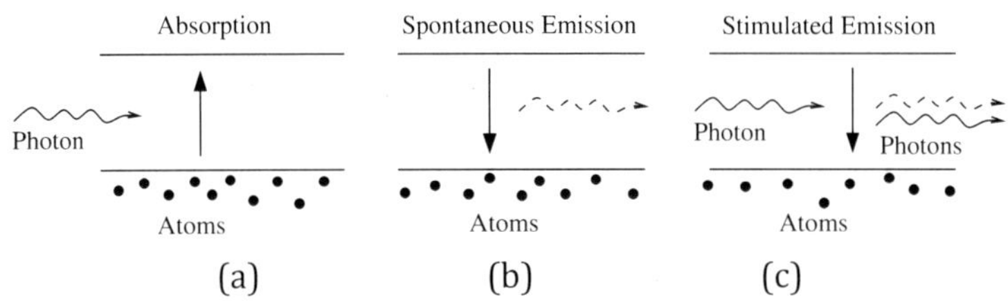

Figure 3.7 Interaction of an atom with electromagnetic radiation: (a) induced absorption; an increase in energy level from E_1 to E_2 yields in absorption of a photon, (b) spontaneous emission; a decrease in energy level from E_2 to E_1 results in emission of a photon, and (c) stimulated emission; the resulting photon has the same phase and frequency as the original photon and travels in the same direction.

3.3.1 *Induced Absorption*

When an atom in the lower energy state encounters radiation of appropriate frequency (corresponding to the energy difference of the two atomic levels), it can absorb the energy of a photon and jump to an excited state (see Figure 3.7a). This process of excitation occurs in the presence of radiation and the rate of absorption depends on the density of radiation at the particular frequency corresponding to the energy separation of the two levels, as well as on the number of atoms in the lower energy state. In photon absorption, the laser material is optically excited to achieve population inversion (see Section 3.3.4) based on Planck's law (see Section 1.1.1), according to which, the change of energy level from the lower energy level, E_1 to the higher energy level, E_2 or vice versa, results in the absorption or emission of photon, respectively.

Let N_1 and N_2 represent the number of atoms (per unit volume) in the levels E_1 and E_2, respectively, i.e., N_1 atoms with energy E_1 and N_2 atoms with energy E_2 per unit volume. Since $E_2 - E_1 = \hbar\omega$, where $\hbar = h/2\pi \approx 1.0546 \times 10^{-34}$ J.s, h Planck's constant, and $\omega = 2\pi\nu$, the absorption process depends on the energy density of radiation at the frequency ω. The energy density, $u(\nu)$, is defined such that $u(\nu)d\nu$, which is the energy incident per unit volume of the system of frequency range ν and $\nu + d\nu$. The rate of absorption is proportional to $N_1 u(\nu)$, thus one writes the number of absorption per unit time per unit volume as equal to $N_1 B_{12} u(\nu)$, with B_{12} as the proportionality constant, called Einstein's coefficient of induced absorption, which is a characteristics of the energy levels.

3.3.2 *Spontaneous Emission*

As stated in Section 1.1, photon is a discrete bundle of electromagnetic energy. The common process for photon creation, known as spontaneous emission of radiation, occurs when an electron jumps from a higher energy state to lower one (see Figure 3.7b). The excited particles, resulting from population inversion (see Section 3.3.4), are unstable. They release their excess energy by non-radiative process, such as collisions (atoms are seldom isolated and there are collisions with the surrounding atoms) with other excited particles or by photon emission, and return to the stable ground state; at room temperature the atoms reside in the ground state. If the energy falls into the material, the outer electrons of the atoms would excite to higher energy levels, e.g., E_2. An electron in such a state can spontaneously relax to the lower energy level, E_1. Such emissions are random and isotropic in nature. The probability per unit time of the atom making a downward transition is independent of the energy density of radiation field, but depends on the levels involved in the transition. The exact time at which the electron chooses to relax is totally random, and therefore the radiation emitted is incoherent. The molecules typically remain excited for no longer than a few nanoseconds, this is often also called fluorescence or, when it takes longer, phosphorescence. To note, the energy levels that result from the molecular binding are found to be quantized in either situation; the molecules also emit photons when the molecules transit from higher energy states to lower ones. However, molecules have energy states that are related to rotational and vibrational motions of the molecules, which lead to energy levels more closely spaced than those associated with electronic transitions. The radiation resulting from transitions between those levels occurs at longer wavelengths (in mid- and far-IR) than that from electronic transition. The rate of spontaneous emission is proportional to the number of atoms in the excited state, N_2, and thus

$$\frac{dN_2}{dt} = -A_{21}N_2, \tag{3.3}$$

where $A_{21}(= 1/\tau_2)$ represents the. proportionality constant, called Einstein coefficient for spontaneous emission and τ_2 the spontaneous emission lifetime of the excited state; the natural process of the decay is by spontaneous emission.

The decay of the population density of an excited state, N_2 as a function of time, t, is indicated through the effective decay time constant, τ_2. The collisions force excited atoms to decay back to their ground state without emitting radiation. However, there are cases where a collision may not decay an atom from its excited state but may interrupt the phase of the radiating atom. This effect will produce a broadening of the emission frequency. In gases, the collisional de-excitation rate depends on the electrons density, the gas pressure and is proportional to the inverse cube of the energy difference between the two levels; higher the pressure, more frequent would be the collisions resulting in the reduced lifetime of the upper laser level and hence also the population.

Atoms may absorb or emit photons over a narrow range of neighboring frequencies ν. The shape is given by the line-shape $g(\nu)$. This width is associated with the characteristic decay time of the radiating levels, as well as with other emission line broadening mechanisms. The main broadening mechanisms for gases are (i) Doppler broadening (see Section 3.3.5.1), (ii) pressure broadening, and (iii) natural lifetime, while for solids they are (i) lattice vibrations (see Section 2.2.1.1), (ii) phonon[a] interactions, and (iii) natural lifetime. The density of the population of the upper energy level due to spontaneous emission is

$$\begin{aligned} N_2(\tau) &= N_2(0)e^{-A_{21}t} \\ &= N_2(0)e^{-t/\tau_2}, \end{aligned} \tag{3.4}$$

which implies that the population of level 2 reduces by $1/e$ in a time τ_2 that is the lifetime of the excited state.

In the case that the upper energy level population is decaying to more than one lower energy levels then more general equations are valid

$$\frac{dN_2}{dt} = \left(\sum_i A_{2i}\right) N_2,$$

therefore, equation (3.4) is derived as

$$N_2(\tau) = N_2(0)e^{-(\sum_i A_{2i})t}, \tag{3.5}$$

with $\tau_2 = 1/\sum_i A_{2i}$, implying that the lifetime of excited energy level is determined by the reciprocal of the sum of all possible decay rates to lower-lying levels.

---

[a]Phonons are quanta of lattice vibrations. Often referred to as a quasi-particle (Schwabel, 2008), a phonon represents an excited state in the quantum mechanical quantization of the modes of vibrations of elastic structures of interacting particles. The phonons are indistinguishable particles like photons (see Section 1.1), they follow Bose-Einstein distribution. These particles play a major role in many of the physical properties of condensed matter, such as thermal conductivity and electrical conductivity. The long-wavelength phonons give rise to sound, while the short-wavelength phonons give rise to heat; the phonon amplitude increases with increasing temperature. In the absence of phonons, all materials would be acoustic insulators. Scattering mechanism (see Section 1.5.2), namely, Brillouin scattering involving acoustic phonons or Raman scattering that involves optical phonons, with higher energies, 1 to 10 eV provides information on phonon spectra. The resulting shifts in frequency can be explained by creation or annihilation of phonons. X-rays can be scattered by phonons as well (Laval, 1958).

The transition probability, A_{21} is related to the oscillator strength of this transition and to the wavelength, λ_{21} according to the relationship

$$A_{21} = \frac{10^{-4} f_{12}}{1.5(g_2/g_1)\lambda_{21}^2}, \tag{3.6}$$

in which f_{12}, g_2, and g_1 are dimensionless numbers.

According to the classical model, the spontaneous emission probability, A_{21} coincides with the decay rate γ. With the inclusion of the collisions, one gets

$$\begin{aligned} N_2 &= N_2(0)e^{-\gamma_2 t} \\ &= N_2(0)e^{-(\gamma_2^r+\gamma_2^c)t}, \end{aligned} \tag{3.7}$$

with $\gamma_2^r = \sum_i A_{21}$ and $\gamma_2^c = 1/T_{21}$ and the decay time

$$\tau_2 = \frac{1}{\gamma_2} = \frac{1}{\gamma_2^r + \gamma_2^c}. \tag{3.8}$$

The collisional decay occurs in various types of media, for instance, (i) binary collision between two atoms or molecules, (ii) interactions of excited atoms with neighborhood, and (iii) phonon collisions in solids.

3.3.3 *Stimulated Emission*

As put forward by Einstein, when an atom is in the excited state, it can make a transition to a lower state through stimulated emission, in which an incident signal of appropriate frequency triggers an atom in an excited state to emit a photon, resulting in amplification of the incident beam (see Figure 3.7c). A photon having the exact energy necessary to cause stimulated emission is made available by the spontaneous emission. These photons from spontaneous emission trigger stimulated emission of other photons resulting in a cascade of stimulated emission. The photons due to such a type of emission are monochromatic, coherent, and appear to originate from a point source. The rate of stimulated emission depends on the intensity of the external field, as well as on the number of atoms in the excited state, which is proportional to $N_2 u(\nu) = B_{21} N_2 u(\nu)$, with B_{21} as the proportionality constant, called Einstein coefficient of stimulated emission. The amplification process due to stimulated transition is phase coherent, i.e., all stimulated photons are identical in respect of phase, wavelength, and direction. At thermal equilibrium, the rate of absorption is equal to sum of the rates of spontaneous emission and stimulated emission, that is

$$N_1 B_{12} u(\nu) = N_2 A_{21} + N_2 B_{21} u(\nu),$$

or,

$$\begin{aligned} u(\nu) &= \frac{N_2 A_{21}}{N_1 B_{12} - N_2 B_{21}} \\ &= \frac{A_{21}}{\frac{N_1}{N_2} B_{12} - B_{21}}. \end{aligned} \tag{3.9}$$

In the absence of collisions, the molecules tend to remain in the lowest energy state available, while the collisions can knock a molecule into a higher-energy state; the higher the temperature, the more this happens. According to the principle in thermodynamics, from Maxwell–Boltzmann's law (see Section 2.2.2.1; equation 2.5), at thermal equilibrium, the ratio of the populations of two states is

$$\begin{aligned} \frac{N_1}{N_2} &= e^{-\Delta E/(k_B T)} \\ &= e^{-h\nu/(k_B T)}, \end{aligned} \tag{3.10}$$

where $k_B (= 1.38 \times 10^{-23}$ J.K$^{-1})$ represents Boltzmann's constant and T the absolute temperature, and thus one may write

$$u(\nu) = \frac{A_{21}}{B_{12}[e^{h\nu/(k_B T)} - 1]}. \tag{3.11}$$

As a result, higher-energy states are always less populated than the ground state, and absorption is stronger than stimulated emission. At thermal equilibrium, the radiation energy density in a range from ν to $\nu + \delta\nu$ emitted by an arbitrary blackbody, is given by Planck's law

$$u(\nu) = \frac{8\pi h \nu^3}{c^3} \left[\frac{1}{e^{h\nu/(k_B T)} - 1} \right]. \tag{3.12}$$

Comparing equations (3.11) and (3.12)

$$\frac{A_{21}}{B_{21}} = \frac{8\pi h \nu^3}{c^3}.$$

As $T \to \infty$, $e^{h\nu/(k_B T)} \to \infty$, so $B_{12}/B_{21} = 1$; hence, $B_{12} u(\nu) = B_{21} u(\nu) = B$, and therefore $\mathcal{P}_{12} = \mathcal{P}_{21} = \mathcal{P}$, which means the probability of induced absorption (B_{12}) is equal to the probability of stimulated emission. From equation (3.12), one gets

$$u(\nu) = \frac{A}{B} \frac{1}{\left[e^{h\nu/(k_B T)} - 1\right]}. \tag{3.13}$$

Equation (3.13) is for the energy density at thermal equilibrium using Einstein's coefficients. If the frequency of radiation is high $h\nu \gg k_B T$, $e^{h\nu/(k_B T)} \gg 1$, and

$$\frac{A}{B} \gg 1 \Rightarrow A \gg B,$$

i.e., the spontaneous emission would be much larger in number than stimulated emission. In the case of low frequency, if

$$h\nu \ll k_B T, \qquad \left[e^{h\nu/(k_B T)} - 1\right] \ll 1,$$

and

$$\frac{A}{B}(u(\nu)) \ll 1 \Rightarrow A \ll B.$$

Hence, for lower frequency, stimulated emission dominate the emission dominate the emission process.

3.3.4 *Population Inversion*

A laser stores energy, surrounded by two mirrors of which a partially reflecting output mirror lets some light out. When the atoms are in the thermodynamic equilibrium, there are larger number of atoms in the lower state, which implies that the number of absorption exceeds the number of stimulated emissions resulting in the attenuation of the beam (Silfvast, 2004). On the contrary, in state of population inversion, in which the number of atoms in the upper energy level is greater than that in the lower energy level, the stimulated photons grow to a very large number resulting in the amplification of the beam. The factor by which an input beam is amplified by a medium is called the gain, G. In order to achieve $G > 1$, stimulated emission must exceed absorption, i.e., $BN_2 I_\nu > BN_1 I_\nu$, where $I_\nu (= vU = c/n_0 U)$ is the intensity representing the energy per unit area per unit time, which is related to the energy density, $U = \int_0^\infty u(\nu) d\nu$, n_0 being the refractive index, and $v(= c/n_0)$ the velocity of the radiation field in the medium; under normal condition, N_2 will always be lesser than N_1. This condition is called inversion, which does not occur naturally (it is forbidden by Maxwell–Boltzmann's distribution). It is inherently a non-equilibrium state

$$\frac{\text{Rate of emission}}{\text{Rate of absorption}} = \frac{A_{21}N_2 + B_{21}N_2u(\nu)}{B_{12}N_1u(\nu)}$$

$$= \frac{N_2}{N_1}\left[\frac{A_{21} + B_{21}u(\nu)}{B_{12}u(\nu)}\right]$$

from Einstein's theory, $B_{12} = B_{21}$

$$= \frac{N_2}{N_1}\left[\frac{A_{21}}{B_{12}u(\nu)} + 1\right]. \tag{3.14}$$

for $h\nu \ll k_\text{B}T$, which makes $\left[e^{h\nu/k_\text{B}T} - 1\right]$ very small, i.e., $\frac{A_{21}}{B_{21}u(\nu)} \ll 1$, thus

$$\frac{\text{Rate of emission}}{\text{Rate of absorption}} = \frac{N_2}{N_1}. \tag{3.15}$$

The method of producing population inversion is known as pumping and the process that leads to emission of stimulated photon due to population inversion is often referred to as lasing. A laser would lase if the beam increases in intensity during a round trip. Additional losses in intensity occur in terms of absorption, scattering, and reflections. However, in order to achieve threshold, the gain should be more than loss; the threshold gain is the gain value that compensates for losses occurring on a round-trip through the cavity. Neglecting spontaneous emission, the stimulated emission without considering absorption is given by

$$\frac{dI_\nu}{dt} = c\frac{dI_\nu}{dz}$$
$$\propto B[N_2 - N_1]I_\nu, \tag{3.16}$$

Considering a collection of atoms and let a near monochromatic beam of frequency, ν be propagating along the z direction, the solution to this equation (3.16) is expressed as

$$I_\nu(z) = I_\nu(0)e^{\sigma[n_2 - N_1]z}, \tag{3.17}$$

with σ as the proportionality constant, which is the absorption/gain cross section.

There can be either exponential gain or loss in intensity. In the case of N_2 being less than N_1, there would be loss (absorption), i.e., the absorption coefficient, $\alpha = [N_1 - N_2]\sigma$. But if N_2 turns out to be more than N_1, there would be gain, i.e., the gain coefficient, $g = [N_2 - N_1]\sigma$. The gain, G, is defined as

$$G = e^{\sigma[N_2 - N_1]L}, \tag{3.18}$$

in which L is the volume of the laser medium, which is perpendicular to the z direction at z and $z + dz$.

3.3.5 *Emission Line-Width*

Atoms may absorb or emit photons over a narrow range of neighboring frequencies. The line-shape function, $g(\nu)$ determines the threshold population inversion. This width is associated with the characteristic decay time of the radiating levels, as well as with other emission line broadening mechanisms. In what follows, the broadening mechanisms are elucidated in brief.

3.3.5.1 Doppler broadening

The Doppler broadening occurs due to the thermal motion of gas atoms. In a gas, atoms are moving in all directions with velocities given by Maxwell–Boltzmann's distribution. If an atom has a z-component of the velocity lying between v_z and $v_z + dv_z$, the probability distribution is

$$\mathcal{P}(v_z)dv_z = \sqrt{\frac{m}{2\pi k_B T}} e^{-mv_z^2/(2k_B T)} dv_z, \tag{3.19}$$

where k_B is Boltzmann's constant, m the mass of the atom, and T the absolute temperature of the gas; one uses the normalizing factor since each atom must be located somewhere

$$\int_{-\infty}^{\infty} \mathcal{P}(v_z)dv_z = 1.$$

The probability $g(\nu)d\nu$ that the transition frequency lies between ν and $\nu + d\nu$ is equal to the probability that the z-component of the velocity lying between v_z and $v_z + dv_z$, in which $v_z = (\nu - \nu_0)c/\nu_0$, thus

$$g(\nu)d\nu = \frac{c}{\nu_0}\sqrt{\frac{m}{2\pi k_B T}} e^{-\frac{mc^2}{2k_B T}\frac{(\nu-\nu_0)^2}{\nu_0^2}} d\nu, \tag{3.20}$$

which corresponds to a Gaussian distribution

$$\mathcal{P}_N(n) = \frac{1}{\sqrt{2\pi \langle\sigma\rangle^2}} e^{-(n-\langle N\rangle)^2/(2\langle\sigma\rangle^2)}, \tag{3.21}$$

with $\langle N\rangle$ is the mean or average, σ the standard deviation (also called RMS deviation).

It may be noted that unlike inhomogeneous broadening where each atom or molecule has a different line-shape, $g(\nu)$ due to its environment, in homogeneous broadening, all atoms behave the same way i.e., each effectively has the same $g(\nu)$. In the frame of the emitter, a homogeneously broadened line-shape is generated, but in the laboratory frame, seen by an external observer, the homogeneously broadened line-shape is frequency shifted by the Doppler effect, i.e., $\nu_o = \nu_0(1 \pm v/c)$, with ν_o as the observed frequency. The relationship between velocity and frequency is

$$\nu = \nu_0 \left(1 \pm \frac{v}{c}\right) \rightarrow v = c\frac{\nu - \nu_0}{\nu_0};$$

$$\nu_o = \nu_0 \left(1 \pm \frac{v}{c}\right) \rightarrow d\nu = \frac{\nu_0}{c} dv \rightarrow dv = \frac{c}{\nu_0} d\nu,$$

therefore, the Doppler broadening is given by

$$\begin{aligned} g(\nu)d\nu &= \sqrt{\frac{m}{2\pi k_{\mathrm{B}} T}} e^{-\frac{m[c(\nu-\nu_0)/\nu_0]^2}{2k_{\mathrm{B}}T}\frac{c}{\nu_0}} d\nu \\ &= \sqrt{\frac{mc^2}{2\pi k_{\mathrm{B}} T \nu_0^2}} e^{-\frac{mc^2(\nu-\nu_0)^2}{2k_{\mathrm{B}}T\nu_0^2}} d\nu. \end{aligned} \tag{3.22}$$

The line-shape function is peaked at ν_0, that is

$$g(\nu_0) = \frac{c}{\nu_0}\sqrt{\frac{m}{2\pi k_{\mathrm{B}} T}},$$

while the full-width at half-maximum (FWHM) is

$$\Delta\nu_D = \frac{2}{\lambda_0}\sqrt{\frac{2k_{\mathrm{B}}T}{m}\ln 2}.$$

The intensity of the emission line as a function of frequency is expressed as

$$I(\nu) = \sqrt{\frac{mc^2}{2\pi k_{\mathrm{B}} T \nu_0^2}} I_0 e^{-\frac{mc^2(\nu-\nu_0)^2}{2k_{\mathrm{B}}T\nu_0^2}}, \tag{3.23}$$

with $I_0 = \int_0^\infty I(\nu)d\nu$.

3.3.5.2 Natural broadening

The homogeneous broadening mechanisms lead to the Lorentzian line-shape emission spectrum, which is the minimum emission line-width. The frequency spectrum associated with spontaneous emission is written as

$$g(\omega) = \frac{1}{2\pi\tau_2} \cdot \frac{1}{(\omega - \omega_0)^2 + 1/4\tau_2^2}, \tag{3.24}$$

where $\tau_2(= 1/A_{21})$ is the spontaneous emission lifetime and $\omega(= 2\pi\nu)$ the angular frequency.

The total effective energy width, ΔE_T, due to the natural broadening is given by

$$\begin{aligned} \Delta E_T &= \Delta E_1 + \Delta E_2 \\ &= \hbar\left(\sum_i A_{2i} + \sum_j A_{1j}\right), \end{aligned} \tag{3.25}$$

with

$$\Delta E_1 = \hbar/\tau_1 = \hbar\gamma_1 = \hbar\sum_j A_{1j}; \qquad \Delta E_2 = \hbar/\tau_2 = \hbar\gamma_2 = \hbar\sum_i A_{2i},$$

γ_1, γ_2 as the respective decay rate at state 1 and at state 2, τ_1, τ_2 the lifetime of the lower and upper states, respectively. $\hbar = h/(2\pi) = \Delta E \Delta t$ (from uncertainty principle), and

$$h\Delta\nu_{21}^N = h(\Delta\nu_1 + \Delta\nu_2) = h\left(\frac{A_1}{2\pi} + \frac{A_2}{2\pi}\right).$$

The natural emission line-width is

$$\Delta\nu_{21}^N = \frac{\sum_j A_{1j} + \sum_i A_{2i}}{2\pi}. \tag{3.26}$$

Taking into account of the total decay rate $\gamma_{21}^T(= \gamma_1 + \gamma_2)$ describing the minimum emission line-width, the line-shape function is given by (Silfvast, 2004)

$$I(\omega) = I_0 \frac{\gamma_{21}^T/(2\pi)}{(\omega - \omega_0)^2 + (\gamma_{21}^T)^2/4}, \tag{3.27}$$

with $I_0 = \int_0^\infty I(\omega)d\omega$ and $I(\omega)$ as the intensity distribution per unit frequency.

To note, if lower state is ground state or an excited state having a long lifetime i.e., a meta stable state, one may take note of the energy width of upper state. The FWHM of the Lorentzian is given by, $\Delta\nu = 1/\tau_2 = A_{21}$. The line-shape function in terms of total decay rate, γ_{21}^T and frequency, ν, is written as

$$I(\nu) = I_0 \frac{\gamma_{21}^T/(4\pi^2)}{(\nu - \nu_0)^2 + \left[\gamma_{21}^T/(4\pi)\right]^2}. \tag{3.28}$$

Therefore including all frequencies, the rate equation from spontaneous emission turns out to be

$$\begin{aligned} \frac{dN_2}{dt} &= -N_2 A_{21} \int_0^\infty g(\nu) d\nu \\ &= -N_2 A_{21}. \end{aligned} \tag{3.29}$$

Here, $\int_0^\infty g(\nu)d\nu (= 1)$ is normalized to unity and $g(\nu)d\nu$ the probability of spontaneous emission in the frequency range ν and $\nu + d\nu$.

3.3.5.3 Collisional broadening

In a gas, the random collision occur between the atoms if they are very close to each other. The energy levels of the atoms change due to their mutual interaction leading to a Lorentzian line-shape function, which is given by

$$g(\omega) = \frac{\tau_0}{\pi} \cdot \frac{1}{(\omega - \omega_0)^2 \tau_0^2}, \tag{3.30}$$

where τ_0 is the mean time between collisions; the FWHM is $\Delta\omega = 2/\tau_0$.

The collisional broadening is a homogeneous broadening mechanism, which is due to the influences from outside. The collisions drives the de-excitation of the atoms from their excited state to the lower state. The increase decay rate due to the collisional interactions of the atoms contribute to the increase of the broadening of the emission line. The increased decay rate γ_2 of an excited level 2 due to collision is expressed as

$$\gamma_2 = \frac{1}{\tau_2} = \sum_i A_{2i} + \frac{1}{T_2'},$$

where T_2' refers to the decreased collisional decay time of excited level 2; a similar equation applying to the decay of the lower level 1 with a collisional decay time, T_1'' is

$$\gamma_1 = \frac{1}{\tau_1} = \sum_j A_{1j} + \frac{1}{T_1''}.$$

The total emission line-width due to natural and T_1' broadening is written as

$$\Delta\nu_{21} = \frac{1}{2\pi}\left[\left(\sum_i A_{2i} + \sum_j A_{1j}\right) + \frac{1}{T_2'} + \frac{1}{T_1''}\right]$$
$$= \Delta\nu_{21}^N + \frac{1}{2\pi}\left(\frac{1}{T_2'} + \frac{1}{T_1''}\right). \quad (3.31)$$

The aforementioned line broadening mechanisms are classified as:

Homogeneous broadening The natural and collision broadening act to broaden the response of each atom in an identical manner, which results in a Lorentzian line-shape. The excited atoms in crystals are generally homogeneously broadened because of its periodic structure. The de-phasing collisions involves a process interrupting the phase of the radiating atoms without destroying the population. Including the natural broadening, the collisional broadening and the broadening due to such collisions, the total homogeneous broadening (Silfvast, 2004) is

$$\Delta\nu_{21}^H = \Delta\nu_{21}^N + \frac{1}{2\pi}\left(\frac{1}{T_2'} + \frac{1}{T_1''} + \frac{2}{T_p}\right), \quad (3.32)$$

where T_p represents the average phase interruption time of excited state or lower state for outside disturbances and equal for both the levels.

In a homogeneously broadened laser, all the atoms contribute to the gain at all frequencies. As the population inversion is reduced, the gain is reduced at all frequencies. The laser would oscillate at the frequency corresponding to the longitudinal mode closest to the gain maximum; if the steady state is reached, one mode would oscillate.

Inhomogeneous broadening In this case, Doppler broadening or the broadening produced due to local inhomogeneities in a crystal lattice (see Section 2.2.1.1) act to shift the central frequency of the response of the individual atoms by different amounts, and thereby lead to an overall broadening of the response of the atomic system; the excited atoms in glasses are inhomogeneously broadened since the host appears to be different at each site. When the effects causing the inhomogeneous broadening are random in origin, the broadened line-shape is Gaussian.

Notably, atoms with slightly different numbers of neutrons within their nuclei exhibit small differences in energy level values. Such energy shifts cause a slight difference in central frequency, ν_0, of the emission lines for the different isotopes, which is known as an isotope shift. The laser gain is modified by a line-shape function, which is a convolution of homogeneous and inhomogeneous terms. Since, the value of $g(\nu_0)$ at the central frequency, ν_0 is approximately the reciprocal of the line-width, $\Delta\nu_0$ for both these types of broadening, the optical gain is inversely proportional to the line-width.

3.3.6 *Principal Components of a Laser*

The requisites of a laser system are as follows:

Active medium The active medium consists of a collection of atoms, molecules, or ions (in solid, liquid, or gaseous form) in a state of population inversion (see Section 3.3.4), which can amplify the input light beam (or spontaneously emitted light) by stimulated emission. The solid-state lasers (see Section 3.3.8) are based on ion-doped crystals or glasses, for instance, cobalt (Co^{3+}), neodymium (Nd^{3+}). The laser transitions commence and end at the inner energy levels of these ions. These levels are shielded from influences by the electrical fields of the host crystals; therefore, the transitions become small. A part of the input energy is absorbed by the active medium in which population inversion occurs at a certain stage. In order to have optical amplification, the medium has to be kept in a state of population inversion, which can be achieved by means of the pump.

Excitation source An excitation source provides energy in an appropriate form for pumping the atoms to higher energy levels. Depending on the external source, the excitation process is called optical pumping, in which the energy input is in the form of light energy (for instance, ruby laser) or electrical pumping, where an AC or DC electrical discharge is used for excitation (for example He–Ne laser).

Optical resonator Often referred to as laser cavity, the optical resonator is a medium with population inversion capable of amplification, which acts an oscillator, a part of the output energy must be fed back into the system. This feedback is brought about by placing the active medium in a resonator consisting of an active medium bound between two mirrors. In most lasers, it is configured as a Fairy-Perot interferometer having a pair of mirrors. These mirrors should be aligned perfectly to ensure that the light waves be reflected exactly along the optical axis. Hence, this axis is the direction in space into which light is emitted by the laser. Also, the spacing between the mirrors must be very small compared to the transverse dimension of the mirrors and is filled by an optically active material. One of the mirrors is completely reflective, while the other is partially transparent. The separation between the mirrors allows selecting a wavelength of radiation to bounce back and forth. The direction of travel of the photon is parallel to the axis of resonant cavity and the repeated bouncing of light back and forth increases the chances of more stimulated emissions. The electromagnetic energy is stored forming standing waves between the laser mirrors. Two types of waves, such as right and left moving waves, exist in such a medium, which interfere producing fringes. A plane wave traveling parallel to the axis from one of the mirrors undergoes diffraction as it reaches the second mirror, and since the mirror is of finite transverse dimension, the energy in the diffracted wave that lies outside the mirror would get lost. The wave reflected from the mirror would again undergo losses when it is reflected from the first mirror. Such a loss is referred to as the diffraction loss. The quality factor of an optical cavity is a quantity that measures the capability of the optical cavity to store electromagnetic energy inside. It is proportional to the ratio of the energy stored inside the standing wave and the energy that is lost from the wave during round trip between the laser mirrors. High quality factor value means that the energy is stored well inside the cavity, while low factor value implies that the energy is emitted from the cavity rapidly. Both the mirrors of the laser cavity should have high reflectivity in order to achieve high value for the quality factor. In order to express the value of the maximum intensity, $I(0)$, in terms of the full optical power of the beam, P, one may integrate $I(r)$ (see equation 3.2)

$$P = 2\pi I(0) \int_0^\infty e^{-2r^2/w^2} r dr = \frac{\pi w^2 I(0)}{2}, \quad (3.33)$$

and therefore

$$I(0) = \frac{2P}{\pi w^2}. \quad (3.34)$$

To note, the radiation coming out of the cavity is linearly polarized. However, although each avalanche of photons at any short time interval consists of photons of the same polarization, it may vary randomly if a longer period of time is involved.

3.3.7 *Laser Systems*

A great variety of lasers are available, which cover the spectral interval from ultraviolet, to mid-infrared. Follwing are the commonly used lasers:

3.3.7.1 Gas lasers

The gas lasers have a gas or a mixture of gases as their light-amplifying substance in a glass (or quartz) tube (Willett, 1974). The gas molecules are excited in an electric glow discharge. These lasers receive their energy input through collisions of gas atoms with high-energy electrons. This energy is provided by applying a high voltage between electrodes located within the gaseous medium to accelerate the electrons to the necessary high energies.

Helium-neon (He–Ne) laser The He–Ne laser the first gas laser constructed by Ali Javan and his associates. This was a continuous-light laser operating on the principle of converting electrical energy to a laser light output. An electrical discharge is given in helium contained in a discharge tube. The optimum gas pressure is determined by the product of the pressure and the tube-bore diameter, so that the optimum pressure is 3.6 to 4.9 torr-mm (1 torr = 0.00131579 atmosphere) with a He:Ne pressure ratio of 5:1

(Silfvast, 2004). A high-power incandescent lamp or flash lamp is used for population inversion. The discharge excites the helium atoms to their first excited level above their ground state. When helium atoms are moving through the lasing medium, the excited energy of the helium atoms is transferred to the neon atoms by resonance transfer due to collisions between them. These excited states decay radiatively to lower energy neon states giving rise to continuous laser emission at $\lambda \sim 632.8$ nm, which generally produce powers in the range of 0.5 to 50 mW in the red, with much lower power on the other region.

CO_2 laser The carbon dioxide (CO_2) lasers operate in both continuous and pulsed modes, and emit light in the 10.6 and 9.4 μm wavelength regions (Silfvast, 2004). It is a molecular gas discharge with a laser action taking place between two vibrational levels of the CO_2 gas. In this laser system, a mixture of CO_2, nitrogen (N_2), and helium gases, typically with CO_2:N_2 ratio of about 0.8:1, is taken in a gas discharge tube and excited electrically with power supply. The transitions may occur between the vibrational states of the same electronic state by achieving population inversion between these states. The nitrogen molecules are excited by collisions with electrons to their first excited vibrational state from which the excitation is resonantly transferred by molecular collisions to excite CO_2 molecules to a particular vibrational state. These molecules, in turn undergo radiative transition to the lower vibrational levels giving rise to laser radiation. The helium gas helps to populate the upper state and assists to empty the lower state of CO_2. The CO_2 laser is one of the powerful lasers, which produces power greater than 100 kW.

Excimer laser Arising from the contraction of the phrase "excited dimer," the excimer lasers use reactive gases, such as fluorine (Fl), chlorine (Cl), bromine (Br), iodine (I) and mixed with inert gases, namely, argon (Ar), krypton (Kr), or xenon (Xe). Excimers are molecules that exist only in the excited state. The population of the ground state of the molecule is always zero, and thus it is relatively easy to obtain population inversion and get high efficiency. Excitation occurs by electron dissociation and ionization of the gas molecule to produce argon, krypton or xenon ions. These ions react with the halide molecules pulling off one of the atoms of that molecule to create an excited diatomic molecule. The excimers are pulsed lasers, which lase in the ultraviolet region. These lasers are commonly used in the production of micro-electronic devices, for example, integrated circuits (IC), and also in eye surgery for medical diagnostics, etc.

3.3.7.2 Dye lasers

The dye lasers use complex organic dyes, for example, rhodamine 6G, in liquid solution or suspension as lasing media, which are tunable over a broad range of wavelengths. Such lasers are an ideal four-level system, and a given dye would lase over a wavelength range of 30–40 nm. The dyes have very broad emission and gain spectrum, which lead to tunable laser output and short-pulse laser output. An important property of dye lasers is the broad emission spectrum (typically 30–60 nm) over which the gain occurs, and combined with a diffraction grating or a prism as the cavity mirrors, the output can be in a very narrow frequency beam ($\sim$10 GHz). The typical dye concentration is a 10^{-4} to 10^{-4} molar solution. There are three main types of dye lasers (Silfvast, 2004), for instance:

(1) **Pulsed dye lasers**: These are pumped by other lasers, such as excimer or frequency-multiplied neodymium doped yttrium-aluminum garnet (Nd:YAG) lasers, which can produce output pulses up to 50–100 millijoules (MJ).
(2) **Continuous wave dye lasers**: Such lasers are pumped by flash lamps producing output up to a few watts.
(3) **Mode-locked dye lasers**: These are capable of producing pulses of 200 fs without prisms or gratings in the cavity.

3.3.7.3 Optically pumped solid-state lasers

The solid-state lasers have lasing material distributed in a solid matrix, in which the population inversions are brought out by implanting impurities (which provide the laser action) within a host material, such as a crystal or a glass and exciting them with a suitable light. The impurity concentration is usually in the range of 0.01 to 3.0%. In most of the lasers, the impurities are in the form of ions in which the energy states are shielded from the surrounding atoms so that the energy levels are narrow.

Ruby laser Historically the first successful demonstration of a laser device was by T. Maiman (1960) using a ruby crystal, which is referred to as ruby laser. Ruby is a crystal of aluminum oxide (Al_2O_3), in which some aluminum ions (Al^{3+}) are replaced by chromium (Cr^{3+}) ions; 0.05% Cr_2O_5 is threshold value of doping. This laser consists of a ruby rod xenon flash tube, a suitable cavity to reflect the light from flash tube to the ruby rod, and a high voltage power supply to give electrical energy to the flash tube. It gives laser radiation on a pulsed length (1 nm). The light from the xenon flash tube was focused by the cylindrical cavity onto the ruby rod, thereby exciting the chromium atoms that were responsible for the laser action. The ruby laser is basically a three-Level system since three energy levels are involved in the process of stimulated emission. The atoms are excited from the ground state (energy level E_1) to an upper state (E_3) through stimulated absorption. Basically E_2 (a meta stable state where the atoms stay for a longer interval of time) and E_3 are very close to each other. From the energy level E_3, the atoms are transferred to energy level E_2 without emitting radiation (non-radiative transfer). The atoms return to the ground state from the meta stable level through the process of stimulated emission giving rise to an intense light at wavelength $\lambda \sim 0.6943$ μm. The laser beam comes out in the form of a pulse of very short duration (about a millisecond).

Nd:YAG laser In this laser, Nd^{3+} ions are present as substitutional impurities on lattice (see Section 2.2.1.1) sites being occupied by yttrium ions; hence, YAG acts as host. The YAG is formed from a mixed oxide system having a composition of $Y_3Al_5O_{12}$. It has the advantage of having a relatively high thermal conductivity to dissipate the heat generated, thus allowing the crystals to be operated at high repetition rates of the order of 100 Hz (Silfvast, 2004). The optical pumping is carried out by the krypton arc lamp or xenon lamp. Typically a Nd:YAG-laser emits light at 1064 nm, in the infrared.

3.3.8 *Semiconductor Diode Laser*

The semiconductor lasers (Kneubühl and Sigrist, 1999) are P-N junction (see Section 2.2.7) devices that convert the electrical energy applied across the junction into optical radiation; the wavelength of the output radiation depends on the energy gap across the P-N junction.

3.3.8.1 Light-emitting diode

A light-emitting diode (LED) transforms electrical energy directly into optical energy (Brennan, 2003). It comprises a semiconductor P-N junction with metallic contact at the ends for the applications of external voltage, whose light output is generated by the spontaneous emission (see Section 3.3.2) due to radiative recombination (see Section 2.2.4.1) of electrons and holes in the active region of the diode under forward bias. At the junction boundary on either side of this diode, the excess minority carriers recombine with the majority carriers and the energy is released in the form of photons. As the forward current is increased, the intensity of emitted light is also increased. In LED, the light energy is transmitted out through the sides of the junction.

An LED emits incoherent, non-directional, and unpolarized spontaneous photons. It starts emitting light as soon as an injection current flows across the junction. The surface emitting LEDs approximate a Lambertian source (see Section 1.2.3). These have a large beam divergence and a radiation pattern approximate a sphere; therefore, most of their total output is not coupled into optical fibers.

The linear dependence of the LED output power on the injected current is valid if the current is less than a certain value, for instance, a few tens of milliamperes on a typical LED. The corresponding wavelengths are present in the visible and IR. However, the spectral properties of monochromatic LEDs are inferior to the laser diodes; the bandwidth of LEDs is $\approx$30–50 nm. The spontaneous emission (see Section 3.3.2) of light due to the radiative recombination from within the diode structure is known as electro-luminescence, the term is used when the optical emission results from the application of an electric field. The amount of radiative recombination and the emission area within the structure depends on the semiconductor materials used and the fabrication of the device.

3.3.8.2 Laser diode

A laser diode is very similar to an LED in structure. However, as compared to an LED in which the output has many phases and is dispersed in different directions, the output from a laser diode is highly coherent and collimated. The first laser diodes were developed in the early 1960s. The laser diodes are used in a wide variety of applications, for example, CD players, CD-ROM drives, laser disk and other optical storage drives, laser printers, bar code scanners, and typesetters. Generally, a laser diode is made of a direct bandgap semiconductor (see Section 2.2.3.1) material, namely, gallium arsenide (GaAs) or other semiconductor substrates with similar optical properties, having a small resonator located in its central part.

Hetero-junction (see Section 5.3.3) concept can be applied in fabricating the laser diode. Two of the materials, $GaAs/Al_xGa_{1-x}As$ (aluminum gallium arsenide) and indium gallium arsenide phosphide, $In_{1-x}Ga_xAs_{1-y}P_y/InP$ (indium phosphide) are used for such a purpose. The active region in the former case is $Al_xGa_{1-x}As$, and depending on the compositional mixing ratio x and its doping, the resulting laser emits at $0.78\ \mu m < \lambda < 0.87\ \mu m$, while in the latter case, the laser diode emits in the $1.1\ \mu m < \lambda < 1.65\ \mu m$ spectral region depending on x and y. The region in the vicinity of the 1.55 μm is favorable since optical fibers with losses as small as $0.15\ dB.km^{-1}$ at this wavelength are available, making it desirable for long distance high data rate optical communication.

When the P-N junction (see Section 2.2.7) is connected to a DC supply in a forward bias, it causes the carriers to cross the junction resulting in the electron–hole recombination, which generates photons with a wavelength proportional to the bandgap energy (see Section 2.2.3.1) at the junction. Hence, the majority carriers are injected across the depletion region in both directions to create a population inversion (see Section 3.3.4) in a narrow active region. The light produced by radiative recombination across the bandgap is confined to this active region. As stated in Section 2.2.4.1, the incident photon causes the radiative recombination; the associated energy emits as a photon of light. This photon, as it propagates, stimulates the recombination of another excited electron with a hole and release another photon. These two photons with the same characteristics are in phase with each other, thus travel together, get reflected at the end face. On their way back, they stimulate more electron–hole recombination with the release of additional photons of same λ and become a part of monochromatic coherent laser beam. This beam gets resonated by traveling back and forth, and finally leaves through the partially reflecting face.

As the current is being passed continuously, more electrons get excited, move to conduction band and new holes generate in the valence band. This maintains the population inversion, while the recombination of electron–hole pairs continues with the generation of laser beam. It may be reiterated that if a P-N junction is forward biased, it raises the electrons from the valence band to the conduction band, but this is an unstable state and within a short span of time ($\sim 10^{-13}$ s), the electrons drop to the lowest level in that band. At the same time, the electrons near the top of the valence band would drop to the lowest unoccupied levels, leaving behind holes. This means that the lowest level of conduction band is full of electrons and the top of the valence band is full of holes, which is the indication for population inversion.

The resonant cavity required for energy amplification, is formed by the front and back faces of the semiconductor material. The cleaned surface is kept perfectly flat and parallel. The back face is made fully reflecting and the front face is made partially reflecting. Although a number of layers are added to the semiconductor, the size of the laser diode is kept to a minimum by arranging the layers in a structure called, quantum well (see Section 5.5.1.1). The main advantage of such a structure is the improvement of the lasing action due to the variation of the energy level distribution of the semiconductor material when its thickness is reduced considerably, for instance, below 20 nm. The high concentration of charge carriers in the narrow strip of the active material reduces the threshold current required to initiate lasing. The high bandgap layers on either side of the active layers perform confinement of charge carriers to the active materials, as well as the confinement of the laser output to a narrow width. An internal photodiode (see Section 5.3.1) can also be used in the feedback loop to stabilize the output power from the laser. The photocurrent (see Section 2.1) is fed to the circuit that drives the laser diode. In semiconductor lasers, the junction forms the active medium, and the reflection end of the laser material provide feedback.

3.4 Fiber-Optics

Optical fibers are transparent glass or plastic fibers, which can carry high speed signals over long distances without employing any repeaters; the propagation is based on the total internal reflection (Tyndall, 1970). In the early 1840s, D. Colladon and J. Babinet demonstrated the usability of such technology. In 1880, A. G. Bell and S. Tainter invented the photophone to transmit voice signals over an optical beam. Although it was an advanced form of telecommunications, it suffered from the atmospheric interferences and became impractical. Modern optical fibers, where the glass fiber is coated with a transparent cladding to offer a more suitable refractive index, appeared in late 1930s (Bates, 2001). Development then focused on fiber bundles for image transmission (Hopkins and Kapany, 1954). In 1963, Nishizawa proposed the use of optical fibers for communications (Nishizawa and Suto, 2004). Kao and Hockham (1966) proposed that the attenuation in optical fibers could be reduced below 20 dB.km^{-1}, making fibers a practical communication medium. They suggested that the silica glass with high purity could be used for fiber, for which the former was awarded the Nobel Prize in Physics in 2009. Kapron et al. (1970) were able to produce silica fibers with a loss of about 17 dB.km^{-1} at a wavelength of 633 nm. By 1985, glass fibers were produced with extremely low losses ($<$0.25 dB.km^{-1}).

3.4.1 *Principle of Operation*

A fiber-optic cable, also known as optical waveguide, is composed of two concentric layers, called the core ($\approx$10 μm in diameter) made

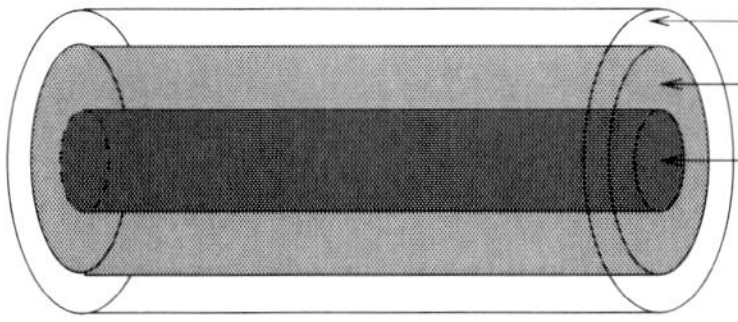

Figure 3.8 Schematic diagram of fiber cross section.

of glass (or plastic) of refractive index, n_1 and the cladding made up of similar material but of slightly lesser refractive index, n_2. The cladding (see Figure 3.8) is enclosed in a polyurethane (a polymer consisting of a series of organic units corroborated by urethane links) jacket that safeguards the fiber against chemical reaction, abrasion, and crushing. One or more of these strands make up the cable as a whole. If the light signal, which enters the cylindrical dielectric core, strikes the interface of the core and cladding at angles of incidence larger than the critical angle, θ_c, it would undergo total internal reflection. Due to symmetry of the structure, the ray undergoes repeated total internal reflection, and hence remains confined within the core. The angle of incidence, θ_i, for which the angle of refraction, θ_r is 90°, is called critical angle, θ_c. When the θ_i exceeds θ_c, there is no refracted ray, one obtains the total internal reflection. From Snell's law one finds, $n_i \sin\theta_i = n_r \sin\theta_r$, therefore, the critical angle is given by

$$\theta_c = \sin^{-1}\left(\frac{n_r}{n_i}\right).$$

in which n_r and n_i are the higher and lower refractive indices, respectively; for example, in the case of glass-air interface, where $n_r = 1.5$ and $n_i = 1.0$, the critical angle turns out to be $\theta_c \sin^{-1}(1.0/1.5) = 41.8^\circ$.

The light that is introduced into the fiber with an angle within the confines of the numerical aperture of the fiber, is transmitted by multiple internal reflections off the interface between the two glasses, provided that its angle of incidence exceeds the critical angle

$$\theta_c = \sin^{-1}\sqrt{n_1^2 - n_2^2}. \tag{3.35}$$

A few parameters are required to be derived, for instance:

3.4.1.1 Numerical aperture (NA)

One of the characteristic parameters of an optical fiber is called the numerical aperture representing light-gathering capability. It is closely related to the critical angle, and is often used in the specification for optical fiber and the components that work with it. Let n_0 be the refractive index of the surrounding medium of the fiber. By applying Snell's law, one derives for refraction at the point of entry of the ray into the core as

$$n_0 \sin\theta_0 = n_1 \sin\theta_i,$$

from which it is expressed as

$$\begin{aligned} \sin\theta_0 &= \frac{n_1}{n_0}\sin\theta_i \\ &= \frac{n_1}{n_0}\sqrt{1-\cos^2\theta_i} \qquad \cos\theta_i = n_2/n_1 \\ &= \frac{n_1}{n_0}\sqrt{1-\frac{n_2^2}{n_1^2}}, \end{aligned} \tag{3.36}$$

in which θ_0 is the waveguide acceptance angle when a ray enters into the core of the fiber-optics and therefore the numerical aperture is given by

$$\mathrm{NA} = \sin\theta_0 = \frac{\sqrt{n_1^2 - n_2^2}}{n_0}. \tag{3.37}$$

If $n_0 = 1$ assuming the medium surrounding the fiber to be air

$$\mathrm{NA} = \sqrt{n_1^2 - n_2^2}. \tag{3.38}$$

The total internal reflection occurs when the angle of incident ray is less than θ_0, i.e., $\sin\theta_i < \theta_0$, which means

$$\sin\theta_i < \sin\theta_0 < \mathrm{NA}.$$

This is the condition for propagation. The minimum focal ratio F_{min} that can be transmitted by the core is dictated by $\sqrt{1-\mathrm{NA}}/2\mathrm{NA}$.

3.4.1.2 Relative core-cladding index difference, Δ

This is defined as the ratio of the refractive index difference between the core and cladding to the refractive index of core of an optional fiber

$$\Delta = \frac{n_1^2 - n_2^2}{2n_1^2}. \tag{3.39}$$

When $n_1 \sim n_2$, i.e., $\Delta \ll 1$, which is true for telecom grade fibers

$$\Delta = \frac{n_1 - n_2}{n_1}\frac{n_1 + n_2}{2n_1} \approx \frac{n_1 - n_2}{n_2} \approx \frac{n_1 - n_2}{n_1}. \tag{3.40}$$

The numerical aperture (NA) is derived from equations (3.37) and (3.40) as

$$\text{NA} = \sqrt{2n_1^2\Delta} = n_1\sqrt{2\Delta}. \tag{3.41}$$

3.4.2 *Types of Fibers*

Depending on the refractive index profile and the number of modes[a] that the fiber can guide, the optical fibers are classified as

Single-mode fiber A single-mode fiber has a small core diameter of 8 to 10 μm, i.e., a few times the operating wavelength in question and overall fiber diameter is standardized as 125 μm. The refractive index of core and cladding is uniform and $n_1 > n_2$. There is a step increase in the value of refractive index from cladding to core (Bass et al., 1995); therefore, its profile takes the shape of a step (see Figure 3.9a).

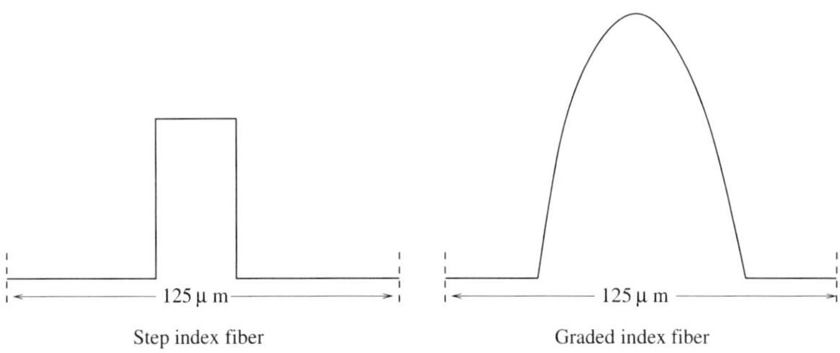

(a) (b)

Figure 3.9 Profiles of (a) step-index fiber and (b) graded index fiber.

[a]Every path a light beam can follow through the fiber is referred to as a mode.

Multi-mode fiber The geometry of multi-mode fiber is similar to that of a single-mode fiber. However, its core has a much larger diameter (62.5 μm in diameter) by virtue of which it is able to support propagation of large number of modes.[a] Different rays take different times to propagate through a given length of the fiber. The number of modes supported for propagation in the fiber is determined through a dimensionless (normalized) frequency parameter, called V-parameter, which is given by

$$V = \frac{\pi D}{\lambda}\sqrt{n_1^2 - n_2^2} = \frac{\pi D}{\lambda} NA, \tag{3.42}$$

where D is the core diameter.

For single-mode fiber, $V \leq 2.405$, while for the multi-mode fiber, the number of modes, $N = V^2/2$ (for step-index fiber) and $N = V^2/4$ (for graded index fiber). The step-index profile fibers exist as single- and multi-mode fibers. They have uniform, albeit different refractive indices in core and cladding. The refractive index profile of a step-index fiber could be similar to that of a single-mode fiber except for its larger core. The core of a graded index multi-mode fiber is such that its refractive index decreases in the radially outward direction from the axis and becomes equal to that of the cladding at the interface. Its index profile has a parabolic shape (see Figure 3.9b), and light is guided along the due to refraction and self-focusing. The refractive index of the cladding remains uniform. This is achieved by using a varying level of dopants in silica across the core diameter so as to yield a parabolic index profile. This light traveling down the center of the fiber experiences a higher refractive index than light that travels further out towards the cladding. Thus, the off-axis propagating light effectively covers shorter paths than on-axis rays covering longer path; the light thus follows a curved trajectory within the fiber. The aim of this is to keep the speed of propagation of light on each path the same with respect to the fiber axis. Thus, energy of an incident signal pulse carried via different mode of the fiber take same amount of time to reach end of the fiber. Figures 3.10a and 3.10b depict the refractive index profiles (RIP) of a

[a]A set of guided electromagnetic waves is called the modes of an optical fiber; a mode by definition is a transverse electromagnetic field distribution, which does not change with propagation except for the phase.

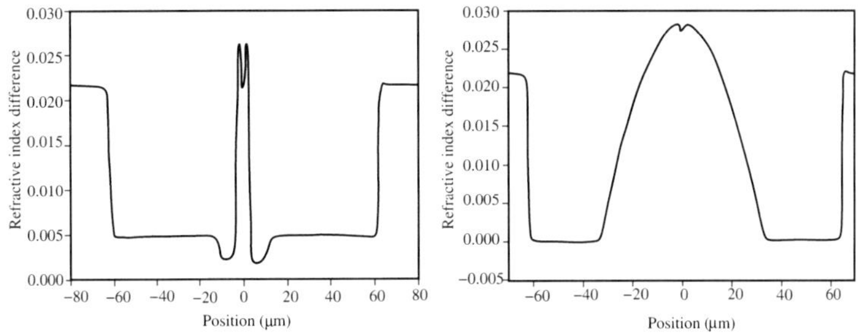

Figure 3.10 Experimental results of refractive index profile (RIP) of (a) a depressed clad single-mode optical fiber; the fiber diameter: 125 μm, core diameter: 7.89 μm, numerical aperture: 0.29, core glass composition: $SiO_2 + F$, $n_{SiO_2} \approx 1.45625$ at 632 nm, and (b) multi-mode optical fiber; the fiber diameter: 125 μm, core diameter: 62.7 μm, numerical aperture: 0.29, core glass composition: $SiO_2 + GeO_2$, $n_{SiO_2} \approx 1.45625$ at 632 nm. Courtesy: S. Bhadra.

depressed clad single-mode optical fiber, as well as of a multi-mode optical fiber, respectively, taken at the Center for Glass and Ceramic Research Institute (CGCRI), Kolkata, India.

Fibers require a laser or an LED as a source of light. However, LEDs are, in general, considered to be unsuitable for injecting light into a single-mode fiber since they emit light over much wider angles than the acceptance cone semi-angle of the fiber. In multi-mode fibers, the intermodal dispersion arises, which is due to different modes traveling with different velocities. The single mode fibers transmit near-IR light in the wavelength range between 1300 and 1500 nm, while the multi-mode fibers are capable of transmitting near-IR light in the wavelength range between 850 and 1300 nm. Some optical fibers, made from plastic, have a large core of about 1 mm in diameter, which transmit visible light at $\lambda = 650$ nm).

3.4.3 *Fiber Laser*

The lasers with a doped fiber as the gain medium are referred to as fiber lasers. In most cases, the gain medium is an optical fiber doped with a rare-earth element, such as erbium (Er^{3+}), neodymium (Nd^{3+}, ytterbium (Yb^{3+}), thulium (Tm^{3+}), or praseodymium (Pr^{3+}).

If such a doped optical fiber is kept between two mirrors, which act as a resonator, with an appropriate pump, one can obtain a fiber laser. The first fiber laser, barium crown glass doped with Nd^{3+} ions, was fabricated by Snitzer (1961), which was transversely pumped by coiling the fiber around a flash-lamp. Later end, pumping was demonstrated making pumping with semiconductor lasers a feasible technique. In more recent times, a precursor to fiber laser has been an Er^{3+}-doped fiber amplifier, whose operating wavelength falls in the range of 1530$\sim$1565 nm. It consists of a semiconductor laser diode (see Section 3.3.8.2) pump laser, a wavelength-selective coupler that allows the pump wavelength to enter the fiber transmission system without disturbing the signal. The Er^{3+}-doped fiber amplifier can be spliced with the fiber signal. An optical isolator placed at the output of the amplifier prevents it from lasing. The laser cavity in fiber lasers is constructed monolithically by fusion splicing different types of fiber. To note, the fiber Bragg gratings, made either directly in the doped fiber, or in an undoped fiber that is spliced to the active fiber, can replace conventional dielectric mirrors to provide optical feedback. A better power-handling capability can be achieved by collimating the light exiting the fiber with a lens and reflecting it back with a dielectric mirror; the intensities on the mirror can be reduced. However, slight misalignment can cause substantial reflection losses, and the additional Fresnel reflection at the fiber end may introduce filter effects. A fiber loop mirror based on a fiber coupler with 50:50 splitting ratio can also be used. The laser diode continually pumps the fiber and as pulses pass through the system they are amplified at the expense of depletion of pump light.

The Er^{3+}-doped fiber laser is useful in long distance communications over fiber-optic networks, particularly in undersea communication links through which amplification of the optical signal is achieved directly by inserting pieces of erbium fiber at appropriate locations in the fiber networks (Silfvast, 2004). Thus, the optically transmitted signal obtains a direct optical amplification as it travels from one continent to other. These amplifiers are also being used in ultrashort pulse production. The fiber lasers, exhibiting high vibrational stability and extended lifetime, may be the possible replacements for distributed feedback semiconductor lasers. Furthermore, the fiber lasers find applications in welding, cutting, drilling, as well as in medical surgery. However, the fiber lasers can suffer from various problems, such as (i) non-linear optical effects, (ii) temperature-dependent polarization evolution, (iii) design difficulties due to strong saturation effects caused by the high optical intensities, (iv) the quasi-three level behavior of nearly all fiber laser transitions, and (v) the complicated pulse formation mechanisms in mode-locked fiber lasers.

3.4.4 *Applications of Optical Fibers*

A spool of fibers, which can be stretched or relaxed to increase or decrease the optical path length, carries information over light waves traveling inside due to the properties of the fiber materials, similar to the light traveling in free space. The light waves are unaffected by other electromagnetic radiation. These fibers are electrically non-conductive, so the information traveling inside is immune to electromagnetic interference. The other advantages of fiber can be envisaged in the form of (i) a large mode area fiber for high throughput, (ii) offering flexibility since the degrees of freedom are located at its entrance and output, (iii) selecting the plane-wave part of a wavefront, (iv) splitting a guided wave into any desired intensity ratio, and (v) combining two guided waves interferometrically.

The fibers are widely used for telecommunication because of their high efficiency in carrying light over long distances ($\sim$100 km). The broadband communication is also probable meaning that audio/video signal and data from a system can be modulated over light carrier wave and demodulated by optical receiver at the other end. Fibers are capable of sensing various physical parameters and generating information. The fiber-optic sensors measure the physical quantity based on its modulation on the intensity, spectrum, polarization, or phase of light traveling through the fiber. The advantages of such sensors are compact size, low cost, multi-functional, accurate and reliability, remote accessibility, and greater flexibility. Unlike electrical sensors, these are immune to electromagnetic interference, and resistant to hazardous and explosive environments. These sensors are broadly classified as intrinsic and extrinsic. In the former case, the physical parameters to be sensed directly alters the properties of the optical fiber, which in turn, leads to changes in a characteristic of light propagating in the fiber. In the case of the latter, an optical fiber, a multi-mode in general, acts as a device to transmit modulated light from an external sensing element connected to an optical transmitter. The sensing element responds to the external perturbation and the change in the characteristics of the sensing element is transmitted by the return fiber for analysis.

3.4.4.1 Fibers used at telescopes

Fibers are also employed to connect telescopes to spectroscopes, which eliminate the need for mirrors or lenses, and alignment required for these elements. Unlike Coudé conventional scheme, where about seven or more reflections cause loss of about 10% light at each reflection are required to bring light from the prime focus to the laboratory, the fiber transmits light with reduced loss of light. The spectrometer, which measures the distribution of brightness (or energy) as a function of wavelength, is, in general, kept on a stable flatform in a temperature and humidity controlled isolated Coudé laboratory that is away from the telescope. The problems of flexure that occur within telescope-mounted spectroscopes can also be reduced substantially as gravitational loads change with the telescope orientations.

Optical fibers are used in imaging optics as well. If the arrangement of the fibers in the bundle is kept constant, the transmitted light forms a mosaic image of the light that struck the end of the bundle. If the relative positions of the fibers in the input and output ends are the same, the bundle is said to be coherent bundle, while in the case of having non-aligned, the bundle is known to form an incoherent bundle. A bright (or dark) spot at the input end of the fiber produces a bright (or dark) spot at the output end. Thus, an image is transmitted in the form of bright (or dark) spots through a coherent bundle of fibers.

3.4.4.2 Fibers used for interferometry

Most of the research performed in recent years is devoted to the development of efficient fibers and waveguides for the infrared. Certain interferometers use fibers, but of very limited length for the beam combining system, so they are not affected by chromatic dispersion or birefringences.

IOTA The Infrared Optical Telescope Array (IOTA; Carleton et al., 1994), was built at Mt. Hopkins, Arizona, USA, which consisted of three 0.45 m collector assembly movable among 17 stations along two orthogonal linear arms. A beam recombiner instrument, called Fiber-Linked Unit for Optical Recombination (FLUOR; Coudé du Foresto and Ridgway, 1992) used fiber, however, fiber combiners are very sensitive to external environmental conditions like temperature and mechanical stresses. This interferometer is closed.

VLTI The Very Large Telescope Interferometer (VLTI), located at a height of 2635 m in Cerro Paranal, Chile, was built by the European Southern Observatory (ESO). It consists of a system of four 8.2 m fixed separate telescopes (unit telescopes; UT). Each of them is a Ritchey-Chretien type and can operate in Cassegrain, Nasmyth, and Coudé focus. These telescopes are accompanied by four mobile Auxiliary Telescopes (AT) of 1.8 m diameter. The complete array is rotated by Earth rotation (Glindemann, 2011, and references therein). This array allows access to baselines between 47 m and 130 m with the UTs, and between 8 m and 202 m with the ATs. The best resolution can be obtained for the large telescopes is 1.5 mas and for the auxiliary telescopes is 1 milliarcsecond (mas). Fibers are used at the following beam combiner instrument of VLTI.

VINCI This instrument, called Very Large Telescope INterferometer Commissioning Instrument (Kervella et al., 2000) is a single-mode optical fiber recombiner, which coherently combines light from two telescopes in the IR K-band (2.2 μm), where the fringe pattern is obtained by temporal modulation of the optical path difference. Light is fed into two single-mode fibers, which converts phase errors across the telescope pupil into amplitude fluctuations in the fiber. Fiber coupler, usually formed through fusion and stretching of two single-mode fibers over a short length, acts as either a beam splitter or a beam combiner for coaxial beam combination. At a coupler, the distance between the cores is a few microns (for fibers that are single-mode in K-band). Due to smallness of the stretched fiber core, the two fibers optically interact through evanescent field, in which light launched into one fiber gets leaked to the other. If the length of interaction is sufficient, leaked optical power could return to fiber 1. If 50% of the launched power remains in the original fiber, it is called a 3 dB power splitter/coupler. It is a reciprocal device.

IONIC An instrument called, Integrated Optics Near-IR Combiner (IONIC; LeBouquin et al., 2004), is used at the VLTI using two auxiliary telescopes to permit operation in the H-band, which is capable of measuring closure phase. Light from each telescope is focused into a single-mode fiber, and three fibers are aligned using a silicon V-groove array mated to the planar waveguides on the integrated optics (IO) device (see Figure 3.11). The optical circuit acts to split the light from each telescope before recombining each telescope pair at three IO couplers. This pair-wise combination scheme leads to six interferometric channels (two for each baselines). The optical path difference (OPD) between the two light beams is modulated to sweep through the fringes, which appears as temporal modulation.

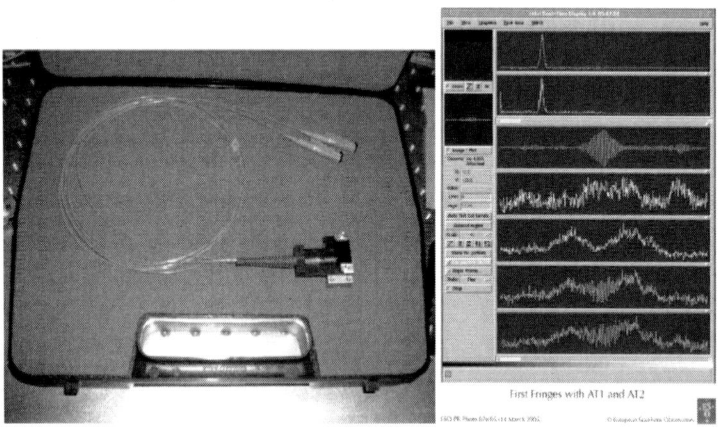

Figure 3.11 Integrated optics beam combiner (IONIC2T-K) at the VLTI (left) using two auxiliary telescopes (LeBouquin et al., 2006). Courtesy: F. Malbet.

A fringe packet centroiding algorithm is applied during observations for removing the instrumental OPD drift.

3.4.5 *Drawback of fibers*

The disadvantage of transmission by optical fibers is the signal attenuation and dispersion that it introduces with propagation and the need for repeaters to reformat and amplify the optical signals after long distances. The degree of attenuation over distance varies according to the wavelength of the light and to the composition of the fiber. The attenuation coefficient, α, can be derived by using Lambert's law, which states that the rate of decrease of intensity of light with distance, L traveled in a homogeneous medium is proportional to the initial power

$$P_{\rm out} = P_{\rm in}e^{-\alpha' L}, \tag{3.43}$$

where α' denotes the fiber loss coefficient in units of nepers.km^{-1}.

In dB, the loss coefficient is expressed as

$$\alpha = \frac{-10}{L}\log_{10}\left[\frac{P_{\rm out}}{P_{\rm in}}\right] \qquad {\rm dB.km}^{-1}, \tag{3.44}$$

in which $P_{\rm in}$ is the input power, while the output power is $P_{\rm out}$, and L the length of the fiber expressed in km; from equations (3.43 and 3.44), it is possible to relate α to α' as $\alpha = 4.343\alpha'$.

The principal sources of attenuation in a fiber waveguide can be classified into two groups, such as absorptive and radiative. The term, α, is generally expressed in terms of its components:

$$\alpha = B + C(\lambda) + E(\lambda) + \frac{A}{\lambda^4}. \tag{3.45}$$

The discussions on each component of the loss spectrum in equation (3.45) are enumerated below.

3.4.5.1 Absorption loss

As stated in Section 1.5.1, an electromagnetic wave propagating in a medium experiences energy loss as part of the signal energy is converted to other forms of energy. The absorption occurs in a fiber-optic when the light beam is partially absorbed by functional groups, namely, SiO, OH^-, etc., which remove energy from the propagation beam. The mode coupling that occurs when a part of the light energy moves back and forth between the core and the cladding and exchange of energy occurs between the core and cladding also lead to the loss of radiation. The absorption loss may be divided into intrinsic and extrinsic losses.

The intrinsic loss may be caused by interaction of the propagation light wave with one or more components of the fiber materials. Such an interaction may eventually lead to quantum transitions between different electronic, as well as vibrational energy levels of the fiber materials. The second term of equation (3.45), namely $C(\lambda)$ represents this loss. Since the bandgap energy (see Section 2.2.3.1) in pure fused silica (SiO_2) is ~8.9 eV, the peak absorption of light due to electronic transition occurs at λ ~0.14 μm in the ultraviolet (UV) region of the electromagnetic spectrum. It is reiterated that the absorption occurs when a photon interacts with an electron and excites it to a higher energy level (see Section 3.3.1). The main cause of intrinsic absorption in the infrared (IR) region is the characteristic vibration frequency of atomic bonds. The vibrational absorption band in SiO_2 is centered at λ ~9.2 μm in the IR-region, with the possibility of weaker absorption bands occurring at $\lambda \approx 3.2$, 3.8, and 4.4 μm due to anharmonic vibrations of silicon-oxygen (Si-O) bonds (Pal, 1992). The interaction between the vibrating bond and the electromagnetic field of the optical signal causes intrinsic absorption. Light energy is transferred from the electromagnetic field to the bond. The absorption bands in the UV and IR-regions decay exponentially resulting in absorption tails that may spill over into the neighboring wavelengths.

The photons are also absorbed by the impurities in the silica glass of which the fiber is made. The extrinsic absorption, represented by the term $E(\lambda)$ in equation (3.45), is caused by the impurities introduced into the fiber material during its manufacture. The presence of tiny traces (in parts per million) of transition metal ions, such as chromium (Cr), cobalt (Co), copper (Cu), iron (Fe), manganese (Mn), and Vanadium (V), etc., as also presence of water, in the form of OH^- ions, dissolved in glass, is noteworthy. The OH^- ion vibrates at a frequency corresponding to 2.7 μm. During the propagation of the signal, the electrons (impurities) absorb the photons and get excited to higher energy levels. Subsequently, these electrons give up their absorbed energy as heat (or light) energy. This energy, in general, is at a different wavelength or phase, which is not useful.

3.4.5.2 Radiative loss

The radiative loss occurs if a part or whole of the guided optical energy is lost through radiation from the fiber.

Scattering loss The scattering (see Section 1.5.2) losses also take place due to defects in the fibers (crystallized region, trapped gas bubbles). The most dominant source of inherent radiation loss in a fiber is due to the Rayleigh scattering (see Section 1.5.2). It is caused by the small scale compositional and density fluctuations that gets frozen into glass lattice at the glass softening point during melting and subsequent cooling. The resulting inhomogeneities lead to attenuation that varies with wavelength λ^{-4}; the light wavelengths toward the blue region suffer greater scattering out of the fiber than those of the red region. Typically, the coefficient, A in equation (3.45) is approximately 0.8 dB-μ m^4.km^{-1} in single mode fibers. The losses in single mode fibers are less than in multi-mode fibers because the relative core-cladding index difference, Δ (see Section 3.4.1.2), is relatively large in the latter. The radiative loss may also be caused by intrinsic fiber imperfections (or deformations) like core-cladding interface irregularities and diameter fluctuations and so on. This explains the presence of the term B in equation (3.45). In the fiber, Mie scattering (see Section 1.5.2) is caused by the imperfections of a size comparable with the wavelength. However, this is not significant concern with modern fibers as recent improvements in manufacturing techniques take care of it.

Bending loss The bending loss is induced by the physical stress on the fiber. During the propagation of the signal in such a fiber, the photon may be scattered because of the sharp changes in refractive index values inside the glass over distances that are small compared to the wavelength of light. Bending occurs in two forms, such as

(1) **Macroscopic bending loss**: Macroscopic bending refers to large-scale bending, which arises when wrapping the fiber on

a spool or during cabling and laying of the cable. The bend curvature creates an angle that is too sharp for the light to be reflected back into the core, and some of it escapes through the fiber cladding, causing attenuation. The light wave suffers loss due to radiation of the evanescent field in the cladding region. As the radius of the curvature decreases, the optical power loss increases exponentially until it reaches at a certain critical radius. Fibers with less core diameter, as well as fibers having larger Δ are more tolerant to sharp bends in terms of fractional power lost due to bending.

(2) **Microscopic bending loss**: Microbends are microscopic distortions along the length of a fiber. The loss occurs due to non-uniformities in the manufacturing of the fiber core by non-uniform lateral pressures created during the cabling of the fiber. If the fiber lay is made to go through a continuous succession of very small bends, the fiber may exhibit a significant rise in attenuation, what is referred to as microbending loss; microbends cause irregular reflections. The power is dissipated through the microbent fiber, because of the repetitive coupling of energy between guided modes and the leaky or radiation modes in the fiber.

Both macro- and micro-bending features create a change in the refractive index locally leading to light getting away from the core into the cladding. In micro-bending, the changes in refractive index occur over minuscule volumes, and hence losses are lesser that caused by macro-bending. In the case of sliced fibers, if they are not joint uniformly, such losses arises. Furthermore, if bundles of fibers inside a jacket are not parallel, this lead to light striking members of a bundle at different angles. Some of the angles fall within the acceptance cone and some outside, thus losses occur, albeit small in magnitude.

3.4.5.3 Dispersion

Dispersion is a phenomenon in which the velocity of propagation of a electromagnetic wave is wavelength dependent. It describes a process by which any electromagnetic signal propagating in a physical medium is degraded because the various wave characteristics (frequencies) of the signal have different propagation velocities within the physical medium. The dispersion in a fiber occurs when the light traveling down a fiber-optic cable, spreads out and becomes longer in wavelength and eventually dissipates. The dispersion coefficient depends on the refractive indices of both the core and the cladding of the fiber.

It is to be noted that for transmission system operation, the group velocity, v_g, which the signal information and energy is traveling down the fiber, is less than the speed of light in the medium. The v_g is related to the phase velocity of a wave in a given medium, $v_p(= c/n)$, in which c is the speed of light in free space and n the refractive index of the medium

$$v_g = c\left(n - \lambda\frac{dn}{d\lambda}\right)^{-1}.$$

In general, the refractive index is a function of the frequency, ν, i.e., $n = n(\nu)$. The group velocity is a function of the frequency of the wave, which results in group velocity dispersion. This causes a short pulse of light to spread in time as a result of different frequency components of the pulse traveling at different velocities. For a wave propagating in a fiber, the phase velocity is given by

$$v_p = \frac{\omega}{\beta},$$

in which $\beta[= 2\pi n(\lambda)/\lambda]$ is the propagation constant in the medium, $\omega(= 2\pi\nu)$ the angular frequency, and ν the frequency.

Hence, the group velocity is expressed as

$$v_g = \frac{d\omega}{d\beta}.$$

The observable delay experienced by the optical signal waveform and energy, when traveling a particular length along the fiber is called group delay

$$\tau_g = \frac{1}{v_g} = \frac{d\beta}{d\omega}.$$

The output is proportional to the delayed version of the input signal, and the delay is inversely proportional to v_g of the wave. Since $\beta(\omega)$ is frequency dependent over bandwidth, $\Delta\omega$, sitting at the center frequency, ω_0, at each frequency, one gets one propagation constant resulting in a specific delay time. The group delay dispersion coefficient, D, is defined as the delay difference per unit length per unit wavelength as

$$D = \frac{1}{L}\frac{d\tau_g}{d\lambda} = -\frac{2\pi c}{\lambda^2}\frac{d^2\beta}{d\omega^2}$$

$$\approx -\frac{\lambda}{c}\frac{d^2n}{d\lambda^2} \quad \text{ps per nm.km.} \tag{3.46}$$

As the output signal is collectively represented by group velocity and group delay, this phenomenon is called intramodal dispersion. This phenomenon arises due to a finite bandwidth of the optical source, dependency of refractive index on the wavelength and the modal dependency of the group velocity. This causes velocity differences among the spectral components of the 1ight source used. A measure of information capacity of an optical fiber for digital transmission is, in general, specified by the product, $\Delta\nu \times L$. The group delay, τ_g, per unit length L is

$$\tau_g = -\frac{\lambda^2 L}{2\pi c}\frac{d\beta}{d\lambda}. \tag{3.47}$$

The dispersion can be categorized into the following types (Kasap, 1999):

Material dispersion The material dispersion leads to chromatic aberration in an optical element, such as a lens. It causes a marked decrease in transmitted power due to the mismatch of the different fibers when light is made to pass through them. It is reiterated that the waves in the guide with different free space wavelengths travel at different group velocities. The waves arrive at the end of the fiber at different times resulting in a broadened output pulse. The refractive index of the material varies as a function of wavelength, $n(\lambda)$. The material induced dispersion for a plane wave propagation in homogeneous medium of refractive index, n is

$$\tau_m = L\frac{d\beta}{d\omega} = \frac{L}{c}\left[n - \lambda\frac{dn}{d\lambda}\right]. \tag{3.48}$$

Therefore, the pulse spread due to material dispersion is

$$\sigma_m = \frac{L\sigma_\lambda}{c}\left|-\lambda\frac{d^2n}{d\lambda^2}\right|$$

$$= L\sigma_\lambda|D_m(\lambda)|, \tag{3.49}$$

where $D_m(\lambda)$ is the material dispersion.

Waveguide dispersion This dispersion is caused by the wavelength dependence of distribution of energy for the fundamental mode in the fiber. It is mainly a problem for single-mode fiber. As the wavelength increases an increasing proportion of the mode energy propagates in the cladding. Since the refractive index of cladding is lower, for a spectral width, σ_λ, the time delay dispersion develops. An important quantity for characterizing an optical fiber is the normalized waveguide parameter, V, which is defined by

$$\begin{aligned} V &= \kappa a\sqrt{n_1^2 - n_2^2} \\ &= \frac{2\pi}{\lambda} a\sqrt{n_1^2 - n_2^2} \approx \kappa a n_2\sqrt{2\Delta}. \end{aligned} \tag{3.50}$$

Δ is the relative core-cladding index difference (see Section 3.4.1.2).

It is convenient to define the dimensionless propagation constant, b as

$$b = \frac{(\beta^2/\kappa^2) - n^2}{n_1^2 - n_2^2} \approx \frac{(\beta/\kappa) - n_2}{n_1 - n_2}. \tag{3.51}$$

The delay time due to waveguide dispersion is expressed as

$$\tau_{\mathrm{w}} = \frac{L}{c}\left[n_2 + n_2\Delta\frac{d(Vb)}{dV}\right]. \tag{3.52}$$

The pulse spread is approximated as

$$\sigma_{\mathrm{w}} = \frac{n_2 L\Delta\sigma_\lambda V}{c\lambda}\frac{d^2(Vb)}{dV^2}. \tag{3.53}$$

The total dispersion is given by the sum of material and waveguide dispersions, i.e., $D_{\mathrm{tot}} = D_{\mathrm{m}} + D_{\mathrm{w}}$.

The mode dispersion, as well as polarization dispersion may also be taken into account. The former is caused by the frequency dependence of the propagating constant of that mode. As the magnitude of mode dispersion is very small, it would be important in single-mode fibers operated near the zero material dispersion region. The latter is caused by the cylindrical asymmetry due to manufacturing, temperature, bends, etc., that lead to birefringence introducing elliptic polarization at the output, when linearly polarized light is injected.

3.4.5.4 Losses due to miscellaneous sources

In addition to the losses mentioned above, there are two more sources of loss in any fiber-optic system. These are

Source-to-fiber coupling loss The light emission pattern from a source is approximated as

$$I(\theta) = I_0(\cos\theta)^m, \tag{3.54}$$

where $I(\theta)$ is the radiant intensity of each ray leaving the surface at an angle, θ, from the surface normal, usually expressed in watts per square meter per steradian, i.e., W.m^{-2}.sr^{-1}, I_0 the intensity of the ray leaving in a direction perpendicular to the surface, and m the directionality of the source radiation pattern; for $m = 1$, the source would correspond to a Lambertian source (see Section 1.2.3).

If a source is directly butted against the fiber, the optical power coupling efficiency, η, is determined by the ratio of power coupled into the fiber, P_1, to the power emitted by the source, P_o, i.e., $\eta = P_1/P_o$. The amount of optical power launched into an optical fiber depends on the radiance (see Section 1.2.1) of the optical source and also on the optical properties of the fiber. If an intervening optics, such as a lens is introduced between the source and the fiber to enhance power coupling efficiency, it would increase over that of the butt coupling case by a factor, $M[= (d_1/d_o)^2]$, in which d_1 and d_o represent the diameters of the fiber and the source, respectively. To note, an intervening optics can enhance the coupling efficiency from a Lambertian source if the source emission area times its emission solid angle is less than the corresponding product for the fiber core. On the contrary, if this product for the source is greater than that for the fiber, most efficient coupling would be a direct butting of the source to the fiber.

Fiber-fiber joint and splice loss Optical fibers are joined either by fusion in an electric arc or by splice. However, before joining them, the opposed cores must be properly aligned, which should be carried out under a microscope. The fiber-to-fiber joint loss is due to:

- change in refractive index from the fiber to the air,
- misalignment at a fiber splice/joint, e.g., the axes of two fibers could be offset in the transverse direction, there could be longitudinal separation and /angular misalignment between them,
- core diameter and numerical aperture mismatch, which has an effect on intrinsic coupling loss,

- reflection loss, called Fresnel reflection, which occurs at every fiber interface caused by a step change in the refractive index that occurs at the fiber joint. The step change in refractive index is caused by the ends of each fiber being separated by a small gap (an air gap). In Fresnel reflection, a small portion of the incident light is reflected back into the source fiber at the fiber interface.

3.5 Light Sources and Illumination Systems

A typical light source emits electromagnetic radiation in the visible spectrum. A perfect optical system produces an image with the same étendue as the source. Étendue (determined by the entrance slit) is the area of the source times the solid angle the system's entrance pupil subtends as seen from the source. It is a constant of the system, which can be viewed as the maximum beam size the instrument can accept. Thermal radiation sources, like incandescent lamps that produce light by means of a filament heated to incandescence by the flow of electric current through it or Nernst rods, have been used in many optical systems, primarily for imaging. The quartz tungsten halogen (QTH) lamps and infrared (IR) emitters are the improved version of the older sources. The former lamp emits at all wavelengths, but transparency of the QTH envelop limits the emission to visible and near-IR ($\approx$2.5 μm). This lamp has an electrically heated tungsten filament positioned inside a transparent bubble made of fused silica and filled with halogen gas. The gas causes a chemical reaction between the tungsten atoms evaporated from the filament and deposited on the bubble wall and the halogen molecules improves the lifetime of the lamp, as well as the transparency of the QTH envelope.

3.5.1 *Lens-Based Light Source*

In order to obtain the light beam from a point source, similar to a star in the sky, an artificial star image can be developed by placing a pair of condensing lenses along with micron-sized pinhole in front of the source (see Figure 3.12). The beam is collimated

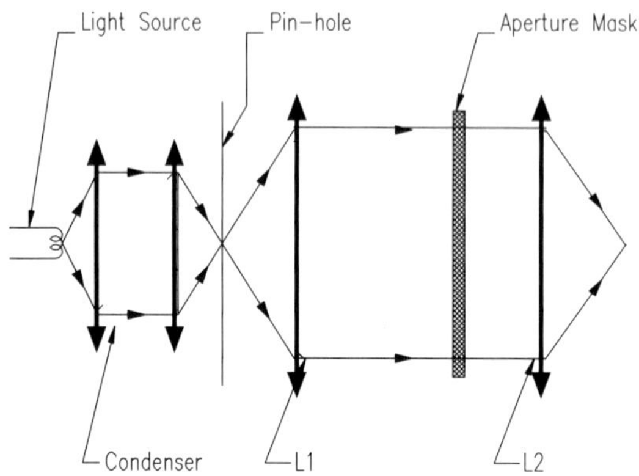

Figure 3.12 Lens-based illuminating system.

with a achromatic lens L_1, which transfers the image of the light source into the entrance pupil of the imaging lens, L_2, which builds the image of the object at the focus. The source in this case is a simple arc-lamp. A laser diode can be employed as source; in such a case, the condensing lens should be replaced with a microscope objective. An aperture mask comprising of several holes arranged in a circular mode at a regular interval is placed in the path of the collimating beam before the imaging lens in order to study the aperture synthesis by Saha et al. (1988).

The aperture synthesis imaging technique with telescope involves observing an object through a masked aperture of several holes and recording the resulting interference pattern in a series of short-exposures. Such a mask introduces a series of overlapping two hole interference patterns projected onto the detector allowing the Fourier amplitude and phase of each baseline can be recovered. By summing up the Fourier phases around each closed triangle, the closure-phase can be obtained. It is a hybrid techniques (speckle interferometry with non-redundant pupils), which allows diffraction-limited imaging from ground-based telescopes.

This concept has been applied by several groups on single telescope (see Saha, 2007, and references therein). In this method, a pattern of holes with sizes $\leq r_0$ in diameter is cut into a plate and placed in the pupil plane of a large telescope. Such a mask introduces a series of overlapping two hole interference patterns on the image plane. The Fourier amplitude and the phase of each baseline can be recovered from the interference fringes. The interference patterns, in a series of short-exposure, contain information about structure of the object at the spatial frequencies from which an image of the same can be reconstructed by measuring the visibility amplitudes and closure phases.

In the presence of these pair of subapertures, the total amplitude and phase of the spectral component of image intensity, I, obtained in the focal plane of the system is the result of addition of all such fringes with frequencies, taking proper account of both amplitudes and their spatial phases. For a large telescope, the aperture, P, can be subdivided into a set of subapertures, p_j. Each pair of them, p_n and p_{m}, separated by a distance (baseline) B would form fringes. The intensity in the focal plane of the telescope, I, according to the diffraction theory (Born and Wolf, 1984), is determined by the expression

$$\begin{aligned} I &= \sum_{n,m} \langle U_n U_{\mathrm{m}}^* \rangle = \left| \sum_{n,m} U_n \right|^2 \\ &= \sum_{n,m} |U_n|^2 + \sum_{n \neq m} \sum_{\mathrm{m}} U_n U_{\mathrm{m}}^*. \end{aligned} \tag{3.55}$$

The term, $U_n U_{\mathrm{m}}^*$, is multiplied by $e^{i\psi}$, where, ψ is the random instantaneous shift in the fringe pattern. The complex disturbance, U, of the field in the telescope plane is obtained by adding the contribution, U_n, of each subaperture. Each subaperture is small enough for the field to be coherent over its extent. The first term on the RHS of equation (3.55) is the sum of the irradiance produced by each subaperture, which does not contain high-resolution information, while the second term that describes the interference through the cross product, contains high-resolution information. The average values of the latter measures the coherence of the wavefront. This expression implies that an image can be reconstructed from sequential measurements of all cross products using pairs of subapertures. For n such apertures, there are $n(n-1)$ independent baselines, with $n-1$ unknown phase errors.

At a point source $O(\boldsymbol{\alpha}) = \delta(\boldsymbol{\alpha} - \boldsymbol{\alpha}_0)$, the mutual intensity of the light, $\mu(\mathbf{u}) = 1$ for all spatial frequencies and the visibility of the fringe pattern is maximum for all baselines, $\mathcal{V} = 1$. The coherent addition of all the fringes provides a sharp concentration at the image position $\boldsymbol{\alpha}_0$. In a circular pupil, Airy pattern is obtained. Since a pair of subapertures produces a fringe pattern, the image produced by the pupil, P, is the sum of fringe patterns, which is the Fourier expansion. Each coherence term is a Fourier component of the expansion. This relationship is known as van Cittert–Zernike theorem, which states that the image of an object is the Fourier transform of the measured visibility or the cross-correlation between the various subapertures.

3.5.2 *Laser as a Light Source*

A light amplifier gets converted to an oscillators, i.e., a laser when a feedback is provided as in case of an electronic oscillator. A laser can be used as a source for testing the focal point of a lens or for aligning the mirrors of any telescope or an interferometer. It can also be used for conducting Ronchi test that is employed for testing of optical surfaces, as well as the aberration of the optical system. If a grating is placed near the center of curvature of a mirror (see Figure 3.13), the image is superimposed on the grating itself, providing a kind of Moiré pattern. Since the shape of these patterns depends on the aberrations of the mirrors, Ronchi applied this phenomenon to the quality testing of the mirrors. Ronchi grating is a specific grating having $d = 2b$, in which d represents the spacing of the grating

Figure 3.13 An experimental setup of Ronchi test using laser as source, in which L_1 represents the collimating lens, L_2 the lens under test, MO the microscope objective, f_0, f_1, f_2 the respective focal lengths of microscope objective, L_1, and L_2.

and b the width of the slit. The spherical aberration is the on-axis aberration, while coma, astigmatism are the off-axis aberrations. The Ronchigrams due to the off-axis aberrations can be obtained by rotating the sample lens in the x–y-plane and rotating the Ronchi grating in y–z-plane.

Holography Another field that attracted attention after the invention of laser was that of the holography, a method that records the amplitude, as well as phase of the light waves; this is accomplished by employing interferometric technique. If the waves leave a scattering medium with uniform amplitudes, they become non-uniform rapidly as the waves propagate. Due to interference of waves from several scattering centers in the aperture the randomly dephased impulse response functions are added, yielding in a speckle and its size is governed by Airy's disk. These speckles were first noticed while reconstructing an image from a hologram that the intensity and directional information of an optical wavefront (Gabor, 1948).

Let a rough surface, illuminated by a laser light, be imaged on the recording plane by placing a lens in the speckle field of the object (see Figure 3.14a). The image appears as random intensity

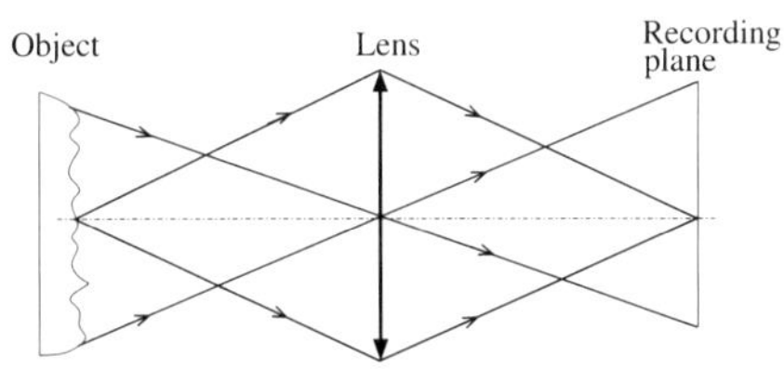

(a)

(b)

Figure 3.14 (a) Schematic diagram for illustrating the formation of speckle pattern and (b) a laboratory setup with laser as a source.

variations. Indeed, every imaging system alters the coherent superposition of waves and hence produces its own speckle size. The fringe separation, θ, is given by, $\theta = 1.22\lambda b/D$, in which b is the distance between the lens and the image plane, and D the diameter of the lens. In terms of aperture speed, $F_{\#}$, of the lens, and magnification, $\mathrm{M}(= b/a)$, with a as the object distance, the speckle size is written as, $\theta = 1.22(1 + \mathrm{M})\lambda F_{\#}$.

In order to record a hologram, a coherent beam from a laser source (see Figure 3.14b) is made to reflect from an object and combined with light from a reference beam at the photographic film. In this process, a series of interference fringes are produced forming a type of diffraction pattern on the photographic film. This recorded interference pattern forms the hologram and contains information about the amplitude and phase of the object wave. By illuminating such a hologram with the same reference beam, diffraction from the fringe pattern on the hologram reconstructs the original object beam. The reconstruction process leads to a virtual and a real image of the object scene.

Wavefront tilt A distorted wavefront can also be generated at the laboratory with a laser source. Figures (3.15a and 3.15b) depict a plane wavefront offering zero volt to the tip-tilt mirror and the wavefront tilt measured with the same source after applying one volt to the said mirror, respectively. These images are grabbed by a CMOS imager-based Shack–Hartman (SH) sensor.

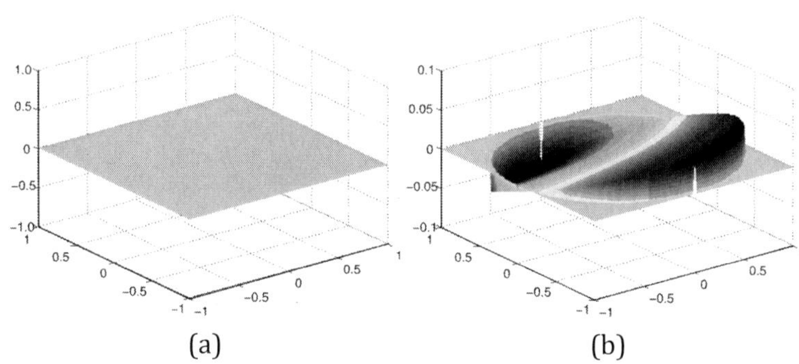

Figure 3.15 Laboratory measurement of wavefront tilt; (a) a plane wavefront and (b) the wavefront tilt.

These plane and tilted wavefronts resemble to the wavefronts arriving to a detector from a distant star before and after passing through the turbulence of the atmosphere, respectively. The laboratory experiment shows only tilt as a major error, while in the case of atmosphere, the wavefront will have complicated contours. Nevertheless, a reverse situation can be created by employing the adaptive optics (AO) systems in order to improve the throughput of the large telescope.

Wavefront correction A laser can be also be used as a source to obtain a corrected image through a telescope in the laboratory. Figure 3.16 displays the images captured by a Peltier cooled electron multiplying CCD (EMCCD; see Section 7.7.1) camera with 10 ms

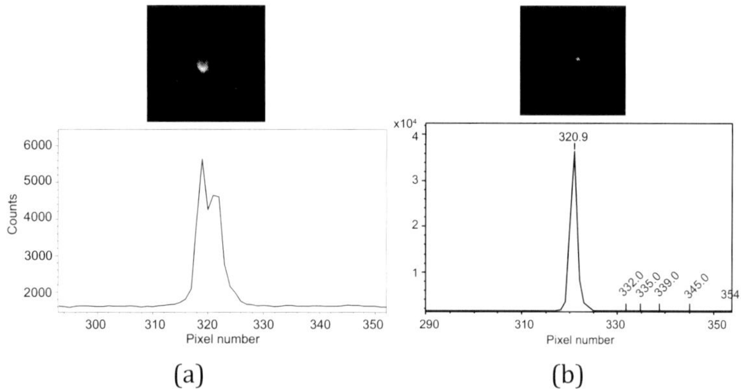

Figure 3.16 (a) Un-corrected image (left panel:top) point source taken with a Cassegrain telescope and its cross section (bottom), (b) corrected image (right panel: top) of the said source with a tip-tilt mirror for tilt error correction and other high frequency errors with a MMDM, and its cross section (bottom); images are twice magnified for better visibility (Courtesy: V. Chinnappan).

exposure time in the laboratory set up using a micro-machined deformable mirror (MMDM). It is found that an aberrated image having 6.4 pixel[a] FWHM can be sharpened to have 3.5 pixel and

[a]A pixel (short for PICture ELement) is a small detecting unit containing a photosensitive area on a solid-state image sensor that takes an image and directly convert it to a digital image. These elements are separated by insulating fixed walls and having no electronic connection together constitute an image on a display

the peak intensity has increased from 5610 counts to 36,500 counts (Chinnappan, 2006).

3.5.3 *Laser Interferometer Gravitational-wave Observatory (LIGO)*

In a classic paper, Einstein (1905b) introduced the concept of special theory of relativity by completing Maxwell's electrodynamics (see Section 1.1). His subsequent work of the general theory on relativity (Einstein, 1916) led to the search for gravitational waves that is defined as the disturbances in a gravitational field triggered by cosmic events, which may cause specific displacements resulting in interference patterns. Such waves travel with the speed of light but with very small amplitudes so they need very sensitive detector for their detection. Key to such detection is the accurate measurement of small changes in distance. The effect of a gravitational wave is an apparent strain in space, transverse to the direction of propagation causing a differential change of the pathlength between the arms of the interferometer, thereby introducing a phase-shift. This shift is due to the changes in the distances between pairs of mirrors that are kept mutually perpendicular in vacuum chambers. Such wave passing through the instrument shortens one arm, while lengthening the other. This relative change in length of the two arms can be measured. One may experience a tiny amplitude-relative oscillations due to gravitational waves.

Like in a classical Michelson interferometer (see Section 3.2.1.1), the laser beacon is divided into two beams by a beam-splitter, oriented at 45° with the direction of the beam. These are reflected at mirrors M_1 and M_2 and return to the beam splitter, where they are recombined before entering the detector via a focusing device. Another passive system can be used by replacing the mirrors by

---

screen. Different optical sensors have different number of such elements, which may be arranged in a single line, what is referred to as line scan or in multiple rows and columns (area scan). There are some defective pixels, which fail to sense the light levels accurately. A dead pixel never displays light, although some stuck pixel always displays light (typically red, blue, or green) on the LCD (liquid crystal display) when the device is active. The size of a typical pixel is very small, say $\sim$5 μm square and the number of electrons per pixels is nearly 1×10^5 to 5×10^5. A pixel may contain the light-sensitive photodiode and a variety of other circuitry.

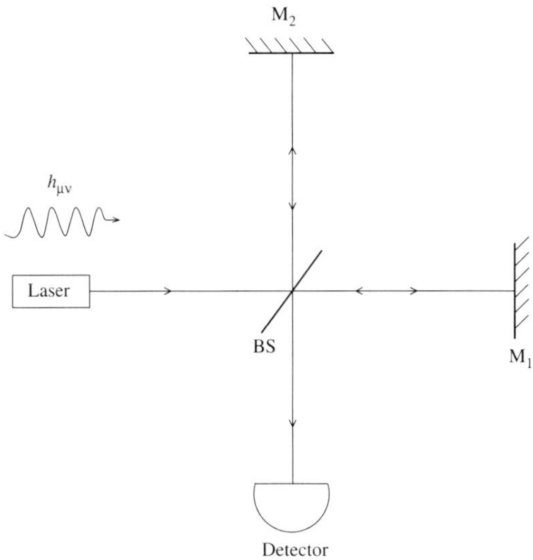

Figure 3.17 Schematic diagram of the LIGO interferometer.

Fabry–Perot cavities in both arms in order to fold the light several times, which lengthens the effective optical pathlengths in each arm.

Advances in modern laser optics technology have made it possible to develop extremely sensitive, kilometer-scale gravity wave detectors. Several ground-based gravity-wave detectors developed. The LIGO instrument is based on the principle of Michelson classical interferometer (see Figure 3.17). It comprises two facilities at widely separated sites, viz., the LIGO Livingston Observatory in Livingston, Louisiana and the LIGO Hanford Observatory, on the Hanford Nuclear Reservation in USA. Both house a 4 km L-shaped interferometer (Sigg, 2006). The interferometer at each site consists of mirrors suspended at each of the corners of the "L". A prestabilized laser emits a beam of up to 35 W that passes through an optical mode cleaner before reaching a beam splitter at the vertex of the "L." The beam splits into two paths, one for each arm, which contains Fabry–Perot cavities that store the beams and increase the effective pathlength.

3.5.4 *Laser Guide Star*

In the field of high-resolution imaging, laser is being used to create a sufficiently bright reference star for monitoring the atmospheric aberrations. Such a system is required in the absence of the natural guide star. Many observatories are currently implementing such an artificial star to monitor the atmospheric aberrations for measuring the wavefront errors, as well as for mapping the phase on the entrance pupil. A laser can produce light from three reflections, such as Mie scattering (see Section 1.5.2), Rayleigh scattering (see Section 1.5.2), and resonance scattering (Foy and Labeyrie, 1985). Unlike Mie scattering by aerosol or cirrus clouds, which may be significant at lower altitudes, but are usually variable and transient, scattering of the upward propagating laser beam is due to Rayleigh scattering; its strength depends on the atmospheric density. Since the density decreases with altitude, this limits the strength of the backscatter at high altitudes. The main drawback is the inadequate sampling of the atmospheric turbulence due to the lower height of the backscatter. The resonance scattering can occur at the layer in the Earth's mesosphere containing alkali metals, for instance, sodium (Na; 10^3 to 10^4 atoms cm^{-3}), potassium (P), calcium (Ca), at an altitude of 90 to 105 km.

In order to produce backscatter light from Na atoms in the mesosphere, a laser is tuned to a specific atomic transition. Sodium atoms scatter beacon strongly from NaD_2 resonance line at 589.2 nm and NaD_1 resonance line at 589.6 nm. The sodium atom absorbs a photon, making the electrons jump to the first energy orbital above ground state, which is followed by the return of the atom to its ground state, accompanied by the emission of single photon. The probability of transition in the former line is higher than that of the latter line. The high altitude of the process makes it suitable for astronomical imaging since it is closer to sampling the same atmospheric turbulence that a starlight from infinity comes through. Figure 3.18 depicts an AO image of θ^1 Ori B with a faint companion, which was acquired with adaptive secondary mirror at the 6.5 m Multi Mirror Telescope (MMT), Mt. Hopkins Observatory, Arizona, USA.

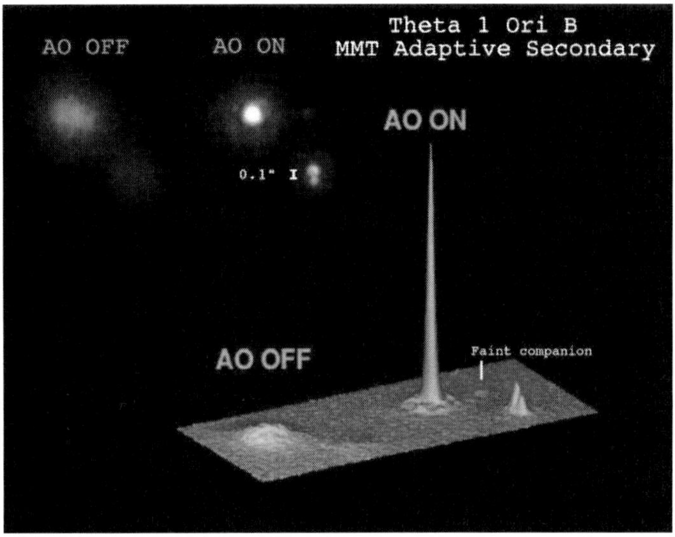

Figure 3.18 AO image of Θ 1 Ori B; without AO, this object appears to be two stars, but with AO turned on it is revealed that the lower star is a close binary having separated by 0.1 arcsec; the brighter one is a laser guide star, and the fainter one slightly to the right (see arrow) is a very faint companion (Courtesy: L. Close).

With a poor beam divergence quality laser, the telescope's primary mirror can be used as an element of the laser projection system, while with a diffraction-limited laser, projection system can be side-mounted and bore-sighted to the telescope (Tyson, 2000). The beam is focused onto a narrow spot at high-altitude in the atmosphere in the direction of the scientific target. Light is scattered back to telescope from high altitude atmospheric turbulence, which is observed and used to estimate wavefront phase distortions. In principle, the LGS should descend through the same patch of turbulence as the target object. Nunez et al. (2014) suggested a modified LGS using subsets of apertures to create an array of artificial stars in the sodium layer and collecting back-scattered light with the same subapertures, which is suitable for a large diluted apertures.

Chapter 4

Photon Detection Process

4.1 Radiation Detectors

A detector is a device that is capable of transforming the energy distribution into a flow of interpretable information. It senses an optical signal within its field-of-view, regardless of its angle of incidence, and converting it into an electrical signal (current or voltage). The process involved here is the quantum process, which is statistical. Given a certain electric field, the number of charges N released is statistically related to the exposure W (wave power multiplied by time) by Einstein's relation, $W = \bar{N}\hbar\omega$, with N as an integer having mean value $\bar{N}$, and obeying Poissonian statistical distribution. The probability of certain number of photons, N_{ph}, being detected in a certain time interval, $\mathcal{P}(N_{\text{ph}})$, in which the average number of photons detected is $\bar{N}_{\text{ph}}$, is given by

$$\mathcal{P}(N_{\text{ph}}) = \frac{(\bar{N}_{\text{ph}})^{N_{\text{ph}}}}{N_{\text{ph}}!}\left(e^{-\bar{N}_{\text{ph}}}\right), \tag{4.1}$$

where the variance for Poisson's distribution is equal to its mean

$$\langle\sigma\rangle^2_{\text{ph}} = \overline{(N_{\text{ph}} - \bar{N}_{\text{ph}})^2} = \bar{N}_{\text{ph}}.$$

As stated in Section 2.1, an incoming photon with energy greater than the bandgap energy (E_g) is absorbed within the semiconductor

High-Resolution Imaging: Detectors and Applications
Swapan K. Saha

ISBN 978-981-4613-27-9 (Hardcover), 978-981-4613-28-6 (eBook)
www.panstanford.com

material. As a result, an electron–hole pair (see Section 2.2.4.1) is created in the detector material, which carries an electric current, called "photocurrent". The probability of converting light into electron–hole pairs depends on the material, the wavelength of the light, and the geometry of the detector.

Several techniques have been developed to convert a single electron into a measurable pulse of multiple electrons. Depending on the wavelength of the radiation, there are incoherent and coherent methods for radiation detection. The incoherent detectors are energy detectors in which current or voltage is generated in direct proportion to received power. This technique is used for detecting higher energy photons (UV, X-ray and Gamma-rays). The particle nature of the photons is utilized for this method, where the incoming radiation is made to interact with the matter. The coherent detectors, in general, optimize frequency and phase in their detection.

Electro-optical detectors These devices, such as single electro-optical cells, photomultipliers (see Section 5.2.2), are based on the photoelectric effect (see Section 2.1), i.e., the emission in the vacuum of an electron from a photocathode by a photon. A photocathode is a very thin light sensitive layer deposited on the inside of the image intensifier tube (see Section 5.2.6) input window that converts the photons into electrons and release them into the vacuum of the tube. All modern such tubes operate under a vacuum of about 10^{-9} to 10^{-10} torr, which is essential to protect the photocathode from oxidation and rapid destruction. However, such devices have several limitations (Morel and Saha, 2005):

(1) The quantum efficiency (see Section 4.1.1.3) is rather low (around 0.1 for multi-alkali photocathodes, 0.2 for gallium arsenide; GaAs photocathodes). It also tends to decrease with time due to the interaction of residual gas molecules with the photocathode (Decker, 1969).
(2) They present false counts due to thermionic emission (random electron emission from the photocathode due to the temperature). Electron emission may also be due to chemical interaction of residual gas molecules with the photocathode (Geiger, 1955).
(3) Residual gas molecules in the tube may be ionized by an electron. In this case, the positive ion will hit the photocathode

and liberate several electrons. This phenomenon called "ion return" causes artifacts in the image that are noticed by bright spots (brighter than those caused by a photoevent).

(4) Their constructions require a high vacuum, implying a fragility of the devices. Also very high voltage for power supply are required for operating an image intensifier, causing problems of electrical insulation.

Solid-state detectors These detectors, viz., photovoltaic (PV) detectors (see Section 5.3), photoconductive (PC) detectors (see Section 5.4) are considered to be the square law detectors since the intensity is proportional to the square of the electric field vector (see Section 1.2); the human eye (see Section 4.2.1) and photographic film (see Section 1.4.2) also exhibit square-law response.

Thermal detectors In these detectors, the incident radiation absorbed by the crystal lattice (see Section 2.2.1.1) leads to a temperature change, which, in turn, alters a physical or electrical property of the detector elements. The thermal detectors have a broad spectral response but are often used for infrared (IR) and sub-mm detection (for more details see Section 8.3).

Coherent detectors The coherent detectors, in which the electric field of the wave is sensed directly and phase information can be preserved, take the advantage of wave-interaction (see Section 4.3.1.4) with a wave with a locally produced field, either before or after conversion of the electromagnetic radiation to an electrical signal. The most common examples are heterodyne detectors (see Section 4.3.1.4), in which a local oscillator (LO) is used at the receiver to down convert the carrier to an intermediate frequency (IF) carrier. The modulation format may include (i) amplitude modulation, (ii) frequency modulation, and (iii) phase modulation. These detectors are used for detecting radio and infrared signals.

4.1.1 *Figure of Merit*

The figure of merit is a quantitative measure of the quality of performance of a device or system. The performance of a detector, which may be described in terms of a specific functional behavior of a combination of a few of the devices or systems in question, depending on their applications or uses, relies on the dark current (see Section 4.1.1.8) of the device and is due to the generation of electron–hole pairs by the effect of temperature, as well as the arrival of photons. Some light can be transmitted through the slab after partial absorption in the material. The intensity of light, I, at any point in the device is given by

$$I = I_0 e^{-\alpha_m x}, \tag{4.2}$$

where I_0 denotes the light intensity at the top surface, x the distance into the material, $\alpha_m (= 4\pi\alpha/\lambda)$ the absorption coefficient of the material (typically in cm^{-1}), which is wavelength dependent, α the extinction coefficient.

The absorption coefficient decreases for longer wavelengths, i.e., the long wavelength photons are absorbed deeper into the semiconductor material than the short wavelengths. Equation (4.2) shows that the intensity of light decreases exponentially throughout the material. The rate of generation provides the number of electrons generated at each point in the device due to the absorption of photons. On differentiating equation (4.2), the generation at any point in the device can be deduced as

$$G_r = \alpha_m N_{ph0} e^{-\alpha_m x}, \tag{4.3}$$

where N_{ph0} is the photon flux at the surface.

Equation (4.3) shows that the generation is highest at the surface of the material. The resolution of a sensor is limited by the lens focal length, pixel size, and detector center-to-center spacing. Both the array size and the spatial resolution together may be gauged by the astronomical capability metric (Bahcall et al., 1991), which is proportional to the number of pixels multiplied by the square of the sensitivity (the output signal in millivolts, mV, per integrated light value in lumen-seconds, lm.s) per pixel. The minimum detectable optical power defines the sensitivity of photodetector; typically, it is given as equivalent electrons RMS. The sensitivity of a photocathode corresponds to the illumination of a blackbody (see Section 1.1.1) at 2856 K. The flux of the emitted electrons is, in general, expressed in microampère (μA) per lumen (see Section 1.2.2). Another way of expressing sensitivity is to use the radiance (see Section 1.2.1), corresponding to illumination by a monochromatic source of specific wavelength. The unit used in this case is often milli-ampère (mA) per watt. This metric indicates the number of positions on the sky that is measured in a given time to a given detection limit. Optical sensors are commonly described by a number of different figures of merit (Dereniak and Boreman, 1996; Rogalski, 2011) that are described below in brief.

4.1.1.1 Spectral bandwidth

The light emitted by an atom in a small spectral band is not strictly monochromatic, but is made up of wave-trains of finite length. A large number of such wave-trains pass at random time intervals during the time necessary to make observations. The coherence of two interfering beams is linked with the duration, called coherence time, τ_c and consequently with the length of the wave-trains, known as coherence length, l_c. The correlated fluctuation can be partially or completely coherent. A polychromatic point source on the sky produces a fringe packet as a function of an applied path length difference. This fringe packet has an extent referred to as the coherence length

$$l_c = \frac{c}{\Delta\nu} = \frac{\lambda^2}{\Delta\lambda} = c\tau_c, \tag{4.4}$$

where $\Delta\nu$ is the effective spectral width and c the velocity of light, and

$$\tau_c = \frac{\lambda^2}{c\Delta\lambda} \sim \frac{1}{\Delta\nu}, \tag{4.5}$$

the coherence time; it is the maximum transit time[a] difference for good visibility of the fringes, a dimensionless number lying between zero and one.

The average of the term, duration of the wave-train or frequency range of the Fourier spectrum satisfy the reciprocity inequality, which is analogous to the Heisenberg uncertainty relation (see Section 1.1.1) in quantum mechanics

$$\tau_c\Delta\nu \geq \frac{1}{4\pi}. \tag{4.6}$$

[a]The transit time is mean time difference between the incidence of the light upon the photocathode and the occurrence of the half-amplitude point on the output-pulse leading edge.

The coherent length determines the bandwidth (optical), $\Delta\nu$, of the radiations emitted by the atoms (Born and Wolf, 1984), for example

$$\begin{aligned}\Delta\nu &= \frac{1}{(\lambda - \Delta\lambda/2)} - \frac{1}{(\lambda + \Delta\lambda/2)} \\ &= \frac{\Delta\lambda}{\lambda^2 - (\Delta\lambda/2)^2} \simeq \frac{\Delta\lambda}{\lambda^2},\end{aligned} \tag{4.7}$$

of the resulting disturbance are related by the order of magnitude

$$\tau_c = \frac{1}{\Delta\nu}. \tag{4.8}$$

Here, the fractional bandwidth is assumed to be very small; therefore, the term $(\Delta\lambda/2)^2$ is ignored. For quasi-monochromatic waves, the frequency interval is $\Delta\nu \ll \nu$. The spectral bandwidth is the wavelength interval in which a radiated light has reached half the maximum value. It is a measure of the extent of the spectrum, defined as full-width at half maximum (FWHM); for a light source the typical spectral widths are 20 to 60 nm for a light-emitting diode (LED) and 2 to 5 nm for a laser diode.

4.1.1.2 Responsivity

The responsivity (or photosensitivity), $\mathcal{R}$, is defined as the ratio of the current produced by a sensor to the amount of light falling on it. It is a function of the wavelength of incident radiation and the bandgap energy, E_g, of the material of which the photosensor is made of. The units of responsivity are amperes (A) per watt (W) or milli-amperes (mA) per milliwatt (mW) for the IR sensor and amperes per lumen for the optical sensors. The convention for infrared (IR) and optical sensors is to speak of a blackbody responsivity, $\mathcal{R}(T, \nu)$ and a spectral responsivity, $\mathcal{R}(\lambda, \nu)$. The responsivity is a spectrally dependent value since a detector is sensitive to the radiation within some finite interval of wavelengths. The wavelength interval, in which $\mathcal{R}_\lambda$, is noticeable, is referred to as the working spectral range of the sensor. The blackbody reference temperature is, in general, 500 K for detectors in the IR wavebands, while it is 2870 K for the detectors in visible wavelengths. The output signal is measured at the frequency, ν, which is the modulation frequency of the incident radiation.

If i_s is the signal output of a detector of area, A_d (cm^2), measured at a frequency, ν in response to incident radiation of root-mean-square (RMS) power, P and irradiance, $\mathcal{E}_e$, modulated at frequency, ν, the responsivity is

$$\mathcal{R} = \frac{i_s}{P} = \frac{i_s}{\mathcal{E}_e A_d}. \tag{4.9}$$

If the entire beam falls within the active area (aperture) of the detector, the responsivity is equal to the ratio of detector response to beam radiant power, while in the case of a detector being placed in a radiation field, which overfills its aperture, it is equal to the ratio of detector response to the irradiance of the field, i.e., $N_d = N_{ph0} A_d / A_i$, in which N_{ph0} is the incident photon flux, N_d the number of photons incident on the detector, and A_d the active area of the detector that is smaller than the area of the image, A_i. Considering Figure 1.6 and using the small angle approximation for parallel rays, one finds

$$\frac{A_o}{s_o^2} = \frac{A_i}{s_i^2},$$

where A_o is the area of the source, s_o is the distance between the source and the entrance aperture of the lens whose area is A_l, and s_i the distance between the lens and the image of the source.

The number of photoelectrons, N_{pe} generated in a solid-state detector is

$$N_{pe} = \int_{\lambda_1}^{\lambda_2} \eta_d(\lambda) N_d(\lambda) d\lambda, \tag{4.10}$$

in which η_d is the quantum efficiency of the detector and λ the wavelength; all the variables are a function of the wavelength.

The sensor responsivity is derived from the relationship between the number and energy of photons arriving at the sensor, the ability of the sensor to convert them into electrons and ultimately into an output voltage. For an incident beam with radiant flux, Φ_λ (see equation 1.75), and photons with energy, $E = h\nu$, in which h is Planck's constant, the number of photons arriving at the detector is $\Phi_\lambda/(h\nu)$. The detector converts photons into electrons with quantum efficiency, η_d, so that the number of electrons, N_e, produced per unit time is given by

$$N_e = \eta_d \frac{\Phi_\lambda}{h\nu}. \tag{4.11}$$

Spectral responsivity The spectral responsivity of an optical detector is a measure of its response to radiation at a specified wavelength (monochromatic) of interest. If each electron contributes to the output signal, then the voltage is $N_e \rightarrow O$, in which O is the product of charge-to-voltage conversion and the internal sensor gain. It is measured in units of volts per electron. The voltage turns out to be

$$V_\lambda = \frac{\eta_d O}{h\nu}\Phi_\lambda, \tag{4.12}$$

and thus the spectral responsivity in terms of quantum efficiency (QE), η_d, is given by

$$\mathcal{R}(\lambda) = \frac{V_\lambda}{\Phi_\lambda} = \frac{\eta_d O}{h\nu}. \tag{4.13}$$

For an ideal detector, the spectral responsivity, $\mathcal{R}(\lambda)$, in amps (A) per watt (W), is given by

$$\mathcal{R}(\lambda) = \begin{cases} \dfrac{\lambda}{\lambda_c}\mathcal{R}_P & \lambda \leq \lambda_c, \\ 0 & \text{elsewhere,} \end{cases} \tag{4.14}$$

in which $\mathcal{R}_P(= \eta_d/E_g)$ denotes the peak responsivity, η_d the detector's quantum efficiency, and λ_c the wavelength at which $\mathcal{R}_P$ occurs.

As stated in Section 2.2.3.1, the bandgap energy, E_g, depends on the crystal structure and the state of the surface of the material, limiting the use of the photoelectric effect in vacuum to detection of wavelengths shorter than a μm. On integrating over the wavelength range the sensor is sensitive to the total responsivity,

$$\mathcal{R} = \frac{\int_\lambda V_\lambda d\lambda}{\int_\lambda \Phi_\lambda d\lambda}. \tag{4.15}$$

4.1.1.3 Quantum efficiency

The quantum efficiency (QE; η_d) is defined as the ratio of the mean number of photoelectrons produced over the mean number of incident photons. It is independent of the intensity and the detection frequency. The QE determines the sensitivity of a sensor at a given light level (Janesick, 2001). A critical factor for designers of photosensors, which is given by

$$\text{QE} = \frac{\text{No. of generated electron, } N_e}{\text{No. of incident photons, } N_{\text{ph}}}.$$

This probability can be determined theoretically from the atomic structure of the semiconductor material and the thickness of the emitting layer. An ideal detector would have a QE of 100% and would be sensitive to all colors of light. However, the QE lies between 0 and 1, i.e., $0 \leq \eta_{\text{d}} \leq 1$; it reduces due to the following reasons (Saleh and Teich, 1991):

(1) When many photons are incident, the QE is the ratio of the flux of generated electron–hole pairs (see Section 2.2.4.1) to the flux of incident photons. As stated in Section 2.2.3.1, the photons possessing energy greater than or equal to the bandgap of the junction, i.e., $E_{\text{ph}} \geq E_{\text{g}}$, are absorbed by the material, while the photons with the characteristic energy less than the intrinsic bandgap, i.e., $E_{\text{ph}} < E_{\text{g}}$, are not absorbed and pass through the semiconductor. However, all incident photons do not contribute in the production of such pairs unless all these photons get absorbed. Because of the probabilistic nature of the absorption process some photons fail to be absorbed; others may be reflected from the top of the surface of the detector. The reflectance that can be reduced by anti-reflection coatings, and the absorption coefficient of the semiconductor material, α_{m}, affect the final shape of the QE curve. Neglecting reflection, the amount of light absorbed by a material depends on α_{m}, as well as on the thickness of the absorbing material. Different semiconductor materials have different absorption coefficients. For example, in a material with a lower value of α_{m}, light is poorly absorbed, while with a higher value of α_{m} more photons get absorbed. The fraction of the photon flux absorbed in the bulk of the material is dictated by

$$\frac{\int_0^d e^{-\alpha_{\text{m}} x} dx}{\int_0^\infty e^{-\alpha_{\text{m}} x} dx} = 1 - e^{-\alpha_{\text{m}} d},$$

in which d is the photodetector depth and x the distance into the material,

(2) There are some electron–hole pairs produced near the surface of the detector, which recombine fast because of the abundance

of recombination centers there; the surface recombination can be reduced by careful material growth.

(3) If the light is not focused precisely onto the active area of the detector, some photons would be lost.

The responsivity and QE are related to each other. The net η_d, of a device is calculated by multiplying the maximum absorption efficiency, η_a at peak wavelength with the PC gain, G, that is

$$\eta_\text{d} = \eta_a \times G.$$

The gain is the average number of circuit electrons generated per photocarrier pair. The QE is wavelength dependent, which is large within a spectral window that is determined by the characteristics of the material. For a large value of wavelength, the QE becomes small, while for small wavelengths, it decreases because most photons are then absorbed near the surface of the device; the recombination lifetime is short near the surface. In a photodetector's response, it is desirable to have the spatial uniformity that is a key factor in image quality, and linearity with respect to the number of photons. The uniformity shows the consistency of response for different pixels under identical illumination. The linearity of a photodetector is defined by the response of the detector being linear, which exhibits the consistency of responsivity over a range of light power. The charge that is generated, collected, and transferred to the sense node should be proportional to the number of incident photons; the response may be linear over a broad range. If the output of the detector is plotted as a function of the input power, there should be no change in the slope of the curve. The noise (any undesired signal; see Section 4.5) determines the lowest level of detectable incident light. The upper limit of the input/output linearity is determined by the maximum current that a detector can produce without becoming saturated. When a photodetector has a linear response, its quantum efficiency, η_d and responsivity, $\mathcal{R}$, are constants that are independent of the power of the input optical signal. However, in reality, such a detector has a finite range of linear response. As the power of the input signal, reaches at a certain level, the response of the detector starts to saturate, thereby deviating from the linearity.

4.1.1.4 Detectivity

The detectivity is a function of operating temperature, detector bias-voltage and cut-off wavelength. The characterization of a detector includes measurements of both responsivity and noise as a function of these parameters. Detection of weak optical signal requires the optimization of a photodetector and its amplification circuitry for a desired signal-to-noise (S/N) ratio. The S/N ratio, which is independent of the bandwidth, Δf, determines the strength of the signal compared to noise. It is defined as the ratio between the signal power and the noise power

$$\mathrm{S/N} = \frac{\overline{|i_\mathrm{s}|^2}}{\overline{|i_\mathrm{n}|^2}}, \tag{4.16}$$

with

$$\overline{|i_\mathrm{s}|^2} = \left(\frac{q\eta_\mathrm{d}}{h\nu}\right)^2 \overline{P_0^2} \tag{4.17}$$

$$\overline{|i_\mathrm{n}|^2} = \left[2q(\bar{i}_\mathrm{s} + \bar{i}_0) + \frac{4k_\mathrm{B}T}{R}\right]\Delta f, \tag{4.18}$$

i_s and i_n the signal and noise current, respectively, i_0 the saturation current that is proportional to the incident flux, h Planck's constant, R the sample resistance, $k_\mathrm{B}(= 1.38 \times 10^{-23}\,\mathrm{JK}^{-1})$ Boltzmann's constant, Δf the measuring bandwidth centered on f, and P_0 the incident radiation power to the detector input of area A_d, η_d the quantum efficiency, and q the electron charge.

The optical power generates a photocurrent with the amplitude equal to that of the total noise. The noise depends on the bandwidth of the detector in question. In this case, the S/N ratio is $S/N = \sqrt{N_\mathrm{ph}}$, with N_ph as the number of photons received. Let $I(t)$ be the intensity of the light at the photocathode, averaged over a few cycles, so that the probability of an emitted electron would be $\eta_\mathrm{d} I(t)dt$, in which the detector quantum efficiency, η_d, is assumed to be constant over the bandwidth, Δf. The mean number of photons, $\bar{N}_\mathrm{ph}$, which is counted in an exposure time, Δt, turns out to be

$$\bar{N}_\mathrm{ph} = \eta_\mathrm{d}\bar{I}\,\Delta t, \tag{4.19}$$

and the variance, $\langle\Delta N_\mathrm{ph}\rangle^2$, for this count would become

$$\langle\Delta N_\mathrm{ph}\rangle^2 = \bar{N}_\mathrm{ph}(1 + \bar{N}_\mathrm{ph}) \qquad \text{when } T \ll 1/\Delta f. \tag{4.20}$$

For weak signals, the S/N ratio should be of the order of 3:1 or 2:1 and as a limiting case, it is chosen to be $S/N = 1{:}1$ (Menn, 2004). The detectivity, D, is given by

$$D = \frac{\mathcal{R}_\lambda}{\sqrt{\langle i_\mathrm{n}\rangle^2}}, \tag{4.21}$$

where i_n is the noise, which is a random fluctuation of time and is characterized by its mean value, $\sqrt{\langle i_\mathrm{n}\rangle^2}$ and $\mathcal{R}_\lambda$ the responsivity.

Specific detectivity, D^* Since the performance of a detector is limited by noise, it is necessary to define the RMS S/N ratio in a 1 Hz bandwidth per unit RMS incident power per square root of its light sensitive area, $\mathtt{A}_\mathrm{d}$. The specific detectivity, D^*, is expressed as

$$\begin{aligned} D^* &= \frac{i_\mathrm{s}\sqrt{\mathtt{A}_\mathrm{d}}}{P}\frac{\sqrt{\Delta f}}{i_\mathrm{n}} \\ &= \left(\frac{i_\mathrm{s}}{i_\mathrm{n}}\right)\left(\frac{\sqrt{A_\mathrm{d}\Delta f}}{P}\right). \end{aligned} \tag{4.22}$$

The unit of D^* is $\mathrm{Hz}^{1/2}\cdot\mathrm{W}^{-1}$ cm.

Noise equivalent power (NEP) The noise equivalent power is defined as the input power required of optical signal for the S/N ratio to be unity at the detector output; one needs to consider the noise that occurs in the detector, as well as in its electric circuit. This radiant power is designated by NEP, whose units is $\mathrm{W.Hz}^{-1/2}$. As seen from equation (4.22), P turns out to be equal to NEP if $i_\mathrm{s} = i_\mathrm{n}$. Hence, from Eqs. (4.17) and (4.18), one finds

$$(\mathrm{NEP})^2 = \Delta f\left(\frac{h\nu}{q\eta_\mathrm{d}}\right)^2\left[2q(\bar{i}_\mathrm{s} + \bar{i}_0) + \frac{4k_\mathrm{B}T}{R}\right]. \tag{4.23}$$

4.1.1.5 Frequency response

The frequency response of a photodetector is the variation of responsivity as a function of the modulation frequency of the incident radiation. It is determined by taking the Fourier transform of the impulse response or by registering the response of the detector at one signal frequency at a time while sweeping the signal frequency. The governing equation is

$$\nu_\mathrm{s} = \frac{\nu_\mathrm{so}}{\sqrt{1 + (2\pi\nu\tau_\mathrm{d})^2}}, \tag{4.24}$$

where ν is the modulation frequency, τ_d the response time and is related to 3 dB frequency by

$$\tau_d = \frac{1}{2\pi\,\nu_{3\ dB}}, \tag{4.25}$$

with $\nu_{3\ dB}$ as the 3 *dB* cut-off frequency.

The 3 dB bandwidth of a photodetector is a function of the combined effect of some physical factors that determine the speed and the frequency response of the detector. For frequencies above $1/(2\pi\,\tau_d)$, the response is falling off significantly and at high enough modulation frequencies, the detector provides a DC output proportional to the average intensity.

4.1.1.6 Response time

The response time refers to the time required to reach ~90% of the final response and it determines the highest signal frequency to which a sensor can respond. It depends on the transit time of photogenerated charge carriers within the photodetector material and inherent capacitance and resistance associated with the device (see Figure 4.1). The response time can be expressed by the frequency response, which is the frequency at which the output of a photodetector decreases by 3 dB.

The response speed of a photodetector is directly related to its frequency response. In order to record an optical signal, a photodetector should have a speed higher than the frequency

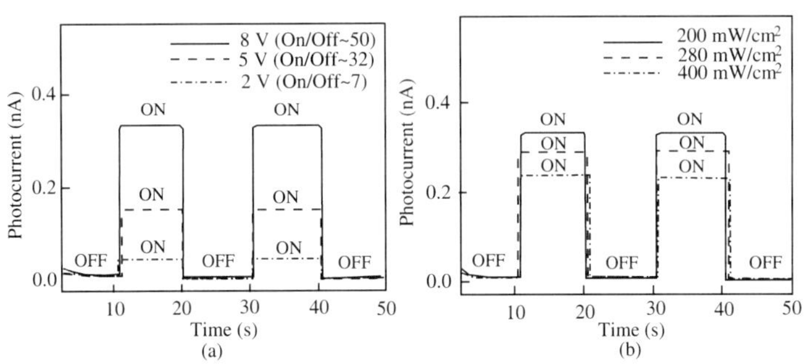

Figure 4.1 Photoresponse of Mn-doped zinc oxide (ZnO) nanowire devices (Prabhakar et al. 2012; Courtesy: K. B. Jinesh).

response that has a bandwidth covering the entire bandwidth of the signal. The notable factors that determine the response speed of a photodetector are the time constant and the carrier transit time in the depletion region. The detection of radiation that varies in time needs fast detector circuit having suitable bandwidth. The radiation is incident on the detector during a very short time interval, Δt. Due to the electrical inertial processes in the detector and in the circuit elements, the generated current has a finite speed of growth characterized by the rise-time, τ_r, as well as a finite speed of decay characterized by the fall-time, τ_f. The rise time is defined as the time interval between the point at which the detector has reached 10% of its peak output and the point at which it has reached 90% of its peak response, when it is irradiated by a very short pulse of light. Also known as the decay time, the fall time is regarded as the time between the 90% point and the 10% point on the trailing edge of the pulse waveform. Both the values are determined with regard to the half the maximum of generated signal and the maximum working frequency, ν_{max}, is given by

$$\nu_{max} = \frac{1}{2(\tau_r + \tau_f)}. \tag{4.26}$$

The response time of a photodetector with its output circuit depends mainly on the three factors:

Drift-time, τ_d The drift-time (or transit-time), of the photocarriers in the depletion region of the photodetector is

$$\tau_d = \frac{w}{v_d}, \tag{4.27}$$

where v_d is the current drift velocity (see Section 2.2.4.2) and w the depletion layer width.

Diffusion-time, τ_{diff} The diffusion-time is defined as the time for carrier generated within a diffusion length from the junction to diffuse into the depletion region

$$\tau_{diff} = \frac{d}{2D}, \tag{4.28}$$

where d is the depth of photoabsorption and D the diffusion constant of the minority carrier.

RC time-constant, τ_{RC} Generally, the overall speed of a photodetector is determined by both its intrinsic bandwidth and its RC circuit-limited bandwidth. The rise-time of the impulse response is determined by the intrinsic bandwidth of a photodetector, while the square-pulse response is determined by the its RC circuit-limited bandwidth. The RC time-constant of the diode-circuit combination that acts like RC low-pass filter after the photodetector with a pass-band, Δf, which is given by

$$\Delta f = \frac{1}{2\pi R_{\mathrm{T}} C_{\mathrm{T}}}, \tag{4.29}$$

with $R_{\mathrm{T}}(= R_{\mathrm{s}} + R_{\mathrm{L}})$ as the sum of the diode series resistance, R_{s} and the load resistance, R_{L}, $C_{\mathrm{T}}(= C_{\mathrm{j}} + C_{\mathrm{s}})$ the sum of photodiode junction, C_{j} and the stray capacitance, C_{s}. As stated in Section 2.2.7, the boundaries of the depletion region act as the plates of a parallel plate capacitor. The junction capacitance is directly proportional to the diffused area and inversely proportional to the width of the depletion region; higher resistivity substrates have lower junction capacitance.

The total rise time, τ_{r}, is derived as

$$\tau_{\mathrm{r}} = \sqrt{\tau_{\mathrm{d}}^2 + \tau_{\mathrm{diff}}^2 + \tau_{RC}^2}. \tag{4.30}$$

In order to achieve a high quantum efficiency, the depletion layer width should, in principle, be larger than the inverse of the absorption coefficient, so that most of the light will be absorbed. At the same time with large width, the capacitance is small and τ_{RC} getting smaller, leading to faster response, but wide width results in larger transit time in the depletion region. Therefore, there is a trade-off between width and QE.

4.1.1.7 Dynamic range

The dynamic range is defined as the peak (maximum) possible signal, the saturation level, to the read-out noise, i.e., $V_{\mathrm{max}}/V_{\mathrm{n}}$, in which V_{max} is the maximum signal voltage and V_{n} the noise voltage; high noise implies low dynamic range. This ratio also gives an indication of the number of digitization levels that might be appropriate for a given sensor. For example, a system with a well depth of 85,000 electrons and a read-out noise of 12 electrons would have a dynamic range $= 20\log(85000/12)$ or 77 dB.

The detector is assumed to be operated in its linear range when the detector current, i_d, is proportional to the incident radiation power. The maximum value of dynamic range is dictated by the saturation of the detector, which means that no additional electrons can be created with the arrival of additional photons at the detector. The minimum value of the dynamic range is governed by the detector noise, which is generally equal to the noise equivalent power (NEP; see Section 4.1.1.4).

4.1.1.8 Dark current

The dark current, i_D, an important factor for sensitivity, possesses an inherent limitation on the performance of the device. It is measured as the signal generated in the absence of the external light and is due to

- thermal generation and diffusion in the neutral bulk substrate,
- arrival of photons within the depletion region of the device that are then swept by the electric field applied to the photodiode, and
- thermal generation of electrons in the surface states.

The dark current continues to flow through the bias circuit. Such a current depends on integration time, but not on the signal level, and also depends on the material used and the manufacturing process, and is given by the relation

$$i_\text{D} = N_\text{e} e^{(E_\text{F} - E_\text{C})/(k_\text{B}T)}, \tag{4.31}$$

where k_B is Boltzmann's constant, E_F the Fermi level of energy (see Section 2.2.5) and E_C the conduction level of energy, N_e the equivalent state of density in the conduction band, and T the absolute temperature.

The generation of dark current is described by a Poisson process

$$\langle\sigma\rangle^2_{i_\text{D}} = 2q\bar{i}_\text{D}\Delta f, \tag{4.32}$$

with Δf as the electrical bandwidth of the measuring instrument in question, since all frequency components are present in equal intensity.

The dark current due to the generation of electron–hole pairs by the effect of temperature can be substantially reduced at lower temperatures.

4.1.2 *Detection of Photoevent*

The arrival of photons is highly localized in space and time, which can be accurately modeled by using Dirac delta function (see Section 1.3.1), $\delta(\mathbf{x}, t)$, the probability of a photon arriving in a region proportional to the irradiance over that region. Thus the probability distribution of photon impacts is the intensity distribution in the image. A detector detects a fraction, $\eta_{\rm d}$, of all incident photons. The event of such detection is called photoevent. An ideal detector array senses each photon impact and measures its position precisely. The fundamental nature of noise of such a detector is produced by the quantum nature of photoelectron. A single photon event provides the centroid location (see Section 7.2.1) with a mean square error equal to the variance of the intensity distribution. For an incoherent light, the semi-classical model for photoelectric light detection in a detector relies on the validity of the assumptions, such as

- the probability, $\mathcal{P}$, of one photon being detected on a small area, dA, which is small compared to the coherence area of the light, during a small interval of dt is given by

$$\mathcal{P}(1, dt, d\mathrm{A}) = \eta_{\rm d} I(\mathbf{x}, t)\, dt\, d\mathrm{A}, \tag{4.33}$$

  in which $I(\mathbf{x}, t)$ is the image intensity, $\mathbf{x}$ the 2-D spatial coordinate on the detector surface, and $\eta_{\rm d}$ the quantum efficiency (QE).
- the probability of more than a single photoevent over the same space and time intervals is negligible, and
- the number of photoevents occurring during non-overlapping space and time intervals is also negligible.

If these conditions are met, the number of photoevents becomes a Poisson random variable (Goodman, 1985) with mean

$$\lambda = \eta \int I(\mathbf{x}, t) dt d\mathrm{A}. \tag{4.34}$$

Since the generation of free charge carriers is dominated by photon absorption (not by thermal excitation), low temperature is needed as the long wavelength threshold, λ_c, of a detector increases, which goes approximately as (Hoffman et al., 2004)

$$T_{\max} = \frac{200\ K}{\lambda_c(\mu\mathrm{m})}. \tag{4.35}$$

4.2 Mechanism of Photon Detection

Modern detectors, such as photomultipliers, charge-coupled devices (CCD), and television cameras, are able to detect individual photons and output a number, which is directly and linearly related to the number of photons that were incident on the detector. However, these detectors suffer from the presence of noise sources that interfere with the detection process. They need to employ a bias voltage and a load resistor in series with the detector. The incident light changes the characteristics of the detector and causes the current flowing in the circuit to change. The output signal is the change in voltage drop across the load resistor. All optical detectors respond to the power in the optical beam.

4.2.1 *The Human Eye*

The human eye provides the sense of sight, allowing to observe the surroundings and interpret the shapes, colors, and the dimensions of objects by processing the light they reflect or emit. It has a spectral response covering the range from ~360 to 700 nm (visible spectrum); the sensitivity of the eye is greatest for light having a wavelength of 555 nm. The eye is a complex and highly non-linear, albeit has a large dynamic range. Once adjusted, it can detect 10^9 difference; the retina at any time has dynamic range of ~15,000. It has a limited depth of field as well. The non-linear response of the eye varies with wavelength and also with the amount of radiant flux, the spatial complexity of the scene being perceived, the adaptation of the iris and retina, etc.

Hand-drawing from eye observations had been used in astronomy since Galileo till the middle of the nineteenth century. Even in the interferometric experiment conducted by Michelson and Pease (1921), which was constructed by installing a 7 m steel beam on top of the 100 inch reflecting telescope at Mt. Wilson Observatory, equipped with four flat mirrors to fold the beams in periscopic fashion, the fringes were observed with naked eye. The outer two flat mirrors allowed them to measure the fringe visibility in the interference pattern formed by starlight at the focal plane. However, the human eye can still be used for searching supernovae and comets; one can discover comets that are close to the Sun.

4.2.1.1 Structure of the eye

The eyeball, nearly spherical in shape, is composed of a dense, imperfectly elastic supporting membrane. The anterior part of the membrane, called the cornea, is transparent having a refractive index $\sim$1.38, which permits light to enter the eyeball. The remainder (white of the eye) is opaque, known as the sclera (Sihota and Tandon, 2004), a protective outer layer, which is connected with the cornea in front and with the sheath covering optic nerve behind. Both cornea and sclera along with the eye lens, a compound lens system (see Section 1.3.2), together form the external covering of the eyeball.

The angular resolution of the eye is dictated by the size of this eye lens, which is about 15 arcsec. Located behind the iris and the pupil, its focal length is 14–17 mm. This lens creates an image on the retina simultaneously in all wavelengths incident on the pupil. It has a refractive index varying from 1.41 at the center to 1.39 at the periphery, whose shape can be changed by muscles attached to it allowing for fine focusing of light entering the eye. The anterior surface of the cornea may be regarded as approximately spherical, the radius of curvature being 8 mm. The centers of curvature of the cornea and the two surfaces of the lens are in the same line, what is referred to as optic axis. The anterior chamber is a space filled with transparent fluid, called aqueous humor with refractive index $\sim$1.34, which is bounded by the iris and the part of the anterior surface of the lens that is exposed in the pupil. The other important anatomical components of an eye (see Figure 4.2) are enumerated below:

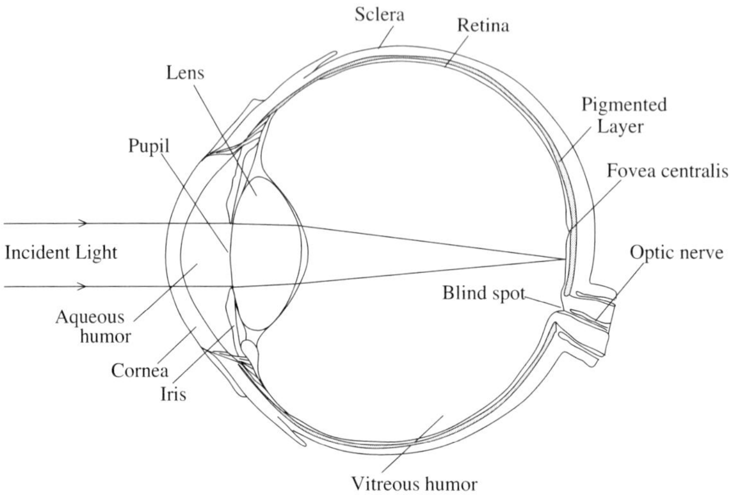

Figure 4.2 Sketch of a human eye.

Conjunctiva The conjunctive is the transparent mucous membrane covering the outer surface of the eyeball except the cornea, and lining the inner surfaces of the eyelids. When this membrane gets irritated, blood vessels beneath it become dilated, resulting in a condition called conjunctivitis.

Extraocular muscles These muscles are strong and efficient although they are relatively small in size. There are six extraocular muscles, which act to turn or rotate an eye about its vertical, horizontal, and antero-posterior axes. These muscles are

- medial rectus (moves the eye inward, toward the nose),
- lateral rectus (moves the eye outward, away from the nose),
- superior rectus (moves the eye upward, rotates the top of the eye toward the nose, and moves the eye inward),
- inferior rectus (moves the eye downward, rotates the top of the eye away from the nose, and moves the eye inward),
- superior oblique (rotates the top of the eye toward the nose, moves the eye downward, and moves the outward), and
- inferior oblique (rotates the top of the eye away from the nose, moves the eye upward, and moves the outward).

Iris Situated behind the cornea, the iris is a thin diaphragm composed mostly of connective tissue and smooth muscle fibers. It controls the amount of light entering the eye by varying the size of the pupillary opening.

Macula A small central area with about 1.5 mm in diameter of the retina surrounding the fovea, called the macula, provides the perfect vision of any location in the retina. Its center is referred to as the fovea centralis, which contains only cone cells.

Optic nerve The optic nerve that acts as a cable connecting the eye with the brain, is a continuation of the axons of the ganglion cells in the retina. There are approximately 1.1 million nerve cells in each optic nerve. As the optic nerve leaves the back of the eye, it travels to the optic chiasm, located just below and in front of the pituitary gland. In the optic chiasm, the optic nerve fibers emanating from the nasal half of each retina cross over to the other side. From there, the nerve fibers become the optic tract, passing through the thalamus and turning into the optic radiation until they reach the visual cortex in the occipital lobe at the back of the brain, where the visual center of the brain is located. The beginning of the optic nerve in the retina is called the optic nerve head. Since there are no photoreceptors (cones and rods) in this nerve head, this area of the retina cannot respond to light stimulation, which is referred to as the blind spot. Any image that falls on this region cannot be seen.

Pupil The pupil of the eye is the circular aperture of about 2–8 mm in diameter at the center of the normal iris. Light entering the eye passes through it and regulate the amount of light passing through to the retina. With the increase of the ambient illumination, the pupil constricts due to the pupillary light response, while under dark conditions, it would dilate.

Retina Situated at the back of the eyeball, the retina is the innermost layer of the eye. It is composed of nerve tissue, which senses the light entering the eye. This complex system of nerves sends impulses through the optic nerve back to the brain. The retina contains two different types of photoreceptors called cones and rods (see Figure 4.5), which produce nerve impulses that are passed on to the subsequent stages of the visual system for processing (Ohta and Robertson, 2005). Seeing is a physiological process originating in photochemical reactions in these photoreceptors. Their tips are embedded in a pigmented layer of cells on the very back of the retina. Absorption of light by the pigment in these cells prevents scattering of any light rays, which are not absorbed by the neural layer. The pigmented layer ensures that the light reaching cones and rods comes through the lens as a focused image. The light causes a chemical reaction with photopigments, such as iodopsin and rhodopsin rods. When light strikes a cone or rod cell, it passes the signal to a bipolar cell, which passes it on to the ganglion cells that perform the first level of information processing.

Figure 4.3 illustrates the cross-sectional information of retina through optical coherence tomography (OCT) imaging (4.3a) represents a normal retina with all retinal layers intact and (4.3b) represents a retinal anomaly with macula detachment. The OCT images manifests the subsurface information based on cross interference between projected signal and reflected signal through classical Michelson interferometer (see Section 3.2.1.1). This imaging allows measurement of various subsurface ocular structures. Ensemble of multiple B-scans accommodates three-dimensional imaging of tissues at histological level (Chauhan and Marshall, 1999; Koozekanani et al., 2000; Bezerra et al., 2009). The major advantage of OCT is that it is a diagnostic tool that can perform tomography/cross-sectional imaging of biological tissues with less than or equal to 10 μm axial resolution using light waves. Since the human retina is easily accessible to external light, OCT finds a prominent role in the diagnosis and monitoring of retinal diseases. The OCT provides information akin to in vivo histopathology of the retina. The uniqueness of OCT is that it is a non-contact, non-invasive device where a broad-band width near-IR light (920 nm) is projected onto the retina. The light gets reflected and scattered differently from tissues with different optical properties. It then compares the echo time-delay of light that is reflected from the layers of the retina to the echo time-delay of the same wavelength that is reflected from a reference mirror at a known distance. The interferometer then combines the reflected pulses from the retina, as well as reflecting mirrors resulting in a phenomenon known as

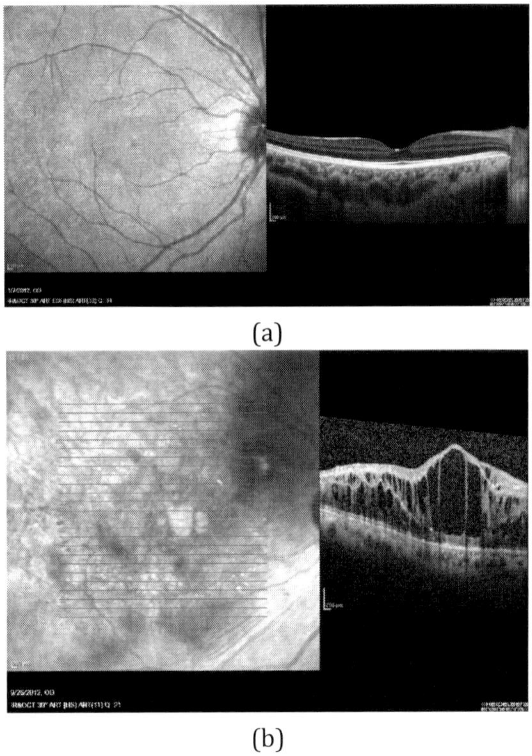

Figure 4.3 The cross-sectional images of (a) normal retina and (b) abnormal retina showing central cystoid macular edema with retinal thickening and cystoid spaces through optical coherence tomography (OCT); a CMOS camera (see Section 6.3) quantizes the reflected optical coherence tomography (OCT) images of retina signals from retina and reconstruct these images. Courtesy: J. Chatterjee and S. Ghosh.

interference (see Section 3.2.1). This interference is measured by a photodetector/CMOS, which determines the distance traveled by various echoes by varying the distance to the reference mirror. This finally produces a range of time delays for comparison. The OCT is also useful for (i) representing anomalies at submicron level, (ii) quantifying morphological abnormality and (iii) demonstration of layer deformations under various pathological conditions. However, it has limitations, namely, (i) depth of penetration is limited due to

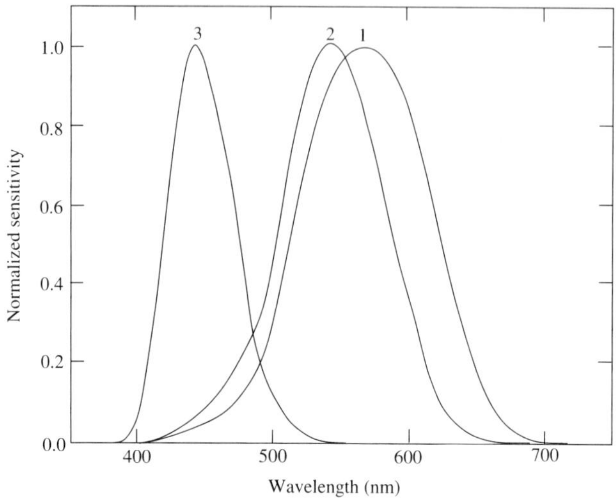

Figure 4.4 Spectral sensitivities of three types of cone pigments: (1) R_λ, cones with erythrolabe, (2) G_λ, cones with chlorolabe, and (3) B_λ cones with cyanolabe.

refractive index of layers, (ii) speckle noise, and (iii) media opacities, which can interfere with optimal imaging.

(1) **Cones**: The cones, which contains a particular type of light-sensitive pigment, iodopsin, are color receptors. Most of the retinal cones are located in the macula and fovea centralis. There are about 6.5 to 7 million cones in each eye. The highest concentration of cones is in the macula. The fovea centralis, at the center of the macula, contains only cones and no rods. There are three types of cone pigments (see Figure 4.4), each one is most sensitive to a certain wavelength of light, such as

   (a) **Erythrolabe**: It has maximum absorptivity in the wavelengths of the red part of the spectrum.
   (b) **Chlorolabe**: It has maximum absorptivity in the green part of the spectrum.
   (c) **Cyanolabe**: This has maximum absorptivity in the blue region of the spectrum.

(2) **Color vision**: The color is a combination of physical effects and the physiology of human sensation. In order to perceive any

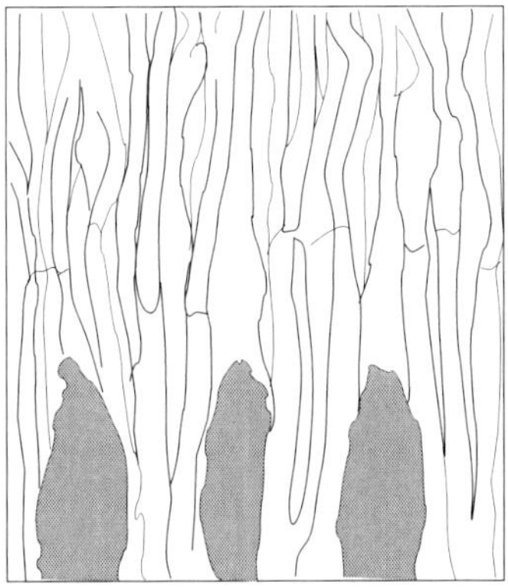

Figure 4.5 Sketch of the cones and rods; the shaded portions are the cones and the thin narrow ones are the rods.

color, the cone (see Figure 4.5) cells must be stimulated by the photons in the electromagnetic waves in a certain spectrum (see Section 1.1.6): violet (400–450 nanometers; nm), blue (450–490 nm), green (490–560 nm), yellow (560–590 nm), orange (590–635 nm), and red (635–700 nm). The color sensation is characterized by three curves, i.e., R_λ, G_λ, and B_λ, representing the spectral sensitivity of these pigments, respectively. The perception of all three stimuli creates the feeling of color of a source. Signals from the cones are sent to the brain, which translates these messages into the perception of color; they work in bright light. The cones are responsible for the day-time color vision and are most sensitive to one of three different colors, such as green, red or blue, while the rods are functioning in the dark. If a light source with spectral radiant flux, Φ_λ, it generates three main stimuli in a human eye, the relative strength of each can be estimated as

$$r = \int \Phi_\lambda R_\lambda; \qquad g = \int \Phi_\lambda G_\lambda; \qquad b = \int \Phi_\lambda B_\lambda. \tag{4.36}$$

(3) **Rods**: Spread over the entire retina except the fovea (a small central region of the retina), the rod cells (see Figure 4.5) contain a particular type of light-sensitive pigment, rhodopsin. They are more numerous ($\approx$20 million) than the cones in the periphery of the retina, decreasing in density up to the macula.

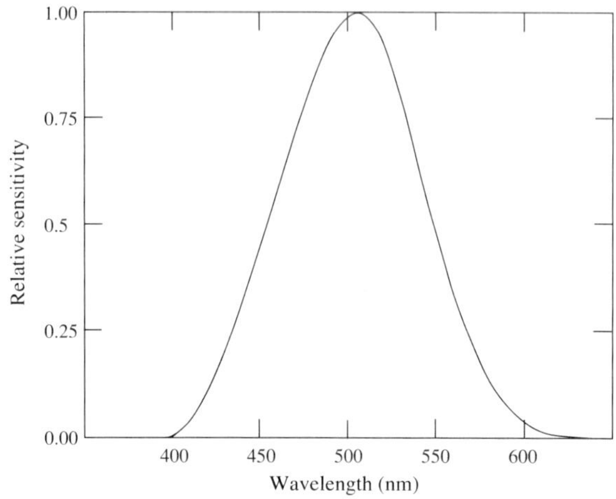

Figure 4.6 Spectral sensitivity of the rods.

> The rods are sensitive to light and dark changes, shape, and movement; they do not detect color. The rod pigment is most sensitive to a wavelength of about 498 nm (see Figure 4.6). The defective or damaged cones result in a color deficiency, while the defective rods result in problems seeing in the dark.

These cones and rods from different parts of the retina are not fully identical, but vary in their morphological structure. However, the density of these cells varies, as a result the sensation of the central part of the retina differs from that of the periphery. The cones are not as sensitive to light as the rods; the quantum efficiency of the eye is 3% (cone; 10% for rod). These photoreceptors are connected through nerve fibers to the optic nerve, which emerges from the back of the eyeball. They receive the optical image and transmit it through the nerve fibers to the brain.

The cones and rods have the same kind of structure as the optical fiber (see Section 3.4), i.e., they consist of dielectric cylindrical rods surrounded by another dielectric of slightly lower refractive index. The core diameters are in the range of a few μm. The light absorbed in these light guides generates electrical signals, which are subsequently transmitted to the brain. The concentration of the rods and cones varies, as a result, the sensation of the central part of the retina differs from that of the periphery. The cones are spread over the entire retina, together with a large concentration within the fovea, which results in the high visual acuity at the center of the field-of-view of the eye. To note, the blood vessels and nerve fibers go around the fovea so that light has a direct path to the photoreceptors.

Vitreous humor The vitreous humor is a transparent thin jelly having a refractive index ~1.34. It makes up about 80% of the volume of the eye and occupies the posterior compartment of the eye, located between the eye lens and the retina. Light initially entering the eye through the cornea, pupil, and lens, is transmitted through this layer to the retina.

4.2.1.2 Operation of the eye

Since the rays enter and leave the refracting system through media of different optical density, the anterior and posterior focal distances are different. The former distance is ~15 mm in front of the cornea, while the latter distance is ~24 mm behind it. The optic axis, produced backwards to meet the retina, cuts it almost exactly at the fovea centralis. Within the layers of the retina the light impulses are changed into electrical signals, which are sent through the optic nerve, along the visual pathway, to the occipital cortex at the posterior of the brain. Light rays from a distant object are converged first by the cornea followed by the lens and passes through the vitreous humor. These rays intersect at a point behind the lens and diverge from that point back to the retina. A real inverted image is formed on the retina, which is re-inverted in the brain.

The eye has a refractive error. If the incoming light from a distant object gets focused before it gets to the back of the eye, such an error is called myopia (near-sightedness), while in the case of not getting focused by the time it reaches the back of the eye, this condition is referred to as hyperopia (far-sightedness). Another error, known as astigmatism,[a] occurs when the surface of the cornea

[a]Astigmatism is an off-axis point wavefront aberration, caused by the inclination of incident wavefronts relative to the optical surface. It occurs when an optical element has different foci for rays that propagate in two perpendicular planes; more de-centered subset of rays may produce astigmatism. Usually centroid shifts with change in focus.

is not spherical. The light rays entering the eye are bent unequally by different meridians preventing the formation of a sharp image focus on the retina. Color blindness, which causes a person to be unable to distinguish certain colors, arises from inherited defects in the pigment in cone cells in the retina.

In humans, the total optical power of the relaxed eye is approximately 60 diopters; a diopter, is a unit of measurement of the optical power of a lens. The optical power, also referred to as dioptric power, P, is the degree to which a lens converges (or diverges) light. Measured in meters (m^{-1}), it is equal to the reciprocal of the focal length of the lens, f, i.e., $P = 1/f$. A convex lens has a positive dioptric value, while a concave lens has a negative dioptric value. When two lenses are in contact, the optical power of the combination is approximately equal to the sum of the optical powers of each lens, $P = P_1 + P_2$, or

$$\frac{1}{f} = \frac{1}{f_1} + \frac{1}{f_2},$$

in which f, f_1, and f_2 are the respective focal distances.

If these lenses are separated by a distance, d, the equivalent focal length f_e of the combination, irrespective of the direction of light, is

$$f_e = \frac{f_1 f_2}{f_1 + f_2 - d}. \tag{4.37}$$

The cornea accounts for approximately two-thirds of this refractive power and the crystalline lens contributes the remaining one-third. In focusing, the ciliary muscle contracts to reduce the tension or stress transferred to the lens by the suspensory ligaments. This results in increased convexity of the lens which in turn increases the optical power of the eye.

The eye is sensitive to radiation over a range from bright sunlight to a flash of light consisting of as little as five or six photons (when adapted to the darkness at least an hour), but is not equally sensitive to all wavelengths of visible light. To note, at 640 K, a target emits a few photons per second. Here, the pupil dilates to about 8 mm ($\mathtt{A}_p = 5 \times 10^{-5}$ m^2, in which $\mathtt{A}_p$ is the area of the pupil) and an observer is able to perceive such an object in completely blackened room. With adaptation to darkness, the spectral response of the eye becomes approximately from 0.38 μm to 0.66 μm. Considering pupil in place of the lens from Figure 1.6 and on using the radiometric equations, the number of photons detected is

$$N_{eye} = \int_{0.38\ \mu m}^{0.66\ \mu m} \frac{A_p}{s_o^2} \frac{\mathcal{M}_q(\lambda, T)}{\pi} A_o d\lambda$$
$$\approx \frac{A_p}{s_o^2} \frac{\mathcal{M}_q(\lambda_0, T)}{\pi} A_o \Delta\lambda, \tag{4.38}$$

where $\mathcal{M}_q(\lambda, T)$ is the spectral photon emittance (see equation 1.78), Λ the wavelength, T the absolute temperature, s_o the distance between the source and the pupil, A_o the area of the source, and $\Delta\lambda = 0.28$ μm; the peak scotopic eye response occurs at 0.515 μm.

The detection capacity of an eye depends on the intensity of the surrounding illuminance, for instance, at 0.01 lx, the object should provide 10 times more flux than the surrounding to be discerned by the eye. As stated in Section 4.2.1.1, fluctuations in the intensity of incoming light change the size of the eye's pupil, which controls the radiation input to the retina by a factor of 12; the remaining adaptation is provided by the cones and rods. The high radiation range, called the photopic range, is mediated by the cones, while the low radiation range, what is referred to as the scotopic range, is mediated by the rods. For example, at low levels of luminance ($< 5 \times 10^{-3}$ lx), the rods operate, while in the case of high light levels ($> 5 \times 10^{3}$ lx), the cones respond. Between these two values, both cones and rods are operating and eye's response is in between the two values. The intermediate range, known as the mesopic range, is mediated by both these elements.

Each human being perceives differently the amount and color of a given beam of radiation. The spectral weighting functions for the photopic and scotopic ranges are known as the spectral luminous efficiency (see Section 1.2.2) functions for photopic and scotopic vision, respectively. The peak scotopic eye response occurs at $\lambda_0 = 0.515$ μm. When the mesopic range requires a weighting function based upon a gradual transition between the aforementioned spectral luminous efficiency functions throughout the mesopic region that depends on the visual adaptation conditions.

4.2.2 *Photography*

Photography was the first technological advancement in astronomical detectors, which had replaced hand drawing over 150 years ago. As stated in Section 1.4.2, in film the chemical grains turn out to be darkened when struck by light. The more photons they absorb, the darker they become leading to the creation of a negative image; the intense light makes the film dark. However, the photographic emulsions can make a permanent record of astronomical images. Direct photometric detections at the focus of a telescope were carried out photographically as early as 1850, when W. Bond and J. Whipple took a Daguerreotype of Vega, although the first Daguerreotype of Moon was made in 1840.

The photography had served as an imaging tool till a few decades ago other than the human eye. By exposing photographic plates for long periods, it became possible to observe much fainter objects than were accessible to visual observations. However, the magnitudes determined by the photographic plate are not, the same as those determined by the eye. This is because the sensitivity of the eye reaches a peak in the yellow-green portion of the spectrum, whereas the peak sensitivity of the basic photographic emulsion is in the blue region of the spectrum; the red sensitive emulsions are also available. Nevertheless, the panchromatic photographic plates may yield photovisual magnitudes, which roughly agree with visual magnitudes by placing a yellow filter in front of the film. The greatest advantage of photography over visual observations was that it offered a permanent record with a vast multiplexing ability. It could record images of hundreds of thousands of objects on a single plate. However, a few percent of the photons reaching the film contribute to the recorded image. Its dynamic range is very low. It cannot record brightness differing by more than a factor of a few hundreds.

Nevertheless, Henry Drapper (1837–1882) photographed the Orion nebula (M 42 or NGC 1976), a diffuse patch of light situated south of Orion's belt, which is visible to the naked eye. Thereafter, many famous astronomical surveys were conducted using photographic plates. The Palomar Observatory Sky Survey (POSS) of the 1950s and 1990s were also taken with these plates.

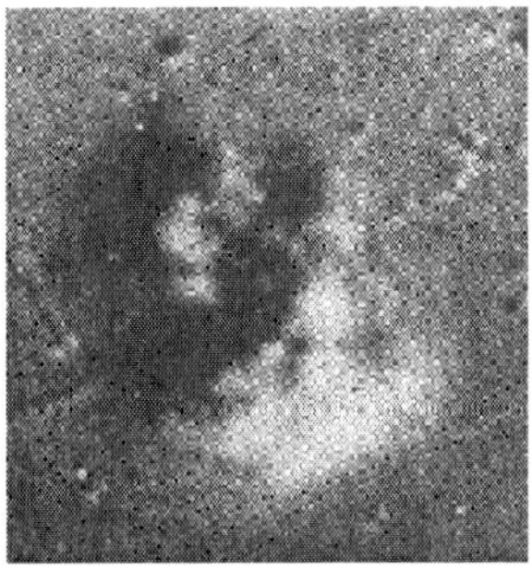

Figure 4.7 Specklegram from an artificial star, which was taken by a photographic film.

Also, many solar system objects were discovered by means of photography. A number of observatories maintain large archives of photographic plates.

During the initial phases of the development in speckle interferometry (see Section 1.5.5), a few observers used photographic films with an intensifier attached to it for recording speckles of various stars (Breckinridge et al., 1979). Saha et al. (1987) used a bare movie camera as a detector in their speckle interferometer to record the fringes and specklegrams of a few bright stars. Figure 4.7 depicts a specklegram obtained in the laboratory, which was recorded with a photographic film; this image was digitized with the photometric data system (PDS) 1010M micro-densitometer.

The photographic plates were also an important tool in early high-energy physics. The sensitivity of certain types of photographic plates to ionizing radiation (usually X-rays) is also a very useful in medical science. Although, the usage of these plates have declined significantly since 1980s, the large format charge-coupled devices (CCD) do not have the detecting area and resolution of most photographic plates.

The photographic technique had many disadvantages, such as (i) non-linearity (see Section 4.4.1.2), (ii) restricted dynamic range (see Section 4.1.1.7), and (iii) low quantum efficiency (see Section 4.1.1.3) of the photographic emulsion (4%), which is a drawback of the photochemical effect (see Section 1.4.2). The threshold of the photochemical effect is governed by the value of the electron excitation energy in AgCl. This limits the sensitivity of plates at the IR boundary of the visible ($\lambda \leq 800$ nm).

Another problem is the "dark background" effect because of the analog nature of the image record on an emulsion. There is a need to use a micro-densitometer to convert analog information into digital information for quantitative processing. The photographic plate is an integrator, since the instant time t is fixed and Δt is the exposure time. The signal is then the density (darkening) of the photographic emulsion, which under certain conditions is proportional to the intensity (see Section 1.2) in the domain of linearity of the emulsion. The signal (or darkening) of the emulsion is defined by the quantity (Lena et al., 2012)

$$X(\mathbf{x}) = \frac{I(\mathbf{x})}{I_0},$$

in which $I(\mathbf{x})$ designates the intensity transmitted by the plate at the point $\mathbf{x}$ when it is illuminated by an intensity, I_0 (densitometry of a photographic image), and $\mathbf{x}(= x, y)$ the 2-D position vector. In the case of observing a faint object, irrespective of the exposure time, the object gets immersed in a background from the plate brighter than the object.

Another difficulty the observer faced was in the case of photometric observation using photographic plate. It was necessary to know both the gamma curve and the spectral response of the process. Comparing two or more pictures was often a difficult task. In addition, the time lag between the recording of an image and its processing was quite long, which reduced observing efficiency significantly.

4.2.3 *Micro-Photometer*

Astronomers faced another problem concerning the measurement of the flux at each point of the plate (or film) whether it represented a field image or a spectrum. In order to determine the exact light level recorded on such a plate, an instrument, called micro-photometer, was developed by Pickering (1910). This instrument measures the relative densities of spectral lines on a photographic plate. The registered signal is measured by the density of the emulsion. If the pixel being examined by such an equipment transmits a fraction I/I_0 of the incident radiation, the density, d, is given by

$$d = -\log_{10} \frac{I}{I_0}. \tag{4.39}$$

The photographic plates were scanned by an instrument which locally measured, using a photoelectric cell, the changes in illumination, which are converted into current variations intensity from the illuminated plate (Carrol and Moss, 1930). With a developed version of such an instrument it became a valuable tool for many investigations besides astronomy. Many branches of applied sciences, such as digital cartography, electron microscopy, medicine, radiography, remote sensing advanced considerably. Later, television cameras, directly mounted on telescopes, were used to replace both the photographic plate and micro-photometer systems (Baum, 1954).

4.3 Photon Detectors

The photon (or quantum) detectors convert incoming photons directly into an electrical signal. They rely on the action of quanta of light energy to interact with electrons in the detectors material and to generate free electron–hole pairs (see Section 2.2.4.1). The photon detectors have wide applications from gamma-rays to the far-IR (McLean, 2008). It is reiterated that the thermal excitation of a carrier from the valence band to the conduction band creates free carriers in both bands. This generation of carriers in the conduction band is applicable in the detection mechanism for both photoconductive and photovoltaic detectors. The photoexcitation is caused by the radiation interacting directly with the lattice sites. Hence, the temperature of the detector must be low enough so that the number of carriers thermally excited across the bandgap is less significant. The electromagnetic radiation can interact with materials in various ways, which are, in general, categorized as: (i) photon effects, (ii) thermal effect (see Section 8.3.1), and (iii) wave interaction effects.

4.3.1 *Photon Effects*

The sensitivity of photon detectors depends on the spectral absorption, as well as on the photoexcitation. The operation of such detectors is associated with the quantum transitions between different energy states. The output electrical signal in a photon detector is determined by the number of the absorbed photons rather than by their energy. The photon effects include interactions of the incident photons with the electrons within the material, whether bound to the lattice (see Section 2.2.1.1) atom or free. In most of the cases the material employed is a semiconductor. The photon effects are important in visible and infrared (IR) detectors. Certain generalities can be made about these effects:

(1) Photon detectors show higher speed of response than the thermal detectors. The signal output in the former case is overshadowed by the noise generated internally to the device at room temperatures. The visible and IR sensors have negative temperature performance coefficients. Since the noise within a semiconductor is partly proportional to the temperature, when cooled, they become more sensitive and their noise generation characteristics reduce significantly. Depending on the composition of the detector, it should be cooled to a certain temperature, below which, leakage current, field emission, and other temperature independent components of the dark current predominate. The present day cooling devices are enumerated in brief:

   (a) **Cryogenic device**: Cryogens are required to be placed inside a dewar, named after its inventor James Dewar who was responsible for liquefying oxygen for the first time. A dewar is a double walled container made out of a suitable material. The space between the two walls is evacuated in order to reduce the conduction of heat from room temperature into the cryogen and power dissipation through electrical leads, reducing convection and gaseous conduction by maintaining vacuum, and limiting radiation losses by reducing emissivity of the cold and warm surfaces that face each other. For astronomical observations, one may require on-chip detector cooling, which are operated at cryogenic temperatures down to the liquid nitrogen (LN_2) temperature at 77 K at the lower wavelengths (≤ 2.5 μm) or the liquid helium (He) temperature at 4.2 to 20 K at the longer wavelengths (Levenduski et al., 1996; Burger et al., 2002; Bhandari et al., 2004). For a device like the CCD (see Section 6.2), the temperature should be maintained between 100 K and 190 K depending on the application, because the charge transfer efficiency, as well as the quantum efficiency drop at very low temperatures. The dark current can approach 0.02 electrons per pixel per hour at 150 K (Janesick et al., 1987).

(b) **Thermoelectric coolers**: The use of aforementioned cryogenic liquids is gradually giving away to closed circuit cryogenic systems (thermoelectric coolers). Also, known as Peltier devices, these are driven by an electric current that pumps heat from the sensor to a heat sink. These devices are lightweight, small and rugged. The heat sink is cooled by passive air, forced air (or liquid, such as water). The final array temperature depends on the amount of the heat generated in the array, the cooling capacity of the thermoelectric device, and the temperature of the heat sink. The present day devices can cool to about 190 K, with water cooling it can cool up to 180 K.

(2) The photon detectors also exhibit a wavelength dependent photosignal per unit incident radiant power, which increases with the wavelength to a long wavelength limit, beyond which the signal falls to zero. Figure 4.8 depicts the output of photon detectors, as well as thermal detectors as a function of wavelength.

(3) The thermal detectors respond to the heat energy delivered by light. Both the thermal conductivity and the coefficient of thermal expansion influence the method of mounting an IR detector in a cooler. Since the thermal detectors rely on the total amount of heat energy reaching the detector, their response is independent of wavelength and extends to longer wavelengths than the response of photon detectors (see Figure 4.8). The

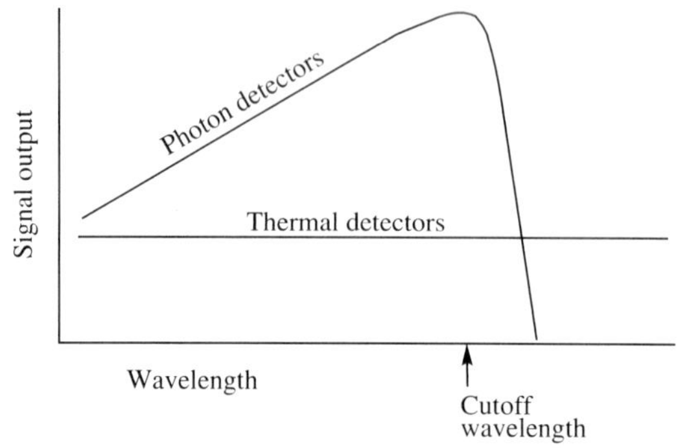

Figure 4.8 Relative output signal (arbitrary units) per unit input for photon detectors and thermal detectors as a function of wavelength (after Rogalski, 2012).

detectors having wavelength beyond 2 μm required to be cooled to minimize the thermal excitation of electrons into the conduction band. With longer limit up to 4–5 μm, they can be cooled to dry-ice (frozen CO_2) temperature, which is of the order of 195 K. The detectors with wavelength limit ranging between 8 and 14 μm require cooling up to 77 K. The lowest practical one is that of a liquid He temperature operating with long wavelength limit beyond 30 μm.

The photon effects can be subdivided into two classes that are elucidated below.

4.3.1.1 Internal photon effects

The photoconduction and junction devices fall under this category. These devices consist of a reverse-biased semiconductor diode whose reverse current is modulated by the electron–hole pairs produced in or near the depletion layer by the absorption of light. Since the charge carriers remain inside the material, it is called internal photoeffect. There are two different modes of operations, such as:

Photoconductivity When radiation is absorbed by a photoconductor, it induces electronic Transitions, which produces a change in electrical conductivity. The rate at which free electrons are generated and the time over which they remain free determines the amount of the increase. The free charges in a semiconductor crystal can occupy different energy levels with an occupancy at an absolute temperature, T (K), which is described by the Fermi distribution function (see equation 2.26). As stated in Section 2.1, for a photoelectric emission, the energies of photons impinging on the material surface provide the extra energy for electrons to overcome the energy barrier. From the energy diagram depicted in Figure 2.11, the transitions can be envisaged as (i) an electron–hole pair may be created by a high-energy photon, (ii) a photon may excite a donor electron into the conduction band, or (iii) a valence electron may go into an acceptor state. The first transition is known as intrinsic excitation, while the last two transitions are referred to as extrinsic excitations. Since the density of states in the conduction and valence bands exceeds the density of impurity states, the photoconductivity is largely due to the intrinsic excitation. The photoconductivity can be categorized into the following:

(1) **Intrinsic conductivity**: A change in photoconductivity takes place when excitation of electrons from the valence to the conduction bands of a semiconductor in which both free electrons and holes change. This process occurs when the photon energy is greater than the intrinsic bandgap energy, E_g (see Section 2.2.3.1). If the wavelength, λ, of the radiation exceeds the critical (or cut-off) wavelength, λ_c, defined by equation (2.11), the energy of the photons becomes less than E_g; hence, such a photon cannot cause a valence electron to enter the conduction band. The concentration of charge carriers, N_{ph0}, integrated for time, Δt, is expressed as

$$N_{ph0} = q \mathtt{A}_d \eta_d \Delta t, \tag{4.40}$$

where $\mathtt{A}_d$ is the area of the detector in m^2, q the electron charge, and η_d the quantum efficiency of the detector with unit electrons/photons; it follows from the law of mass action

$$N_{ph0} = T^{3/2} e^{-E_g/(2k_B T)}.$$

For $k_B T > 2 \times 10^{-2} E_g$, in which k_B denotes Boltzmann's constant, and T the temperature, the current is dominated by thermal excitation of the electrons near the conduction band, and not by photoionization (see Section 5.2). Hence, a photoconductor is required to be cooled to a temperature, $T \ll E_g/k_B$, and the temperature must decrease as the energy of the quanta decreases.

The semiconductor materials that are completely free of impurities and defects (see Section 2.2.6.1) are used for photon detection, which are known to be intrinsic detectors. Pure crystals of germanium (Ge) and silicon (Si) and compounds containing two or more components in stoichiometric proportions, for instance, lead sulfide (PbS), indium antimonide (InSb), indium arsenide (InAs), mercury cadmium telluride (HgCdTe), lead tin telluride (PbSnTe), are intrinsic semiconductors. Although there are advantages in such detectors, for example, (i) easy bandgap tailoring, (ii) good material and dopants, and (iii) possible monolithic integration, the disadvantages are many, for example, (i) non-uniformity over large area, (ii) high thermal expansion, (iii) high cost in growth and processing, (iv) surface instability, (v) hetero-epitaxy with large lattice mismatch, and (vi) long wavelength cut-off limited to 7 μm at 77 K.

(2) **Extrinsic conductivity**: The ionization of impurities in semiconductors provide free electrons and holes, which lead to photoconductivity. The extrinsic detectors are based on silicon or germanium doped with impurities (see Section 2.2.6.2). Examples of such detectors are germanium doped with gold, mercury, cadmium, copper or zinc. The major difference between an intrinsic and extrinsic material is that in the former case, the electron–hole pair (see Section 2.2.4.1) is the charge carrier whereas in the latter, either of the two is the charge carrier. The extrinsic detectors have a low absorption coefficient, irregular performance, and poor radiation hardness with extreme noise due to dark current. In order to reduce the dark current, an intrinsic layer of the Si-semiconductor is introduced directly below the top contact. This thin layer of silicon blocks the impurity band current while allowing the excited electrons into the conduction band, and get collected at the contact. This type of detector is called blocked impurity

band detector with a higher S/N ratio. However, this needs to be kept at a low cryogenic system (<10 K), which makes the device expensive. Also, it is difficult to adjust the condition, for Space Observatory in particular. The extrinsic blocked impurity band devices are, in general, used for astronomy in the 20–200 μm wavelength range. Among the disadvantages of such a system, (i) high thermal generation, (ii) extremely low temperature operation, and (iii) complicated design are noteworthy.

Quantum wells are another type of PC detectors (see Section 5.5.1.1) that have inter sub-band transitions leading to photoconductivity. The free carrier conductivity occurs when radiation is absorbed by free electrons in a semiconductor, especially at a very low temperature where the coupling between the electrons and lattice becomes weak. This process leads to a change in electron mobility and hence a change in conductivity. Examples of such materials are AlGaAs/GaAs and Si/SiGe.

Photovoltaic effect The photovoltaic (PV) effect (Becquerel, 1839) is created by the voltages or electric current in a material upon exposure to light. The electrons are transferred between different bands, i.e., from the valence to conduction bands, within the material, resulting in the buildup of voltage between two electrodes. If two charge carriers, an electron and a hole, are released by an incident photon on the unbiased junction, the voltage across it changes by an amount

$$\Delta V = \frac{q\delta x}{\epsilon \mathtt{A}}, \tag{4.41}$$

in which δx designates the thickness of the junction, $\mathtt{A}$ its area, q the electron charge, and ϵ the electrical permittivity of the material.

The change in voltage is proportional to the incident radiation. As stated in Section 2.2.7, the electrical contact, established between two semiconductors of the same base material, albeit different impurity so as to make one P-type and the other N-type, would produce energy. The energy levels in the P-region is displaced upwards to the N-type region because of space-charge effects at the boundary. When photons are absorbed at the boundary region, which results in the production of free electron–hole pairs that are separated by the electric field. This produces a voltage across the junction, which causes a closed circuit current. The spectral response is determined by the forbidden gap energy, E_g.

4.3.1.2 External photon effects

The non-amplifying phototube and the amplifying photomultiplying tube (see Section 5.2.2) fall under this category. The photoemissive (PE) effect involves when a photon of a particular frequency impinges on the surface of a semi-cylindrical metal plate with an emitting surface, called photocathode, transfers the energy to the surface electrons. Some of them leap out by overcoming the binding force, and are attracted to anode. A photocathode that serves as a negative terminal is generally maintained at a lower potential with respect to the anode. As a consequence of the electron transport between these elements, a photocurrent (of the order of $\sim\mu$A) proportional to the photon flux, is ensued in the external circuit. As discussed in Section 2.1, the conventional photocathodes (metal or semiconductor photocathodes) are characterized by a photoelectric work function, ϕ_0, representing the energy required to raise an electron from the Fermi level, E_F, to a level above the potential barrier at the surface, thereby permitting the electron to escape from the material.

4.3.1.3 Other photon effects

In addition to the aforementioned photon effects, there are a few other photon effects, which are enumerated below:

Photoelectromagnetic (PEM) effect This effect may be described as the Hall effect of a photodiffusion current. If an electric current, $\mathbf{i}$, passes through a specimen (conductor or semiconductor) placed in a magnetic field, $\mathbf{B}$, a potential proportional to the current and to the magnetic field is developed across the material in a direction perpendicular to both the current and to the magnetic field, which is referred to as Hall effect (Hall, 1879). The charge carriers experience a Lorentz force, $q\mathbf{v} \times \mathbf{B}$, in which q is the charge and $\mathbf{v}$ the velocity of the carriers, which would deflect them toward one side of the slab. A build-up of charge at the sides of the specimen would balance this magnetic influence, producing a measurable voltage between the two sides of the specimen, i.e., $V_H = Bi/(\rho w)$, in which ρ is the charge density and w the width of the specimen in the direction of the magnetic field. The photo-Hall effect is an effect in which the illumination of a semiconductor in a magnetic field produces a change in its Hall resistance. Light impinging perpendicularly on one surface of a semiconductor is absorbed. The excess electron–hole pairs produced at the illuminated surface, set up a concentration gradient and diffuse downwards. These oppositely charged carriers moving in the same direction are separated longitudinally under the influence of the magnetic field in the z-direction. This separation of charge produces an output voltage that fluctuates according to the fluctuation of photons.

Photo-Dember effect This effect consists in the formation of a charge dipole in the vicinity of a semiconductor surface after the fast photogeneration of charge carriers (Dember, 1931). If the sample thickness and carrier diffusion lengths are large compared to the absorption depth of radiation, the carrier would diffuse away from the front surface to the back. If the carrier mobilities and diffusion lengths are unequal, a charge separation occurs, giving rise to an electric field between the front and back surfaces. Thus, a photocurrent flows in the external circuit connecting these surfaces. The photo-Dember effect is present in most semiconductors but it is particularly strong in narrow bandgap semiconductors, such as InAs and InSb owing to their high electron mobility.

4.3.1.4 Wave interaction effects

The wave interaction effects arise from the interaction of the electromagnetic field of the incident radiation with the sensing material. There exist two such effects:

Optical heterodyne detection The heterodyne detection technique helps to detect radiation by non-linear mixing with radiation of a reference frequency. Such a technique depends on the interaction of the electric field vector of the signal radiation with that from the reference source. The excitation rate of free carriers in a semiconductor depends upon the rate of absorption of photons, which is a measure of the intensity of the absorbed radiation.

As stated in Section 4.1, a photovoltaic (PV) or photoconductive (PC) detector is a square law device, an alternate way of viewing photoexcitation is that the semiconductor acts as a mixer element, beating electric field vector against itself in a homodyne fashion. The mixing action occurs if two coherent sources of radiation having frequencies, ω_1 and ω_2 are superimposed upon a semiconductor. Let $\mathbf{E}_1$ and $\mathbf{E}_2$ be the amplitudes of the two waves; therefore, the resultant intensity is

$$\begin{aligned} I &= \mathbf{E}^2 = [E_1 \cos\omega_1 t + E_2 \cos\omega_2 t]^2 \\ &= E_1^2 \cos^2\omega_1 t + E_2^2 \cos^2\omega_2 t + E_1 E_2 \cos(\omega_1 - \omega_2)t \\ &\quad + E_1 E_2 \cos(\omega_1 + \omega_2)t. \end{aligned} \tag{4.42}$$

If ω_1 and ω_2 are close, the frequency of I is low enough to fit in the bandwidth of the detector and its electronics and I carries the phase information from the radiation of the source. The electron lifetimes in semiconductors are orders of magnitude too large to permit photoexcitation at the frequency of $\sim 10^{14}$ Hz. However, if the difference frequency is less than 1 GHz, the photoexcitation at difference frequency can occur if the response time is sufficiently short, for example, in the nanosecond range. This indicates that the wavelength of two sources are nearly equal. The performance of a heterodyne coherent detection system can be evaluated either on the basis of carrier-to-noise (C/N) ratio at the intermediate frequency (IF) filter output or by the S/N ratio at the low-pass filter output of the envelope detector. To note, a heterodyne receiver has a second level of detection at base-band in the form of an envelope detector (linear or square-law).

Optical parametric effect The parametric mixing is the mixing of electromagnetic waves to form waves with frequencies linearly related to the frequency of incident radiation in a medium possessing optical non-linearities. This is employed to enhance the intensity of weak signal and to change the optical frequency to a spectral interval where it is more easily detectable by a photocell having a high spectral sensitivity. The parametric up-conversion is important. The crystals useful for such conversion should exhibit non-linear optical properties, such as birefringent and optically transparent at the frequencies of interest.

4.3.2 *General Requirements of an Ideal Detector*

A photodetector is, in general, viewed as a two port device with an optical input and an electrical output. As a linear system, it is characterized by its impulse or frequency response, whose performance depends on two parameters, namely, bandwidth and quantum efficiency. In a circuit of a photodetector consisting of a capacitor, C and a resistor, R, the bandwidth, Δf, is limited by the response of the current source and the response of the overall circuit. The response of the current source applies to all photodetectors and is determined by the velocities of the photogenerated electrons and holes and the distances they travel. If the carriers are uniformly generated across the depletion region and the carrier velocities are constant and equal, The transit bandwidth limitation is dictated by

$$\Delta f = 0.55 \frac{v_\text{d}}{w},$$

in which v_d is the drift velocity (see Section 2.2.4.2) and w the thickness of the depletion region, which is the distance traveled by the carriers; a large bandwidth requires a thin depletion region. The RC bandwidth limitation is given by

$$\Delta f = \frac{1}{2\pi RC}, \tag{4.43}$$

in which $C(= \epsilon \mathtt{A}/w)$, ϵ is the dielectric constant of the depletion layer and $\mathtt{A}$ its area.

4.3.2.1 Photocurrent

The high electric field present in the depletion region causes photogenerated carriers to separate and be collected across the reverse-biased junction. This gives rise to a photocurrent flow in an external circuit, i_ph. When a light source illuminates directly such a detector, the i_ph becomes a function of the detector sensitivity and the detector area, the distance from the source to the detector and the source intensity. Let the modulated optical power signal, $P(t)$, falls on the photodetector with the form of

$$P(t) = P_0[1 + ms(t)], \tag{4.44}$$

where P_0 designates the incident radiation power to the detector input of area A_d, $s(t)$ the message electrical signal, and m the modulation index.

Since the measured photovoltage distribution is not exactly symmetric (Diament and Teich, 1992), following equation (3.21), it is approximated as a Gaussian function, which is given by

$$\mathcal{P}(V) = \frac{1}{\sqrt{2\pi \langle \sigma_V \rangle^2}} e^{-(V-\bar{V})^2/(2\langle \sigma_V \rangle^2)}, \tag{4.45}$$

with V as the voltage, $\mathcal{P}(V)$ the probability density at V, $\bar{V}$ the mean voltage, and $\langle \sigma_V \rangle^2$ the variance in voltage distribution; the voltage is related to the photocurrent by the detector load resistance of the detection circuit.

4.3.2.2 Gain

The gain, G, is defined to be the power amplification in decibels (dB). The primary photocurrent, i_{ph0}, defined as the current passing through the photodetector, when all the photogenerated carriers are separated in the electric field and are swept out without replenishment. The optical power absorbed, $P(x)$ in the depletion region can be written in terms of incident optical power, P_0:

$$P(x) = P_0 \left(1 - e^{-\alpha_s(\lambda)x}\right), \tag{4.46}$$

with $\alpha_s(\lambda)$ as the optical power absorption coefficient, which depends on wavelength and x the distance into the material.

The i_{ph0}, resulting from absorption is

$$i_{\mathrm{ph0}} = \frac{P_o}{h\nu} \left[1 - e^{-\alpha_s(\lambda)w}\right] (1 - R_f), \tag{4.47}$$

where P_o denotes the output optical power, w the depletion layer width, h Planck's constant, ν the frequency, and R_f the entrance face reflectivity (Fresnel reflectivity), which is an essential condition to attain a high device quantum efficiency, a low reflection coefficient at the top surface of a photodetector is required, which can be realized by the special dielectric coatings.

The primary photocurrent (for P-I-N photodiode $M = 1$) is given by

$$i_{ph0} = I_{DC} + I_{AC}(t) = \frac{\eta_d q}{h\nu} M P(t), \quad (4.48)$$

in which I_{DC} conveys the direct current (DC), I_{AC} the alternate current (AC; the flow of electric charge periodically reverses direction), η_d the quantum efficiency, q the electron charge, and the multiplication factor, M

$$M = \frac{1}{1 - (V/V_{br})^{\gamma}}, \quad (4.49)$$

V the applied voltage, γ the constant depending upon the semiconductor material, doping profile, and radiation wavelength, and V_{br} the breakdown voltage.

The photocurrent autocorrelation function, Γ, is defined as

$$\Gamma(\tau) = \langle i_{ph}(t) i_{ph}(t + \tau) \rangle,$$

where $i_{ph}(t)$ is the instantaneous photocurrent at time t, τ a time delay, and the angle brackets, $\langle\,\rangle$, stands for expectation values.

If all the carriers are swept out before recombining and if there is no replenishment by re-injection, the photocurrent, $i_{ph} = i_{ph0}$. The i_{ph} is related to the incident optical power as

$$i_{ph} = q\eta_d \frac{P_0}{h\nu}.$$

On accounting various mechanisms for increasing or decreasing the actual photocurrent, an additional factor called gain, G, is considered; hence, the photocurrent becomes

$$i_{ph} = q\eta_d \frac{P_0}{h\nu} G. \quad (4.50)$$

The bandwidth-efficiency product is an important figure of merit for photodetectors. The general form for external quantum efficiency of such a detector is

$$\eta_d = \zeta \left[1 - e^{-\alpha_s w}\right] (1 - R_f), \quad (4.51)$$

with ζ as the fraction of electron–hole pairs (see Section 2.2.4.1) that contribute to the detector current.

A means of obtaining high bandwidth-QE product is to choose a material having a large product of carrier velocity and absorption coefficient at the wavelength range of interest.

4.4 System Analysis

The usefulness of the entire equipment for any particular application needs to be determined by system analysis. This method includes the determination of all boundaries and constraints the interdependences and the interactions of various parts of the system, and also noise, which would limit the overall dynamic range of the system. A system analyst attempts to determine its performance from the knowledge of the performance of the individual building blocks; a certain amount of compromise is required. The system that yields the specific performance estimate within acceptable limits with the least expenditure of the resources may be pursued.

4.4.1 *Principal Performance Functions*

Although, in Section 1.4.1, the determination of final image fidelity from a knowledge of the input/output relationships of the subsystems, for example, transfer function, was emphasized, the following subsections highlight some of the limiting functions encountered in photoelectronic imaging. The imaging systems may introduce a certain amount of distortions or artifacts in the signal. There are several techniques that can evaluate the image quality. However, in the involvement of the human element in photoelectron system, the tolerance of the human observer to varying degrees of some image impairments need to be considered.

4.4.1.1 Limiting resolution

The resolution of an imaging system is defined as its ability to distinguish object detail. It is expressed as the total number of lines that may be discerned within a specified unit of length, lines per picture height or in terms of line-pairs (a sequence of one black line and one white line) per millimeter (mm). This measure is also known as frequency; the inverse of which yields the spacing in millimeter between two resolved lines. Under bright light level, the image is virtually free of photon granularity. The optical resolution in line-pairs/mm is constant, which is independent of illumination intensity. These working conditions are referred to as resolution limited regime. The limiting resolution is described by means of the modulation transfer function (MTF; see Section 1.3.1), which is the ratio of output signal contrast between black and white to input image contrast, and the contrast transfer function (CTF). The CTF of an optical system is a function of the spatial modulation frequency of the input signal. The input must be 100% modulated and may be an image of parallel black and white line-pairs. The output turns out to be an image of darker gray and lighter gray line-pairs. The contrast depends on the spatial frequency; the finer the input pattern, the smaller the level difference between white and black for the image signal becomes. The human eye is able to notice a contrast down to 3%. To note, the MTF is dependent on both the lens and the image sensor. If mirrors are used to reflect light onto an imaging sensor, this affects the MTF as well. Thus, one reads the MTF of such a sensor as the product of the lens MTF and the sensor MTF. In the measurement of the MTF, a sinusoidal input signal is applied, while in the case of CTF measurement, a square wave function is used. The limiting resolution of certain components depends on the various factors, for example, in the case of a photodetector, it depends on the acceleration voltage between photocathode and the micro-channel plate (MCP; see Section 5.2.7), while in the case of phosphor screen, it is given by the voltage between the MCP and the phosphor screen, and also by the thickness of the phosphor layer.

4.4.1.2 Non-linearity

The non-linearity is the variation in the ratio of the change in photocurrent to the same change in light power. The detectors are characterized by a response that is linear with incident intensity over a broad range. However, most of the detectors have non-linear intensity transfer functions (ITF), for instance, photographic plates, the effects of the dead time at high count rates in photon-counting systems (see Section 7.2), and the low light level non-linearity of image dissector scanners. Many of the elements contained in the electro-optical systems are also non-linear. The ITF is a plot of the output signal of an element as a function the applied signal. While plotting on log-log coordinates, the linear slope of such a system defines the gamma of the system. The output signal is written as

$$E_{\mathrm{o}} = K E_{\mathrm{i}}^{\gamma}. \tag{4.52}$$

In reality, γ of a conventional cathode ray display is typically ranges between 2 and 3. The CCDs are, in general, nearly linear compared to the most other detectors used in astronomical observations, although in particular pixels or regions, CCDs are suffer from deviating signal zero points.

4.4.1.3 Spectral response

Depending on the frequency of the incident radiation, the spectral response describes the wavelength range over which a sensor responds. In a phototube, the spectral response is represented in units of amperes per watt of radiant energy versus wavelength. Let P_λ be the input spectral radiant flux to the detector. The total radiant flux, P_0 is given by

$$P_0 = \int_0^\infty P_\lambda d\lambda. \tag{4.53}$$

If the output of the detector varies with the wavelength of the input radiation, the expression for the output power, P_{o} is

$$P_{\mathrm{o}} = \int_0^\infty s(\lambda) P_\lambda d\lambda, \tag{4.54}$$

with $s(\lambda)$ as the spectral response of the detector.

4.4.1.4 Field-of-view

The field-of-view (FOV) of an optical system is defined as the maximum angular size of the object as seen from the entrance pupil. It is determined by magnification and the focal lengths of the objective and eyepiece lenses; the more magnification the less FOV. The FOV of a sensor on an optical instrument, such as a telescope, is critical when selecting targets for imaging. It is a determining factor in establishing the S/N ratio of the system when limiting the radiant energy collected and transmitted to the photosurface. The focal length of a lens determines the angular field that can be envisaged by a given sensor. At infinity, the angular FOV for an object is

$$FOV = \frac{\tan^{-1} D}{2 f_{\rm e}}, \tag{4.55}$$

in which D is the photosurface diagonal and $f_{\rm e}$ the effective focal length of the lens.

The relative aperture of a lens system, $F_{\#} = 1/(2n \sin\alpha)$, in which n denotes the refractive index in the image space and α the angle subtended at the axial point of the image by the semi-diameter of the circular exit pupil of the lens at a given diaphragm setting, is a measure of its ability to collect radiant energy and direct this energy toward an image point. The image irradiance, $I_{\rm e}$ is related to the object irradiance, I, by

$$I_{\rm e} = \frac{\pi t I}{4 F_{\#}^2 (\mathtt{M}^2 + 1)}, \tag{4.56}$$

where t is the lens transmission and $\mathtt{M}$ the magnification of the image or object.

4.4.2 *Limits on Radiation Detector Sensitivity*

One of the important problems, one faces in the context of interferometric experiment is the non-availability of a detector linearly sensitive to the electromagnetic field in the optical domain, which forbids post-detection amplification and imposes the requirement to encode information prior to detection. As stated (see Section 1.2), the light intensity is transformed into electrical signal by a detector. The minimum detectable power for a photodetector is limited by noise (see Section 4.5). The fundamental noise for an ideal amplitude interferometer that detects both polarizations, arises due to thermal radiation incident on the sensor. For an ideal sensor, this produces a noise power fluctuation of

$$N_{\rm d} = h\nu \left[\frac{2\Delta f}{t} \frac{1-\epsilon}{e^{h\nu/k_{\rm B}T} - 1}\right]^{1/2}, \tag{4.57}$$

in which k_B is Boltzmann's constant, $h\nu$ the quantum energy, h Planck's constant, T the temperature of optics and atmosphere through which the signal is received (in K), Δf the bandwidth, t the post-detection averaging time (s), and ϵ the fractional transmission of light reaching the telescope.

Equation (4.57) is valid for a photodiode. An amplitude interferometer cannot determine the phase of a wave (phase is complementary to energy or number of quanta); hence, noise due to the uncertainty principle is absent. A significant limitation comes from this bandwidth of the instrument, for instance, an interferometer in which the time resolution of the detectors should reach 1 ms, the effect of which is to produce a frequency dispersion.

4.5 Noise

The noise is envisaged as an intensity variation with time that exists within the reproduced image of a photoelectronic system. It occurs due to fluctuations in the number of quanta in the received light, which is characterized by statistical parameters, such as mean value, mean square value, standard deviation, correlation function, etc. The noise may arise within the detector, in the radiant energy to which the detector responds, or in the electronic system that follows the detector. Most of the present day detectors are made from semiconductors, the identical block of such material should meet the optical and electrical requirements. These requirements impose limitations on the performance of imaging system, under low-light level in particular. In what follows, the various kinds of noise associated with such detectors are enumerated briefly.

4.5.1 *Radiation Noise*

The performance of the optical detection is reached when there is no noise generated internally, no amplifier noise, and there is no electromagnetic interference and no radiating background against which the signal must be detected, the events produced within the detector are due to signal photons. Under these conditions, the electric pulses generated by the light detector are due to signal photons. The radiation noise includes two categories, for example:

4.5.1.1 Photon signal fluctuation limit

As the arrival of photons is randomly distributed over time, shot-noise (see Section 4.5.2.3) is introduced into the light signal. This limit is known as the signal fluctuation limit, since the random nature of the rate of arrival of signal photons in a given observation time determines the minimum power that can be detected. The signal fluctuation limit is operative for ultraviolet and visible detectors. Consider a photodiode viewing a monochromatic source of wavelength, λ. The short-circuit signal, i_{sc}, current is given by

$$i_{\mathrm{sc}} = \eta_{\mathrm{d}} q N_{\mathrm{ph}}(\lambda), \tag{4.58}$$

where q is electron charge, $N_{\mathrm{ph}}(\lambda)$ the number of absorbed photons, and η_{d} the quantum efficiency of the detector.

Neglecting all the internal noise, as well as the background radiation falling on the detector, the detector output short-circuit noise current, i_{n}, is expressed by

$$i_{\mathrm{n}} = \sqrt{2qi_{\mathrm{sc}}\Delta f}, \tag{4.59}$$

with Δf as the bandwidth of the system.

Therefore, the S/N ratio (see Section 4.1.1.4) is

$$\frac{i_{\mathrm{sc}}}{i_{\mathrm{n}}} = \sqrt{\frac{\eta_{\mathrm{d}} N_{\mathrm{ph}}(\lambda)}{2\Delta f}}. \tag{4.60}$$

If $P_0(\lambda)(= N_{\mathrm{ph}}(\lambda)h\nu)$ is the incident power of wavelength, λ, the number of absorbed photons, $N_{\mathrm{ph}}(\lambda)$ turns out to be $\lambda P_0(\lambda)/(hc)$, and so

$$P_0(\lambda) = \frac{2hc\Delta f}{\lambda\eta_{\mathrm{d}}}\left(\frac{i_{\mathrm{sc}}}{i_{\mathrm{n}}}\right)^2. \tag{4.61}$$

Since the minimum detectable signal is the signal for which the output S/N ratio is unity, that is

$$P_0(\lambda, \min) = \frac{2hc\Delta f}{\lambda\eta_{\mathrm{d}}}. \tag{4.62}$$

For any real situation, however, the observer can wait only a finite length of time for a photon to be detected. There must be a reasonably high probability that for a given average rate of arrival, at least one photon would arrive during the observation time, Δt, for a given average photon flux onto the detector. Here Poisson's distribution function (see equation 4.1) must be employed to describe the detection probability, since the average number of photons, $\bar{N}_{\text{ph}}$ being detected within the observation interval, Δt, is assumed to be 1, which is small. In a practical sense, this probability must be at least 0.99, i.e., there is a one percent probability that no photon would be detected; hence, $\mathcal{P}(0) = e^{-\bar{N}_{\text{ph}}} = 0.01$ or $\bar{N}_{\text{ph}} = 4.61$. Thus, the minimum detectable power is

$$P_0(\lambda, \text{min}) = \frac{\bar{N}_{\text{ph}} h\nu}{\Delta t} = \frac{4.61 hc}{\Delta t \lambda \eta_{\text{d}}}, \tag{4.63}$$

with $\Delta f = 1/(2\Delta t)$ and hence

$$P_0(\lambda, \text{min}) = \frac{9.22 hc \Delta f}{\lambda \eta_{\text{d}}}. \tag{4.64}$$

4.5.1.2 Background photon fluctuation limit

Often the detection limit is not the signal fluctuation limit but the background fluctuation limit, also known as the photon-noise limit or background limited performance (BLIP) limit, which is operative for IR detector. Since the background level fluctuates about the average value due to the random nature of the photoemission, the signal radiation is compared to the fluctuations, or noise in the background radiation. The background fluctuation limit arises from fluctuations in the background radiation that is incident on the detector. This characterizes the Earth, most tactical targets, and some of the internal surfaces of the detector housing. The radiating background arises generally from objects at an ambient temperature of about 295 K; the thermal emission from a body having such a temperature peaks at 10 μm. The background limit for a photon detector depends on (i) the spectral distribution of the target, (ii) the spectral distribution of the background, (iii) the spectral response of the detector, (iv) the temperature of detector, (v) the mode of operation of the detector, and (vi) the field-of-view within which the detector receives background radiation. The average rate of arrival of background photons of frequency, ν_{b} at a detector of sensing area,

$\mathtt{A_d}$ is given by

$$\bar{N}_{\text{ph, b}}\mathtt{A_d} = \frac{M(\nu_b, T_b)}{h\nu_b}\mathtt{A_d} d\nu_b, \tag{4.65}$$

where T_b is the background temperature

$$M(\nu, T_b) = \frac{2\pi h\nu_b^3}{c^2[e^{(h\nu_b/k_B T)} - 1]}. \tag{4.66}$$

Planck's distribution function (see Section 1.1.1), and k_B Boltzmann's constant; the background is assumed to emit blackbody radiation of temperature, T, into the viewed solid angle, $\pi \sin^2(\theta/2)$, with θ as the full cone angle, of the detector.

The modulation frequency dependence of the mean square fluctuations in the rate of generation of current carriers due to the arrival of detectable photons is

$$\begin{aligned} P_{N_{\text{ph}}}(f) &= \mathtt{A_d} \int_{\nu_0}^{\infty} \eta(\nu_b)\langle\sigma\rangle^2 d\nu \\ &= \mathtt{A_d} \int_{\nu_0}^{\infty} \eta(\nu_b)\frac{M(\nu_b, T)e^{(h\nu_b/k_B T)}}{h\nu_b[e^{(h\nu_b/k_B T)} - 1]} d\nu, \end{aligned} \tag{4.67}$$

with $\nu_0 = hc/\lambda_0$,

$$\langle\sigma\rangle^2 = \bar{N}_{\text{ph, b}} \frac{e^{(h\nu_b/k_B T)}}{e^{(h\nu_b/k_B T)} - 1}. \tag{4.68}$$

as the variance in $\bar{N}_{\text{ph, b}}$, and λ_0 the long wavelength limit of the detector, which is assumed to be equal to the wavelength of the monochromatic signal source.

The detection limit is higher since the quantum efficiencies of the real detectors are smaller than one. Such a limit is further enhanced by the thermal noise of the detector load resistance (see Section 4.5.2.1), as well as the shot-noise (see Section 4.5.2.3) of the dark current, i_D (see Section 4.1.1.8), of the detectors, and the equivalent noise resistance of the signal amplifier (see Section 4.5.3). This results in an increased noise equivalent power (NEP; see Section 4.1.1.4; Heering, 1998). The rate of generation of carriers excited by photons from a monochromatic source of power, $P_s(\lambda)$ incident on the detector is written as

$$N_{ph,s} = \frac{\eta_d(\nu_s)P_s(\lambda)}{h\nu_s}, \tag{4.69}$$

in which $\eta_d(\nu_s)$ is the quantum efficiency for photons of frequency, ν_s.

The RMS fluctuations in the bandwidth, Δf, are given by the square root of the quantity $2\Delta f$ times the frequency dependence of the mean square fluctuations. Hence, for the signal equal to noise

$$N_{\mathrm{ph},s} = \sqrt{2\Delta f P_{N_{\mathrm{ph}}}(f)}. \tag{4.70}$$

On combining Eqs. (4.67), (4.69), and (4.70), one gets

$$P_{\mathrm{s,\,min}} = \frac{h\nu_s}{\eta_d(\nu_s)}\sqrt{2A_d\Delta f}\left\{\int_{\nu_0}^{\infty}\frac{\eta_d(\nu_b)2\pi\nu_b^2 e^{(h\nu_b/k_B T)}}{c^2[e^{(h\nu_b/k_B T)}-1]^2}d\nu\right\}^{1/2}. \tag{4.71}$$

As stated in Section 4.1.1.4, the reciprocal of NEP is the detectivity. The specific detectivity, $D^*(\lambda)$, is derived as

$$D^*(\lambda) = \frac{\eta_d(\nu_s)}{h\nu_s}\left\{\int_{\nu_0}^{\infty}\frac{4\pi\eta_d(\nu_b)\nu_b^2 e^{(h\nu_b/k_B T)}}{c^2[e^{(h\nu_b/k_B T)}-1]^2}d\nu\right\}^{-1/2} \tag{4.72}$$

$$= \frac{\sqrt{A_d\Delta f}}{P_{\mathrm{s,\,min}}}. \tag{4.73}$$

For $h\nu_b/(K_B T) \gg 1$, Eqs. (4.71) and (4.72) simplify to

$$P_{\mathrm{s,\,min}} = \frac{h\nu_s\sqrt{2A_d\Delta f\bar{N}_{\mathrm{ph,\,b}}}}{\eta_d(\nu_s)} = \frac{hc\sqrt{2A_d\Delta f\bar{N}_{\mathrm{ph,\,b}}}}{\lambda_s\eta_d(\nu_s)}, \tag{4.74}$$

and

$$D^*(\lambda) = \frac{\eta_d(\nu_s)}{h\nu_s\sqrt{2\bar{N}_{\mathrm{ph,\,b}}}}. \tag{4.75}$$

The fluctuations depend on the following aspects:

Mode of operation For photoemissive (PE) and photovoltaic (PV) detectors, where the signal is developed as the excess carriers move across the boundary at the photocathode or the P-N junction, the fluctuations in the carrier generation rate due to fluctuations only in the rate of arrival of photons to cause the fluctuations in the signal from the detector. In the photoconductors, the signal arises from the change in the concentration of the carrier resulting in a change in resistance. Fluctuations in both generation and recombination (GR) rates cause fluctuations in the instantaneous value of the carrier concentration. Since noise power due to recombination fluctuation equals that due to generation fluctuations, i.e., $P_{N_{\text{ph}}}(f)$ of equation (4.67) should be doubled and $D^*(\lambda)$ is reduced by a factor of $\sqrt{2}$.

Operating temperature If the sensitive element of the detector is at the same temperature as the background, it would receive radiation from both the forward and reverse hemisphere. The calculations above assumed photons from the former hemisphere. Although the rear side of the detector element is mounted on a substrate, the radiation would enter either through or from the substrate. For PV detectors, the backside is much farther from the junction than the sum of the optical absorption depth and the carrier diffusion length. Therefore, most of these detectors have a preferred surface, and the background limit does not depend on the operating temperature.

To note, the photoconductors respond to the number of photoexcited carriers, irrespective of where they are excited in the materials. Thus, contributions from both back and forward hemispheres, reduce the specific detectivity, $D^*(\lambda)$, by the factor of $\sqrt{2}$. The sensors with highest detectivity are photomultipliers, in the ultraviolet and short range of visible range in particular. They are less detective than an ideal detector. The PE detectors (see Section 5.2) having translucent photocathodes would respond to radiations from these hemispheres, while opaque cathodes do not. By cooling (see Section 4.3.1) the detectors below the background the radiation from back hemispheres can be made very small. If the temperature, T, is brought to zero, according to equation (4.62), the expression (4.71) turns to the lower limit, because the dark current, in general, vanishes at absolute zero. Equation (4.71) implies a S/N ratio of 1. The signal fluctuation limit of an ideal detector can be reached by selected cooled charge coupled devices (CCD; see Section 6.2). As stated in Section 2.3.2, the metal-oxide semiconductors capacitors are made of silicon (Si), the gate electrodes of which are separated from the semiconductor by a silica layer. The noise of its dark current effects a marginally higher detection limit than that of an ideal detector. The noise equivalent charge of such a device can be down to a few electron charges (Heering, 1998).

4.5.2 *Intrinsic Detector Noise*

Since the incoming photons have photon (shot) noise (see Section 4.5.2.3), due to which the signal contains an intrinsic noise containing a contribution of detector noise. Also it has several other sources, for instance, (i) Johnson (thermal) noise (see Section 4.5.2.1) found in all resistive materials including semiconductors, (ii) generation-recombination (GR) noise (see Section 4.5.2.2) caused by fluctuations in the rates of thermal generation and recombination of free carriers in semiconductors, (iii) flicker ($1/f$) noise (see Section 4.5.2.4) being associated with the presence of potential barriers at the contacts or surface of semiconductors, and (iv) pattern noise resulting from the local variation of the dark current, i_{D}, photoresponse and the cut-off wavelength. The real detection limit is often considerably higher than the signal fluctuation limit. While $1/f$-noise and GR-noise have a power spectrum, which falls to higher frequencies as $1/f^n$ with $0.8 < n < 1.5$, respectively $n = 2$, shot-noise and thermal noise exhibit a white noise (a random signal with a flat frequency spectrum) spectrum (Van der Ziel, 1968).

In the detector output, the additive noise, which superimposed on the signal by the statistical addition and exemplified by the thermal agitation noise of an amplifier being added to the signal, is prominent; the photoconductors have more noise by a factor of $\sqrt{2}$. Another type of noise is the product noise exemplified by mechanism, such as photon-noise, phosphor grain, and target grain noise and is characterized by the fact that it disappears or is reduced significantly reduced upon the removal of the input signal.

4.5.2.1 Thermal noise

The thermal noise, also called Johnson noise or Nyquist noise, is the electronic noise generated by the thermal agitation of the electrons inside an electrical conductor at equilibrium. Regardless of any applied voltage, it arises from the thermal fluctuations in the electron density within a conductor or a semiconductor. It can also be the random voltage noise generated in the external resistor either at the input of an amplifier or at the output of the multiplier by thermally driven Brownian motion (random walk; Einstein, 1905d) of charge carriers. An idealized resistor (in series with a voltage source) is assumed to contain a voltage generator causing a fluctuating EMF at the terminals; therefore, in terms of voltage measured on the load resistor, explained by Nyquist (1928), is

$$\sqrt{\overline{V_{\text{th}}^2}} = \sqrt{4R_{\text{L}}k_{\text{B}}T\,\Delta f},$$

where k_{B} is Boltzmann's constant in joules per kelvin and T the resistor's absolute temperature in kelvin, Δf the bandwidth, R_{L} the load resistor of the detector circuitry, and the bar represents a time average.

The load resistance is not a unique quantity. Its magnitude is often determined by the value of the detector resistance, which is quite often temperature and background dependent. In order to obtain maximum signal voltage at the input of a preamplifier, it is necessary to match the load resistor to the impedance of the photoconductor, which is critically dependent on the material from which it is fabricated and the operating temperature. The voltage source has a Gaussian distribution with a mean value zero. Since all frequency components are present in equal intensity, within the frequency band, Δf, the thermal noise is estimated as

$$i_{\text{th}} = \sqrt{\overline{i_{\text{th}}^2}} = \sqrt{\frac{4k_{\text{B}}T\,\Delta f}{R_{\text{L}}}}, \tag{4.76}$$

and the variance due to thermal noise is given as

$$\langle\sigma\rangle_{\text{th}}^2 = \frac{4k_{\text{B}}T\,\Delta f}{R_{\text{L}}}.$$

4.5.2.2 Generation-recombination noise

As stated in Section 2.2.6.1, in an intrinsic semiconductor, the number of free electrons is equal to the number of holes. However, thermal agitation continues to generate new electron–hole pairs (see Section 2.2.4.1) per unit volume per second, while other such pairs disappear as a result of recombination. The free electrons fall into empty covalent bonds, resulting in the loss of a pair of mobile carriers. An electron or a hole would exist for their respective mean lifetime τ_e or τ_h. The generation-recombination (GR) noise (van Vliet, 1999) is related to the statistical fluctuations in the population of charge carriers due to the random transitions of charge carriers between states in different energy bands, for instance, between conduction band and localized levels in the energy gap, conduction and valence bands, etc., in a semiconductor. It represents a noise source in the semiconductor materials, in which the carrier concentration may vary considerably. Thus, the carrier concentrations fluctuates, giving rise to the fluctuation in the electrical resistance of the sample. When a bias current flows through it, this is observed as a fluctuating voltage across the sample:

Extrinsic photoconductor Under the assumption of the thermal ionization of very few extrinsic centers, the GR current noise, i_{GR}, is expressed as

$$i_{GR} = 2i_B\sqrt{\frac{\tau\Delta f}{N(1+\omega^2\tau^2)}}, \tag{4.77}$$

where τ is the lifetime of the free carriers, N the total number of free carriers in the sample, i_B the bias current through the device, and ω the angular frequency.

Near intrinsic semiconductor Under this case, the GR current noise, i_{GR}, is given by

$$i_{GR} = 2i_B\left(\frac{\beta+1}{\beta N_e + N_h}\right)\sqrt{\left(\frac{N_e N_h}{N_e + N_h}\right)\frac{\tau\Delta f}{1+\omega^2\tau^2}}, \tag{4.78}$$

with $\beta(= \mu_e/\mu_h)$ as the ratio of the electron and hole mobilities (see Section 2.2.4.4), N_e and N_h the respective total number of free electrons and holes in the sample; τ is assumed to be the same for electrons and holes.

The level of GR-noise is proportional to the square of the average of DC current through the device.

4.5.2.3 Quantum (Shot) noise

The incoming photons have an inherent noise, known as photon-noise (Poisson or shot-noise). The photon-matter interactions

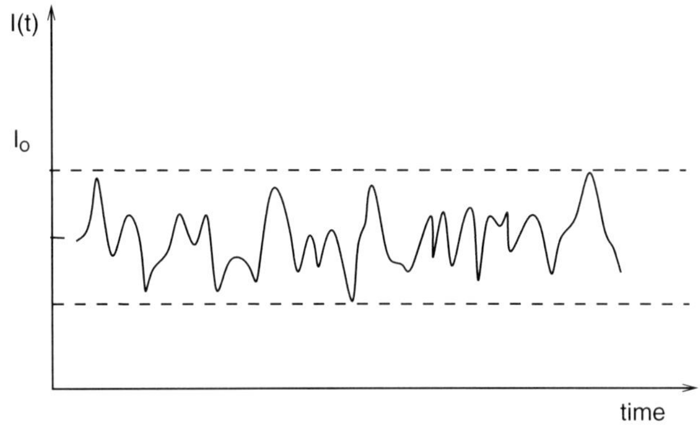

Figure 4.9 Noise fluctuations.

occurring in light detectors are random giving rise to noise in an imaging system. The amount of photons emitted by the source exhibits detectable statistical fluctuations. This is known as Schottky effect (Schottky, 1918). The devices emitting discrete charged particles, for example, electrons, at an average rate proportional to the direct current (DC), but with emission event being statistically independent, can generate these fluctuations. If the illumination is constant and the rate of photoelectric events is large, the resulting current is a superposition of many waveforms initiated at random times with a long time average rate. This gives rise to a fluctuating current (see Figure 4.9). The fluctuating component of such a current is called shot-noise, i_Q, which is fundamental to the photoelectric detectors. All incoming photons have an inherent noise, which is referred to as a photon shot-noise. It is a form of signal dependent noise due to the random arrival time of the signal carriers. The magnitude of fluctuations depends on the magnitude of the charges on the individual carriers. Shot noise is due to the discrete nature of electricity, which arises even at the initial state of the photocathode even in the presence of constant luminous flux. The shot-noise, i_Q, is given by the following equation:

$$i_Q = \sqrt{\overline{(I(t) - I_{DC})^2}}, \tag{4.79}$$

where $I(t)(= \int I(\omega)e^{i\omega t}d\omega)$ is the instantaneous current.

The number of photons, $N_{\rm ph}$, arriving at the detector during a particular time interval, Δt, varies from one time interval to another. During an interval of time, Δt, the DC is given by

$$I_{\rm DC} = \frac{q\bar{N}_{\rm ph}}{\Delta t}, \tag{4.80}$$

in which q is the electron charge, $\bar{N}_{\rm ph}$ the mean number of the charge carriers emitted in time, Δt.

Similarly, the instantaneous current is expressed as

$$I(t) = \frac{qN_{\rm ph}(t)}{\Delta t}, \tag{4.81}$$

with $N_{\rm ph}(t)$ as the instantaneous number of charged particles emitted during the time interval, Δt.

The independent emission events having a constant mean, $\bar{N}_{\rm ph}$ and variation, $\langle\sigma\rangle^2_{\rm ph}$ imply that $N_{\rm ph}(t)$ is described by a Gaussian distribution of probabilities (see equation 4.45). In a certain electric field, the number of charges released is statistically related to the exposure Δt (wave power multiplied by time) by Einstein's relation, $\Delta t = \bar{N}_{\rm ph}\hbar\omega$. For example, if $\bar{N}_{\rm ph} \gg 1$, and where most of the time during the exposure, Δt, there is no charge carrier arriving the detector, Gaussian probability distribution becomes a special case, such as Poisson's statistical distribution (see equation 4.1). The number of photons impinging on the detector, $N_{\rm ph}$, limits the number of statistical tests of generated charge carriers. In such a case, the statistics obeys Bernoulli's distribution

$$\mathcal{P}(N_{\rm e}) = \frac{N_{\rm ph}!}{N_{\rm e}! \times (N_{\rm ph} - N_{\rm e})!} p^{N_{\rm e}}(1-p)^{N_{\rm ph}-N_{\rm e}}, \tag{4.82}$$

with the variance expressed as

$$\langle\sigma\rangle^2_{\rm e} = \overline{(N_{\rm e} - \bar{N}_{\rm e})^2} = \bar{N}_{\rm e}(1-p),$$

with $N_{\rm e}$ the number of electrons and p the probability of a single event, which in this case is the quantum efficiency of the detector, $\eta_{\rm d}$.

Each photon induces a free electron, during the time Δt, the number of electron produced is, $\bar{N}_{\rm e} = \eta_{\rm d}\bar{N}_{\rm ph}$. Hence, for the total fluctuation of carriers caused by both the processes, the variation of incoming photons, as well as the variation of the generated electrons for each given number of photons, one gets

$$\begin{aligned}\overline{(N_\mathrm{e}-\bar{N}_\mathrm{e})^2_{tot}} &= \overline{(N_\mathrm{ph}-\bar{N}_\mathrm{ph})^2}\times\eta_\mathrm{d}^2+\overline{(N_\mathrm{e}-\bar{N}_\mathrm{e})^2}\\ &= \bar{N}_\mathrm{ph}\eta_\mathrm{d}^2+\bar{N}_\mathrm{ph}\eta_\mathrm{d}(1-\eta_\mathrm{d})=\bar{N}_\mathrm{e}. \end{aligned}\tag{4.83}$$

The fluctuation in the number of carriers is related directly to the detector current noise, i_Q, by the following equation:

$$\begin{aligned}\overline{i_\mathrm{Q}^2} &= \overline{(N_\mathrm{e}-\bar{N}_\mathrm{e})^2_{tot}}\times\left(\frac{q}{\Delta t}\right)^2\\ &= \left(\frac{\bar{N}_\mathrm{e}q}{\Delta t}\right)\frac{q}{\Delta t}=i_\mathrm{d}\frac{q}{\Delta t},\end{aligned}\tag{4.84}$$

where i_d is the detector current.

For a frequency band of $\Delta f(\simeq 1/2\Delta t)$, the shot-noise current is estimated as

$$i_\mathrm{Q}=\sqrt{\overline{i_\mathrm{Q}^2}}=\sqrt{2qi_\mathrm{d}\Delta f}.\tag{4.85}$$

To note, the detector current i_d must include the signal and the dark current.

4.5.2.4 Flicker noise ($1/f$-noise)

Flicker noise (also called $1/f$-noise or excess noise since it exceeds shot-noise at frequencies below a few hundred cycles; see Figure 4.10) arises in several phenomena, where a measurement is a result of certain statistical phenomena. Usually, it comes at low frequency regimes (Sze and Ng, 2007). Therefore, $1/f$-noise appears not only in physical systems determined by thermodynamics, but also chemical, biological, economic, and even psychological systems. In semiconductor devices, such as detectors, $1/f$-noise comes from the statistical behavior of the electron occupancies in allowed energy levels. Generally, electrons in the valence band of a semiconductor are excited to the conduction band, and their distribution in the conduction band levels are determined by Fermi–Dirac statistics (see Section 2.2.5) when the excitation rate is large (at room temperature), the number of electrons excited to the conduction band by thermal agitation is quite small. In this case, the electrons follow Maxwell–Boltzmann distribution law. As per the Fermi–Dirac statistics, the occupation probability of electrons at an energy level

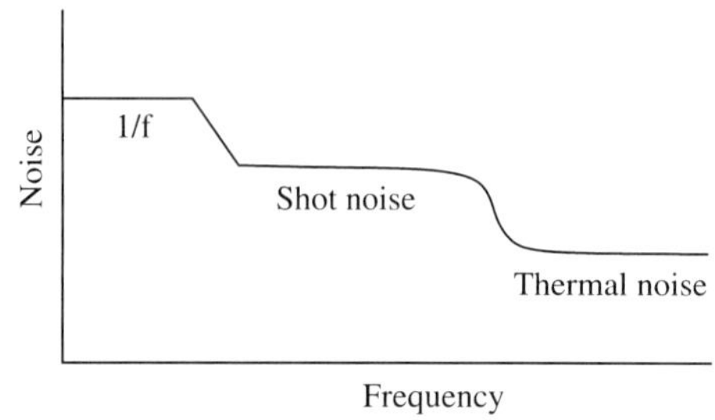

Figure 4.10 Sample of flicker ($1/f$) noise.

E at a given temperature T, is given by equation (2.26). When, this reduces to a similar form of Maxwell–Boltzmann statistics, the statistical mechanics of the thermally excited electrons in a detector and the resultant generation-recombination mechanism can produce flicker noise (Schroder, 2006).

The $1/f$-noise was discovered by Johnson (1925) from data in an experiment designed to confirm Schottky effect in vacuum tubes. An intermediate between the white noise with no correlation in time and random walk (Brownian motion; Einstein 1905d) noise with no correlation between increments, it is a signal with a frequency spectrum such that the spectral density is inversely proportional to the frequency. The Brownian noise refers to a power density that decreases 6 dB per octave with increasing frequency over a frequency range that does not include DC value. In the electrical domain, such density is defined as the Fourier transform of the autocorrelation function that is given by (Mandel and Wolf, 1995),

$$\Gamma(\tau) = \langle i(t)i(t+\tau)\rangle,$$

where $i(t)[= \mathcal{R}P(t)]$ stands for the instantaneous photocurrent at time, t, which is proportional to the instantaneous optical power, $P(t)$, $\mathcal{R}$ the detector responsivity (see Section 4.1.1.2), τ the time delay, and $\langle\,\rangle$ the expectation values; the signal current, $i_s(\equiv \langle i(t)\rangle$, can be regarded as the a product of the detector responsivity, $\mathcal{R}$ and the total optical power, P_T.

In terms of power at a constant bandwidth, $1/f$-noise falls off at 3 dB per octave (density proportional to $1/f$). The noise spectral density is important at low frequency. The flicker noise current is given by

$$i_{\mathrm{FN}} = \sqrt{\frac{K_1 i_{\mathrm{B}}^{\alpha} \Delta f}{f^{\beta}}}, \tag{4.86}$$

where K_1 denotes the proportionality constant that depends weakly on temperature, i_{B} the bias current, Δf the bandwidth, α the constant ranging between 0.5 to 2, and β the constant close to unity.

Flicker noise is important in the heterogeneous structure of the device; the more heterogeneous the structure, higher the $1/f$-noise. Any inhomogeneity in the materials, as well as contamination during processing, etc., may increase such a noise. It is low in a device with homogeneous structure. This noise causes the specific detectivity, D^*, to decrease for low frequencies. For high frequencies, it may also be small due to the decreasing sensitivity of the detector.

4.5.2.5 Dark current noise

The dark current noise, i_{n}, is proportional to the square root of the dark current, i_{D} (see Section 4.1.1.8), which can be reduced by cooling the device. It is created by the false pulses resulting from thermally generated electrons (dark signal). A false alarm may arise when noise alone exceeds the threshold and is interpreted as signal. Also, there is a probability of missed detection when the signal along with noise does not exceed threshold. The noise arising from the dark charge is given by Poisson's statistics as the square root of the charge arising from the thermal effects. This noise is the combination of bulk dark current, $\langle i_{\mathrm{D}}^2 \rangle$, which is due to thermally generated electrons and holes in the P-N junction, and the surface dark current due to surface defects, bias voltage and surface area. This process is dependent on temperature by equation (4.31). The surface dark current noise, $\langle i_{\mathrm{DS}}^2 \rangle$, is the same as given in equation (4.32) and the generation of bulk dark current noise is expressed as

$$\langle \sigma \rangle_{\mathrm{DB}}^2 = 2q i_{\mathrm{D}} \Delta f \langle M^2 \rangle F(M). \tag{4.87}$$

with $\langle M^2 \rangle$ as the mean-square value of the internal gain and $F(M)$ the noise factor or equivalently the noise figure (see Section 4.5.5).

4.5.3 *Amplifier Noise*

The amplifier noise consists of two components, such as flicker noise (see Section 4.5.2.4) and white noise. A photoelectric substance gives a saturation current, i_0, upon being illuminated by a luminous flux, Φ_ν (see Section 1.2.2). As stated in Section 4.1.1.3, a photodetector converts the incident photons into electrons at a rate determined by the quantum efficiency, η_d. A signal voltage, V_s, is created when the electrons are transferred to the capacitance, C. The current being weak needs support of multipliers or amplifier having wide bandwidth, and large gain. A voltage, V_s, is created when the electrons are transferred to the sense node capacitance, C, which is given by (Holst, 1996)

$$V_s = N_e \frac{qG}{C}, \tag{4.88}$$

where N_e is the total number of electrons in the charge packet, which includes both photoelectrons and dark current electrons, q the charge, G the source follower amplifier gain that is, in general, is less than one, Δf the bandwidth, and qG/C the device output conversion gain, the unit of which is volts/electron with values typically ranging from 0.1 μV per electron to 10 μV per electron.

In a linear optical amplifier, according to the semi-classical picture, the amplifier output is considered as a mixture of an amplified signal and amplified spontaneous emission. The former is a coherent field with Poisson's statistics, while latter may be described as a thermal field. The amplifier noise is described as

$$\langle i_{\text{AMP}} \rangle^2 \approx 2qG\Delta f F(M), \tag{4.89}$$

in which $F(M)$ describes the additional noise introduced by the random nature of the impact ionization.

Considering Figure 1.6 without any imaging system and assuming a Lambertian source (see Section 1.2.3), the number of photoelectrons, N_{pe} is given by

$$N_{pe} = \frac{1}{q} \frac{\mathtt{A}_i \mathtt{A}_d}{s_o^2} \int_{\lambda_1}^{\lambda_2} \mathcal{L}_e(\lambda) \mathcal{R}(\lambda) \Delta t d\lambda, \tag{4.90}$$

where q denotes the electron charge, Δt time interval, $\mathtt{A}_d$ the active area of the detector, $\mathtt{A}_i$ the area of the image, s_o the object distance,

$\mathcal{L}_e(\lambda)$ the spectral radiance (see Section 1.2.1), and $\mathcal{R}(\lambda)$ the spectral responsivity.

From Eqs. (4.88) and (4.90), one may derive the array output voltage, V_s

$$V_s = \frac{G}{C}\frac{\text{A}_i\text{A}_d}{s_o^2}\int_{\lambda_1}^{\lambda_2}\mathcal{L}_e(\lambda)\mathcal{R}(\lambda)\Delta t d\lambda. \tag{4.91}$$

Expressing the responsivity in the form

$$V_s = \langle\mathcal{R}\rangle\left[\frac{1}{\text{A}_d}\frac{\text{A}_i\text{A}_d}{s_o^2}\int_{\lambda_1}^{\lambda_2}\mathcal{L}_e(\lambda)\mathcal{R}(\lambda)\Delta t d\lambda\right] \quad \text{V.J}^{-1}.\text{cm}^{-2}, \tag{4.92}$$

where $\langle\mathcal{R}\rangle$ designates the average responsivity, which depends on the source characteristics and the spectral quantum efficiency, and combining Eqs. (4.91) and (4.92) yields

$$\langle\mathcal{R}\rangle = \frac{G}{C}\text{A}_d\frac{\int_{\lambda_1}^{\lambda_2}\mathcal{L}_e(\lambda)\mathcal{R}(\lambda)d\lambda}{\int_{\lambda_1}^{\lambda_2}\mathcal{L}_e(\lambda)d\lambda}. \tag{4.93}$$

Since the spectral radiance, $\mathcal{E}_e(\lambda)$, and the spectral radiant emittance, $\mathcal{M}_e(\lambda)$, are functions of wavelength (see equation 1.78), these quantities may be considered as constants for a very small wavelength. Hence, equation (4.93) is approximated as

$$\langle\mathcal{R}\rangle \approx \frac{G}{C}\text{A}_d\frac{\mathcal{M}_e(\lambda_0)\mathcal{R}(\lambda_0)\Delta\lambda}{\mathcal{M}_e(\lambda)\Delta\lambda} \approx \frac{G}{C}\text{A}_d\mathcal{R}(\lambda_0), \tag{4.94}$$

which is related to the quantum efficiency of the detector, η_d, by $\mathcal{R} = q\eta_d/h\nu$; therefore,

$$\langle\mathcal{R}\rangle \approx \frac{Gq}{C}\frac{\lambda_c}{hc}\text{A}_d\eta_d(\lambda_c) \approx \frac{G}{C}\frac{\lambda_c}{1.24}\text{A}_d\eta_d(\lambda_c), \tag{4.95}$$

where h is Planck's constant, c the speed of the light, and q the electron charge.

4.5.4 *Read-Out Noise*

In an optimized read-out procedure, the intrinsic noise is limited by the electronic noise in the amplifiers and the minimum detectable optical power; the minimum signal that can be detected is the square of the read noise. A detector may measure at regular intervals the photocharges accumulated since the last measurement. Each charge transfer is accompanied by a fluctuation, whose standard deviation is measured in numbers of electrons per reading. This number is the read-out noise, a component of the noise on the signal from a single pixel, which is independent of the signal level. It occurs due to

- if the conversion from an analog-to-digital (A/D) number is not perfectly repeatable and
- if the electronics of the camera introduces spurious electrons into the process with unwanted random fluctuations in the output.

A value for gain, G, is defined by

$$G = \frac{C_\mathrm{m}}{\langle\sigma\rangle^2 - \sqrt{\langle i_\mathrm{ro}\rangle^2}}, \tag{4.96}$$

with C_m as the mean counts, $\langle\sigma\rangle^2$ the variance, and $\langle i_\mathrm{ro}\rangle^2$ the read-out noise.

For a standard CCD (see Section 6.2), the S/N ratio, S/N, is given by

$$\mathrm{S/N} = \frac{i_\mathrm{s}}{\sqrt{i_\mathrm{s} + i_\mathrm{ro}^2}}, \tag{4.97}$$

where i_s is the signal current.

High S/N ratio is, therefore, achieved either by reducing the read-out noise or by increasing the signal. The read-out noise is presently limited by the device manufacturing capabilities while the input signal can be increased either by integrating the signal longer or multiplying the signal before read-out. Another important origin of noise in the read-out procedure is that it results from the dark current caused by thermally generated electrons in each elements of the detector array. The read-out noise RMS value limits the low end of the registered radiation power. This value is used to deduce the dynamic range of the detector array. It is measured as the number of electrons per read-out sequence. Both the dark current and read-out noise can be reduced by employing an effective cooling system.

4.5.5 *Noise Figure*

The noise figure, F, is one of the most important characteristics of the optical amplifiers, which is defined as the ratio of equivalent noise power developed at the input to that generated by thermal noise in the source resistance, which is expressed in decibel (dB; Alexander, 1997). A low noise figure is essential for determining the transmission performance of an optically amplified system; an ideal amplifier would have a noise figure of 0 dB. Following equation (4.16), the S/N ratio is written as

$$\mathrm{S/N} = \frac{|\bar{i}_{\mathrm{s}}|^2}{\langle\sigma_{\mathrm{T}}\rangle^2}, \tag{4.98}$$

where $\langle\sigma_{\mathrm{T}}\rangle^2$ is the photocurrent variance, and for the Gaussian distribution described by equation (4.45), it is expressed as

$$\mathrm{S/N} = \frac{\overline{V}^2}{\langle\sigma_V\rangle^2}.$$

In order to determine the amplifier noise factor, the $(S/N)_{in}$ is required to be known (Kweon, 2003). However, it should be measured for a shot-noise limited optical source with an ideal photodetector with unit quantum efficiency, η_{d}. The comparison between the thermal noise and the shot-noise (equations 4.76 and 4.85) is derived as

$$\gamma = \frac{\langle i_{\mathrm{Q}}^2\rangle}{\langle i_{\mathrm{th}}^2\rangle} = \frac{i_{\mathrm{d}}R_{\mathrm{L}}}{2(k_{\mathrm{B}}T/q)} = \frac{i_{\mathrm{d}}R_{\mathrm{L}}}{5\times 10^{-2}}. \tag{4.99}$$

For successful determination of the signal, both the thermal noise and shot-noise should be low. The voltage drop at the input of the amplifier (in case of photoelectric cell with an amplifier) should be at least 0.05 volts and hence, the flux sensitivity of the setup would be very poor. To do away with it, multiplication is used so that how feeble the photocathode current may be, the output current is always large enough to produce 0.05 volts at the output resistance. If they are present simultaneously in the system, the optimal load resistance is defined as the one that causes both the noise components to be equal.

In order to calculate the total noise presented in photodetector, one may sum up the root mean square of each noise current by assuming that those are uncorrelated. The total RMS photodetector noise current, $\langle i_N^2\rangle$, is determined by

$$\begin{aligned}\langle i_N^2\rangle &= \langle\sigma_N\rangle^2 = \langle i_Q^2\rangle + \langle i_{\mathrm{DB}}^2\rangle + \langle i_{\mathrm{DS}}^2\rangle\\ &= 2q(i_{\mathrm{s}} + i_{\mathrm{D}})\Delta f M^2 F(M) + 2qi_{\mathrm{d}}\Delta f,\end{aligned} \tag{4.100}$$

where $\langle i_Q^2 \rangle$ is the noise generated by the photon signal,

$$i_{\mathrm{Q}} = \sqrt{\eta_{\mathrm{d}} N_{\mathrm{ph}}},$$

η_{d} the quantum efficiency, and N_{ph} the number of photons, i_{D} the bulk dark current, $\langle i_{\mathrm{DB}}^2 \rangle$ the bulk dark current noise, $\langle i_{\mathrm{DS}}^2 \rangle$ the surface current noise, M the multiplication factor (current gain), and

$$F(M) = \frac{\langle M^2 \rangle}{\langle M \rangle^2}$$

the excess noise figure; for a P-I-N photodiode, $M^2 F(M) = 1$.

It is pertinent to note that the actual information that is the optical signal, $\langle i_{\mathrm{OP}} \rangle$, created by the light must be more than the total noise (addition of the variances of all these noises, such as shot-noise, thermal-noise, and amplifier noise)

$$\langle i_{\mathrm{OP}} \rangle^2 \gg \langle i_{\mathrm{Q}} \rangle^2 + \langle i_{\mathrm{th}} \rangle^2 + \langle i_{\mathrm{AMP}} \rangle^2. \tag{4.101}$$

Having obtained the signal and total noise, the S/N ratio is expressed as

$$S/N = \frac{\langle i_{\mathrm{d}}^2 \rangle M^2}{2q(i_{\mathrm{s}} + i_{\mathrm{D}})\Delta f M^2 F(M) + 2qi_{\mathrm{d}}\Delta f + 4k_{\mathrm{B}} T \Delta f / R_{\mathrm{L}}}, \tag{4.102}$$

where i_{d} is the detector current.

Since the noise figure $F(M)$ increases with M, there always exists an optimum value of M that maximizes the signal-to-noise ratio, S/N. For a sinusoidally modulated signal, M is equal to 1, and the noise figure, $F(M) \approx M^{\chi}$ for $0 \leq \chi \leq 1.0$, χ is the excess noise index:

$$M_{opt}^{\chi+2} = \frac{2qi_{\mathrm{d}} + 4k_{\mathrm{B}} T / R_{\mathrm{L}}}{xq(i_{\mathrm{s}} + i_{\mathrm{D}})}. \tag{4.103}$$

In a photoemissive device, such as a photomultiplier (see Section 5.2.2), the noise limitation is, in general, due to the noise equivalent current of the load (Johnson noise) rather than the inherent noises in the device due to the dark current or emitted electrons. It is possible to obtain higher signal than the noise level introduced by loading circuitry using secondary emission phenomenon. These introduce shot-noise (see equation 4.85) in the emitted stream of electrons due to the fluctuations in the secondary emission at the dynodes. The avalanche multiplication process is statistical in nature and there is an associated excess noise. The thermal noise in the final stage is due to the finite temperature of the device.

4.5.5.1 Noise accumulation in photoelectric system

It is reiterated that noise is any intensity variation that exists within the reproduced image of a photoelectronic system does not exist in the original scene. Noise imposes the final limit on the total dynamic range, which may be reproduced. It imposes the final constraint on the amount of electronic correction as well in the form of high frequency boosting that may be applied for a modulation transfer function (MTF; see Section 1.3.1) fall-off. The noise may be injected and accumulated, at a number of points throughout the photoelectronic imaging system (see Figure 4.11). There exist two kinds of noise, for instance:

Additive noise This is superimposed upon the signal itself by statistical addition and exemplified by the thermal agitation noise of an amplifier being added to the signal.

Product noise It is exemplified by the mechanism, such as photon-noise, phosphor grain, and target grain noise and characterized by the fact that it disappears or is significantly reduced upon the removal of the input signal.

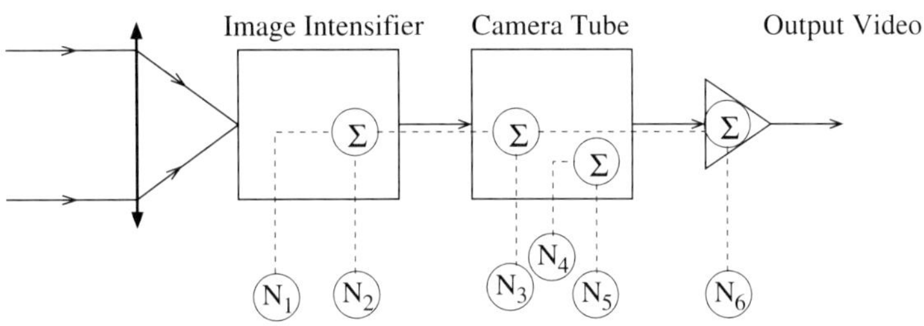

Figure 4.11 Accumulation of noise in intensified television camera. Here N_1 is the photoelectron noise, N_2 the phosphor noise, N_3 the camera photoelectron noise, N_4 image section partition noise, N_5 the reading beam shot-noise, and N_6 the preamplifier noise.

4.5.5.2 MTF measurements in a photoelectric system

As stated in Section 1.3.1, the modulation transfer function (MTF) is a means of characterizing the image quality of the optical system. An optical or electro-optical system can be characterized by analyzing the image produced by a narrow slit object or pulse in space or to a sinusoidally varying test object and the result may be interchanged by a Fourier transform. Any frequency dependent phase changes introduced by the system would result in the image as lateral shifts of the sinusoidal frequency components comprising the image (Steward, 1983). This phenomenon is called phase transfer function (PTF; see Section 1.3.1) and its relation with the optical transfer function (OTF) is shown mathematically as

$$\mathrm{OTF} = \mathrm{MTF}\, \mathrm{e}^{\mathrm{i}\,(\mathrm{PTF})}. \tag{4.104}$$

The contribution of this phase transfer function is often negligible. The normalized form of $\widehat{\mathcal{T}}(\mathbf{f})$ is

$$\widehat{\mathcal{T}}_{\mathrm{n}}(\mathbf{f}) = \frac{\widehat{\mathcal{T}}(\mathbf{f})}{\widehat{\mathcal{T}}(0)}. \tag{4.105}$$

For a perfect optical imaging system with a uniformly illuminated circular pupil of diameter D, the pupil function may be expressed as

$$P(\mathbf{x}) = \begin{cases} 1 & \text{inside the pupil} \\ 0 & \text{otherwise,} \end{cases} \tag{4.106}$$

The frequency region in which the system operates is, in general, well corrected for the PTF. Hence, it does not contribute significantly to the determination of transfer function. There are two notable methods for MTF measurements:

PSF or LSF analysis A series of 2-D convolutions, first with the image and the lens followed by the result of that procedure with the sensor, and so on through all of the components of the system take place. Though the measuring arrangement is simple in the case of point spread function (PSF) or line spread function (LSF) analysis, it loses its accuracy due to low photometric value and signal-to-noise (S/N) ratio. These difficulties can be avoided in case of sine wave analysis. In this case, a slit is employed as object and the image is scanned by a sinusoidal object pattern or vice versa.

Spatial frequency domain Each of the components of the system may be transformed into spatial frequency domain, and then multiply the 2-D results. A system response may be determined without reference to an object. This method is computationally easier.

Chapter 5

Photodetectors

5.1 Photodetector Elements

In both the eye (see Section 4.2.1) and a camera, the light coming from a source is focused with a lens onto a detection plane. In the former case, the detection plane is the retina where cells, which are sensitive to changes in light intensity, encode the image and send it to the brain, while in the latter case, the detection plane contains either a photographic film (see Section 1.4.2) or an electronic detector, in which light intensity is converted to an electric signal. In the electronic camera, the photodetectors are required to detect light, which put out an electrical signal whose amplitude is proportional to the photons striking the surface. Such cameras can be either analog or digital. In the former case, the detectors coupled acquisition and display together and provide higher resolution but with limited dynamic range and fixed detector contrast, while in the latter case, the detectors have separated acquisition and display system providing higher dynamic range and noise-limited contrast but with lower resolution.

Since the photo-electric effect (see Section 2.1) demonstrates the quantum nature of light, which can actually detect a single

High-Resolution Imaging: Detectors and Applications
Swapan K. Saha

ISBN 978-981-4613-27-9 (Hardcover), 978-981-4613-28-6 (eBook)
www.panstanford.com

photon. The quantum outputs measured for each wavelength in a given space provide the spectral response of the photocathode. The interaction of photons with materials changes the conductance of the material. The energy that has to be supplied by the absorbed light quanta can be expressed in terms of the work function, ϕ_0, as well as in terms of bandgap energy, E_g, and electron affinity (see Section 2.3.2) for semiconductor.

The photocathodes are, in general, coated with photoelectric material on the inner side. For high photoemissive efficiency, the absorption coefficient of the material should be high. The photoelectric surfaces with high efficiency are not generally obtainable with metals, but rather with semiconductors. For most metals, the reflectivity (in the visible and near-IR region of the spectrum) is very high (90–99%). They have the same work function value determined by thermionic emission and photoelectric emission. On the other hand, the semiconductors generally have 25–50% reflectivities over the range of interest, and absorbed photon energy is taken up by excitation of free electrons. Their thermionic work function may differ considerably from their photoelectric work function though semiconductors are effective photoemitting surfaces. In what follows, the salient features of photon detectors, namely, (i) photoemissive (PE), (ii) photovoltaic (PV), and (iii) photoconductive (PC) devices are enumerated.

5.2 Photoemissive Devices

A photoemissive detector is based on the photoelectric effect, in which incident photons release electrons from the surface of the detector material. The free electrons are collected in an external circuit. In this a photocathode and an anode are kept inside a sealed vacuum tube, what is referred to as phototube (photocell). The main characteristic of this cathode is its quantum output that is to say that the PE effect (see Section 4.3.1.2) manifests itself. Operation of a phototube is based on the photoelectric effect obeying Einstein's equation (see equation 2.1). The photocells can be divided into the following two groups:

High vacuum tube The photoelectric cell, the first device to make practical use of this effect, comprising an evacuated glass bulb with two electrodes was devised in 1890 by Elster and Geitel. The cathode consisted of a thin film of a metal, such as potassium and the anode was a wire sealed in to complete the circuit. This anode was maintained at a positive potential in order to collect the negative charges released by light from the cathode. This cell emitted electrons when stimulated by light, and at a rate proportional to the intensity of the light. Hence, a current could be obtained from the cell. They observed that the photoelectric currents fall off gradually, which was caused by oxidation of the cathode by the air; modern cathodes are enclosed in sealed vessels, so that such problems do not arise. Following its success, Elster and Geitel (1893) suggested the construction of a photoelectric photometer for measuring the intensity of sunlight.

Gas-filled device The possibility of detecting photoelectrons individually was envisaged by Elster and Geitel (1916), based on the design of the α-particles counter by Rutherford and Geiger (1908). It consists of a bulb filled with gas containing a photocathode and an anode. The voltage between the electrodes was regulated so that the presence of a photoelectron could cause, by ionization of gas, a discharge (measured by an electrometer), and therefore detected.

Photoionization A high energy ($\gtrsim$10 eV, ultraviolet or X-ray regions) photon may interact with an ionic electron leading to removal of that electron from the ion. This process is termed as photoionization. The released electrons are accelerated by an external electric field, and cause further ionizations by collision. The charges collected on the accelerating electrode are measured by an electrometer. The operating limit is reached when the multiplier effect leads to a discharge in the gas; the response is linear below this limit. The photoionization probability is typically described in terms of a cross section defined as the photons absorbed per unit time by the incident photon flux. The photoionization cross section in a gas is a function of the incident energy, E, varying as $E^{-7/2}$.

Geiger-Müller gas detector The counting of particles issued by radioactivity with the system of Geiger-Müller (1928), based on the

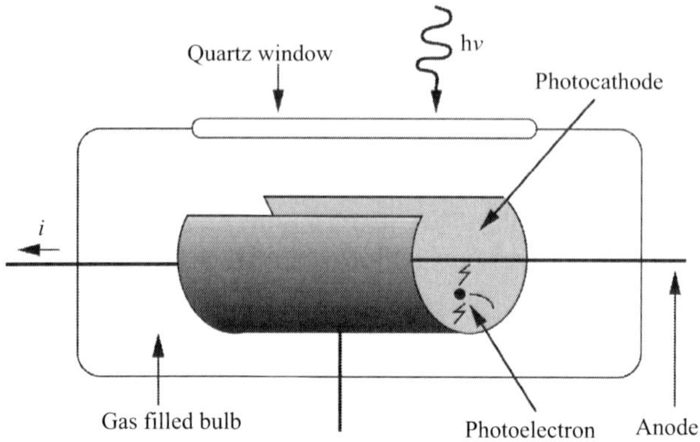

Figure 5.1 Principle of a Geiger-Müller gas detector (after Morel and Saha, 2005).

system represented before, was adapted later by Locher (1932) to count visible photons. It accomplished several counters composed of a gas bulb, in which a photocathode had a three-quarters of cylinder shape and the anode was a simple wire aligned on the cylinder axis (see Figure 5.1). Several alkaline materials, associated with hydrogen, were tried for the photocathode. The most sensitive was potassium, but it was more noisy; the dark noise was 122 per minute, as against only 5.8 per minute for cesium, and 4.7 per minute for sodium. At any rate, the flux maximum rate, limited by the reading electronics, did not exceed 300 counts per second. Because of their weak precision, gas detectors as stellar photometry sensors were abandoned in the 1940s. There are some applications for which PE devices are uniquely suited, for example,

- the detection of low intensity signals, either weak in total available power or diffuse at the focal plane,
- high-speed detection of low level signals, and
- the acquisition of high spatial resolution information (imaging).

A major advantage of these detectors is the case with which fast, high-gain (see Section 4.3.2.2), low-noise amplification can be incorporated within the detector by use of an integral electron multiplier. Another advantage is the case with which uniform, large-area detector surfaces can be fabricated. These surfaces are a prerequisite for successful imaging devices, including both direct-view intensifiers and image tubes (Csorba, 1985) with scanned electron beam read out. The electron beam integration technique provides an advantage of such devices, namely, extremely high detector element density, which permits high spatial resolution (for imaging); the electron multiplier gain is available in these devices. However, the major limitation of PE devices is the restricted range of the spectrum over which response can be obtained. Owing to transmission limitations of common window materials, in the ultraviolet region of the spectrum, the standard PE materials do not respond beyond $\sim$0.15 μm, while in the IR-region, it may not respond beyond 1.5 μm.

5.2.1 *Photoemissive Surfaces*

A material that can exhibit photoelectric phenomena (see Section 2.1) is said to be photoemissive; an electric field is required to be applied to collect the photoelectrons. For a given metal, there is a minimum frequency of incident electromagnetic radiation below which the photolytic effect does not take place; hence, no photoelectrons are emitted. The property of a PE substance causing it to emit electrons when radiation in the visible region is incident on it is called photoemissivity. The work functions of some of these surfaces are listed in Table 5.1.

5.2.1.1 Classical photoemissive surface

The photoemitter is a thin evaporated layer compound containing an alkaline metal, such as sodium (Na) and potassium (K), or metals (almost always including cesium; Cs), one or more other metallic elements from Gr.VB of the periodic table (e.g., antimony, Sb) and possibly also oxygen and /or silver; Cs is very unstable because it oxidizes readily. These metals, therefore, can be broadly used in association with some antimony for the manufacture of photocathode. A semiconductor like cesium antimonide (Cs_3Sb; Glover, 1941) as a photoemitter, which has a bandgap of 1.6 eV and an intrinsic photoelectron threshold of 2.05 eV, due to a low electron affinity, has found wide application in commercial photodetection devices. Doping this material with electrons that

Table 5.1 Work function of certain photoemissive surfaces

Metal	Symbol	Work function, eV
Aluminum	Al	4.06–4.26
Calcium	Ca	2.87
Cesium	Cs	2.14
Copper	Cu	4.53–5.10
Lead	Pb	4.25
Lithium	Li	2.93
Mercury	Hg	4.49
Platinum	Pt	5.12–5.93
Potassium	K	2.29
Rubidium	Rb	2.26
Silver	Ag	4.52–4.74
Sodium	Na	2.36

go to states close to the Fermi level (see Section 2.2.5) leads to a photoelectric threshold close to the work function, which is 1.25 eV. This corresponds to a photon wavelength in near-IR (~1 μm). Tubes with Cs_3Sb on several semi-transparent and opaque substrates, and with various window materials, have been given several "S" numbers, for example, S 4, S 5, S-11, S-13, S-17, and S-19; other photoemitters include S-10 (BiAgOCs), S-20 (Na_2KSbCs; tri-alkali material). The "S"-numbers denote the total spectral response, including the effect of the input window; they do not identify specific types of cathode or cathode materials, or absolute sensitivities. Unlike Cs_3Sb photocathode whose spectral response barely extends through visible, the Na_2KSbCs photocathode has high sensitivity in the longer wavelength of the visible range (red) and near-IR.

However, the quantum output of photocathodes based on alkaline metals is often too weak for expulsions with longer wavelength. To note, the direct bandgap (see Section 2.2.3.1) emitters are generally preferable to indirect ones because they allow the region of photon absorption and electron generation to be kept thin, and for a given thickness they also exhibit a more constant yield above the threshold. For visible wavelength, the photoelectric effect occurs when the Fermi energy is of the order of 1.5 eV, which corresponds to that of alkaline metals (Na, K, Cs).

The photoemission of electrons from a solid is a three step process, such as (i) photoabsorption, (ii) transport of excited electrons to the surface, and (iii) emission of electron. In order to maximize the quantum efficiency, it is required to provide attention to each one of these steps. Large excitation results from materials with large density of electrons states close to the vacuum level, for example, high electron density and low photoelectron threshold. The conditions for efficient photoemission from a solid are:

(1) The optical absorption coefficient should be high, so that incoming photons can be absorbed near the surface and excited electrons can escape before making too many collisions with the lattice or with other electrons.
(2) Incoming photons should excite electrons high into the conduction band, well above the vacuum level (A rather than B in Figure 5.2), so as to enable them to escape even after losing energy in collisions.
(3) The probability of energy losing collisions should be minimized. By employing a semiconductor, electron–electron collision effects (C; see Figure 5.2) can be avoided.
(4) The excited electrons may lose energy during transport to the surface and so their flux is attenuated. The attenuation is less for materials where inelastic electron scattering is hindered, like insulators or wide bandgap semiconductors. A low surface barrier is required for maximum escape. The electron affinity, χ, should be low, to increase the number of electrons able to escape. In particular, the condition that $\chi \ll E_g$ must be

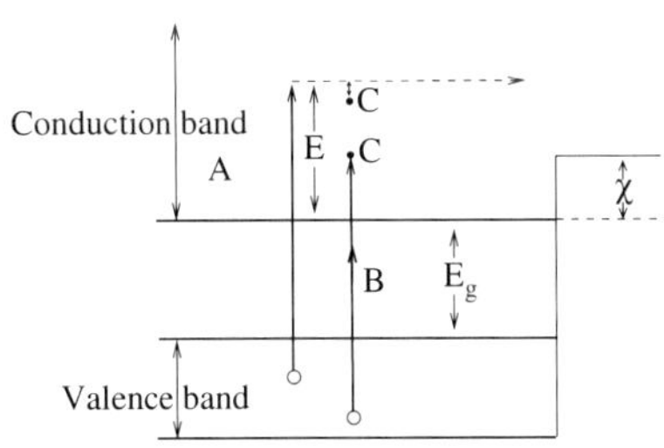

Figure 5.2 Photoemission from a positive affinity material; E_V and E_C are the valance and conduction bands, respectively and E_g the bandgap.

satisfied or else an electron having an energy, E, with respect to the bottom of the conduction band (see Figure 5.2), much higher than E_g can excite electrons from the valence band into the conduction band with loss of energy. Thus, the electrons lose most of their energy quickly until it becomes smaller than E_g. The electrons for which $E < E_g$ can lose their energy by colliding with the lattice and exciting lattice vibrations. Since the energy loss per collision in this process is small, which is of the order of $h\nu_p$, in which h is Planck's constant and ν_p the frequency of the lattice vibrations, the electrons having $E < E_g$ can come from a greater depth, suffer many collisions on their way to the surface, and still be able to escape when they arrive at the surface. To note, the PE threshold energy, E_T, for electron emission in a metal is generally of the order $\sim \phi_m$ and in a semiconductor, $E_T = \chi + E_g$. A metal should contain a very large number of free electrons, of the order of 10^{22} per cm^3, which may be chosen for a given frequency of the light.

5.2.1.2 Negative electron affinity (NEA) photoemissive surface

Semiconductors with negative electron affinity surfaces are used for photoemitters and secondary emitters (Ng, 2009). The NEA utilizes a photoconductive single crystal semiconductor substrate with a very thin surface coating comprising Cs and a small quantity of oxygen in general. By suitably treating the surface of a P-type semiconductor material, the band levels at the surface can be bent downward so that the effective electron affinity is actually negative. The energy level of the electron at the bottom of the conduction band in the bulk of the semiconductor should be greater than the zero energy level of an electron in the vacuum (see Figure 5.3). If an electron that is excited to the conduction band within the bulk travels to the activated surface without first recombining, it can energetically fall out of the photocathode into the free space. Thermalized electrons in the conduction band are, in general, repelled by the electron-affinity barrier. In the case of secondary emission, the secondary electrons can be created at greater depths in the material and still escape, thus providing a much greater secondary-emission yield.

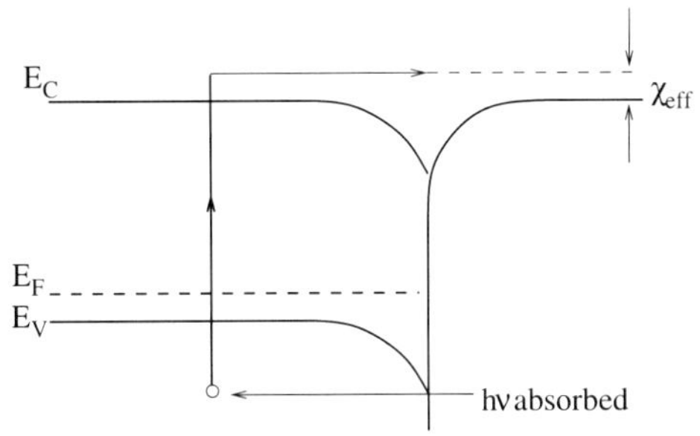

Figure 5.3 Photoemission from a negative electron affinity P-type direct bandgap semiconductor (see Section 2.2.3.1) material; E_V, E_C, and E_F are the valance band, conduction band, and Fermi level energies (see Section 2.2.5), respectively.

By selecting appropriate semiconductor and coatings, it has been possible to achieve extended-red and near-IR sensitivities greater than those obtainable with classical surfaces; the quarternary compound, indium gallium arsenide phosphide (InGaAsP), can reach 1.5 μm (see Table 2.1) with quantum yield of the order of 2.5×10^{-6}. The advantages of NEA photoemissive surfaces include:

- increased quantum efficiency in the regions of the spectrum,
- substantial extension of response into the near-IR region of the spectrum ($\lambda > 1\,\mu m$),
- possibility of good resolution in image sensors (this results from the narrow energy distribution of the emitted cold electrons),
- low temperature dark current, and
- capability of uniform absolute sensitivity over an extended region of the spectrum.

However, semiconductor doping may not be uniform, the surface doping may differ from the bulk one, the band bending occurs. This effect may reduce the difference between the vacuum level and the bottom of the bulk conduction band, which may become negative. The quantitative measure of the degradation caused by the lateral leakage of charge in a photoconductor requires the specification of its point spread function (PSF). This can be determined by a potential spread on its vacuum side caused by a point source illumination. Scheer and van Laar (1965) conceived this idea and made negative affinity photocathodes. They showed the way to overcoat the surface of selected P-type semiconductors with an evaporated layer of a low work function material such that the resulting structure has a NEA. In this case, the incident photon energy must be equal or exceed the energy gap of the semiconductor for emission to occur.

5.2.1.3 Types of photocathodes

Both classical and NEA emitters can be fabricated in two modes, namely:

Semi-transparent mode The photocathodes of this kind are deposited on the inside of the input window. The light enters at the rear side of photoemissive layer and is absorbed throughout its thickness. In this case the generated electrons diffuse through the bulk to the side (non-illuminated side) from which they are emitted toward the collecting electrodes. The cathode can be large (up to a few hundred millimeters in diameter) and the window on which it is deposited can be flat or curved.

Reflection mode (opaque) Such photocathodes are deposited on a metal electrode inside the tube, in which the electrons are emitted from the illuminated side. The light is incident through the tube envelope directly on to the side of the photocathode from which electron emission occurs. The area is about a few square centimeters because of the size of the focusing electrodes.

Although both forms of emitter provide similar electron collecting efficiency, the semi-transparent structure is preferred when imaging or interrogation by a scanning electron beam is used. However, the improper thickness of the semi-transparent device would limit both the range of useful spectral response and overall quantum efficiency. The total number of excited electrons per unit time, t, $N' = N/t(= i/q)$, at which photoelectrons having charge, q, are emitted from a PE surface is determined from the photoelectric current, i

$$i = \frac{Nq}{t}, \tag{5.1}$$

where N the total number of excited electrons; in order to determine the rate N', it is required to calculate the probability of exciting an electron.

The widely used photocathode materials are cesium-silver-oxygen (CsAgO), cesium-antimony (CsSb), and the bi-alkali and tri-alkali compounds, such as KCsSb, RbCsSb, and Na_2KCsSb; the CsSb offers high gain but has poor high temperature stability. The spectral sensitivity characteristics are limited at long-wavelengths by the photoemission threshold of the material, and at short-wavelengths by the transmission of the window. In PE devices, one of the main causes of the pulse component of the dark-current is the thermionic emission obeying Richardson's law (see equation 2.3) from the photocathode surface. Although the thermionic work function and the photoemission threshold are separate quantities, within a given family of tubes, there exists a statistical correlation that makes high red and IR sensitivity incompatible with low dark current. At room temperature, the thermionic emission of photocathodes with maximum sensitivity in the blue region of the electromagnetic spectrum (300 to 500 nm) is between 10 and 1000 electrons per cm^2 per second; it increases as the sensitivity extends toward the longer wavelengths. A means of reducing thermal dark emission is to cool (see Section 4.3.1) the photocathode. However, cooling of such an element has its limitations that are due to residual tube structure leakage, high energy particle interference (from the window, the environment, and extra-terrestrial sources), and mechanical stress considerations. Cooling can raise the resistance of the active layer of most classical devices as well. The higher resistance causes lateral cathode voltage drop during emission, which requires that emission be reduced to maintain uniformity.

5.2.2 *Photomultiplier*

The important photoemissive device specifications, other than dark current, are determined mainly by the internal electron-multiplier amplifier, for instance, the photomultiplier tube (PMT; see Figure 5.4a). A highly sensitive detector of light, the PMT is an evacuated glass vessel incorporating an electron multiplier to enhance the photocathode output before it is coupled to the external electronics.

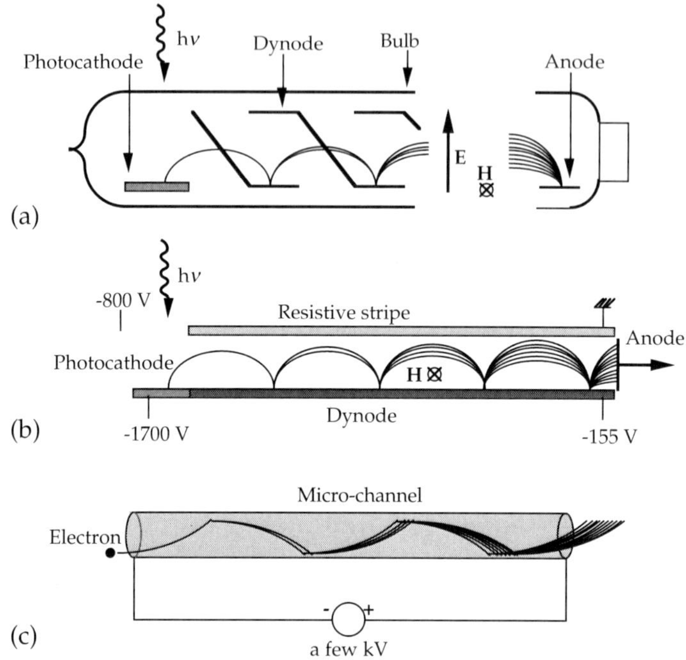

Figure 5.4 Schematic diagram of a (a) photomultiplier tube, (b) continuous dynode photomultiplier, and (c) micro-channel (after Morel and Saha, 2005).

In a photocathode, light enters from one side and the electrons are emitted from the other side into the vacuum, which are further amplified by a chain of secondary electrodes called dynodes (see Figure 5.4b), possessing the property of emitting many more electrons than they receive under electron bombardment. Ejected by a photon, an electron leaves the photocathode, having the energy of the incoming photon minus the work function of the photocathode. Using an electric field, each electron is accelerated from the photocathode toward the first dynode to gain a sufficient energy to be able to eject more than one electron. The mechanism of gain (see Section 4.3.2.2) at each dynode is secondary emission of electrons whose number depends on (i) the nature of the surface, (ii) the energy of the primary, and (iii) the incident angle of the primary electrons. The ratio of the average number of secondary electrons emitted by a target to the primary electrons bombarding the dynode is characterized by the "secondary emission ratio," δ. The dynodes are maintained at a certain electrical potential (typically 1–3 kV), which increases from dynode to dynode in order to accelerate the electrons through the PMT to the anode, where they are absorbed generating an output signal in the form of an electrical pulse. The peak output current, signal gain, bandwidth and rise-time, and noise figure (see Section 4.5.5) are functions of the dynode material used for each gain stage and of the geometrical design of the multiplier structure including the number of stages. The material used for each dynode has lower dark current and residual noise than the cathode limiting photomultiplier noise to either photon-noise (see Section 4.5.2.3) or residual cathode dark current noise. In order to avoid field emission, the dynode surfaces must have some conducting property, i.e., the surface oxide layer must have some metal in excess.

A dynode material is chosen from the behavior of its secondary emission ratio, δ versus accelerating potential. The oxides of alkali metals prove to be good materials, for instance, (i) AgMgO that offers high gain and better stability (with both temperature and high emission density) and (ii) CuBeO, which is similar in gain and easily reactivated after exposure to air. However, the dynodes have upper current limits due to defocusing of traveling electron cloud, anode heating, noise, and space-charge effects, which limit stable gain to a maximum of about 10^7. Among the emitted electrons three groups of electrons are distinguishable:

- primary electrons that are elastically reflected without loss of energy (high energy species),
- primary electrons that are back diffused, low energy species in comparison, having continuous distribution of energy, and
- true secondary electrons (low energy species).

If the primary energy is sufficiently large, the energy distribution becomes independent of primary energy. Because, for highly energetic primaries a large number of secondaries are produced deep inside the material, and hence, most of them are unable to escape, resulting in a dip of the number of secondaries as primary energy goes very high.

The secondary electron emitted by the first dynode can be directed onto the second dynode, which functions in the same manner as the first. Each dynode is held at a more positive voltage than the previous one. The same phenomenon recurs between the first and the second dynode and so on until the electrons hit the anode whose output current is large. At each stage, electron flux increases with a gain G; as it is possible to make a very large amplification of the electric signal to an extent that it is able to detect the arrival of a single photon. The system of dynodes should satisfy the conditions (Hiltner, 1962):

- each dynode surface receives the largest possible fraction of secondaries emitted by the preceding dynode,
- secondary electrons are emitted into an accelerating electric field,
- system is insensitive to perturbing fields, such as the Earth's magnetic field, and
- ionic feedback is eliminated, and electron cold field emission is avoided.

If a photomultiplier, containing N dynodes, each of same amplification factor, δ, provides an overall amplification of δ^N; therefore, one finds

$$i = i_0 \delta^N, \tag{5.2}$$

where N is the number of stages and assuming after each stage the noise (see Section 4.5) is likely to be the same Schottky form $(2qi_0\Delta f)$ and gets multiplied by the same way as current. so

$$\langle i_s \rangle^2 = 2q\delta i_0 \Delta f + \delta^2 (2q\delta i_0 \Delta f). \tag{5.3}$$

After first stage, $i_1 = \delta i_0$; hence, after N^{th} stage, one gets

$$\begin{aligned} \langle i_N \rangle^2 &= 2qi_0 \Delta f \delta^N \left[1 + \delta + \delta^2 + \cdots + \delta^N \right] \\ &= 2qi_0 \Delta f \delta^N \left[\frac{1 - \delta^{N+1}}{1 - \delta} \right]. \end{aligned} \tag{5.4}$$

The signal-to-noise (S/N) ratio (see Section 4.1.1.4) at the input (ip) divided by the S/N ratio at the output (op) can be specified as

$$\frac{(\mathrm{S/N})_{\mathrm{ip}}}{(\mathrm{S/N})_{\mathrm{op}}} = \left[\frac{1 - \delta^{N+1}}{1 - \delta} \right]^{1/2} = \mathcal{A}. \tag{5.5}$$

Thus $\mathcal{A}$ is small when δ is greater, while it is small if n is small. But for all practical purposes, the noise introduced by increasing n is negligible when $\delta > 2$.

The gain (see Section 4.3.2.2), G, of a photomultiplier is given by

$$G = \frac{\Phi_\nu}{\Phi_{\rm m}}, \tag{5.6}$$

in which $\Phi_{\rm m}$ is the minimum luminous flux detectable with an ideal photomultiplier and Φ_ν the minimum flux detectable by the photocathode that is without multiplication (directly coupled with an amplifier).

Since a photocathode yields a current, i_a, that contains thermionic current and current due to background light flux (sky background light) along with the true signal, $i_{\rm m}$, one defines a modulation factor, Γ

$$\Gamma = \sqrt{2}\frac{i_{\rm m}}{i_a}. \tag{5.7}$$

Now one obtains

$$G = \frac{i_\nu}{i_{\rm m}} = \frac{\Gamma}{C\rho}\sqrt{\pi k_{\rm B} T C}, \tag{5.8}$$

where ρ is the S/N ratio defined by $i_{\rm m}/i$, C the capacitance, and i_ν can be expressed by

$$i_\nu = \rho\sqrt{\left(\frac{4k_{\rm B}T\,\Delta f}{R}\right)}. \tag{5.9}$$

Thus for $C = 20$ pF and $\rho = 2$, for room temperature, $G = 1600\Gamma$. For a 100% modulated signal, $\Gamma = 1$ and $G = 1600$, i.e., the multiplier can detect 1600 times fainter flux than the amplifier. In an ideal case, PMT should be cooled and it should have no background flux; the PMTs are usually operated in low temperatures using dry-ice. But for a non-ideal situation, efficiency of multiplier rapidly goes to the amplifiers. If an impedance transfer mechanism or electronic amplifier is used at the output of the multiplier, in order that the shot-noise (see Section 4.5.2.3) is greater than the Johnson noise (see Section 4.5.2.1)

$$M^2 \geq \frac{5 \times 10^{-2}}{i_0 R}, \tag{5.10}$$

so that high gain is required for cooled (small i_0) multiplier.

There is an upper limit to the last dynode current $\sim 10^{-7}$ A. Thus, a multiplier with a higher number of stages reaches the highest cut-off flux situation more quickly than the one with smaller stages. The latter can also be used to measure very weak flux using high amplifier input resistance. The S/N ratio of the multiplier is given by

$$\mathrm{S/N} = \frac{Mi_\phi}{M\left[2q(i_\mathrm{e} + i_\phi)\Delta f\right]^{1/2}} = \frac{i_\phi}{\left[2q(i_\mathrm{e} + i_\phi)\Delta f\right]^{1/2}}, \qquad (5.11)$$

where i_ϕ and i_e are the signal and extraneous components of the photocathode current, respectively.

Let i_Φ and i'_e, respectively, be the signal component of the multiplier current and the extraneous component of the output current, one may write

$$i_\Phi = Mi_\phi; \qquad i'_\mathrm{e} = Mi_\mathrm{e}. \qquad (5.12)$$

Equation (5.12) suggests that the S/N ratio is independent of the amplification, M, and increases with when

- i_ϕ is large, i.e., for a given light flux, the quantum efficiency of the photocathode is large, thus the measurement of weakest light flux requires only photocathode of higher quantum efficiency (QE),
- i_e is smallest; it tends to very less thermionic current and background flux, and
- Δf is small, i.e., the S/N ratio of the instrument can be improved by the time constant of the measurement.

The anode current is related to the photocathode current as

$$i_a = i_\mathrm{ph} G_\mathrm{n}. \qquad (5.13)$$

The use of PMT is a limiting factor to obtain higher spatial resolution and robustness, as they are bulky, fragile, and susceptible to magnetic field. Also, the bandwidth or anode pulse rise time of such a tube is limited by the spread in electron times of flight (rather than the time of flight).

5.2.3 *Development of PMT*

A photomultiplier is useful in low intensity applications, such as fluorescence spectroscopy, television cameras, burglar alarms, and light sensors. Applications in the imagery of this detector regarded mechanical scan television using Nipkow's disk, invented by P. G. Nipkow in 1884 although J. L. Baird is generally considered to be the creator of television, which required only a detector "mono-pixel." It was a spinning disk, with a spiral pattern of apertures arranged to raster scan the specimen with the light source, illuminating through the apertures. When the disk is spinning, numerous small points of light scan the specimen simultaneously. The spinning disk can be made out of any metal, with a series of equally spaced circular (or square) apertures of equal diameter arranged in a spiral around its edge. Each aperture would provide a single line of video signal. While one aperture was scanning the projected picture, several samples were being taken, representing the horizontal resolution of the signal. In this method, the light intensities of small portions of an image are analyzed and transmitted. As the disk rotated, the image was scanned by the perforations in the disk, and light from different portions of it passed to the selenium[a] photocell. The number of scanned lines was equal to the number of perforations and each rotation of the disk produced a television frame. Although Nipkow's equipment lacked the means of amplifying the impulses, this method was an important stepping stone for the development of television.

However, the magnetic system of MS-10, even if it is left for photomultipliers today, have been the first stage the miniaturization of the multipliers of electrons. Heroux and Hinterreger (1960) resumed the idea of the magnetic photomultiplier, but simplified it by using only one dynode, which consisted of a coating on a plate. The electrons bounced and were multiplied on this plate during their travel from the cathode to the anode (see Figure 5.4b). Goodrich and Wiley (1961) fabricated an identical system, providing

[a]Selenium (Se) was discovered by J. J. Berzelius in 1817. It is metalloid in the oxygen group with electrical properties. It was in 1873 when J. May found that some selenium rods had the photosensitive properties, which led to the possibility of converting light waves into electrical impulses.

a 10^7 electron gain, a few millimeter thick. The process of miniaturization was therefore already commenced.

The first cathode-ray tube (CRT) oscilloscope was invented by K. F. Braun in 1897, and a two-diode thermionic valve (an electronic valve in which electrons are emitted from a heated cathode) was introduced by A. Fleming a few years later. The history of the photomultiplier begins with the discovery of the secondary emission, the first implement of which to use was "Dynatron," a system with negative resistance used for oscillators invented by Hull (1918). Later in 1924, J. L. Baird demonstrated a mechanically scanned television system. His 30 line images were the first demonstrations of television by reflected light. The concept of the electronic camera came from P. T. Farnsworth (1906–1971). In 1927, he succeeded in transmitting a television image comprising 60 horizontal lines from one room to another. He went on to invent many different devices that include equipment to convert an optical image into an electrical signal, amplifier, cathode-ray tube, vacuum tube, electrical scanners, electron multipliers and photoelectric materials.

A few years later, V. Zworykin developed the iconoscope, a tube for television transmission used in the first cameras; an improved CRT, called Kinescope, was also developed by him in 1929, which was used to demonstrate a television system with all the features of a picture tube. Later, Zworykin (1936) conceived in a multiplier of electrons the 12 elements of which in secondary emission were made of a mixture of silver, zirconium, and cesium. Subsequently, the Société Radioelectrique company perfected a tube called photomultiplier "MS-10" (Coutancier, 1940). It featured 10 dynodes of composition $AgCs_2O$, providing a gain from 4000 to 12,000. This tube used a magnetic field between 10 and 20 milliteslas (mT), in order to apply to electrons a Lorentz's force associated with the electric field, making them bounce from an element to the next (see Figure 5.4a).

5.2.4 *Astronomical Applications of PMTs*

Early applications of the photomultiplier were in astronomy and spectroscopy, because of the high quantum efficiency, which was an order higher than that of photographic film. Furthermore, the output current of such a tube is linear with incident radiation power. The PMT can also be used directly in photometric and spectro-photometric instrument.

5.2.4.1 Photometer

The development of photoelectric photometry had made considerable contributions in the field of observational astronomy. The light gathering power of a small telescope using such an instrument had improved substantially; one could measure radiation from a star of magnitude m_V $\sim$20 or fainter. It could be achieved with off-set photometers refrigerated photomultipliers and good telescope drives that enable the use of small diaphragms. A photometer measures the apparent brightness of a stellar object by directing its light on to a photosensitive cell, such as a PMT. The additional requirements are (i) a field lens (Fabry lens), and (ii) a set of specialized optical filters. Also, a small diaphragm is kept in the focal plane to stop down a star and minimize background light from the sky and other stars; it has several openings ranging from a large opening for initially centering the star to the smallest. In order to achieve proper centering, an assembly consisting of a movable mirror, a pair of lenses, and an eyepiece for viewing the star in the diaphragm, is required. Figure 5.5 depicts a schematic layout of a photoelectric photometer. When the mirror is swung into the light path, the diverging light cone is directed toward the first lens. The focal length of this lens is equal to its distance from the diaphragm. The second lens is a small telescope objective that re-focuses the light. The eyepiece gives a magnified view of the diaphragm. Once the star is centered, the mirror is swung out of the way and light passes through the filter. The choice of the filter is dictated by the spectral region to be measured. The Fabry lens refracts the light rays onto a photocathode of the PMT. This lens spreads the light on the photocathode and minimizes the photocathode surface variations. The photocathode is located, in general, exactly at the exit pupil of Fabry lens so that the image of the primary mirror on the cathode is in good focus. A detector, usually a photomultiplier tube, is housed in its own subcompartment with a dark slide. The output current is

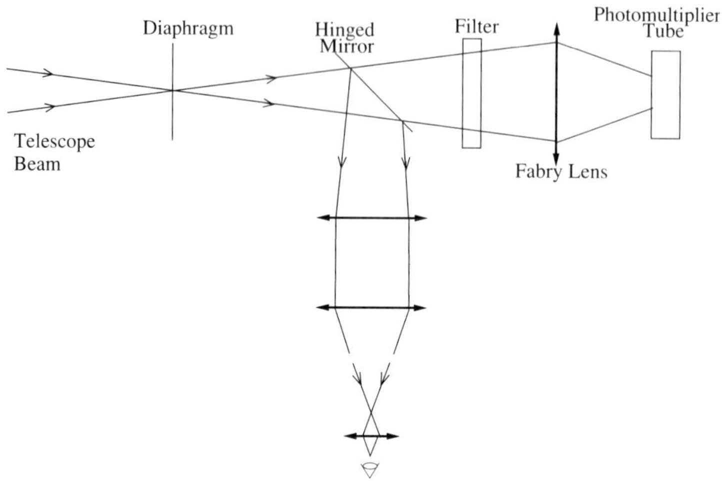

Figure 5.5 Schematic diagram of a photometer.

intensified further by a preamplifier, before it can be measured and recorded by a device, such as strip chart recorder or in digital form on disk.

The photometer is calibrated on the detector spectral comparator that is used, in general, for spectral transmittance measurements and other characterizations requiring a known optical beam and has an absolute spectral responsivity, $\mathcal{R}(\lambda)$ (see Section 4.1.1.2). Let $\mathcal{Q}_\lambda$ represent the general radiometric quantity (spectral radiant energy) and $\mathcal{Q}_\nu$ the general photometric equivalent function (luminous energy), following equation (1.85) the relationship is given by

$$\mathcal{Q}_\nu = K_{\rm m} \int_\lambda \mathcal{Q}_\lambda V(\lambda) d\lambda, \tag{5.14}$$

where $V(\lambda)$ is the spectral luminous efficiency (CIE, 1986) for photopic vision (see Section 4.2.1) and $K_{\rm m}$ the luminous efficacy; the range of the integral has practical limits restricted to the region of non-zero values for $V(\lambda)$.

A source is placed at a distance r from the detector system whose input is determined by the amount of optical power entering the precision aperture of area $\mathtt{A}$. The uniform irradiance of the source, $\mathcal{E}_\lambda$ on this aperture produces a current, i_0 from the detector system

$$i_0 = \mathtt{A} \int_\lambda \mathcal{E}_\lambda \mathcal{R}(\lambda) d\lambda. \tag{5.15}$$

The responsivity of the photometer to the light in terms of luminous flux (equation 1.85) is expressed as

$$i_0 = \mathtt{A}\mathcal{R}_{\nu,f} K_\mathrm{m} \int_\lambda \mathcal{E}_\lambda V(\lambda) d\lambda, \tag{5.16}$$

with

$$\mathcal{R}_{\nu,f} = \frac{\int_\lambda \mathcal{E}_\lambda \mathcal{R}(\lambda) d\lambda}{K_\mathrm{m} \int_\lambda \mathcal{E}_\lambda V(\lambda) d\lambda}, \tag{5.17}$$

as the luminous flux responsivity and if the luminous flux (see Section 1.2.2), Φ_ν, is uniform over the aperture, $\mathtt{A}$, one finds

$$\Phi_\nu = \mathcal{E}_\nu \mathtt{A}, \tag{5.18}$$

in which $\mathcal{E}_\nu$ is the illuminance, and hence, the illuminance responsivity, $\mathcal{R}_{\nu,i}$, turns out to be $\mathcal{R}_{\nu,i} = \mathtt{A}\mathcal{R}_{\nu,f}$.

For a point source, for instance, an unresolved star, the illuminance is related to the luminous intensity by the inverse square law (see Section 1.2.1)

$$\mathcal{E}_\nu = \frac{\mathcal{I}_\nu}{r^2} = \frac{i_0}{\mathcal{R}_{\nu,i}} = \frac{i_0}{\mathtt{A}\mathcal{R}_{\nu,f}},$$

therefore, the luminous intensity, $\mathcal{I}_\nu$ is expressed as

$$\mathcal{I}_\nu = \frac{i_0 r^2}{\mathtt{A}\mathcal{R}_{\nu,f}}. \tag{5.19}$$

The photometer is generally placed at the Cassegrain focus behind the primary mirror of a telescope. In a differential photometry, a second star of nearly the same color and brightness as the target star, is used as a companion star that should be very close so as to enable the observer for switching over rapidly between these stars. The extinction (see Section 1.1.5) correction can often be ignored, since both stars are seen through identical atmospheric layers. Photon-counting with adequate sky subtraction ensures the high quality observations. To note, the wide-band photomultiplier tubes enables observations from U to I to be done with the same detector.

A photometer can also be used for airglow, observations. A ground-based filter photometer was developed to measure (i) the density of atomic oxygen at the lower thermosphere and (ii) the temperature of dominant OH level at the mesopause (Saha, 2001).

The intensities of OI [557.7], OI [630.0], OH(7-2) bands Σ R [683.8], and $P_1(3)$ line of OH [692.28] from the zenith sky together with the background continuum at 712.0. The filter wheel containing these filters with the respective half-power bandwidths of interference filters are 1.003 nm, 1.150 nm, 2.253 nm, 0.807 nm and 4.0 nm, took 12 min to complete a rotation; the calibration was carried out with ^{14}C source.[a]

Airglow Airglow is a faint photoluminescence caused by the chemical reactions of incoming solar energetic radiation with atoms and molecules, such as atomic oxygen (O), nitrogen molecules (N_2) that are predominant in the thermosphere (above 100 km altitude), and other atmospheric components, including hydroxyl radicals (OH). The molecular oxygen (O_2) and OH emissions are bands of many closely spaced wavelengths (Meinel bands; Meinel, 1950) since the transitions involve changes in vibrational energy together with smaller changes in rotational energy. The wavelengths of emission from these elements span bands ranging from about 380 to 490 nm (violet-blue) and about 650 to 700 nm (red), respectively. The brightest emission is green (557.7 nm) light from oxygen atoms in the lower thermosphere; atomic oxygen also emits at 630.0 nm wavelength. The reaction rates for the production of O I(557.7) radiation was suggested by Chapman (1931), that is

$$O(^3P) + O(^3P) + O(^3P) \xrightarrow{k_1} O(^1S) + O,$$

in which k_1 is the respective reaction rate.

The temperature at the lower thermosphere region can be determined from OH(7-2) night airglow observations. The ratio of intensities of $P_1(3)$ line (6922.8) and ΣR branch (6838) of OH(7-2) band is measured experimentally. The same ratio can be

[a]The radiocarbon dating, ^{14}C is based on the radioactive decay of natural radiocarbon that every organism takes up during its lifetime. The radioactive half-life for a given radioisotope is the time for half the radioactive nuclei in any sample to undergo radioactive decay. The radioactive decay law is dictated as

$$A(t) = A_0 e^{-(\ln 2.t)/T_{1/2}},$$

where A is the amount of radioactive material compared to the original amount, A_0, $T_{1/2}$ the half-life of the radioactive isotope, and t the time; for ^{14}C, the half-life is 5730 ± 40 years (Taylor, 2005).

determined theoretically by using the following equation at different temperatures from synthetic spectrum of OH(7-2) band:

$$I' = C_{\nu}^{4} S_j e^{-F(j)hc/k_B T},$$

where k_B is Boltzmann's constant, I' the intensity of rotational line, S_j the line strength, j the upper level rotational quantum number (see Section 1.1), T the rotational temperature, C an arbitrary constant.

The function $F(j)$ is defined by $B_\nu j(j+1)$ where B_ν is the effective rotational constant. The rotational temperature of OH is considered to be the true temperature of the atmospheric level where OH is dominant. The temperature can be estimated from other hydroxyl bands too, for instance, 8-3 band of OH; in fact, the ratio of intensity of any two lines from a given vibration rotation band would provide temperature.

The nightglow (nocturnal phenomenon) is feeble in the visible bands but strong in the near-IR region; it varies through the night. The airglow is strongest over low- and mid-latitudes. However, it is not discernible during the daylight due to the scattered light from the Sun. Of course, with the advent of new technology especially involving fast electronics and using a mask system, it has become possible to measure the dayglow over low and equatorial latitudes (Vineeth et al., 2005, 2007; Hussain et al., 2010).

5.2.4.2 Interferometric applications of PMTs

The intensity interferometry, in visible waveband, introduced by Hanbury Brown and Twiss (1956a, b), has made a major impact to the development of the subject of quantum optics. However, like a two-telescope amplitude interferometer, the measurements from intensity interferometer was useful to derive the diameter and limb-darkening coefficients of stars, but unable to produce aperture synthesis images since the visibility phase information is not preserved. The degree of partial coherence represents the Fourier transform of the energy distribution on the source. The correlation depends on the square of the intensities and it does not provide the degree of coherence directly, but provides the square of the modulus of the complex degree of coherence. From the second-order correlation function, it is clear that by measuring correlations with two detectors, no information on the phase of the complex degree of coherence, γ, can be retrieved. Of course, the phase behavior may be recovered with multi-element intensity interferometry. The advantage of such an instrument is its independent ability to measure the distance to each source by searching for the maximum signal. Sensitivity improvements of an intensity interferometer can result from observing the same star with an array of telescopes, providing measurements over several baselines simultaneously.

The PMTs were employed as the detectors to demonstrate the interference that could be detected in the intensity of light, despite the fact that light was detected as a stream of photons (see Section 1.1). Both these detectors measure fluctuations in the intensity of light from a particular star. The intensities are transformed into electrical currents in the photodetectors at the focii (see Figure 5.6). The current output of each detector is proportional to the instantaneous intensity of the incident radiation. The fluctuations in the current output by these detectors are partially correlated. This correlation gradually changes as the detectors are moved apart from each other. The Fourier transform of the size and shape of the source along the projected baseline is obtained as a function of separation. The correlation between the two signals (see detailed explanations in Mandel, 1963, and Saha, 2010) is made after the detection of photons at the two photocathodes for coherent incident light.

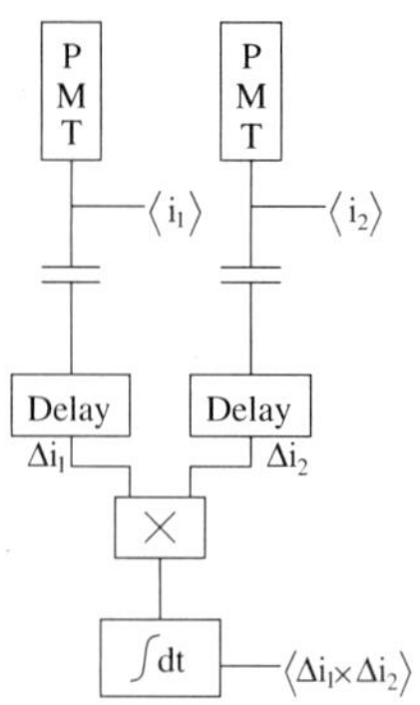

Figure 5.6 Principle of an intensity interferometer.

A correlator measured the correlation between the fluctuations in the anode currents of the photomultiplier detectors at the focii of the two telescopes. It multiplied the fluctuations in the two channels together and the correlation was a unidirectional output superimposed on random noise. This type of observation can be explained by statistical fluctuation of the instantaneous intensity, $I(t)$, having a probability distribution

$$\mathcal{P}(I)dI = \frac{I}{\bar{I}}e^{-I/\bar{I}}dI. \tag{5.20}$$

For a mean irradiance, $\bar{I}$ recorded in a time interval of δt, the mean number of emitted electrons $\bar{N} = \bar{I}\eta_{\rm d}\delta t/\hbar\omega$, in which $\eta_{\rm d}$ is the quantum efficiency (QE) of the detector, and ω the angular frequency. The mean value of $\bar{N}$ depends on the mean irradiance and the time over which the averaging is carried out. The term, $\bar{I}(\mathbf{r}_1)$ is measured by the multiplier followed by an integrator. Therefore, one obtains the degree of partial coherence, $|\gamma(\mathbf{r}_1, \mathbf{r}_2, 0)|$ and the angular profile of the stars.

The correlation, $\mathcal{C}(\mathbf{B})$, which is proportional to the square of the degree of coherence of the radiation at the two detectors is given by

$$\begin{aligned}\frac{\mathcal{C}_{\rm n}(\mathbf{B})}{\mathcal{C}_{\rm n}(0)} &= \frac{\langle \Delta i(\mathbf{r}_1, t)\Delta i(\mathbf{r}_2, t)\rangle}{\langle i(\mathbf{r}_1)\rangle\langle i(\mathbf{r}_2)\rangle} \\ &= |\gamma(\mathbf{r}_1, \mathbf{r}_2, 0)|^2 = \mathcal{V}^2,\end{aligned} \tag{5.21}$$

in which $|\gamma(\mathbf{r}_1, \mathbf{r}_2, 0)|^2$ is the complex degree of coherence, $\Delta i(\mathbf{r}_1, t)$ and $\Delta i(\mathbf{r}_2, t)$ are the fluctuations in the two currents, $\mathbf{r}_1$, $\mathbf{r}_2$ the position vectors, $\langle\,\rangle$ the long-term averaging, $\mathcal{V}$ the corresponding fringe visibility, $\mathcal{C}_{\rm n}(\mathbf{B})$ the normalized correlation, $\mathbf{B}$ the baseline vector, and $\mathcal{C}_{\rm n}(0)$ the constant of the equipment.

Equation (5.21) indicates that the covariance of the time-integrated product of the current fluctuations, $\langle \Delta i(\mathbf{r}_1, t)\Delta i(\mathbf{r}_2, t)\rangle$, is the squared modulus of the covariance of the complex amplitude. This correlation is proportional to the square of the modulus of the degree of coherence, $|\gamma(\mathbf{r}_1, \mathbf{r}_2, 0)|^2$, of the light at the two detectors.

Intensity interferometer In the mid-1960s, the development of the Narrabri intensity interferometer (Hanbury Brown, 1974, and references therein) having a pair of light collectors having diameter of 6.5 m with 11 m focal length on a circular railway track spanning

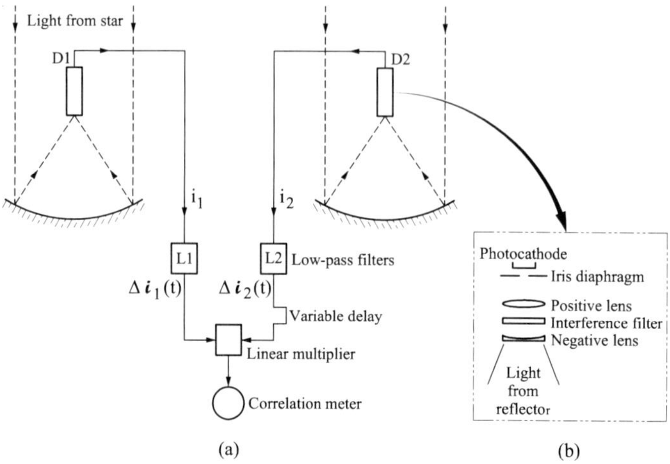

Figure 5.7 (a) Schematic diagram of Narrabri intensity interferometer and (b) the optical system at the focus.

188 m, provided the measurements of 32 southern binary stars with angular resolution limit of 0.5 milli-arcseconds (mas). In this arrangement (see Figure 5.7), starlight collected by two concave mirrors is focused on to two PMTs and the correlation of fluctuations in the photocurrents is measured as a function of mirror separation.

The DC component of the signals was measured and recorded while the AC components were sent to the input of the correlator, which measured the correlation between the fluctuations in the anode currents of the photomultiplier detectors at the focii of the two telescopes. The DC component to be measured on the output of the multiplier is very small in comparison with the random fluctuations. On the output of the multiplier, the phase switching frequencies were amplified and demodulated before the signal was integrated, thus removing any significant contamination due to offsets. The measurement of correlation for different baselines, B, of the two photomultipliers (PMT) were taken. The degree of coherence $|\gamma(\mathbf{r}_1, \mathbf{r}_2)|$ at the two photocathodes is varied in a manner that was calculable from the geometry of the arrangement. The experimental curve obtained in this way approaches the theoretical curve that provides $|\gamma(\mathbf{r}_1, \mathbf{r}_2, 0)|^2$ as a function of B, from which one derives $|\gamma(\mathbf{r}_1, \mathbf{r}_2, 0)|$, and hence, the angular diameter of the star.

The requirement of opto-mechanical accuracy for the intensity interferometry is less stringent, since it depends on the electrical bandwidth of the detectors, but not the wavelength of light. The alignment tolerances are extremely relaxed since the pathlengths need to be maintained to a fraction of $c/\Delta f$, in which c is the speed of light in free space and Δf the electrical bandwidth. It requires control of the light paths to an accuracy determined by the light coherence time ($\sim$100 ns), therefore it is much less sensitive to atmospheric phase fluctuations; the signals input to the correlator follow roughly the same path through the atmosphere, for at least small separation between the two telescopes, although the significant effect comes from scintillation induced by the atmosphere. Such an instrument eliminates the need for optical delay-lines as well. However, the sensitivity of this interferometer was found to be very low; it was limited by the narrow bandwidth filters.

5.2.5 *Use of Photoemission Devices for High Energies*

The concepts on radiation detectors in X-ray and gamma-ray energies are fundamentally different from those used in optical/IR. These radiations are of high energy and a quanta of photon interacts by photoelectric effect (see Section 2.1), Compton scattering or pair-production and gives all or a substantial part of its energy to one electron: this electron is a fast-moving electron and gives multiple electron–hole pairs/ multiple photons (Leo, 1994; Knoll, 2000). Barring the photographic plates (see Section 1.4.2), most of the current X-ray/ gamma-ray detectors are individual photon counting devices. The concept, terminology, and language are those used in atomic and nuclear physics.

In the case of X-ray radiation, photoelectric effect is the dominant mode of interaction. Here, complete or part of energy, $h\nu$ of the incoming photon ejects an electron from an atomic shell of the atom; the ejection of the electron is given by equation (2.1). This fast-moving electron loses energy by ionization (see Section 5.2) and a series of electron–ion and electron–hole pairs are created (see Section 2.2.4.1). Most detectors use these charged pairs to identify the incoming radiation and to measure its energy (which is related to the number of charges produced).

Most X-ray and low energy gamma-ray observatories employ detectors like proportional counters, semiconductor detectors, and scintillators to measure the flux and the energy spectrum of astrophysical sources. In these detectors, photoelectric effect is the primary mode of interaction. The direction of the initial fast electron can also give information about the polarization of the incoming radiation. This property, however, is not yet exploited in any Space Observatory.

Compton Scattering For soft gamma-ray band and beyond ($>$100 keV), the incoming photons interacts with the electron of the absorbing material and part of its energy is transferred to that electron, which gets scattered. This phenomenon is called Compton scattering. A. H. Compton (1892–1962) observed the scattering of X-rays from electrons by solid materials (mainly graphite) and found the shift of the wavelength of the scattered photon, which could be explained by assuming the photon having momentum given in equation (1.20). This is termed as Compton effect (Compton, 1923), for which he was awarded Nobel prize in 1927. Figure 5.8 illustrates Compton scattering. The kinetic energy of the scattered electron is given by

$$\begin{aligned} E &= h\nu - h\nu' = h\nu - \frac{h\nu}{1+\alpha(1-\cos\theta)} \\ &= h\nu\frac{\alpha(1-\cos\theta)}{1+\alpha(1-\cos\theta)}, \end{aligned} \tag{5.22}$$

with

$$\alpha = \frac{h\nu}{m_e c^2}, \tag{5.23}$$

where the rest mass is $m_e c^2 = 511$ keV.

Since the energy given to the electron is only a part of the incident radiation, detectors having Compton scattering as the major form of interaction are rarely used as X-ray/ gamma-ray detectors in astronomy. There are, however, some attempts to use a thick total absorber below a Compton scatterer to measure the energy and the direction (limited to a cone angle) of the incident radiation

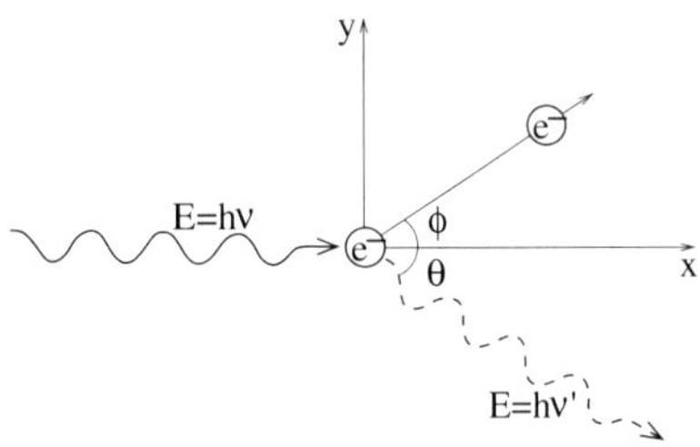

Figure 5.8 Schematic diagram of Compton scattering; the angle θ is that between the directions of the incident and scattered photons and ϕ is that between the directions of the incident photon and the recoil electron. Energy and momentum are conserved in such an event, as a result the scattered photon has less energy (longer wavelength) than the incident photon.

in the MeV range. Since the direction of the scattered photon is strongly related to the polarization angle of the incident radiation, the Compton scattering properties can be exploited to measure the degree of polarization of the incoming radiation.

Pair production In this process, called pair production, a high energy photon interacts with the electric field of a heavy nucleus to create electron–positron (production of e^+–e^-) pair. For radiation with energy higher than ~5 MeV, pair-production becomes significant. In this phenomenon, an electron–positron pair is formed by absorbing the complete energy $h\nu$ of the incoming photon. Thus

$$h\nu = (E_{e^-} + m_e c^2) + (E_{e^+} + m_e c^2), \tag{5.24}$$

in which E_{e^-} and E_{e^+} are the kinetic energies of the electron and positron, respectively.

Stacked detectors to track these high energy e+/ e− pairs have been successfully used in gamma-ray astronomy in the energy range of 10 MeV to 100 MeV. The kinematics can be worked out to measure the energy and direction of the incident photons.

5.2.5.1 PMTs with scintillators detectors and medical applications

Achievement of two major milestones in the nuclear medicine development in the 1950s has opened up the path to use functional imaging procedure in cancer diagnosis. The first one is the development of technetium-99m (^{99m}Tc) radionuclide generator system by a group at Brookhaven National Laboratory where ^{99m}Tc is eluted from the much longer-lived Molybdenum-99 (^{99}Mo; half-life ~66 hours), which ensured the availability of ^{99m}Tc. The second one is the development of the gamma camera (Anger, 1952, 1958) for the medical imaging. For the application of such an imaging system, in general, the photomultiplier tubes are used to locate a small lesion or tumor.

In a conventional X-ray mammography, a process where doses of ionizing radiation (emission of energy in the form of waves or particles) are used to create image, the presence of such abnormalities at the early stage of cancer are not definitely detected because of the complex anatomy of the breast tissues. A word of caution, the increased use of such an instrument or a computed tomography has raised concerns about potential carcinogenic side effects (Heyes et al., 2006; Brenner and Hall, 2007). The breast-specific gamma imaging is preferred in this respect.

In a gamma ray mammography, a radioactive isotope combined with a suitable compound is injected into the blood stream or ingested orally by the patient. The radioactive material disintegrates and gamma rays are ejected from the locations of abnormalities. A collimator absorbs and blocks all gamma-rays except in the forward direction and these gamma-rays interact in a large-area scintillation crystal and produce a burst of photons (see Figure 5.9). The PMTs

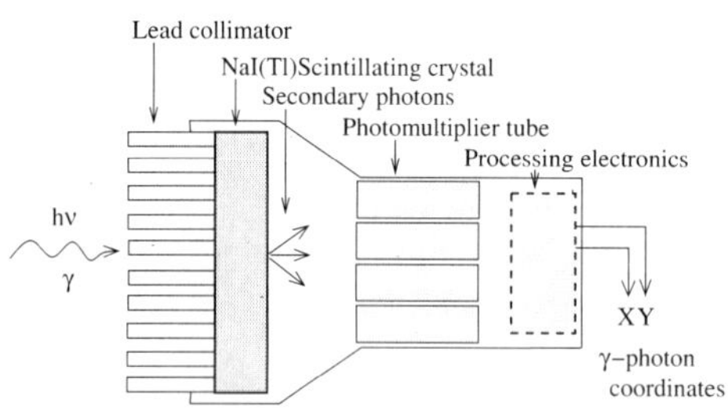

Figure 5.9 Anger gamma camera. The Na I(T1) scintillators convert each gamma photon into a burst of visible photons, the secondary photons spread on the photocathodes of the PMTs. The combination of the analog signals given by the PMTs provides the (x, y) coordinates of the gamma photon.

arranged in a hexagonal array behind the crystal detect these photons and the relative amplitude of the signals in the PMTs is used to pin-point the location of the gamma-ray interaction accurately. This method of photon position gamma imaging was originally proposed by Anger (1952). Counting is continued until a few million counts are obtained and the organ in question is satisfactorily delineated. The location of each scintillation is represented by a point on a cathode-ray-tube presentation.

The advantages of scinti-mammography are the following: (i) It improves evaluation of positive mammograms, (ii) differentiates between benign and malignant tissue, (iii) detects small tumors in cases where mammograms are difficult or improbable to read, (iv) could reduce the need for biopsy, and (v) allows images where none were possible before. However, the low-resolution image produced by conventional scinti-mammography does not differentiate between very small benign and malignant tissues.

5.2.5.2 Use of PMTs for gamma-ray astronomy

Gamma-rays are a form of electromagnetic radiation, which occurs during nuclear fission and during the interaction of ultra-relativistic particles with magnetic and photon fields. Narrabri interferometer telescopes were employed to detect atmospheric Cerenkov radiation produced in the atmosphere, light from extensive air showers by Grindlay et al. (1975); they were looking for sources of very high energy ($E > 300$ GeV) gamma rays. The cosmic rays and gamma rays reaching the Earth's atmosphere trigger extensive particle showers, which produce Cerenkov light. The Cerenkov radiation is polarized light generated by a high-energy charged particle moving through a dielectric with a velocity faster than the velocity of light in the dielectric. To note, the cosmic rays are not traveling faster than the speed of light, but travel faster than the phase velocity of light in the medium. These particles emit faint flashes of blue light, known as Cerenkov radiation. The light is emitted only in directions inclined at a certain angle to the direction of the particles' motion depending on their momentum; a conical wavefront is formed with the particle at its apex. The half angle of the conical wavefront is $\theta = arc\cos[c/(nv)]$, in which c is the speed of light in free space,

v the speed of the particle, and n the index of refraction of the dielectric.

Since Cerenkov radiation is very weak, fast photomultipliers are used in the focal planes of telescopes and signal is examined in time domain (nanoseconds) to extract the Cerenkov pulse information from the background. In recent times, multiple photomultipliers are used in the focal plane to get rough imaging. Use of multiple imaging telescopes gives a stereoscopic view of the path of the charge particle or gamma-ray interaction. The fact that gamma-ray tracks from a given source are parallel to each other while the charge particle tracks are randomly oriented can be exploited to reduce the background and improve sensitivity.

HAGAR An instrument called High Altitude Gamma ray telescope ARray (HAGAR) is installed at the Indian Astronomical Observatory (IAO), India. Situated at an altitude of 4300 m, it aims to detect atmospheric Cerenkov radiation of a threshold energy of $\sim$50 GeV with an atmospheric Cerenkov instrument. This instrument offers an advantage in lowering the threshold of energy of gamma ray observation, which has seven telescopes, each with seven mirrors of total area of 4.4 m^2. The total light gathering area of such an instrument is thus 31 m^2. The telescopes are deployed on the periphery of a circle of radius with one telescope at the center. Each of the seven mirrors in each telescope would be looked at by a UV sensitive PMT.

5.2.6 *Image Intensifiers*

The term, image intensifier, refers to a vacuum optical tube that amplifies (or intensifies) the number of incident photons internally by several orders of magnitude (Csorba, 1985; Brennan, 2003), as a miniature electron multiplier with 2-D resolution. It comprises a photocathode input, which is a coating of multi-alkali or semiconductor layers on the inside of the input window, and a phosphor screen, which is a fluorescing phosphor coating on the inside of the output window of the intensifier tube, usually fiber-optics (see Section 3.4). The phosphor screens emit green light and are made of rare earth oxides or halides, for instance, gadolinium

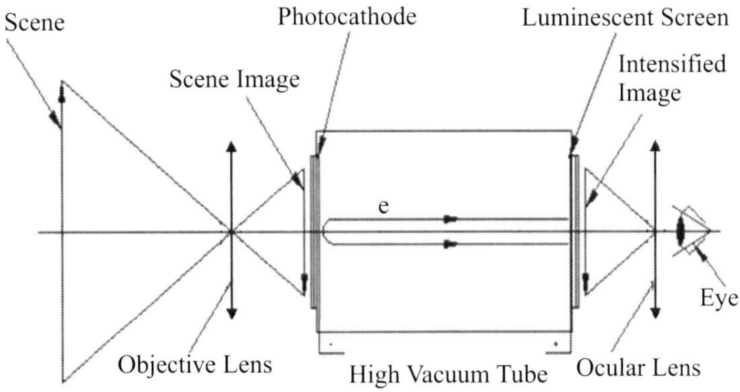

Figure 5.10 General principle of an intensifier.

(Gd), lanthanum (La), yttrium, with decay times of a few hundred nanoseconds to a few milliseconds. The image intensifiers sense an image from photons and energetic particles providing an output image by converting the electronic image into an intensified visual image by using a phosphor coating as the rear electrode to provide electroluminescent. Figure 5.10 depicts the schematic diagram of an intensifier.

The photocathode current, i_P, is given by

$$i_\mathrm{P} = k_\mathrm{P} \mathtt{A}_\mathrm{P} \mathcal{I}_\mathrm{P}, \tag{5.25}$$

where k_P is the photocathode sensitivity, $\mathtt{A}_\mathrm{P}$ the photocathode area, and $\mathcal{I}_\mathrm{P}$ the irradiance at the photocathode.

The power carried by the electrons to the screen, P_s, is

$$P_\mathrm{s} = i_\mathrm{P} V = k_\mathrm{P} \mathtt{A}_\mathrm{P} \mathcal{I}_\mathrm{P} V_a, \tag{5.26}$$

in which V_a is the accelerating voltage.

So, the flux radiated by the screen, Φ_s, is

$$\Phi_\mathrm{s} = \eta_\mathrm{d} P_\mathrm{s} = \eta_\mathrm{d} k_\mathrm{P} \mathtt{A}_\mathrm{P} \mathcal{I}_\mathrm{P} V_a, \tag{5.27}$$

with η_d as the luminous efficiency.

If the efficiency image conversion is characterized by the radiant flux conversion coefficient, $\Gamma_\mathrm{P} (= \Phi_\mathrm{s}/\Phi_\mathrm{P})$, in which $\Phi_\mathrm{P} = \mathcal{I}_\mathrm{P} \mathtt{A}_\mathrm{P}$, it can be expressed as

$$\Gamma_\mathrm{P} = \eta_\mathrm{d} k_\mathrm{P} V_a. \tag{5.28}$$

The image intensification co-efficient, Γ_B, defined by the ratio of the image luminance to the object illuminance, describes the system performance, which is related to Γ_P as

$$\Gamma_B \approx 0.25t\left(\frac{D}{f}\right)^2 \frac{\Gamma_P}{M_s^2}, \tag{5.29}$$

where D is the diameter of the entrance pupil, f the focal length of the lens, t the transmission of the lens, and M_s the magnification of electron–optical system.

The photocathode being subjected to an electrical field when the intensifier works, equation (2.3) becomes

$$\mathbf{J} = A_R T^2 e^{-(\phi_0 - \Delta\phi_0)/k_B T}, \tag{5.30}$$

with $\Delta\phi_0 = \sqrt{q^3\mathbf{E}/(4\pi\epsilon_0)}$, ϵ_0 as the permittivity in vacuum, q the elementary charge, and $\mathbf{E}$ the electrical field in the photocathode.

In order to bring the image intensity above the dark background level of the photographic plate, Lallemand (1936) introduced an archetype new imaging device, using a monitoring screen, generally known as "phosphor," onto which the energy of each accelerated electron from the photocathode was converted into a burst of photons (spot). Such a device consists of a 35 cm glass tube, with a potassium photocathode at one end, and 8 cm diameter zinc sulfide monitoring screen at the other end (which may be replaced with a photographic plate for recording the image). The focusing of electrons was performed by an electrostatic lens made by an inner silver coating on the tube and by a magnetic lens consisting of a 10 cm coil fed with a 0.5 A current. The accelerating voltage inside the tube was 6 kV, providing intensified images by the collision of accelerated photoelectrons onto the screen or the plate. This electronic camera could operate at low levels of intensity and did not exhibit the threshold effect of the normal photographic plate. The time integration capacity for the long-exposure was good, which was useful for imaging the faint astronomical objects. Figure 5.11 depicts the schematic diagram of a Lallemand tube used for astronomy.

The description of the first operational Lallemand tube on a telescope dates from 1951 (Lallemand and Duchesne, 1951). Similar tubes have been used until the beginning of the 1970s for faint object imaging. However, the sensitivity of Lallemand tubes did not reach

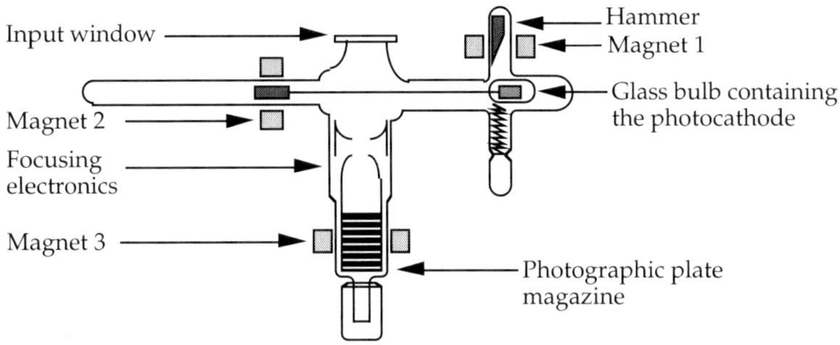

Figure 5.11 The Lallemand tube. The ring magnet 1 is employed to move the hammer and break the glass bulb containing the photocathode. The ring magnet 2 is used to bring the photocathode behind the input window, while the ring magnet 3 is used to change the photographic plate.

the quantum limit because of the dark threshold of the photo plates. Moreover, such tubes are made of glass and are, hence, very fragile and inconvenient for operation on telescopes.

A similar version was also developed by Holst et al. (1934), who used proximity focusing without electronic or magnetic lenses, in which the photocathode and the phosphor were separated by a few millimeters. To note, the proximity focus intensifiers of new generation are of compact mechanical construction with their length being smaller than their diameter. They are completely free of geometric distortion and feature high resolution over the photocathode's useful area. The image magnification is 1:1. The other advantages include (i) their immunity against electrical and electromagnetic stray-fields and (ii) ability to function as fast electronic shutters in the nanosecond range. This tube, in spite of its poor resolution owing to its structure, was constructed in large numbers during the Second World War for observation in infrared (Pratt, 1947). The industrial production of first-generation (Gen I) image intensifiers began from the 1950s; they were developed in most cases for nocturnal vision. The tubes in this category feature high image resolution, a wide dynamic range, and low noise.

A common type of detector is based on the television (TV) camera. A photoelectron accelerates under 15 kV about 900 photons by striking a phosphorus of type P-20, where they form an image

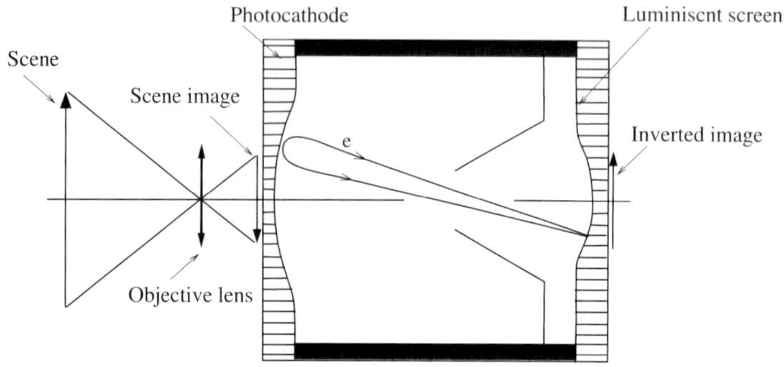

Figure 5.12 Schematic of a Gen I intensifier; an image is formed by a lens on photocathode, which is inverted by electrostatic and magnetic field.

in the form of an electric charge distribution. Following exposure, the charge at different points of the electrode is read by scanning its surface with electron beam, row by row. This produces a video signal that can be transformed into a visible signal on a TV tube. The information can be stored in digital form. In order to overcome the problem of photon gain limitation of these intensifiers, cascades of Gen I intensifiers (see Figure 5.12) were used for high sensitivity cameras. The Gen I intensifiers employed electrostatic focusing and electron acceleration to achieve signal gains up to 150. The main drawbacks of these devices are image distortion, short-lived components, and the large size. Also, in cascaded Gen I tube, three intensifiers are connected in series, which makes the unit longer and heavier. This equipment produced an image that was a clear at the center of the field-of-view, but suffered from large optical distortion around the periphery. These intensifiers are obsolete at present.

5.2.7 *Micro-Channel Plate*

Goodrich and Wiley (1962) achieved a major breakthrough by inventing the "micro-channel," a simple glass tube in which the inner side is coated by a secondary emission semiconductor. These authors pointed out that the electron gain of such a tube does not depend on the diameter, but on the ratio of the length to the diameter, in a proportional manner. With such dimensions, the parallel assembling of micro-channels in arrays, with an intention of making the enhancement of image became realistic. However, the gain of micro-channels is limited by the positive charges left by the secondary electron cascade, which goes against the electric field that is applied at the ends of the micro-channel. The maximum electron gain of a micro-channel is a few 10,000.

Further development on how to build arrays of micro-channels, known as micro-channel plates (MCPs; see Figure 5.4c), was carried out in the 1960s. It is an intensifier, exploiting the photoelectric effect (see Section 2.1) in solids. The photoelectrons are guided and multiplied in a channel with accelerating voltage gradient. The MCP is a conductive glass wafer consisting of parallel array of millions of channels fused together with pore size in the range 4–12 μm, which is fabricated from two compartment, lead silicate glass tube and acid soluble glass rod. The channel array is formed by fused fiber-optics technique and acid etching technique. The electrons entering the channel collide with coated walls and produce secondary electrons. Each channel behaves as a continuous dynode, when an electron enters and hits the channel wall, commences a cascade of electrons avalanche propagating the channel. The interior walls of each capillary (hollow tube) are coated with a secondary-electron-emissive material. Both faces of the plate are coated with thin metal films that act as electrodes and a voltage is applied across them.

A potential difference of a few kilo-volts (kV) is applied to ends of the tube in order to cause the multiplication of electrons. The incident photon releases a photoelectron, which is accelerated by the electric field, and multiplied by successive electronic collisions. The phosphor is excited at the exit and emits a large number of photons. The proximity focused MCP has photomultiplier like ultraviolet (UV) to near-IR response. Such a MCP reduces the image distortion; it has response ranging from UV to near-IR. The local photon flux in an image can, therefore, be rapidly converted into a substantial electron flux and back again.

The operational MCPs, known as "second generation" (Gen II) image intensifiers (see Figure 5.13a) were ready to be mounted at the telescopes in 1969 (Manley et al., 1969). The photocathodes are, in general, made of photoalkali halides. The wafer tube with over ten million holes (see Figure 5.13b) is a geometrical distortion free

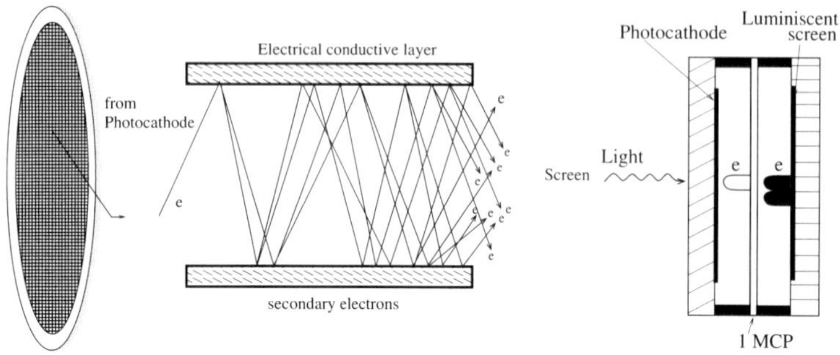

(a) (b)

Figure 5.13 Schematic of a Gen II intensifier: (a) path of a single electron multiplying through one of the millions of channels in such a tube; with each bounce, the electron multiplies and accelerates and (b) cross-sectional view of the Gen II wafer tube.

device. It operates by use of an S 25 (extended red) photocathode, MCP plate for internal multiplication, and a fiber-optic twist for image erection. The photoelectrons are accelerated into a channel of the MCP releasing secondaries and producing an output charge cloud of about 10^3 to 10^4 electrons with 5–10 kV potential. With further applied potential, these electrons are accelerated to impact a phosphor, thus producing an output pulse of $\sim 10^5$ photons.

The MCPs devices are small and have high resolution and produce no image distortion. Also, they offer a larger gain compared to Gen I image intensifiers (see Figure 5.12), but a smaller quantum efficiency, due to the fact that some electrons, ejected from the photocathode by a photon, do not enter any micro-channel. The channels of these MCPs had a 40 micron diameter. Several electronic read-out techniques have been developed to detect the charge cloud from a high-gain MCP. However, the shortcomings of the MCPs are notably due to its local dead-time, which essentially restricts the conditions for use of these detectors for high spatial resolution applications. These constraints are also related with the luminous intensity and the pixel size.

Third generation (Gen III) image intensifiers are similar in design to Gen II intensifiers, with a gallium arsenide (GaAs) Photocathode, which offers a larger QE ($\sim$50%) than multi-alkali photocathodes (Rouaux et al., 1985); the Gen II image intensifier photocathodes mostly lie below 20%. They offer a larger QE (~0.3) than multi-alkali photocathodes. Such tubes employ proximity focus and have a luminous sensitivity of approximately 1200 μA.lm^{-1}. The main advantage is their high sensitivity in the red and near-IR; the GaAs is extremely photosensitive in the near-IR region above 800 nm, but is relatively poor in the blue/green region of the visible band. However, the high infrared sensitivity makes these tubes more susceptible to high thermal noise (see Section 4.5.2.1).

Ion-poisoning effect The photocathode in Gen III intensifier was improved by coating oxidized cesium (Cs:O) on an $Al_{1-x}Ga_xAs/GaAs$ hetero-structure (see Section 5.3.3) substrate to form a negative electron affinity (NEA; see Section 5.2.1.2) thin layer. If these ions and natural gas molecules are allowed to escape from the channels of MCP and impact onto the photocathode, they can cause the NEA layer suffering a physical and chemical damage and a rapid decay of cathode sensitivity. This phenomenon is referred to as ion-feedback or ion-poisoning (Estrera et al., 2000; Kulov et al., 2005; Nevshupa et al., 2008). The amount of poison species in the image intensifier can be identified by the following equation developed by Sandel et al. (1977):

$$\frac{G}{G_0} = \frac{1}{1 - \beta\left[1 - e^{Q/Q_H}\right]},$$

with $\beta = 1/(1 - n_1/n_2)$, n_1 as the source species population responsible for the photoemission, n_2 the poisoning species population, G the gain, G_0 the initial gain, Q the total charge per unit of MCP area extracted from the MCP, and Q_H the half-life charge density characterizing the decay.

In order to minimize ion-poisoning effect, the Gen III intensifiers (see Figure 5.14) utilize high-resolution MCP plates (6 μm diameter channels) and dielectric films called ion-barrier films (IBF), which is typically either Al_2O_3 or SiO_2 with thickness ranging between 25 Å and 100 Å. The IBF is coated on the input surface of MCP, which suppresses the ions and molecules from exiting the MCP and damaging the photocathode. The application of such an film enhances the mean time to failure of Gen III intensifiers. However, the IBF has the drawback of scattering the photoelectron impinging on the MCP, thus preventing some low energy photoelectrons

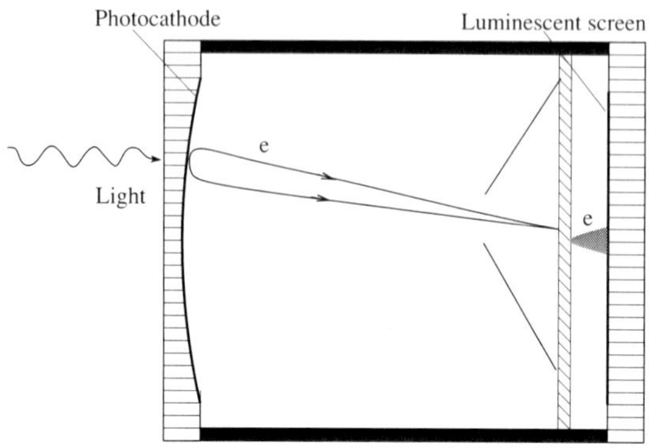

Figure 5.14 Schematic of a inverter MCP intensifier.

reaching the channel causing a significant degradation in signal-to-noise (S/N) ratio and impairment in resolution of these intensifiers. Extensive efforts have been made to reduce the thickness of the film or remove it in order to improve S/N ratio, while maintaining sufficient lifetime (Estrera et al., 2000; Floryan et al., 2003).

Gen III Pinnacle development Efforts have been made to develop filmless Gen III intensifier tube, but unable to achieve the desired reliability standard. However, using thin film (10,000 times thinner than a human hair), the enhanced version of Gen III tube was dubbed appropriately the "Pinnacle" because it represented the apex in technological achievement. Such a tube has exceeded the same reliability standards for gain and S/N ratio, while reducing the halo effect by minimizing the spacing between the photocathode and MCP as compared to the existing intensifier tubes. When operating in dynamic light conditions, images provided through image intensifier tubes show halos around the brightest spots in the scene, which lower the image quality and whiting part or the entire image; the smaller the halo, the better the image quality. In Gen III tubes, the halo diameter is minimized from 1.25 mm to 0.7 mm. The primary cause of a halo is the fraction of photoelectrons generated by the photocathode when they hit the surface between holes of the MCP input and the backscatter. Another cause for the halo is generated by optical reflection at the input gap and backscattered electrons on the anode side. The diameter of the halo is approximately equal to four times the gap between the photocathode and the MCP.

Auto-gated power supply Auto-gated power supply applied between the photocathode and MCP, while extending the dynamic range of image intensifier, also limit the ion-feedback of MCP, prolong the lifetime of image intensifier (Estrera and Saldana, 2003). The auto-gating is an electronic solution that reduces the duty cycle of the cathode voltage by rapidly switching the voltage on and off. It is a proprietary means to preserve the high modulation transfer function (MTF; see Section 1.3.1) and resolution that is obtained at the full operating voltage while reducing the strain of high illumination levels on the image intensifier tube, which can be done by lowering incoming current from the photocathode.

Performance The performance of an image intensifier is determined by the certain factors: (i) signal-to-noise (S/N) ratio (see Section 4.1.1.4), (ii) limiting resolution (see Section 4.4.1.1), which can be derived from the MTF curve, (iii) photocathode sensitivity, ($\mu A.lm^{-1}$; see Section 4.1.1), (iv) Gain ($cd.m^{-2}.lx^{-1}$; see Section 4.3.2.2), and (v) equivalent background illumination (lx; see Section 4.5.1.2). In the low-light level regime, the information density is determined by the light level. However, from a certain light level onwards the image quality is no longer determined by the input light level but the image transfer characteristics of the image intensifier, which is expressed by the MTF parameter that provides the contrast at the output when a 100% modulation is applied. This contrast at the output is a function of the spatial frequency.

Of course, an alternative to the MCP is the microsphere plate (Tremsin et al. 1996) comprising a cluster of glass beads whose diameter is about 50 μm each. These beads have a secondary emission property. Electrons are, therefore, multiplied when they cross a microsphere plate. Compared to MCPs, microsphere plates require a less drastic vacuum, a reduction of ion return and a faster response time (about 100 ps). The drawback is a poor spectral resolution. Hence, they can be used for PMTs, but not for image intensifiers.

5.3 Photovoltaic Devices

Soon after the appearance of the photoemissive cell, the photovoltaic cells came into existence. These are used extensively in wide range of electronic systems, such as fiber-optics (see Section 3.4) communication arrays. The PV devices generate electricity from incident light, commonly by exciting electron–hole pairs (see Section 2.2.4.1) in the material. Also, there are electrochemical cells, which produce electricity by redox reaction of certain electrolytes. Generally, solid-state PV devices are similar to the reverse-biased diodes (see Section 2.2.7), with a wide depletion region generated by the drift of the electron and holes to the respective sides of the P-N junction. The current–voltage profile of the PV cells in dark is, therefore, similar to a diode, with low dark current at reverse biasing and exponential increase in current while forward biased. Light falling in the depletion region excites electron–hole pairs, which are separated by the electric field across the device. The reflective coating on the bottom of such a device provides the double chances, for instance, injection and reflection, of photon absorption. Since the PV sensor works at reverse-biased mode, the current and thus the power dissipation are minimized. Also, these are low-noise detectors since the reverse-bias diode junction is depleted of minority carriers. However, applying such a bias would increase the dark and noise currents. Commonly, the PV devices are, in general, fabricated from silicon (Si), germanium (Ge), gallium arsenide (GaAs), indium antimonide (InSb), and mercury cadmium telluride (HgCdTe). There are two types of junction devices, for example, non-amplifying and amplifying devices. The depletion layer non-amplifying photodiode family include P-N junction, P-I-N diode, Schottky barrier, and hetero-junction, while amplifying photodevices are avalanche photodiode and phototransistor.

5.3.1 *P-N Junction Photodiode*

A P-N junction photodiode utilizes the effect of photons onto the charge carriers in the depletion layer of a P-N junction diode. As stated in Section 2.2.7, application of a reverse bias can greatly improve the speed of response and linearity of the devices, which is due to increase in the depletion region width, and consequently decrease in junction capacitance. The displaced carriers establish an electric field across the junction, setting up an intermediate region depleted of free carriers. The depletion layer is, therefore, both charged and highly resistive. The strength of the electric field and the size of this layer is increased by applying an external electric field (reverse bias). When an incident photon is absorbed by the photosensitive material (quantum energy of light should be larger than the energy gap between the valence band and the conduction band; see equation 2.11), it creates a free electron–hole pair (see Section 2.2.4.1) in the transition region. The electrons flow to the N-region and holes flow to the P-region, so that a full electron charge flows through the external circuit that is connected directly to the input of a current-to-voltage converter. The two photogenerated charges drift in the material and would recombine. If the mobility of the charge carrier (a hole in an N-type material) is high enough, it would reach the depletion region before recombining and be swept to the other side of the junction by the electric field. There, in the P-type region, it would recombine with one of the electrons, causing a reduction of the voltage across the junction. This change of voltage is proportional to the number of photogenerated charges providing a direct measure of the photons captured on each pixel.

The average distance over which the holes are collected by the P-region is equal to the hole diffusion length, $L_{\mathrm{h}}(=\sqrt{D_{\mathrm{h}}\tau_{\mathrm{h}}})$, in the N-region and the average distance over which the electrons are collected by the N-region is equal to electron diffusion length, $L_{\mathrm{e}}(=\sqrt{D_{\mathrm{e}}\tau_{\mathrm{e}}})$, in the P-region. Here D_{h}, D_{e} are the respective diffusion coefficient of the holes and electrons and τ_{h} τ_{e} the lifetime of holes and electrons, respectively. The photogenerated carriers within the volume of the photosensitive region of the P-N photodiode ($L_{\mathrm{h}}+w+L_{\mathrm{e}}$, in which w is the width of the transition region) give rise to a current, which is proportional to the photon flux, Φ (Kasap, 1999).

Short-circuit current, i_{sc} The short-circuit current is the output current, which flows when the load resistance is zero and is nearly proportional to the active area of the device. If the P-N junction is short-circuited by an external conductor, current would flow in the circuit when the junction is illuminated. The short-circuit

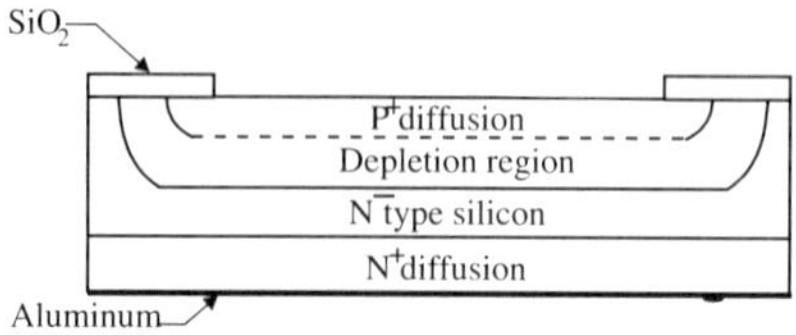

Figure 5.15 Planar diffused photodiode.

photocurrent, i_{sh}, is given by

$$i_{sh} = qN_0(L_h + w + L_e), \tag{5.31}$$

with N_0 as the number of electron–hole pairs generated per second per unit area of the surface.

The carriers absorbed here must diffuse to the intrinsic layer before they recombine. Slow diffusion can lead to slow "tails" in the temporal response. A planar diffused photodiode is fabricated by growing a layer of oxide over a slice of high-resistivity silicon (Si), etching a hole in the oxide and diffusing boron into the silicon through the hole (see Figure 5.15). This structure leads to the devices with high breakdown voltage and low leakage current. The theory of the production of a photovoltage at a P-N junction follows the relationship between the current, i_s, and voltage, V, is

$$i_s = i_0\left[e^{V/(\eta V_T)} - 1\right], \tag{5.32}$$

with $V_T(= k_B T/q)$ as the volt equivalent of temperature, T the absolute temperature in kelvin, $k_B(= 1.38 \times 10^{-23}$ J/K) Boltzmann's constant, i_0 the reverse saturation current, i.e., the current in the junction in the reverse-biased condition, V the applied voltage across the terminal of the diode in volts, and the parameter, η, is unity for germanium and 2 for silicon.

Open-circuit voltage, V_{oc} The current increases exponentially with increasing forward bias. The potential barrier gets reduced under such bias. There is a reduction in the drift current (see Section 2.2.4.2) and increase in the diffusion current (positive net current; see Section 2.2.4.3). If a reverse bias is applied, due to the thermally generated minority carriers, a small reverse saturation current appears, which is related to the dark current (see equation 4.31). As discussed in Section 2.2.7, these minority carriers fall down the potential hill at the junction, whereas such a barrier does not permit majority carriers to cross over the junction. In the case of P-N junction being illuminated, additional electron–hole pairs are generated, which are proportional to the number of incident photons. Under reverse bias, the potential barrier is increased; the drift current dominates (negative net current). Hence, the current under large reverse bias is $i_s = i_{sh} + i_0$, so that the short-circuit current varies linearly with light intensity. The change of voltage, ΔV, in output voltage is due to the incident light turns out to be $\Delta V = i_{sh} R_L$, in which R_L is the load resistance. The volt-ampere characteristics (Millman and Halkias, 1972) is written as

$$i_s = i_{sh} + i_0 \left[1 - e^{V/(\eta V_T)}\right]. \tag{5.33}$$

As the applied reverse bias increases, there is a sharp increase in i_s. The photodiode should be operated below the breakdown voltage (V_{br}; see Section 2.2.7). Such voltage varies from one photodiode to another and is measured generally for small areas at a dark current of 10 μA.

The voltage, at which zero resultant current is obtained, is known as PV voltage (Rappaport, 1959; Loferski and Wysocki, 1961; Loferski, 1963). Also, known as the open-circuit voltage, V_{oc}, that is developed when the load resistance is infinite exhibiting a constant value independent of the active area of the device. Under open-circuit condition, there appears a voltage equal to the amount by which the barrier potential is reduced across the open terminals of the P-N junction. If the current, $i_s = 0$, from equation (5.32) the open-circuit voltage is derived as

$$V_{oc} = \eta V_T \ln \left(1 + \frac{i_{sh}}{i_0}\right). \tag{5.34}$$

If $i_s/i_0 \gg 1$, V_{oc} increases logarithmically with i_s and, hence, with illumination. When the illumination is parallel to the junction, i_{sh} in equation (5.34) is given by equation (5.31), while in the case of transverse illumination, the derivation leads to equation (5.33) where the short-circuit photocurrent, $i_{sh}(= qN_0)$ is under the assumption $L_h \ll L_e$. If N_λ represents the number of photons of wavelength, λ, absorbed per unit area of the sample per unit time, N_0 turns out to be $N_0 = \eta_d N_\lambda$, with η_d is the quantum efficiency (QE; Section 4.1.1.3), i.e., the number of excess carriers

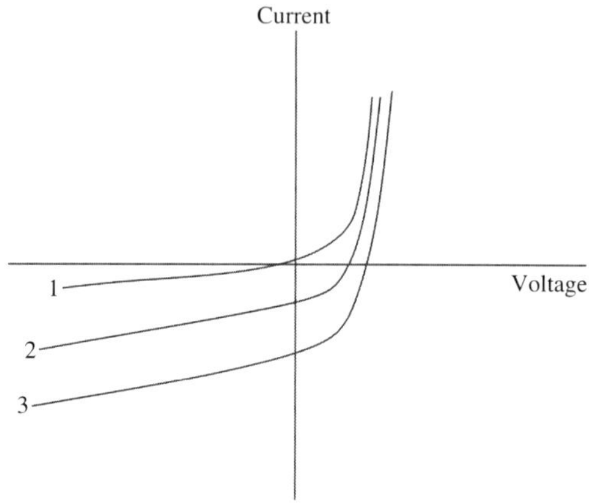

Figure 5.16 Current–voltage characteristic of a photovoltaic sensor: (curve 1) in absence of light, which is similar to a rectifying diode (see Figure 2.5). This curve shifts to a different position, for instance, at low light level, such a curve moves to curve (2) and on increasing further the amount of incident light, it shifts to position curve (3).

produced per absorbed photon. Figure 5.16 depicts the current–voltage characteristics of a junction photodiode at different light levels.

Sensitivity The sensitivity, expressed by the term called photo-sensitivity (or responsivity; Section 4.1.1.2), $\mathcal{R}$, and the quantum efficiency, η_{d}, of a photodiode have the following relationship:

$$\eta_{\mathrm{d}} = \frac{\mathcal{R}_\lambda(\text{observed})}{\mathcal{R}_\lambda(\text{ideal})} = \mathcal{R}_\lambda \frac{hc}{\lambda q} = 1.24\frac{\mathcal{R}_\lambda}{\lambda}, \tag{5.35}$$

with h as Planck's constant, c the speed of light in free space, $\mathcal{R}_\lambda$ the responsivity in A.W^{-1}, and λ the wavelength in nm.

It is reiterated that the responsivity of a photodiode is a measure of the effectiveness of the conversion of the light power into the current, which varies with the applied reverse bias, wavelength, and temperature. The non-uniformity of response is the variations of responsivity observed over the surface of the photodiode active area with a tiny spot of light, which is inversely proportional to the spot-size; the smaller the spot-size, larger the non-uniformity. If the generated photocurrent increases linearly with incident light power,

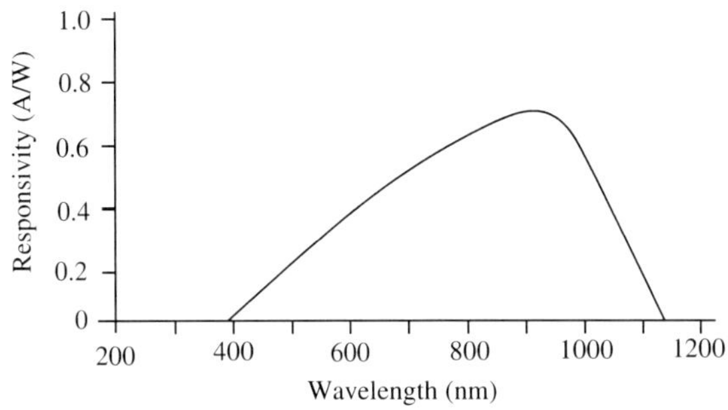

Figure 5.17 Responsivity as a function of wavelength for a typical Si photodiode; a Si photodiode responds over the spectral range of 400–1100 nm.

a photodiode is considered to be a linear. Figure 5.17 depicts the spectral responsivity of a typical silicon photodiode.

Noise In a photodiode, two sources of noise, such as (i) thermal noise, i_{th} (see Section 4.5.2.1) and (ii) shot (quantum) noise, i_Q (see Section 4.5.2.3) can be identified. The total noise current generated in such a device is determined by

$$i_{\text{tot}} = \sqrt{i_{\text{th}}^2 + i_Q^2}. \tag{5.36}$$

A typical value for the noise equivalent power (NEP; see Section 4.1.1.4) of a photodiode is 10^{-14} W.Hz$^{-1/2}$.

The advantages of the photodiodes are (i) excellent linearity of output current as a function of incident light, (ii) large spectral response, (iii) low noise, (iv) compact and lightweight and also low cost, (v) longer lifetime, and (vi) high quantum efficiency, typically 80%. However, the disadvantages are (i) area being small, (ii) having no internal gain (except avalanche photodiodes, typically 102–103), (iii) lower overall sensitivity, and (iv) slow response time. One may use it for the purpose of photon-counting but with cooling system incorporated.

5.3.2 *P-I-N Photodiode*

Another type of semiconductor structure used in photodiodes is the P-I-N photodiode (see Figure 5.18; upper panel). The device

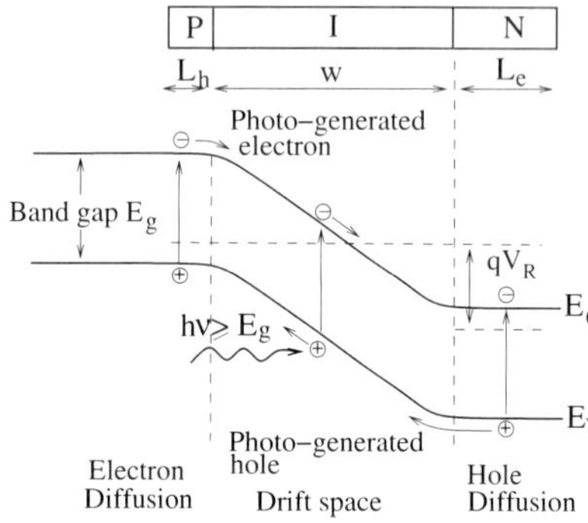

Figure 5.18 Top panel: Cross-sectional view of a P-I-N photodiode and Bottom panel: P-I-N photodiode energy band diagram under reverse bias; V_R is the reverse voltage and q the electron charge.

has a layer of an intrinsic region, I, sandwiched in-between a heavily doped P-type and a lightly doped N-type regions. In this, the depletion layer can be tailored to optimize the frequency response and the sensitivity by confining the generation of electron–hole pairs in the depletion region exclusively. By making the surface region thin compared to the optical absorption depth, the incident radiation would penetrate into the I-region, which produces electron–hole pairs provided that the photon energy is high enough. Upon applying a reverse bias, the intrinsic region gets fully depleted, and the generated carriers are swept across the region with high velocity and are collected in the heavily doped regions due to the strong electric field. The I-region takes control of the depletion region width and makes it possible to control the wavelength sensitivity of the junction of a particular photodiode. Under steady-state condition, the photogenerated carriers move by drift and diffusion, and therefore the total current density, **J**, through the reverse-biased depletion region is expressed as

$$\mathbf{J} = \mathbf{J}_{\mathrm{d}} + \mathbf{J}_{\mathrm{diff}}, \tag{5.37}$$

where $\mathbf{J}_{\mathrm{d}}$ is the drift current (see Section 2.2.4.2) density due to carrier generation in the depletion region and $\mathbf{J}_{\mathrm{diff}}$ the diffusion current (see Section 2.2.4.3) density due to carrier generation outside the depletion region in the bulk semiconductor and diffusing into the reverse-biased region.

The depletion layer in P-N diodes is relatively thin such that the incident light is not completely absorbed; a near complete absorption of light may be achieved by using a thick intrinsic absorption layer. Figure 5.18; bottom panel depicts the energy diagram of P-I-N photodiode. According to Lambert–Beer's law, the electron–hole generation rate, $G(x)$, per unit area decreases exponentially, that is

$$G(x) = \eta_{\mathrm{d}} G_{\lambda} e^{-\alpha_{\mathrm{s}}(x)}, \tag{5.38}$$

where G_{λ} is the photon flux density of wavelength λ, $\alpha(x)$ the absorption coefficient, and η_{d} the quantum efficiency of the detector.

One may derive the expression for the total current density under the assumption that the thermal generation current is neglected and that the surface layer is much thinner than optical absorption depth. If all the carriers are swept out by the electric field in the depletion region, the drift current density, $\mathbf{J}_d$, is

$$\mathbf{J}_{\mathrm{d}} = -q \int_0^w G(x)dx = q\eta_{\mathrm{d}} G_{\lambda} \left(1 - e^{-\alpha_{\mathrm{s}}(w)}\right), \tag{5.39}$$

with w as the width of the depletion layer that is approximately the same as the thickness of the I-region.

For $x > w$, the minority carrier (electron) density in the bulk semiconductor is determined by the 1-D diffusion equation, that is

$$D_{\mathrm{e}} \frac{\partial^2 N_{\mathrm{h}}}{\partial x^2} - \frac{N_{\mathrm{h}} - N_{\mathrm{h}0}}{\tau_{\mathrm{e}}} + G(x) = 0, \tag{5.40}$$

in which D_{e} is the diffusion coefficient for electrons, τ_{e} the lifetime of excess carriers, N_{h} as the concentration of holes, and $N_{\mathrm{h}0}$ the equilibrium electron density.

Under the boundary conditions, $N_{\mathrm{h}} = N_{\mathrm{h}0}$ at $x = \infty$ and $N_{\mathrm{h}} = 0$ at $x = w$, the solution of equation (5.40) is

$$N_{\mathrm{h}} = N_{\mathrm{h}0} - \left(N_{\mathrm{h}0} + Ce^{-\alpha_{\mathrm{s}}(w)}\right) e^{(w-x)/L_{\mathrm{e}}} + Ce^{-\alpha_{\mathrm{s}}(x)}, \tag{5.41}$$

with $L_{\mathrm{e}}(=\sqrt{D_{\mathrm{e}}\tau_{\mathrm{e}}})$ the electron diffusion length in the P-region, and the constant

$$C = \frac{\eta_{\mathrm{d}} G_{\lambda} \alpha_{\mathrm{s}} L_{\mathrm{e}}^2}{D_{\mathrm{e}}(1 - \alpha_{\mathrm{s}}^2 L_{\mathrm{e}}^2)}. \tag{5.42}$$

The diffusion current density (Grundmann, 2010) at $x = w$ is

$$\mathbf{J}_{\text{diff}} = -q D_{\text{e}} \left(\frac{\partial N_{\text{h}}}{\partial x} \right)_{x=w}$$
$$= -q \eta_{\text{d}} G_{\lambda} \frac{\alpha_{\text{s}} L_{\text{e}}}{1 + \alpha_{\text{s}} L_{\text{e}}} e^{-\alpha_{\text{s}}(w)} - q N_{\text{h}0} \frac{D_{\text{e}}}{L_{\text{e}}}. \quad (5.43)$$

Under normal operating conditions, the term involving, $N_{\text{h}0}$ is much smaller, so that the total photocurrent ($\mathbf{J}_{\text{d}} + \mathbf{J}_{\text{diff}}$) is proportional to the photon density flux. In addition, for a large response, it is needed to have $\alpha_{\text{s}}(w) \gg 1$ and $\alpha_{\text{s}} L_{\text{e}} \gg 1$. However, for $w \gg 1/\alpha_{\text{s}}$, there may be considerable transit time-delay. The expression for the total current density is

$$\mathbf{J} = -q \eta_{\text{d}} G_{\lambda} \left(1 - \frac{e^{-\alpha_{\text{s}}(w)}}{1 + \alpha_{\text{s}} L_{\text{e}}} \right) - q N_{\text{h}0} \frac{D_{\text{e}}}{L_{\text{e}}}$$
$$\approx -q \eta_{\text{d}} G_{\lambda} \left(1 - \frac{e^{-\alpha_{\text{s}}(w)}}{1 + \alpha_{\text{s}} L_{\text{e}}} \right). \quad (5.44)$$

The external quantum efficiency, η'_{d}, described in equation (4.51), of the diode is expressed as

$$\eta'_{\text{d}} = \frac{\mathbf{J}/q}{P_0 / h\nu} = (1 - R_f) \left(1 - \frac{e^{-\alpha_{\text{s}}(w)}}{1 + \alpha L_{\text{e}}} \right), \quad (5.45)$$

where P_0 denotes the input optical power and R_f the entrance face reflectivity.

The P-I-N diode can be optimized to have the diffusion component as small as possible for faster response; the speed of response is limited by the parameters given in equations (4.27–4.29). The frequency response (see Section 4.1.1.5) can be very high, of the order of 10^{10} Hz, which is higher than the frequency response of P-N junctions without the intrinsic region. In this, the junction capacitance is kept small by (i) limiting the diode area and (ii) reducing the doping density of the I-region; absorption on the top can be minimized by using higher bandgap material (Si_3N_4) that is transparent to optical radiation. The shot-noise current, i_{Q} (see Section 4.5.2.3), for such a device is given by

$$\langle i_{\text{Q}} \rangle^2 = 2q \Delta f i_{\text{D}}, \quad (5.46)$$

where Δf is the bandwidth and i_D the dark current.

5.3.3 *Hetero-junction Photodiode*

Composite semiconductor structures consisting of two or more layers of different materials, one grown on another, are referred to as hetero-structures, which have been used in semiconductor devices. The structures were mostly grown by the techniques of liquid phase epitaxy or chemical vapor deposition techniques. A depletion layer photodiode can be realized in a hetero-junction, where the junction is, in general, formed between two lattice-matched semiconductors of different bandgaps. It is to be noted that in a homo-junction, the energy bands remain continuous and smooth across the junction because the semiconductors on the two sides of the junction have the same bandgap (see Section 2.2.3.1). In the former case, at the junction where the two semiconductors are joined together, the disparity in their bandgaps results in a discontinuity of ΔE_C in the conduction-band edge and a discontinuity of ΔE_V in the valence-band edge, i.e., $\Delta E_C + \Delta E_V = \Delta E_g$, in which ΔE_g is the difference between the bandgaps of the two dissimilar semiconductors (see Figure 5.19). When junction is formed, the Fermi level (see Section 2.2.5) should coincide and the vacuum level is everywhere parallel to the band edges and is continuous. The discontinuity in the ΔE_C and ΔE_V is invariant with doping when the bandgap energy, E_g and the electron affinity, χ, are not functions of doping (non-degenerate semiconductors).

The main attractive feature of such a junction is that if barrier to holes is larger than for electrons, the current would consist entirely of electrons, which makes the device unipolar. The large bandgap material ($E_g > h\nu$) can make use of its transparency

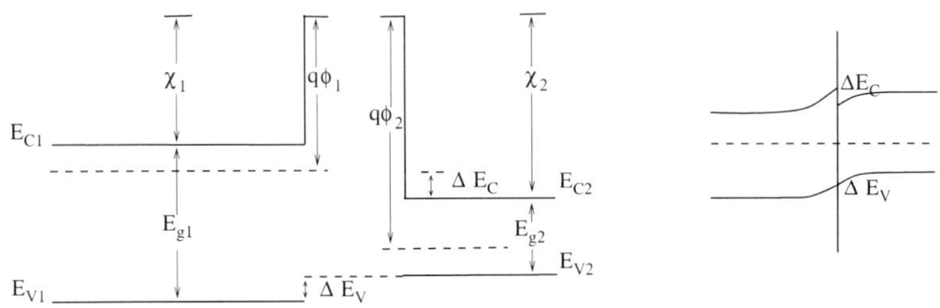

Figure 5.19 Energy relation and formation of a hetero-junction.

to minimize optical absorption outside the depletion region. However, the mismatch in lattice constants (see Section 2.2.1.1) and thermal expansion coefficients severely limit the potential of the device. The problem of preparing hetero-junction with electrical and optical properties conforming to the ideal caused a decrease in development of this area. However, there remains some interest in hetero-junctions prepared from alloy semiconductors (Sze and Ng, 2007), for example, III–V material systems. These III–V semiconductors have a direct bandgap and this property makes them suitable for optoelectronic applications. The wavelength of absorption of the material can be changed by mixing two similar (direct) compound semiconductors with different bandgaps. The group III–V element technology has high production yield, low cost, high speed, radiation hardness, and very long wavelength capability. Most III–V compounds are completely miscible in ternary or even quaternary crystals.

Binary semiconductors A binary compound, such as gallium arsenide (GaAs) has a constant lattice structure and, hence, a constant bandgap, in which Ga is a tri-valent material and As a penta-valent material (see Section 2.2.3.1).

Ternary semiconductors The lattice mismatch can be minimized by the materials, such as the following:

(1) $Al_xGa_{1-x}As$ /GaAs: Aluminum gallium arsenide, $Al_xGa_{1-x}As$, is a compound with properties that interpolate between those of aluminum arsenide (AlAs) and GaAs; the subscript x indicates the fraction of the Ga atoms in GaAs that are replaced by aluminum (Al) atoms; an additional element like Al is added to a portion of the existing element. This modifies the lattice structure and, hence, the bandgap; thus one obtains different wavelength outputs. As the Fermi level on either side and bandgaps are different, for example, if the bandgap of one of these compounds is higher than the other, it provides a trap for the electrons and holes, and thus their recombination is enhanced. In this case, the bandgap energy, E_g, for this material varies between 1.42 eV for GaAs and 2.16 eV for AlAs, as x varies

between 0 and 1. The operating wavelengths are in the visible (mainly red) and near-IR, depending on the molar fraction, x.

(2) $In_xGa_{1-x}As$: Indium gallium arsenide ($In_xGa_{1-x}As$) is a direct bandgap material (see Section 2.2.3.1), which operates in the near-IR region of the spectrum.

(3) $Hg_xCd_{1-x}Te$ /CdTe: Mercury cadmium telluride ($Hg_xCd_{1-x}Te$) is useful in the mid-IR region of the spectrum. The mercury telluride (HgTe) and cadmium telluride (CdTe) have nearly the same lattice parameter, and can therefore be lattice matched at nearly all compositions. This material provides a compositionally tunable bandgap that operates in the wavelength range between 3 and 17 μm.

Quaternary semiconductor The quaternary semiconductors offer more flexibility for fabricating materials with desired properties by virtue of an additional degree of freedom than do ternary compounds. The material, namely, $In_{1-x}Ga_xAs_{1-y}P_y$/InP based systems can be lattice-matched to InP substrates, which offer more flexibility for fabricating materials with desired properties since the fourth element provides an additional degree of freedom that permits lattice matching to be achieved for different compositionally determined values of E_g. The bandgap energy in $In_{1-x}Ga_xAs_{1-y}P_y$ varies between 0.36 eV (indium arsenide; InAs) and 2.26 eV (gallium phosphide; GaP) as the compositional mixing ratios x and y vary between 0 and 1.

5.3.4 *Schottky Barrier Photodiode*

The Schottky barrier photodiode, similar to the Schottky diode (see Section 2.3.1.2), is a metal–semiconductor diode and can be used as a high-efficiency photodetector. In metals, the energy barrier toward vacuum is measured from the Fermi level and its height is called the work function, $q\phi_m$, that is the minimum energy required to release an electron from atomic binding (see Section 2.1). When the metal and the semiconductor are not in contact, the work function becomes $E_0 - E_F = -q\phi_m$, in which q is the electron charge and E_F the Fermi energy (see Section 2.2.5), and the semiconductor is

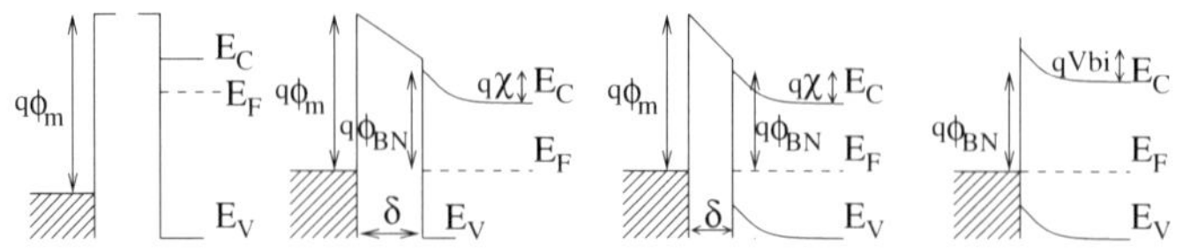

Figure 5.20 Energy band diagrams of metal–semiconductor contacts.

characterized by its electron affinity, χ; $E_0 - E_C = -q\chi$, E_C the near-surface conduction band edge.

For an N-type semiconductor, the energy difference between the Fermi level and the conduction band is denoted as $-qV_N = E_C - E_\mathrm{F}$ ($V_N < 0$ for non-degenerate semiconductors); the subscript N stands for the contact on a N-type semiconductor. Thus, the position of the semiconductor Fermi level is given as $E_\mathrm{F} = E_0 + q(\chi + V_N)$. The potential difference $\phi_\mathrm{m} - (\chi + V_N)$ is known as the contact potential. Let an intimate contact of a metal with an N-type semiconductor be considered. Figure 5.20a depicts the energy diagram of metal and semiconductor before the contact. If the metal and semiconductor are connected to each other through a wire, the Fermi levels would equilibrate (see Figure 5.20b; Grundmann, 2010); the electrons would flow from the semiconductor to the metal. The negative surface charge of the metal is compensated by a positive charge in the semiconductor in the vicinity of the surface. As the gap, δ, decreases (see Figure 5.20c), an increasing negative charge is built up at the surface of the metal; an equal and opposite charge (positive) must exist at the semiconductor surface. Owing to the low carrier concentration of the semiconductor, this positive charge is distributed over a barrier region near the semiconductor surface. When δ becomes comparable to inter-atomic distances, the gap becomes transparent to electrons, and thus one obtains the limiting case as depicted in Figure 5.20d. The limiting value of the barrier height, $q\phi_\mathrm{BN}$ (neglecting the Schottky lowering), is given as

$$q\phi_\mathrm{BN} = q(\phi_\mathrm{m} - \chi_\mathrm{m}). \tag{5.47}$$

A typical Schottky barrier structure along with the equilibrium energy band diagram is shown in Figure 5.21. The presence of Schottky barrier, which forms at the interface, makes this

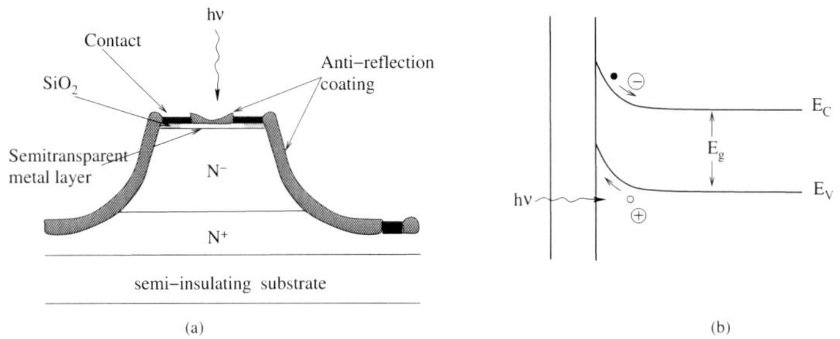

Figure 5.21 (a) Schematic metal–semiconductor photodiode and (b) band-to-band excitation of electron–hole pairs; the symbol • and ○ represent electron and hole, respectively.

metal–semiconductor junction rectifying like a P-N junction. These detectors are responsive to photon energies greater than the Schottky barrier height. The Schottky photodiodes can be fabricated from the materials, namely, gold (Au) on N-type Si, which operates in the visible, and platinum silicide (PtSi) on P-type Si, which operates over a range of wavelengths stretching from the near-UV to the IR. In order to avoid large reflection absorption losses when the diode is illuminated through the metal contact, a thin semi-transparent metallic film (~100 Å), made of a metal–semiconductor alloy, together with an anti-reflection coating is used.

A thin gold coating is sputtered onto the N-doped semiconductor material layer to form such a diode. The junction is fabricated by oxidation of the Si-surface and subsequently etching of a hole in the oxide, followed by the evaporation of a thin transparent and conducting gold layer. For $h\nu > E_g$, the radiation generates electron–hole pairs in the semiconductor. When the reverse bias across the diode is small compared to the breakdown voltage, the electrons and holes are swept out of the depletion region and the general characteristics of these diodes are similar to those of P-I-N photodiodes.

However, the breakdown effects around the edge of the metalized area in the presence of high electric fields that are present around the edge of the plate, as well as the leakage effects may be noticed. In order to overcome such problems, a guard ring of P^+-type semiconductor fabricated using a diffusion process is used along with an oxide layer around the edge. The guard ring operates by driving this region into avalanche breakdown before the Schottky junction is damaged by large flow of reverse current during transient events. If the N^+ is omitted and the Al is deposited directly upon the N-type Si, an equivalent P-N structure is formed, which yields in a metal–semiconductor diode (Millman and Halkias, 1972). The metal side acts as the anode and N-type semiconductor acts as the cathode of the diode.

5.3.5 *Avalanche Photodiode*

The avalanche photodiodes (APD; Ingerson et al., 1983) are the solid-state photodiodes with internal current gain, which are made using more complex structures (see Figure 5.22a). They are operated at high reverse-bias voltage close to the breakdown voltage, V_{br}, where the avalanche multiplication (see Section 2.2.7) takes place. However, the avalanche multiplication should be spatially uniform across the entire light sensitive area of the diode. The gain (see Section 4.3.2.2) is a function of the APDs reverse voltage, V_r, which varies with applied bias. By applying a high reverse-bias voltage (typically 100–200 V in silicon), an APD show an internal current gain effect (around 100), since it multiplies internally the primary photocurrent before entering to the following circuitry. They are based on the ionization of a high-voltage $P\,N^+$ junction, triggered by a single photoelectron (see Figure 5.22b). The current gain-bandwidth product of the multiplication can be as high as

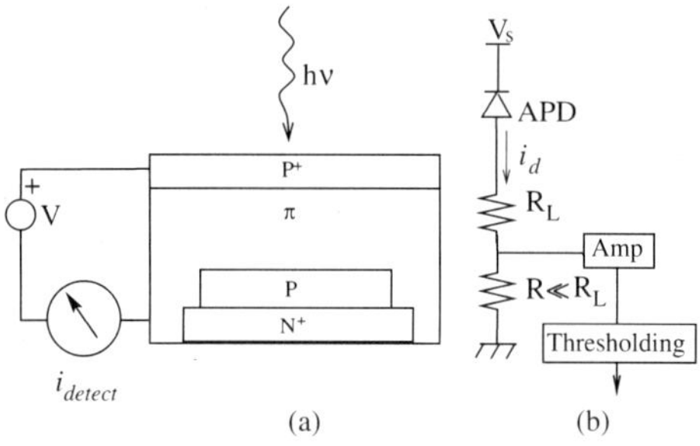

Figure 5.22 (a) Schematic of an APD device and (b) Passive quenching of an APD.

100 GHz, so that substantial gain may be achieved at microwave frequencies.

On applying a reverse voltage to the P-N junction, there is a reverse current across the depletion layer. The electrons drift toward the N^+ side due to the electric field developed across this junction; the drift speed depends on the electric field strength. As the field strength is increased to a certain level, the carriers are undergoing collision with the crystal; therefore, the drift speed turns out to be saturated at an average speed. The photogenerated primary carrier pairs within high electric fields impart sufficient energy (greater than the bandgap, E_g) to ionize bound electrons in applied voltage, V. Upon colliding with the lattice atoms, they provide energy to create an electron–hole pair. This process of pair generation is known as impact ionization. With each collision produces additional electron–hole pairs in a process like a chain reaction. The newly created carriers in the presence of high electric field result in more ionization called avalanche effect. Such an effect in the high-field region of an APD multiplies the number of photoexcited carriers by the avalanche gain. This raises the signal level, which may be useful for raising low signal levels above the amplifier noise (see Section 4.5.3).

This internal amplification process enhances the responsivity of the detector. Let $\mathbf{J}_e$ and $\mathbf{J}_h$ be the electron and hole current densities, respectively. The incremental change of the hole current density due to electron–hole pairs generated along a line element, dx, is given by (Saleh and Teich, 1991; Grundmann, 2010)

$$d\mathbf{J}_h = (\mathbf{J}_h\alpha_h + \mathbf{J}_e\alpha_e)\,dx, \tag{5.48}$$

where α_e and α_h denote the respective electron- and hole-ionization coefficients.

Assuming that at $x = 0$, hole current density $\mathbf{J}_{h0}$ enters the depletion layer. In the high field region, this current increases and reaches the value at the end of the depletion layer ($x = w$), $\mathbf{J}_h(w) = M\mathbf{J}_h(0)$. Similarly, an electron current density is increased to $\mathbf{J} - \mathbf{J}_h$ on its way from w to $x = 0$, in which w is the multiplication layer width. The total current in the depletion layer is $\mathbf{J} = \mathbf{J}_h + \mathbf{J}_e$, which is constant under steady-state condition; therefore,

$$\frac{d\mathbf{J}_h}{dx} - (\alpha_h - \alpha_e)\,\mathbf{J}_h = \alpha_e\mathbf{J}. \tag{5.49}$$

The solution of the first order differential equation is

$$\mathbf{J}_{\mathrm{h}}(x) = \mathbf{J}\frac{1/M + \int_0^x \alpha_{\mathrm{e}} e^{-\int_0^x (\alpha_{\mathrm{h}}-\alpha_{\mathrm{e}})dx'} dx}{e^{-\int_0^x (\alpha_{\mathrm{h}}-\alpha_{\mathrm{e}})dx'}},$$

in which the amplification factor for all carriers generated in the photodiode is

$$M = \frac{(1-k)e^{\alpha_{\mathrm{e}} w(1-k)}}{1-ke^{\alpha_{\mathrm{e}} w(1-k)}}, \tag{5.50}$$

$k(= \alpha_{\mathrm{h}}/\alpha_{\mathrm{e}})$ the ionization ratio, a parameter to indicate device noise.

For APD, the photocurrent, i_{ph}, at low voltage is $i_{\mathrm{ph0}} + i_{\mathrm{D0}}$; therefore, the amplification factor, M, is

$$M = \frac{i_{\mathrm{ph}} + i_D}{i_{\mathrm{ph0}} + i_{\mathrm{D0}}} = \frac{1}{1-(V - iR/V_{br})^{\gamma}}, \tag{5.51}$$

where i_{ph} is the component due to photocurrent, i_D the dark current component, V the applied voltage, γ the exponent that varies with the semiconductor material, doping profile, temperature, and wavelength, V_{br} the breakdown voltage, i the average value of the total output current, $V - iR$ the applied voltage across the device, and R the sum of the series resistance and the space-charge resistance.

Under the conditions that $i_{\mathrm{ph0}} \gg i_{\mathrm{D0}}$ and $iR \ll V_{\mathrm{br}}$, the maximum value of the photomultiplication is given by

$$(M_{\mathrm{ph}})_{\mathrm{opt}} = \sqrt{\frac{V_{\mathrm{br}}}{\gamma i_{\mathrm{ph}} R}}. \tag{5.52}$$

Equation (5.52) indicates that the maximum photomultiplication depends inversely on the square-root of i_{ph}. In order to achieve high carrier multiplication, as well as sensitivity, the primary dark current should be kept as low as possible. For $x = w$, one gets

$$1 - \frac{1}{M} = \int_0^w \alpha_{\mathrm{e}} e^{-\int_0^x (\alpha_{\mathrm{h}}-\alpha_{\mathrm{e}})dx'} dx. \tag{5.53}$$

Avalanche breakdown is reached for $M \to \infty$, when $\int_0^w \alpha_{\mathrm{e}} e^{-\int_0^x (\alpha_{\mathrm{h}}-\alpha_{\mathrm{e}})dx'} dx = 1$. If $\alpha_{\mathrm{h}} = \alpha_{\mathrm{e}} = \alpha$, equation (5.53) simplifies to

$$\int_0^w \alpha dx = 1. \tag{5.54}$$

An APD has a diffused P-N junction, with surface contouring to allow high reverse-bias voltage without surface breakdown. A large internal electric field leads to multiplication of the number of charge carriers through ionizing collisions. As stated in Section 2.2.7, if the electrons are swept across the P-N junction to ionize bonds, the sufficient energy to ionize additional bonds is created causing a regenerative breakdown and current flow. If electrons are injected into the field region at $x = 0$, the multiplication factor for electrons is

$$M_e = e^{\alpha_e w}, \tag{5.55}$$

for $\alpha_h = 0$. If the electron and hole impact ionization coefficients are the same, the multiplication factor, M, for electrons and holes is given by

$$M = \frac{1}{1 - \alpha w}. \tag{5.56}$$

The APD differs from the P-I-N photodiode (see Section 5.3.2) by proving internal photoelectronic signal gain through avalanche multiplication. The avalanche effect in the high-field region of an APD multiplies the number of photoexcited carriers by the avalanche gain. By applying a high reverse-bias voltage (typically 100–200 V in silicon), APDs show an internal current gain effect (around 100). The signal is increased, so is detectivity, provided that the limiting noise is not from background radiation, which may be useful for raising low signal levels above the amplifier noise. The RMS value of the current noise is

$$\langle i_s \rangle^2 = 2q(i_{ph} + i_B + i_D)\Delta f \langle M^2 \rangle F(M), \tag{5.57}$$

where i_{ph} is the photocurrent, i_B the bias current, i_D the dark current, Δf the bandwidth, $\langle M^2 \rangle$ the mean-square value of the internal gain, $F(M)$ the noise figure (see Section 4.5.5) that is a measure of the increase in shot-noise (see Section 4.5.2.3) compared to an ideal noise-less multiplier, and q the electron charge.

To note, the charge generated in APD during the photoelectric conversion process needs to be converted to a voltage pulse in the pre-amplifier stage. However, the design of such a pre-amplifier is the most critical since it determines electronic noise, and thus energy resolution. There are a few commercially available monolithic amplifier integrated circuits (IC; see Section 2.3) that are suitable for scintillation pulse amplifier.

Another consideration is the amount by which the noise is increased through the avalanche process. APDs must have low dark currents, since the latter might be multiplied in the device along with any photocurrent. Since an APD works in reverse mode, sensitivity at low light level is limited by the shot-noise and the APD leakage current. Since the ionization carriers is not uniform, the excess noise term $2qi_{\mathrm{D}}\Delta f$ is added during the multip1ication process. Since the maximum photomultiplication depends inversely upon the square root of i_{s}, it is essential to keep the dark current as low as possible in order to achieve high carrier multiplication and sensitivity. The noise current generated by the multiplication process in the white noise part of the spectrum is expressed as

$$\overline{i_{\mathrm{n}}} = \sqrt{2qi_{\mathrm{s}}\Delta f M^{\chi}} \qquad 2 \leq \chi \leq 4, \tag{5.58}$$

with χ as excess noise index.

The responsivity of APD can be calculated by considering the current gain

$$\mathcal{R}_{e(APD)} = \frac{\eta_{\mathrm{d}} q}{h\nu} M = \mathcal{R}_{\mathrm{e}0} M, \tag{5.59}$$

where η_{d} the quantum efficiency.

Although the P-I-N system gives higher bandwidth and bit rate, the APD system provides higher sensitivity. Table 5.2 provides a comparative study between these systems.

Three types of APDs are available commercially (Laforce, 2009), for example, (i) silicon (Si) APDs that operate between 400 to 1100 nm, (ii) germanium (Ge) APDs that detect infrared up to a wavelength of 1800 nm, but has high multiplication noise, and (iii) indium gallium arsenide (InGaAs) APDs, which detect up to a wavelength of 1700 nm and less multiplication noise than Ge. Although InGaAs has higher quantum efficiency for P-I-N sensor, Ge has higher gain (M) for APD system.

The InGaAs is normally used as the absorption region of a heterostructure diode, most typically involving indium phosphide (InP) as a substrate and as a multiplication layer. Although the photomultipliers (see Section 5.2.2) offer a high electronic bandwidth of over 1 GHz and a high gain, it suffers from a relatively low quantum

Table 5.2 Typical characteristics of P-I-N and avalanche photodiodes

Parameter	Symbol	Unit	Type	Si	Ge	InGaAs
Wavelength	λ	nm		400–1100	800–1650	1100–1700
Responsivity	$\mathcal{R}$	A/W	P-I-N	0.4–0.46	0.8–0.87	0.5–0.95
QE	η_d	%	P-I-N	75–90	50–55	60–70
APD Gain	M		APD	–	50–200	10–40
Dark current	i_D	nA	P-I-N	1–10	50–500	0.5–20
			APD	0.1–1	50–500	1–5
Rise time	τ_r	ns		0.5–1	0.1–0.5	0.05–0.5
			APD	0.1–2	0.5–0.8	0.1–0.5
Bandwidth	Δf	GHz	P-I-N	0.125–1.4	0–0.0015	0.0025–40
			APD		2–10	1.5–3.5
Bit-rate	B_R	Gbit/s	P-I-N	0.01	–	0.1555–53
			APD	–	–	2.5–4
Reverse voltage	V_R	V	P-I-N	50–100	6–10	5–6
			APD	200–250	20–40	20–30
K-factor	K_A		APD	0.02–0.05	0.7–1.0	0.5–0.7

efficiency (QE); a photomultiplier with gallium arsenide phosphide (GaAsP) photodiode has QE-curve peaking close to 50%. The solid-state photodetectors with quantum efficiencies reaching over 70% seem very attractive. The Geiger mode of operation provides an excess noise much smaller than in standard photomultipliers. However, they have the inconvenience of a temperature sensitive gain. They are also non-linear under high illuminations as the detector element dead time is responsible for a decrease in the number of active elements when the photon rate increases. Short-wavelength mercury cadmium telluride (HgCdTe) APDs are able to exploit a favorable property in the band structure of the alloy when the bandgap is about 0.90 eV (Norton, 2002). These photodiodes are attractive for low flux and high-speed applications, such as active and hyperspectral imaging (Rothman et al., 2008), since its material properties favor uniform avalanche gain across an array.

5.3.6 *Phototransistor*

The bipolar transistors (see Section 2.2.8) can be made to conduct by exposure to light, since absorption of photons in the base region generates a photocurrent that acts as a base current. Such devices have a transparent window in the package and are referred to as the phototransistors. A phototransistor is more sensitive than a photodiode. It is connected in a common-emitter configuration with the base open, and radiation is concentrated on the region near the collector junction. If the emitter junction is slightly forward biased and the collector junction is reverse biased, i.e., the transistor is biased in the active region, and assume that there is no radiant excitation, the minority carriers are generated thermally, and the electrons crossing from the base to the collector, as well as the holes crossing from the collector to the base, constitute the reverse saturation collector current, i_{C0}. The collector current is given by equation (2.40), with the base current, $i_{\mathrm{B}} = 0$, that is

$$i_C = (\beta + 1)i_{C0}. \tag{5.60}$$

In the case of light being turned on, additional minority carriers are photogenerated, which contribute to the reverse saturation current in a the same manner as they do the thermally generated minority charges. The total collector current is given by

$$i_C = (\beta + 1)(i_{C0} + i_{\mathcal{L}}), \tag{5.61}$$

$i_{\mathcal{L}}$ is the component of the reverse saturation current due to the light.

Due to the transistor, the current caused by the radiation is multiplied by the factor $\beta + 1$. The semiconductors that have been employed in phototransistors are germanium (Ge) and silicon (Si). The former material has a limited IR response, while the latter responds in the visible and near-IR. However, neither has found application when high performance is required.

5.3.7 *Photovoltaic Devices for High Energies*

Photovoltaic devices are commonly used at high energies (X-ray and gamma-ray region). The interaction mechanism produces a bunch of electron–ion or electron–hole pairs and since the requirement is to use them as individual photon counting devices, either some internal charge amplification mechanisms (like PMTs or proportional counters) or low noise amplifications are the essential prerequisites of using photovoltaic devices at high energies. In the following, some representative examples are discussed.

5.3.7.1 Modern gamma-ray observatories

During the past decades, a renewed enthusiasm for the revival of intensity interferometry by which to achieve diffraction-limited imaging over a square-kilometer synthetic aperture is envisaged (LeBohec and Holder, 2006; LeBohec et al., 2008). Using a precise electronic timing of arriving photons within intense light fluxes, combined with digital signal storage, may enable digital intensity interferometry with high degrees of freedom. These telescopes can be kept fixed and the electronic delays can enable the tracking of stars across the sky. the phase behavior may be recovered with multi-element intensity interferometry (Saha, 2010, and references therein).

As stated in Section 5.2.4.2, Cerenkov radiation can be detected with large telescopes equipped with fast photodetectors and electronics. Today, several gamma ray detection observatories, such as (i) collaboration of Australia and Nippon for a gamma ray Observatory in the Outback (CANGAROO), (ii) high energy stereoscopic system (HESS), (iii) major atmospheric gamma imaging Cerenkov (MAGIC), and (iv) very energetic radiation imaging telescope array system (VERITAS), are potentially making available large collecting area mirrors having diameters ranging from 10 to 17 m.

These instruments may contribute more to the measurements of the stellar diameter wavelength dependence. Predictions from three-dimensional dwarf star atmospheric models can also be checked from limb-darkening curves observed at different wavelengths. The other possible stellar programs that can be made are (LeBohec et al., 2008): (i) angular diameters and limb-darkening of single stars, (ii) resolving pre-main-sequence (PMS) stars, (ii) determining atmospheric structures of super-giant stars, including large-scale stellar convection, (iii) calibration of Cepheid period-luminosity relation, (iv) oblateness of fast-rotating stars, (v) binary stars, double-lined spectroscopic binaries in particular, and (vi) emission-line observations.

An important aspect is to be noted that since the time of the Narrabri interferometer (see Section 5.2.4.2), many advances in photodetection (Morel and Saha, 2005; Saha, 2007) have been made, which could benefit the implementation of a modern intensity interferometer. Until recently, large-area photodiodes, required for practical use in intensity interferometry did not offer competitive electronic noise and bandwidth when compared to photomultiplier tubes, which provide an electronic bandwidth of over 1 GHz. As stated in Section 5.2.2, the photomultipliers offer a high bandwidth and a high gain, but suffer from a relatively low quantum efficiency (typically less than 30%). Since the sensitivity of an intensity interferometer is proportional to the quantum efficiency, arrays of Geiger-APDs on a single substrate can be used as detector, which offer detection efficiencies close to 60%. These devices are now becoming available with high signal bandwidth and gains from 10^5 to 10^6, which are being considered for ACT (Atmospheric Cerenkov Telescope) high energy imaging cameras.

5.3.7.2 Proportional counters and ASTROSAT LAXPC/ SSM

The ionization of gases is the basis of proportional counters. These counters are mainly used to detect low energy X-ray radiation. A proportional counter is basically a Gas-filled chamber, wherein the charge produced by the ion-electron pair created due to the incident photon, gets amplified. This amplified signal is proportional to the energy of the original photon. The gas that is filled inside the chamber is an inert gas, which gets ionized by incident radiation. The electric field inside the chamber is kept low in order to maintain it as an ion-chamber. But it should be sufficient for the ionization process to take place. For this the chamber geometry and the applied voltage should be optimized. This field drifts the positive ions toward the cathode (outward toward the wall of the chamber) and the electrons toward the anode (which forms the axis of the tube, inside).

Because of the increasing field, at the vicinity of the anode, Townsend avalanches are produced. Thus each electron produces a multiplication effect in this avalanche region. In the outer volume of the chamber, near the cathode, ions gets drifted and it interacts with the inert gas molecules, producing multiple ion pairs proportional to the incident photon energy. This region is called ion-drift region. For a chamber with a cylindrical geometry, the electric field at a radius,

r from the central axis, is given by

$$E_{cg} = \frac{V}{r \ln(r_c/r_a)}, \tag{5.62}$$

where, V is the applied voltage between the cathode and the anode and r_a and r_c the anode wire radius and the cathode inner radius, respectively.

From equation (5.62), it is clear that field would be higher at low values of r, which would be at the vicinity of the anode wire, where the electrons get drifted. The field gets reduced rapidly toward the cathode. This limits the multiplication region to a very small volume. This creates a uniform multiplication for all ion pairs formed from the initial radiation interaction. The parameter that characterizes the proportional counter is the gas multiplication factor, M, which is expressed as

$$\ln M = \frac{V}{\ln(r_c/r_a)} \frac{\ln 2}{\Delta V} \left(\ln \frac{V}{pr_a \ln(r_c/r_a)} - \ln K \right), \tag{5.63}$$

where, p is the gas pressure and ΔV the potential difference between successive ionizing events, through which an electron moves. K is equal to the minimum value of E_{cg}/p, below which multiplication cannot occur.

The statistical limit of the energy resolution of a proportional counter, assuming that non-linear effects are absent, is given by

$$\frac{\sigma_Q}{Q} = \sqrt{\frac{W(F+b)}{E}}. \tag{5.64}$$

Here, Q is the charge developed in a pulse, and σ_Q its standard deviation, W the required energy to form one ion pair, F the Fano-factor, b, the parameter characterizing the avalanche statistics, whose value is between 0.4–0.7, and E the energy deposited by the incident radiation; for a given gas, that is filled, the factor, $W(F+b)$ is constant.

ASTROSAT A multi-wavelength space mission, called ASTROSAT (ASTROnomySATellite), is planned to be launched soon (ASTROSAT Handbook, Agrawal, 2006). This mission contains three pointed X-ray instruments and a ultraviolet telescope, namely

- large-area xenon proportional counters, covering the medium energy X-rays from 3 to 80 keV,
- soft X-ray telescope with conical foil mirrors, and X-ray CCD detector,
- Cadmium Zinc Telluride Imager (CZTI) for hard X-rays (10–100 keV), and
- twin UltraViolet Imaging Telescopes (UVIT) based on Ritchey–Chrétien design with an aperture of $\sim$375 mm.

There is also an X-ray sky scanning monitor consisting of three 1-D position-sensitive proportional counters with coded masks (Seetha et al., 2006). The UVIT, which has a field-of-view (FOV; see Section 4.4.1.4) of about 0.5° with roughly $1''$ spatial resolution, permits simultaneous monitoring of targets with precise timing accuracy in three wavelength channels. One of the twin telescopes has far-UV (130–180 nm) capability, while the other would be used for near-UV (180–300 nm) and blue optical range (320–530 nm) by means of a dichroic beam splitter. Each channel has a filter wheel containing interference or long-pass filters. The UV channels have gratings with a resolution in the range 100–150 and the optical channel has a neutral density filter.

5.3.7.3 X-ray CCDs and ASTROSAT SXT

Photovoltaic devices with internal amplification like proportional counters or scintillators with PMTs have inherently poor energy resolution due to the large amount of energy needed to generate one electron–ion pair and the resultant low number of charge carriers. Semi-conductor devices like Si and Ge have low bandgap and they are generally used in the X-ray and gamma-ray regions due to their excellent energy resolution.

At the focal plane of imaging X-ray telescopes, X-ray CCDs are one of the commonly used devices. Unlike the CCDs used in the optical region, X-ray CCDs need to be configured to have a large depletion region (to have sufficient cross section to absorb the X-rays) and to use as individual photon counting devices, they need to be cooled to low temperatures (-80°C) and coupled to low-noise amplifiers.

The Soft X-ray Telescope (SXT) of ASTROSAT uses a nested grazing incidence telescope and at the focal plane an X-ray CCD (Kothare et al., 2009), sensitive in the energy range of 0.3–8.0 keV, is used. It is a three-phase frame transfer device with an open electrode structure. With a fine pixel size of 40 microns, in conjunction with the somewhat poor spatial resolution (PSF of 2 arc minutes), the SXT will provide very good quality X-ray spectra for bright X-ray sources.

5.3.7.4 New-generation semiconductors and ASTROSAT CZTI

The photoelectric cross section of Si is quite low at energies higher than about 10 keV and Ge detectors require more complex cryogenics for operation. In the past couple of decades, there have been a lot of efforts to develop semiconductors with a slightly higher bandgap than Si and Ge so that they can be used at near room temperatures. Cadmium zinc telluride (CZT) is a promising candidate with a bandgap of 1.4 eV. It can be configured into a large-area mosaic and operated at room temperatures, and hence, they are ideally suited for hard X-ray (above 10 keV) operations. Figure 5.23a depicts a significant Compton scattering probability above 150 keV and Figure 5.23b demonstrates the ability to measure Compton scattered events and hence the polarization of incident X-rays.

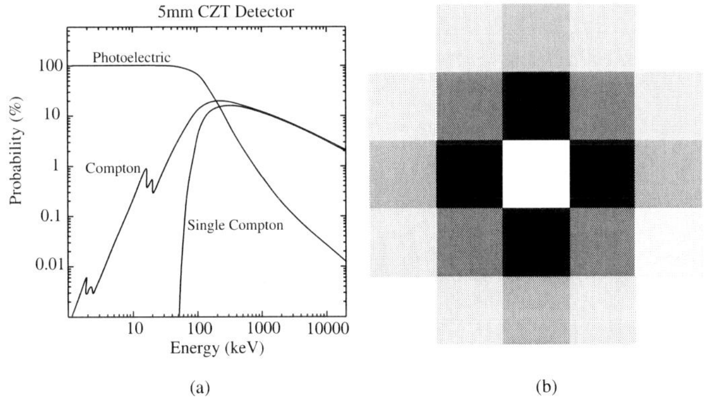

Figure 5.23 (a) Absorption properties of CZT detectors used in the ASTROSAT CZT-Imager and (b) observed distribution of double events in the neighboring pixels for 122 keV X-rays (Courtesy: A. R. Rao).

The ASTROSAT CZT-Imager (CZTI) covers a broad energy range of 10–100 keV and it has sufficient all sky monitoring capability above 100 keV (Rao et al., 2010). A large detection area (1024 cm^2) ensures good sensitivity and it uses an imaging process known as the coded aperture mask. In order to achieve low background a passive shield, collimator, and an active shield in the form of a scintillator detector are used. The coded aperture mask provides moderate imaging so that multiple X-ray objects in the field-of-view (FOV) could be distinguished.

One of the exciting new features of the CZT-Imager in the ASTROSAT satellite is the possibility of using as a X-ray polarimeter above 150 keV. At these energies, Compton scattering becomes the significant mode of interaction and the scattered photons could be captured by the neighboring pixels in an CZT detector. By studying the azimuthal distribution of these scattered photons, polarization properties of the incident radiation can be measured. From laboratory tests, it is estimated that polarization in bright X-ray and gamma-ray sources could be studied. Figure 5.24 displays

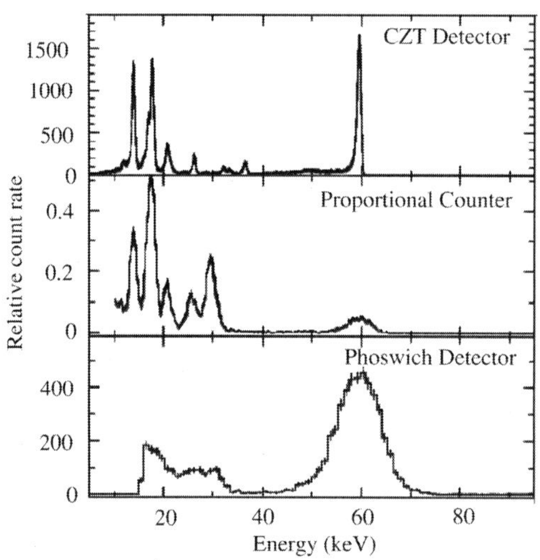

Figure 5.24 Relative pulse height distribution for X-rays from the radioactive source 241-Am (which has a predominant emission at 59.54 keV) for different X-ray detectors demonstrating the superior energy resolution of CZT detectors (Courtesy: A. R. Rao).

the relative pulse height distribution (PHD; see Section 7.2.1) for X-rays from a radioactive source.

5.4 Photoconductive Detectors

In a photoconductive detector, the incoming light produces free electrons carrying electrical current so that the electrical conductivity of the detector material, semiconductor in general, changes as a function of the intensity of the incident light. As stated in Section 4.3.1.1, the PC devices operate on the basis of internal photoeffects in which the photoexcited carriers remain within the semiconductor. They operate by registering either the photocurrent, i_{ph} or the voltage drop across a load resistor, R_L placed in series with the circuit. The radiant energy ionizes covalent bonds and the electron–hole pairs, in excess of those generated thermally, are created. These increased current carriers reduce the resistance of the material; the addition of released electrons reduce its resistance from several megaohms (MΩ) in the absolute darkness to less than 100 Ω in sunlight.

The photogenerated carrier pairs move in response to the applied voltage across the absorbing region, which leads to an increase in the current flowing in the circuit resulting in a temperature change, which in turn, changes the physical or electrical property of the detectors responsivity; changes in output current are about one milli-ampere (mA) per lumen. The conductivity of a semiconductor, made of intrinsic semiconductor material (see Section 2.2.6.1), that has electron and hole concentrations of N_{e} and N_{h}, respectively, is

$$\sigma = q(N_{\mathrm{e}}\mu_{\mathrm{e}} + N_{\mathrm{h}}\mu_{\mathrm{h}}), \tag{5.65}$$

with q as the charge of the electron, μ_{e} and μ_{h} the electrons and holes mobilities, respectively (see Section 2.2.4.4), N_{e} as the magnitude of free electron density, and N_{h} the magnitude of hole density.

In the absence of optical illumination, the conductivity, known as the dark conductivity, σ_D, is given by

$$\sigma_D = q(N_{e0}\mu_e + N_{h0}\mu_h). \tag{5.66}$$

The electron and hole concentrations in this situation are the equilibrium concentrations, N_{e0} and N_{h0}, respectively. When the sample is illuminated with light of a sufficient photon energy ($h\nu > E_g$), the electron concentration is increased by ΔN_e and hole concentration by ΔN_h. The photoconductivity is the additional conductivity contributed by these photogenerated excess carriers. The change in conductivity, according to Ohm's law, is expressed as

$$\Delta\sigma = \sigma - \sigma_D = q(\Delta N_e\mu_e + \Delta N_h\mu_h), \tag{5.67}$$

in which $\Delta N_e (= N_e - N_{e0})$ and $\Delta N_h (= N_h - N_{h0})$ are the photogenerated excess electron and hole concentrations, respectively.

The change in conductivity implies either a change in concentration or mobility. The increase in conductivity is proportional to the photon flux. Let n electron–hole pairs be generated per second as a result of photoexcitation. If τ_e and τ_h are the respective lifetimes of electrons and holes, one finds $\Delta N_e = n\tau_e$ and $\Delta N_h = n\tau_h$. Thus the increase in conductivity is

$$\Delta\sigma = Nq(\mu_e\tau_e + \mu_h\tau_h). \tag{5.68}$$

In the case of extrinsic photoconductors (see Section 4.3.1.1), if either an N-type or a P-type semiconductor is irradiated with light quanta having $h\nu > E_i$, in which E_i is the intermediate energy level close to the conduction band, the electrons may be raised from the donor energy level to conduction band or from valance band to the acceptor energy level leading to the creation of electrons and holes, respectively. Consequently, the free carrier concentration increases over the thermal equilibrium value yielding in a change in the conductivity of the sample. The aforementioned derivations hold with the substitutions that $h = 0$ for N-type and $e = 0$ for P-type semiconductors.

The device exhibits an internal gain, which arises because the recombination lifetime and transit time generally differ. Suppose that the electrons travel faster than holes and that the recombination lifetime is large. As the electron and hole are transported to opposite sides of the photoconductor, the electron completes its trip sooner than the hole. Hence, another electron moves faster to complete its trip before the hole reaches. This process continues until the electron recombines with the hole. A single photon absorption can, therefore, result in an electron passing through the external circuit several times. The gain, G, of a photoconductor can be derived as

$$G = \frac{\tau}{\tau_d}, \tag{5.69}$$

where τ is the excess carrier recombination lifetime and $\tau_d (= w/v_e)$ the electron transit time across the sample, which depends on the properties of the photoconductor, the contacts, and the applied voltage, w the width of the transition region, $v_e (= \mu_e \mathbf{E})$ the velocity of electron, and $\mathbf{E}$ the electric field.

The gain factors may range from values many orders of magnitude below unity (τ is very small) for insensitive photoconductors to values several orders of magnitude above unity for very sensitive photoconductors (τ is very large) provided that the applied voltage is sufficiently large. If the velocity of hole, $v_h \ll v_e$ and $\tau_e = w/v_e$, the photocurrent, i_{ph}, is given by (Saleh and Teich, 1991)

$$i_{ph} \approx q\eta_d \frac{\tau\Phi}{\tau_e}, \tag{5.70}$$

in which η_d is the quantum efficiency of the detector.

The spectral sensitivity of the PC detectors is governed by the wavelength dependence of η_d. Unlike photoemissive detectors, the PC devices can operate into the infrared (IR) region on band-to-band transitions. The substances, for example, lead sulfide (PbS), lead selenide (PbSe), and lead telluride (PbTe) are sensitive to IR wavelengths. A selenium (Se) cell is sensitive throughout the visible end, toward the blue end of the spectrum in particular. Other semiconductor materials used as photoconductors are cadmium telluride (CdTe), indium antimonide (InSb), indium phosphide (InP), germanium (Ge), silicon (Si), gallium arsenide (GaAs), gallium phosphide (GaP), gallium nitride (GaN), and mercury cadmium telluride (HgCdTe); the cadmium sulfide (CdS) cell is also used. The intrinsic photoconductors (see Section 4.3.1.1) detects the light of wavelength dictated by equation (2.11), while the extrinsic photoconductors can be used as detectors at longer wavelengths.

The best PC devices are made using nearly pure semiconductor materials having uncompensated extrinsic carrier concentration below 10^{15} cm^{-3}. The extrinsic photoconductivity operates on transitions involving forbidden-gap energy levels. It takes place when the photon interacts with a bound electron at a donor site, producing a free electron and a free hole. However, these detectors require cooling (see Section 4.3.1).

The Photoconductors can be used in two ways: (i) to measure electric current (photoconduction) and (ii) to measure pulses. In the former case, a flux is measured (photometry), while in the latter case, the energy of the quanta is measured (spectrometry). Unlike photodetectors, the solar cells generate electrical power. The light-emitting diodes (LEDs; see Section 3.3.8.1) and laser diodes (see Section 3.3.8.1) convert electrical energy to optical energy. A PC cell can be employed in photography as the light meters used to measure the intensity of illumination. It is also used to record a modulating light intensity.

5.5 Detector Arrays

A detector array (combination of several single detectors) is a linear array of discrete detectors on an integrated circuit (IC; see Section 2.3) chip, which can simultaneously register the photon fluxes (as functions of time) from many spatial points. Development of such an array pave the way to produce images containing hundreds of pixels at the same time. There exist two kinds of arrays, namely, line detectors (1-D arrays) and area sensors (2-D arrays). In spectroscopy, such a detector is placed at the image plane of a spectrometer to permit a range of wavelengths to be measured simultaneously. The incident light is dispersed by a grating (Berns, 2000); a full spectrum can be acquired in milliseconds or less. The detector arrays having numbers of elements (pixels) ranging from 128 to 2048 can be fabricated using semiconductor materials and have two parts, such as the following:

(1) A mosaic of closely spaced detector elements, which convert the incident electromagnetic radiation into electrical signals.

As stated in Section (2.2.4.1), the incident photon creates the electron–hole pairs and the electrons migrate to the nearest P-I-N junction (see Section 5.3.2); in the visible, a single such pair is created for each photon absorbed. The holes diffuse to the substrate and recombine, the electrons are collected within a pixel and give rise to the signal charge. At X-ray and far-UV wavelengths, each photon has enough energy to create several such pairs.

(2) A read-out circuit that relays and multiplexes the electrical signal from each detector element (or pixel) to a few output amplifiers. After a fixed integration time, the charge at each element is sequentially read to generate the detector response as a function of linear distance along the array.

5.5.1 *Focal Plane Arrays*

A focal plane array (FPA) is a general term for a two-dimensional (2-D) detector array that is hybridized with a read-out integrated circuit (ROIC). These arrays are also called staring arrays as they do not require scene scanning, which is necessary if a single element or a liner array is used. Usually, the optics part of an opto-electronics imaging device is limited to focusing of the image onto the detectors array. The FPAs use either photon or thermal detectors. The photon detectors (see Section 4.3) respond to the photons having an energy greater than the bandgap of the semiconductor material. The detector response is proportional to the rate of absorption of these photons. These photon detector materials can be used in three basic types of structures, such as

- photoconductive mode in which the absorbed photons cause a change in conductivity (see Section 4.3.1.1),
- photovoltaic mode where the absorbed photons create a change in voltage at a potential barrier either produced by a P-N junction or induced by electric fields with a metal-insulator-silicon (MIS) structure (see Section 4.3.1.1), and
- photoemissive modes (see Section 5.2), in which, for instance, silicide Schottky barrier device (see Section 5.3.4), the photons are absorbed in the silicide electrode and a small fraction of

the excited carriers are emitted over the Schottky barrier into the semiconductor; the charge generated at the electrode is transferred to a read-out structure.

An important PC material system for area arrays is doped extrinsic silicon (Si). As stated in Section 4.3.1.1, the extrinsic photoconductors are the impurity band conduction detectors using Si:As and Si:Ga (Stetson et al., 1986), which reduced recombination noise (see Section 4.5.2.2) and longer spectral response than standard extrinsic devices due to the higher dopant levels; the Si:Ga and Si:In were the first mosaic focal plane array PC detector materials. These detectors have a combination of PC and PV characteristics, including extremely high impedance, photovoltaic like noise, linear photoconductive gain, high uniformity, and stability (Kozlowski and Kosonocky, 1995).

The photodiode array detector makes an ideal sensor for an entire spectrum of wavelengths stretching from the near-UV to visible dispersive spectro-photometer. The platinum silicide (PtSi) is a PV-Schottky barrier detector. An array of tiny PtSi photodiodes on P-type Si is sensitive to such wavelengths range, which corresponds to the Schottky barrier height of about 0.2 eV. The quantum efficiency, η_d, ranges between 35% and 60% in the UV and visible regions, where the photon energy exceeds the bandgap of Si. At these wavelengths, the light transmitted through the PtSi film generates large quantity electron–hole pairs in the Si substrate. It is noted that for wavelengths less than 1100 nm, the choice is restricted to silicon detectors operating in a PV mode either as photodiodes or as MOS capacitors (see Section 2.3.2).

Further advancement in imaging systems requires solutions for some fundamental and technological issues related to wide-FOV, resolution, pixel pitch, optics, multicolor, and low cost (Dhar et al., 2013). The conventional detector arrays are designed for a narrow band of wavelengths due to inadequate absorption and charge collection from photons with varying wavelengths. The broad-band absorption is usually inadequate due to quantum efficiency (QE) roll-off. The development of nanotechnology and science (Logeeswaran et al., 2011) of emerging materials and material designs has significant impact on sensors and many other electronic devices. Nanostructures offer very thin absorption layers due to many intricate designs, for example, plasmonics and meta-materials to concentrate photons and enhance electric field. The advantages of thinner absorber in a photodetector means shorter carrier transit time, thus high-speed detectors and greater structural compatibility with ever-shrinking electronic devices (Tohyama et al., 2006).

5.5.1.1 Quantum well devices

A quantum well is a double hetero-junction structure, in which the active region is formed by using two materials having compatible crystal structure and lattice space formation, with different bandgaps, out of which one material with shorter bandgap is sandwiched between the materials of larger bandgap providing the rise to the potential well for charge carrier in the conduction band, as well as in the valence band (Esaki and Tsu, 1970). For example, in aluminum gallium arsenide/gallium arsenide (AlGaAs/GaAs), quantum well transition takes place among the energy levels of the conduction band (Nag, 2002 and references therein). The offset in the conduction band between these two materials create a confining place. Thus, the so-formed well confines the electronic wave functions and hence quantizes the energy levels. While depositing well layer, at the center of the region a dopant is added. Quantum wells are of two classifications, for instance, N-type that is doped with silicon (Si) and P-type doped with Be. In the case of the former, electrons are transferred to GaAs and confined to this region by the potential profile of GaAs/AlGaAs, in which the former acts as well and the latter as a barrier.

The component of momentum in the direction perpendicular to the interfaces being quantized, the absorption spectrum of such structures depict peaks corresponding to transitions from the hole levels to the electron levels at near liquid helium temperature as illustrated in Figure 5.25. The band-edge potential varies from layer to layer as a result of the difference in the bandgaps and a periodically varying potential is produced in the structure with a period equal to the sum of the widths of two consecutive layers. For layer thicknesses of the order of 10 nm, the wavelength, as well as the mean free path of the electrons extend over several layers and

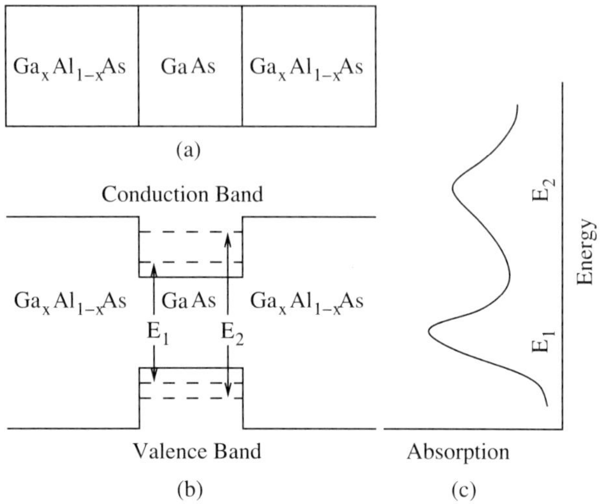

Figure 5.25 Schematic of absorption spectrum of a quantum well: (a) the hetero-structure, (b) quantized energy levels, and (c) absorption coefficient for different photon energy (after B. R. Nag, 2002).

the periodic potential transforms the energy bands of the host lattice into mini-bands. Phenomena like Bloch's oscillation (Bloch, 1928; see Section 2.2.3.2) and low-field negative differential resistance may be produced by the electrons in such minibands.

In order to grow a quantum well in semiconductor material, the method like molecular beam epitaxy (MBE) is used. The MBE is a kind of ultrahigh vacuum evaporation in which the beam of atoms or molecules containing the desired atoms are directed from effusion cells to a heated substrate. The atoms on arriving at the substrate combine at the lattice sites. The flux of the incident beams are determined by the temperature of the effusion cells. The composition of the grown layer may be controlled by opening and closing the shutters in front of the effusion cells, which are kept at predetermined temperatures. Other techniques, such as metal organic chemical vapor phase deposition (MOCVD), molecular organic vapor phase deposition (MOVPD), chemical beam epitaxy (CBE), and Metal organic molecular phase deposition (MOMPD) can also be employed.

5.5.1.2 Super-lattice structure

A super-lattice structure is a device made of a repeating sequence of thin layers of different materials. Hetero-structures for quantum well devices are constructed in three forms, for instance, (i) the single-junction structures (often referred as hetero-structures), (ii) the double-junction structures (mostly referred as quantum wells), and (iii) multi-junction structures called super-lattices (see Figure 5.26). Typical calculations of ground (E_0) and excited (E_1) require basic quantum mechanics (see Section 1.1) and boundary conditions linking envelope wave functions of the well and barrier sides of the hetero-interface (Levine, 1993; Gunapala and Bandara, 1995). For a quantum well with finite barrier, one may write exact wave functions for the ground and excited state wave functions in both well and barrier regions, and evaluate the absorption peak and oscillator strength.

The multi-layered structures of different semiconductor materials that interchange with each other are called multi-quantum well structures, which can be fabricated such that the energy bandgap varies with position in any number of ways. If the energy barriers between the adjacent wells are sufficiently thin, in a quantum mechanical sense, so that electrons can readily tunnel through the barriers between them, the discrete energy levels broaden into miniature bands in which case the multi-quantum well structure is also referred to as a super-lattice structure (Nag, 2002, and references therein). In crystalline thin film deposition technique like MBE (see Section 5.5.1.1), the atoms are deposited atom by

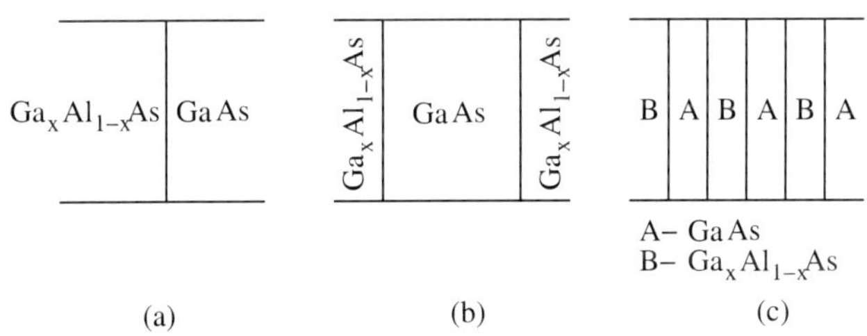

Figure 5.26 Schematic diagrams of hetero-structures: (a) single-junction hetero-structure, (b) double-junction hetero-structure, and (c) super-lattice structure (after B. R. Nag, 2002).

atom and layer by layer. This repetition of atomic layers brings about the formation of Brillouin zone center (see Section 2.2.1.1) about which the crystal is built. Thus, super-lattice (SL) is formed, in which a new Brillouin zone is created, compared with the other two dimensions, having a reduced wave vector range and narrow energy bands since the layer thickness is of few angstrom. In N-type semiconductor materials, the super-lattice potential is created by the discontinuities in the conduction-band profile due to the sandwich effect of the second material of two semiconductors. The mean free path of electrons along the SL direction should be greater than the period of the interfacing potentials. The band offset ratio is the ratio of the difference in conduction band energies to the difference in valence band energies.

In the case of a detector, the bandgap energy corresponds to the desired cut-off wavelength. When two semiconductors are brought in contact, there are various ways the valance and conduction bands can be aligned. If both the valence and the conduction band edge of the second material are above the band edges of the first material, it is referred to as a broken type-II band alignment. The strain of the individual layers is a key feature of type-II SL as it suppresses Auger generation of carriers (see Section 2.2.4.1; Auger, 1923), in which the conduction band of one component is lower in energy than the valence of the other. The staggered band alignment of this type creates a situation, where the bandgap energy, E_g, of the SL can be adjusted to form either a semi-metal (for wide InAs and GaInSb layers) or a narrow bandgap (for narrow layers) semiconductor material. The bandgap of the SL is determined by the energy difference between the electron miniband and the first heavy hole state at the Brillouin zone.

5.5.1.3 Focal plane array architecture

Development in detector focal plane array (FPA) technology has revolutionized many kinds of imaging (Norton, 2003). From gamma rays to the infrared and even radio waves, the rate at which images can be acquired has enhanced by more than a factor of a million in some cases. A number of architectures are used in the development of IR focal plane arrays (FPA; Scribner et al., 1991). The architectures of such devices are enumerated below:

Scanning system The scanning linear FPA consists of a row of detectors. An image is generated by scanning the scene across the strip using a mechanical scanner. At standard video frame rates, at each pixel (detector), a short integration time has been applied and the total charges are accommodated. This system does not include multiplexing functions in the focal plane. A typical example of such a detector is a linear photoconductive array (lead salt detectors, such as PbS, PbSe, and HgCdTe). In this, an electrical contact for each element of a multi-element array is brought off the cooled focal plane to the outside of dewar, where there is a single electronic channel at ambient temperature for each detector element. The HgCdTe arrays employ 60, 120, or 180 PC element depending on the applications.

Staring array The staring arrays having 2-D array of detector pixels are scanned electronically by circuits integrated with the arrays, which provide increased sensitivity and gain; there are more than 10^6 elements on the focal plane. These ROICs include, pixel deselecting, anti-blooming on each pixel, subframe imaging, and output pre-amplifiers.

Time-delay and integration Also, the intermediary systems exist, which are fabricated with multiplexed scanned photodetector linear arrays with time-delay and integration (TDI) functions. For instance, HgCdTe multi-linear 288×4 arrays are fabricated for both 3–5 μm and 8–10.5 μm bands with signal processing in the focal plane. To note, in signal processing, one needs to find out the characteristic points of a light spot (maxima or minima of the intensity distribution). The TDI imagers operate like line-scan imagers, but with a large number of lines; as the image of the object moves past each line, each line captures a snapshot of the object (see Figure 5.27). These imagers combine multiple exposures synchronized with the motion of the object. The CCD TDIs combine signal charge, while the CMOS TDIs combine voltage signals.

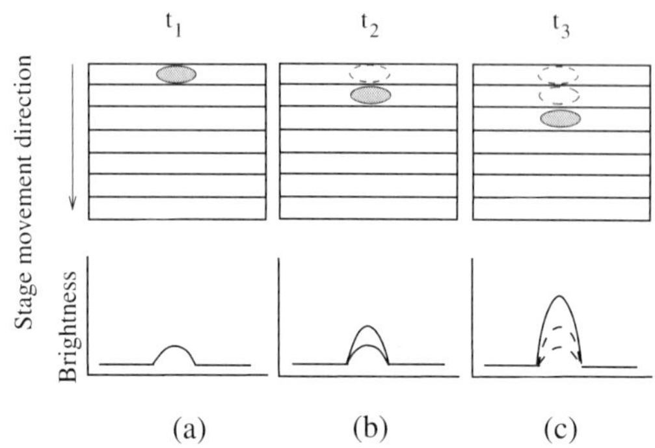

Figure 5.27 Principle of time-delay and integration (TDI) imager.

5.5.2 *Broad Classifications of Detector Array*

There are broadly two types of detector arrays available:

Monolithic In this approach, some of the multiplexing is done in the detector material rather than in an external read-out circuit. The basic element of a monolithic array is a metal insulator semiconductor structure. There exist three kinds of monolithic FPA:

(1) **Completely monolithic**: In this, the detector array and the read-out multiplexer are integrated on the same silicon substrate.
(2) **Partially monolithic**: This includes narrow-band detector arrays of HgCdTe (Roberts, 1983) and InSb (Gibbons and Wang, 1983) integrated on the same substrate with the row and column read-out from a 2-D detector array. In this, the multiplexing of the detected signal is completed by additional silicon integrated-chips packaged on the imager focal plane.
(3) **Vertically integrated**: In this, a silicon read-out multiplexer is used with the narrow-bandgap HgCdTe detectors; HgCdTe chips are attached to a silicon multiplexer wafer, followed by the fabrication of the HgCdTe photodiodes.

Hybrid systems The hybrid FPA detectors and multiplexers are fabricated on different substrates and mated with each other by flip-chip bonding interconnection. The indium bump bonding of read-out electronics provides for multiplexing the signals from thousands pixels onto a few output lines. This simplifies the interface between the vacuum-enclosed cryogenic sensor and the system electronics. Two hybridization approaches that are used at present are

- indium bumps are formed on both the detector array and the ROIC chip, which are aligned and force is applied to cause the indium bumps to cold-weld together, and
- indium bumps are formed on the ROIC chip; the detector array is brought into alignment and proximity with the said chip, the temperature is raised to cause the indium to melt, and contact is made by reflow.

The detector array can be illuminated from either the front-side, in which the photons passing through the transparent silicon multiplexer or backside where the photons passing through the transparent detector array substrate; in the latter approach, the multiplexer would have the areas of metallizations and other opaque regions, which can reduce the effective optical area of the structure.

Chapter 6

Charge Transfer Devices

6.1 Introduction

Modern photodetectors provide quantum efficiency, η_d, approaching unity. However, for terrestrial applications, the only way to significantly enhance the focal plane sensitivity without sacrificing resolution is to increase the number of detectors used by the imaging system. Presently, the number of detectors used in the imaging systems is limited by the requirement to connect each detector to its preamplifier. The charge transfer device, a semiconductor structure where the discrete charge packets are removed, may eliminate these shortcomings.

Evolved from small analog shift-registers to the imaging detectors, it has a simple parallel plate metal-oxide semiconductor (MOS; see Section 2.3.2) or a metal–insulator–semiconductor (MIS). As capacitor like structure, such a device is composed of many independent sensors, called pixels, which can collect, store, and transfer localized packets of minority carriers along the semiconductor-insulator interface. The charge is stored in such a way that the charge pattern corresponds to the irradiation pattern. These devices can be linear or two-dimensional. According to the method used to detect the charge pattern, two types of charge transfer devices can be

High-Resolution Imaging: Detectors and Applications
Swapan K. Saha

ISBN 978-981-4613-27-9 (Hardcover), 978-981-4613-28-6 (eBook)
www.panstanford.com

distinguished (Streetman and Banerjee, 2000), such as the charge-coupled device (CCD) and the charge injection device (CID). Both these systems are being used in a number of fields of chemistry, the fields of Raman spectroscopy, molecular fluorescence, molecular absorption, atomic emission and X-ray detection (see Section 5.3.7).

The CCDs were initially expected for use in analog systems (Hobson, 1975) but soon their light registration capabilities were discovered. Following the introduction of the concept of charge-coupled semiconductor, Fairchild Inc. introduced the first commercial CCD with 100×100 pixel resolution in 1973. As main optical detectors, CCDs are in use at every observatory for imaging, spectroscopy, and photometry since their introduction in the late 1970s (Monet, 1988). Also, the CCDs can be directly mounted on telescopes replacing both photographic plate and micro-photometer system. They provide an order of magnitude increase in sensitivity and have revolutionized astronomical imaging enabling the study of very faint objects (Saha, 2009). Today, CCDs are commonly used in other scientific fields also, such as optical and UV spectroscopy, remote sensing, and biomedical science. However, in the presence of atmospheric turbulence, a conventional CCD cannot track the image of a point source, since it can shift charge along one direction. As discussed in Section 1.5.6, in an adaptive optics (AO) system, the low frequency tilt error can be removed by a tip-tilt mirror, which leads to a better resolution. However, such a system suffers from the complexity and the optical loss. In order to achieve this capability electronically within an image sensor, an orthogonal transfer CCD (OTCCD) that shifts charge in all directions in order to track image motion, was also developed (Burke and Tonry, 1997).

6.2 Charge-Coupled Device

W. S. Boyle (1924–2011) and G. E. Smith (1930–) invented an imaging semiconductor circuit, called charge-coupled device (CCD) sensor, in the late 1960s at the Bell laboratories, for which they shared the Nobel Prize in Physics, along with Charles Kao (1933–) in 2009. Such a device refers to a light-sensitive semiconductor chip made up of P-type silicon (Si). The chip consists of bi-dimensional

Table 6.1 Typical performance of CCD

Characteristics	Range
Charge transfer efficiency (CTE)	99.995%
Dynamic range	$\lesssim 10^6$ charges pixel^{-1}
Linearity	0.1
Output amplifier sensitivity	6.0 μV.e^{-1}
Peak signal	200 ke . pixel^{-1}
Quantum efficiency (back-illuminated)	~90%
Read-out noise at 20 KHz	4–6 e RMS
Spectral range	400–1100 nm (peak at 550 nm)
Temporal response	~50 ms
Threshold	one quantum
Uniformity of response	5

array of sensors. Each sensor has a set of MOS structures (Boyle and Smith, 1970). Although the CCD was envisioned as a new kind of computer memory circuit, which is adopted from the magnetic bubble memory[a] technology, because of its ability to transfer charge and the photoelectric interaction with light, it has grown up to become the dominant process for other applications, for example, signal processing and imaging. A CCD gives the best performance in terms of resolution, sensitivity, and other parameters given in Table 6.1. It also provides low terminal capacitance and good geometric stability.

6.2.1 *Principle of Operation*

The operating principle of a CCD is based on the photoelectric effect (see Section 2.1). Unlike a photomultiplier where the photoelectrons leave the substratum in order to produce an electric current, this

[a]Bubble memory uses a thin film of a magnetic material to hold small magnetized areas, known as bubbles, which each store one bit of data. Conceptually, it is a stationary disk with spinning bits. The unit, a couple of square inches in size, contains a thin film magnetic recording layer. Globular-shaped bubbles (bits) are electromagnetically generated in circular strings inside this layer. In order to read or write the bubbles, they are rotated past the equivalent of a read/write head. One of the limitations of bubble memory was the slow access. A large bubble memory would require large loops, so accessing a bit requires cycling through a huge number of other bits first.

device uses a quantity of electrical charge to represent the light intensity, which is sampled at discrete times. A CCD samples the inherent image by collecting incoming photons into a sample grid. The grid is made-up of tiny square (or rectangle) shaped pixels. When a photon is collected by a pixel, information about the exact location of that photon is the location of the pixel. Each pixel has a brightness value that is assigned a shade of gray color by the display routine. Since the pixels are square, the edges of features in the image would have a stair step appearance. The more pixels and shadows of gray that are used, the smoother the edges would be. The memory function arises from shifting these charges simultaneously down a row of sensing cells in discrete time. The CCD allows the photoelectrons to remain where they are released, thus creating an electronic image, analogous to the chemical image formed in a photographic plate. The CCDs have three functions (Holst, 1996), such as

- the generation and collection of charges (magnetic bubble formation),
- the transfer of charge by manipulating the electrode gate voltages (bubble propagation by an external rotating magnetic field), and
- the conversion of charge into a measurable voltage (bubble detection as either true or false).

A CCD operating in the inversion mode is an array of (MOS); the energy band structure of such a semiconductor for both N-type and P-type is illustrated in Figure 2.9. As discussed in Section 2.3.2, the insulator is formed by direct oxidation of the semiconductor (SiO_2 on silicon). The applied potential difference creates an electric field, which modifies the charge distribution in the semiconductor in the form of a potential well (see Figure 2.10) that is capable of collecting many photogenerated electrons. The chip is a thin wafer of silicon consisting of millions of photo sites each corresponding to a MOS-capacitor. The capacitor is referred to as gate and MOS-capacitor is called photogate. Under the control of an external circuit, each capacitor can transfer its electric charge to one or other of its neighbors. For charge transfer, the depletion regions, which are gradients, should overlap.

The CCD register comprises a series of electrodes, each electrode has its own control voltage, which is varied as a function of time. This voltage is a square wave and is known as the clock signal. In the case of low electrode voltage, it acts as a barrier, while when the voltage is high, charge can be stored.

The number of MOS-capacitor structures in a single pixel is determined by the number of phases (Φ) in operation. To note, the set of pixels is organized in a 2-D array forming the detector array of $N \times M$ pixels specifying the $N \times M$ format of the CCD. In general, the CCDs are made with format 1024×1024 (usually denoted by $1\text{k} \times 1\text{k}$). The large array CCDs beyond $4\text{k} \times 4\text{k}$ format are made by placing small array CCDs side by side, what is known as mosaic CCD. The mosaic CCDs are useful imaging large size celestial objects. Of course, there is a dead space between the CCDs. In a three-phase CCD, three MOS-capacitors are placed very close to each other in a pixel. The center electrode is biased more positively than the other two and the signal electrons are collected under this phase, which is called the collecting phase. The other two phases are called barrier phases.

The whole CCD array can be conceived as shift-registers arranged in the form of columns close to each other, which move charges that are collected between capacitive bins in the device. Highly doped P-regions, called channel stops, are deposited between these columns so that the charges do not move across the columns; the depletion regions overlap in one direction only. The channel stops are parallel to, and exclusive of, the channel, or charge carrying regions. Every third electrode in these shift-registers is connected to the same potential. The electrodes of each pixel in a column are connected to the corresponding electrodes in other columns also. By manipulating the voltages on these electrodes, the charges can be shifted along the columns. This array is the imaging area and is referred as the parallel register. At the end of columns, a similar kind of CCD shift-register is arranged at right angle to the imaging area, which is called output register or serial register (multiplexer).

Multiplexer (or MUX) A multiplexer (or logic circuit) is a device that selects one of n input (analog or digital) signals and direct it to an output. A 2^n to 1 multiplexer routes one of 2^n input lines to a single output line. It has two sets of inputs, for instance, (i) 2^n data input lines and (ii) n select lines, to pick one of the 2^n data inputs. Its output is a single bit, which is one of the 2^n data inputs. The multiplexers can implement arbitrary functions. A de-multiplexer is basically a multiplex, where it takes a single input signal and allows one of many output-lines.

Charge transfer mechanism Soon after the completion of the exposure, one should discharge the CCD keeping the memory of the charge of each active element. The charge transfer mechanism for three-phase CCD is illustrated in Figure 6.1. By applying proper

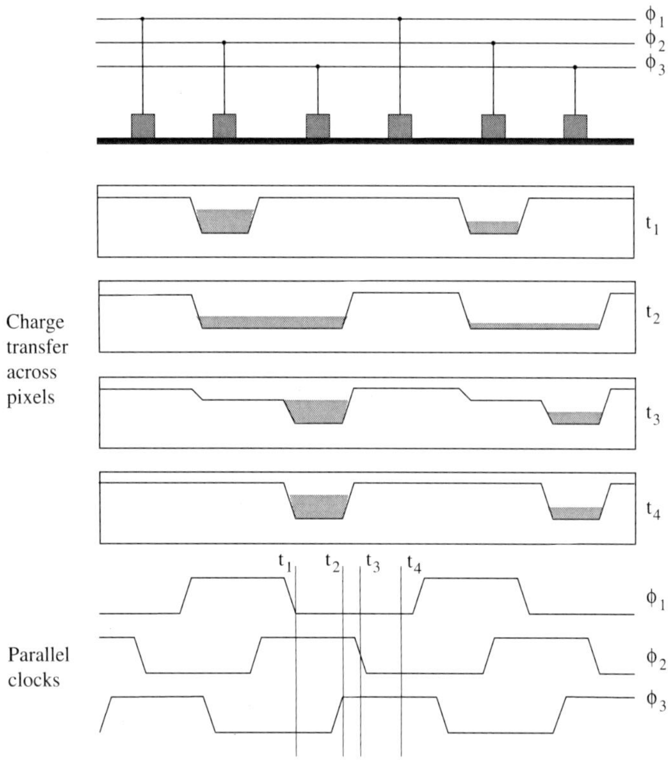

Figure 6.1 Sequence of charge transfer from Φ_2 to Φ_3: (1) charges under Φ_2 well, (2) charges shared between Φ_2 and Φ_3 wells, and (3) charges under Φ_3.

phase voltage pulses to a set of electrodes on the surface, the potential well may be caused to travel in a selecting direction, which transfers the packet of charge with it. During a specified exposure-time, the central electrode in each pixel is kept at a more positive potential than the other two (three-phase CCD). The charge collected under this electrode should be measured. First, the charges should be shifted vertically along the parallel register (the columns) onto the output register. After each parallel shift, the charges should be shifted along the output register horizontally onto the output amplifier. Hence, there should be n serial shifts after each parallel shift, where n is the number of columns.

At time t_1, only phase two is positive and hence all the electrons are under phase two. At time t_2, both the phases two and three are at the same positive potential and the electrons are distributed equally under these phases. At time t_3, phase two potential is going lower whereas phase three is positive and hence electrons start leaving phase two and cluster under phase three. At time t_4, when the phase three is alone positive, the electron are fully under phase three. The electrons that were collected under phase two are now under phase three by this sequence. The repetition of this clock sequence results in the transfer of electrons across the columns onto the output register. The instant in which an electron reaches the output register characterizes the position of the element on the array and its intensity is amplified and digitized. This is done for all the arrays simultaneously so that one obtains a matrix of numbers that represents the distribution of intensities over the entire field.

6.2.2 *CCD Architectures*

Two modes of operation have been implemented in CCD construction:

Surface channel operation When a positive potential is applied to the electrode that is separated by thin SiO_2 from the silicon lattice, the carriers move at the Si–SiO_2 interface. Thus, the signal electrons are collected and transferred at the interface between the substrate and the SiO_2 insulating layer. Hence, they are referred to as surface channel CCDs. However, the surface layer has many crystal irregularities and defects (see Section 2.2.2) in the crystal lattice, which can trap charge but not release it instantly.

Buried channel structure In order to alleviate the problems caused by surface irregularities at the interface of the oxide and semiconductor, the buried channel structure is used at present. A thin N-type is silicon layer is deposited on the surface of the P-type substrate (just below the SiO_2 insulating layer). This N-type layer creates a complex potential well with potential maximum generated at slightly below the Si–SiO_2 interface, where the signal electrons are collected and transferred. This is known to be as buried channel CCD; the band diagram is illustrated in Figure 6.2. Since this process takes place inside the bulk of the silicon, the charge transfer is very efficient as the trapping sites become much less.

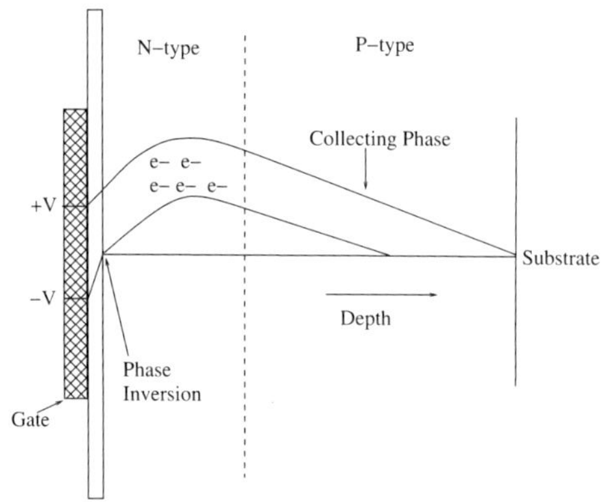

Figure 6.2 Energy band structure for buried-channel CCD device.

A CCD used for imaging has an array of cells to capture a light image. The packets of charge move from cell to cell by the coupling and decoupling of potential wells within the semiconductor. At the end of the line, the charges from all the pixels are converted into electrical signals. Several CCDs can be formed into a mosaic. A CCD may be designed with the following methods:

6.2.2.1 Linear array CCD

A linear CCD is composed of a photosensitive row having a range of a hundred to several thousand pixels. It is designed with a single

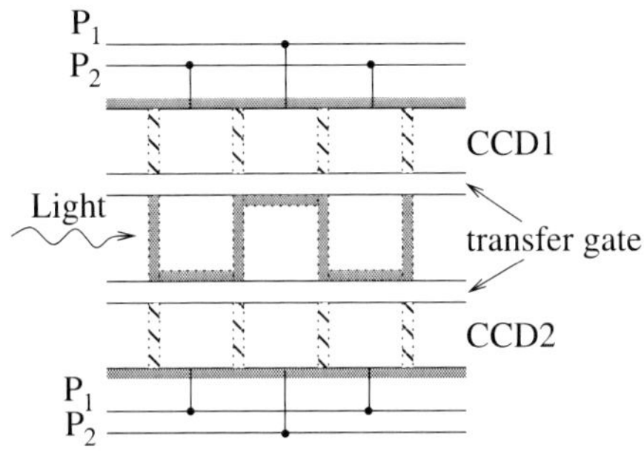

Figure 6.3 Linear imaging.

line of pixels, which is achieved by shifting the signals between stages within the device one at a time. The device captures single column of images of moving objects at a time and adds row after row until the frame is complete. Located next to each sensor is the CCD shift-register (see Section 6.2.1), which is light sensitive covered with a metal light shield. For a linear imaging device, the ratio of the integration time to read-out time must be large to avoid superimposed but time delayed multiple images. The practical approach is to use light-sensitive storage regions. At the end of the integration period, the charges are read out simultaneously into one or more charge-coupled channels (see Figure 6.3), the reason for which is to match the minimum periodicity of the CCD. The size of the pixel is limited by the size of the gate, for instance, for a three-phase system, the pixel width is three times the width of a single gate. For a fixed gate size, the active pixel width can be reduced with a bi-linear read-out. Since the resolution is inversely related to pixel pitch (sensor-to-sensor spacing) has twice the resolution. For fixed pixel size, the bi-lincar read-out enhances the charge transfer efficiency (see Section 6.2.5.1).

6.2.2.2 Area array CCD

The area array CCDs utilize the pixels to create a rectangular or square image capture surface. There exist four possible read-out organizations of which two are depicted in Figure 6.4. Devices with X-Y addressed organization (see Figure 6.4a) are being pursued

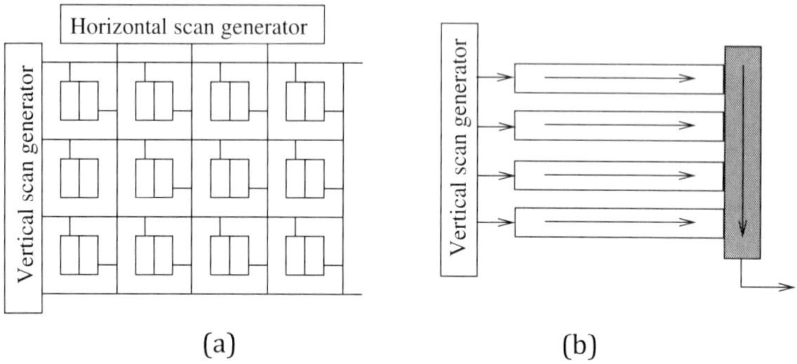

Figure 6.4 Read-out organizations of area image sensors.

under the name of charge injection devices. An area sensor can be formed in a straight by assembling simple linear sensors with illuminated charge transfer registers in a 2-D array (see Figure 6.4b), following which this array is provided with an addressing circuit that switches the clock pulses to a certain line when its read-out is desired. The resolution of an area image sensor is typically described by the number of pixels in the horizontal direction by the number of pixels in the vertical direction, for example, 2k×2k, the total resolution of which is 4 megapixels. The area image CCDs are classified by their shifting techniques into:

Full frame transfer CCD In a CCD, the signal charge is collected in a potential well in the photosensitive area during the signal integration time, following which the image pixels are read out line-by-line (see Figure 6.6a), and transferred to the output section through the multiplexer. All charges must be clocked out of this register before the next line can be transferred. The photosites are continually irradiated during read-out, which results in a smeared image. When a photo site is subjected to excessively strong illumination, the accumulated charges can become so numerous that they spill on to adjacent photo elements. A saturated pixel produces a characteristic diffusion similar to the halo surrounding bright stars on a photographic plate. In addition, the number of charges accumulated in a saturated well can be such that its contents cannot be emptied in one or more transfers. A trail starting at the saturated point then appears in the direction of the transfer of rows. This

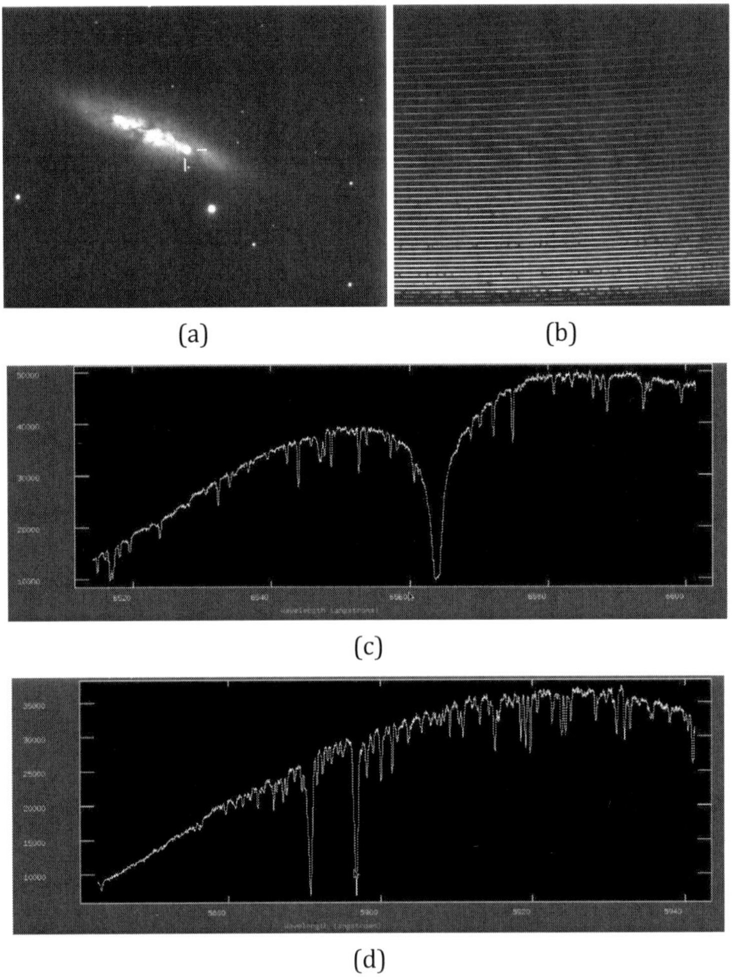

Figure 6.5 (a) Image of the supernova SN 2014 J in the nearby galaxy M 82, obtained through the R-band filter (see Table 1.1) using the Himalaya Faint Object Spectrograph Camera (HFSOC) instrument on the 2 m HCT at the IAO; the field-of-view of 10×10 arcmin is imaged on the central $2k \times 2k$ pixels of the E2V CCD with $2k \times 4k$ pixels, each 15 μm (Courtesy: S. Srivastav and G. C. Anupama), (b) image of the echelle spectrum of HR 1865 from 2.3 m VBT echelle spectrometer displayed in $2k \times 4k$ CCD, (c) tracings of two orders showing Hα, and (d) Na I D lines. Courtesy: N. K. Rao.

effect, called blooming, is generally caused by the presence of a bright object in the scene being imaged. If it is not controlled, the resultant image would suffer from a large over-exposure. The smear takes place in the direction of the charge transport in the imaging part of the array. In order to avoid such a smear, an external shutter is used for shielding the array. If the image integration time is longer than the read-out time, the smear is considered to be insignificant.

The full frame transfer CCDs are used for scientific applications, in conventional astronomy in particular. Figure 6.5a shows the image of the supernova SN 2014 J in the nearby galaxy M 82, taken at the 2 m Himalayan Chandra Telescope (HCT), Indian Astronomical Observatory (IAO), India. Figure 6.5b shows the high-resolution two-dimensional (2-D) image of a star, HR 1865, spectrum having 62 orders covering wavelength range 4000–10,000 Å, with small gaps in between the orders. The numerous narrow dark lines are absorption lines superposed on the continuum. The respective tracings of two orders showing Hα and Na I D lines of the said star are also presented in Figures 6.5c and 6.5d. These spectra, as well as the image are taken with the fiber-coupled echelle[a] spectrometer of 2.34 m Vainu Bappu Telescope (VBT), Vainu Bappu Observatory (VBO), India (Rao et al., 2004). This spectrometer transmits light through an optical fiber (see Section 3.4) from the said telescope focus to the Coudé laboratory.

Frame transfer CCD A frame transfer CCD (see Figure 6.6b) is composed of two parallel registers, of which one is kept in the photosensitive section, known as image array and the other is in the storage section, which is shielded with an opaque plate, namely, aluminum, to prevent any exposure to the light for data storage, and a single serial register. After the image array exposed for a specified integration time, the entire image area is transferred as fast as possible to the storage register array. The transfer time to the shielded area depends on the size of the array but is typically less than 500 μs. The smear is limited to the time it takes to

[a]An echelle (stairs or ladder) grating is characterized by a relatively low groove density and is used at high angles in high diffraction orders. The echells are used in applications where high dispersion and resolution are essential, such as atomic absorption spectroscopy, laser tuning, and astronomy.

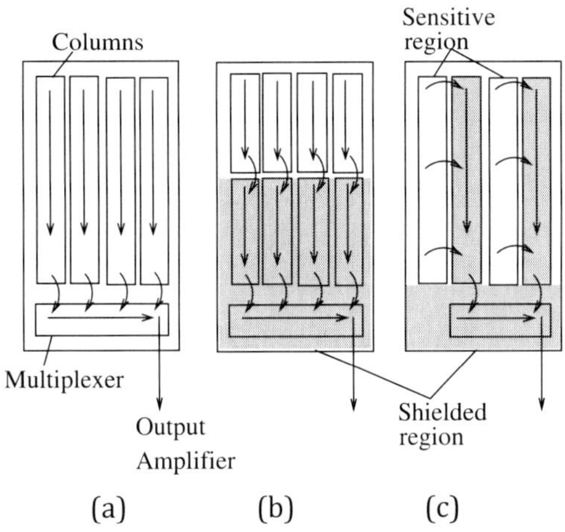

Figure 6.6 CCD read-out mechanisms: (a) line transfer, (b) frame transfer, and (c) interline transfer.

transfer the image to the storage area, which is much less time than that is required by the full frame CCD device. While the storage array is read out (in analog or digital format), the image array continues to capture the next frame. The frame transfer CCDs are usually operated without shutter at television frame rates. Since the storage area is placed away from the image area, these devices are able to capture close to 100% of the light that falls on each pixel, which makes them useful for low-light level imaging. However, such CCDs are twice the size and may contain a few dummy charge wells between the active area and storage area. These wells collect any anticipatory light leakage. Also, due to the distance between the image array and the storage area, these devices appear to be slower than the interline transfer CCDs. These are more susceptible to smearing. Some arrays do not have the integral light shield, which can be operated either in the full frame mode (e.g., 1k$\times$1k) or in the frame transfer mode (e.g., 1024$\times$512). By removing the opaque plate on the storage array, this can be used as full frame imager by clocking the parallel gates of the two arrays together. The frame transfer CCD is useful for certain applications, such as image or wavefront reading in times less than the coherence time of atmospheric turbulence (see Section 1.5.3).

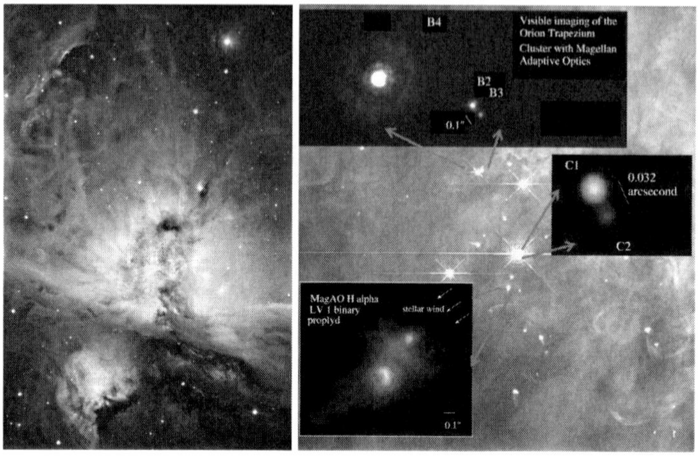

(a) (b)

Figure 6.7 (a) Image of the Trapezium Cluster and (b) Top inset: MagAO's photo of the "mini-cluster" of young stars in the θ 1 Ori B group (B1-B4). There is a clear evidence of relative motion of these stars around B1. The lowest mass member (B4) will likely be ejected in the future. Middle inset: Highest resolution astronomical photo of the θ 1 Ori C1-C2 pair. Bottom inset: LV 1 binary young star pair shaped by the wind from θ 1 Ori C (in the visible light of hydrogen gas at 656 nm). Courtesy: L. Close and Ya-Lin Wu. The background image is a Hubble Space Telescope (HST) Orion Trapezium Cluster visible image (NASA, C. R. O'Dell and S. K. Wong).

Figures 6.7a and 6.7b depict the wide-field view image of the Orion Trapezium,[a] a cluster of young stars still in the process of forming, which is taken using the Schulman telescope atop Mt. Lemmon, USA, and high-resolution images of θ 1 Ori B, respectively. The detector used was a compact 1k$\times$1k frame transfer CCD (E2V CCD47) device. The high-resolution images were acquired with

[a]The Trapezium system, θ^1 Ori ABCD, comprising massive O-type and early B-type stars, are located at the center of the brightest diffuse Orion nebula, M 42. They range in brightness from magnitude 5 to magnitude 8; two fainter stars (E and F) can also be envisaged with a moderate telescope. Both the θ^1 Ori A1 and θ^1 Ori B stars are the eclipsing binary systems. The former is known as V 1016 Ori, some part of its light is blocked off by its companion in about every 65.43 days, while the latter has a period of 6.47 days with a magnitude range of 7.96 to 8.65. The θ^1 Ori C is a massive star having 40 $M_\odot$ with a temperature of about 40,000 K. It has the power to evaporate dusty discs around nearby new stars.

adaptive secondary mirror (ASM) at the 6.5 m Magellan telescope adaptive optics (MagAO) system (Close et al., 2013). The adaptive secondary mirror (Saha, 2007, and references therein), which makes relay optics obsolete that are required to conjugate a deformable mirror (DM) at a reimaged pupil, as well as to minimize thermal emission (Bruns et al., 1997), can change its shape at 585 points on its surface 1000 times a second.

Recently, Crouzier et al. (2013) used a split frame transfer device, CCD39-01 from E2V, for the Nearby Earth Astrometric Telescope (NEAT) mission and measured the pixel positions with an accuracy of about 1×10^{-4}. In order to optimize the dynamic range, the sensitivity is maximized by combining back-illumination technology with large pixels and non-anti-bloomed architecture. The CCD is read at 1 kHz, the Poisson-noise (see Section 4.5.2.3) from the dark current at ambient temperature is found to be 5 electrons per pixel, which is smaller than the read-out noise (see Section 4.5.4) of 20 electrons. This CCD has two buffer zones (masked pixels) allowing a rapid transfer and high frame rate. The high quantum well size and frame rate allow a very fast integration.

The Hubble Space Telescope The Hubble Space Telescope (HST), a space-based observatory, named after Edwin P. Hubble, is a cooperative program of the European Space Agency (ESA) and the National Aeronautics and Space Administration (NASA). HST, in orbit at an elevation of 575 km, circles the Earth every 97 min. Although its primary mirror is only 2.4 m, it has revolutionized astronomy by providing unprecedented deep and clear views of the Universe, ranging from our own solar system to extremely remote galaxies. One of the key projects of HST on the extragalactic distance scale is to determine the Hubble Constant, H_0. In addition, This telescope is used to measure the age of the Universe and to find evidence of dark energy.

NEAT astrometric mission The Nearby Earth Astrometric Telescope (NEAT), a space-based mission, is proposed to the European space agency (ESA) with the objectives of detecting Earth-like exo-planets in the habitable zone of nearby solar-type stars (Malbet et al., 2011, 2013).

Interline transfer Another development in the CCD sensor is the construction of interline transfer CCD devices with greatly reduced cell size. It has been the major factor in the successful production of compact, low-cost, and high-quality image capturing equipment including video cameras, digital still camera, etc.; it reduces substantially smear, dark current, infrared response. An interline transfer CCD consists of photodiodes (see Section 5.3.1), which create the charge and the storage element as part of each pixel, separated by vertical shift-registers that are covered by an opaque metal shield (see Figure 6.6c). Most devices use buried photodiodes with a low dark current. With silicon photodiodes arrays, each absorbed photon creates an electron-hole pair. Either the electrons or holes can be stored and transferred. A vertical shift-register contains segmented storage areas for each active pixels and connects to the active area by means of a transfer gate (an analog switch). A horizontal shift-register at the bottom of each column collects the information from each vertical shift-register. The signal charge that is collected under the image registers is transferred to the opaque storage register at read-out time. The serial register lies under the opaque register. The arrival of information is timed so that it feeds sequentially to the processor, yielding in a continuous feed of image packets for processing. The read-out procedure is similar to the full frame CCD.

Frame-interline transfer The frame interline transfer CCD is configured with an added storage section. In order to solve the smear problem, such a system transfers the signal charge to the storage section at a high speed soon after the receiving the signal charge from the photodiode by the vertical shift-register. This ensures a reduction of smearing as compared with the aforementioned interline CCD.

6.2.2.3 Scanning images

Scanning refers to how the image is drawn on an electronic display monitor. There exist two basic ways in which CCD images can be displaced:

Interlaced images A cathode ray tube (CRT) image is created by drawing lines of video information to the screen from left-to-right and top-to-bottom. An interlaced image is made up of two fields, such as (i) odd field consisting of all odd lines on the monitor screen and (ii) even field consisting of all even lines. This mode scans 1, 3, 5, $\cdots$ to bottom, then proceeds to fill in the 2, 4, 6, $\cdots$ lines, which enables to obtain two images simultaneously. This occurs at a rate of 30 FPS (frames per second), meaning the entire process takes 1/30 of a second. Although it is limited by the spatial resolution to skip every other line, it improved temporal resolution. Also, the interlaced scan signals have the reduced bandwidth. Since, an interlaced system collects fields in sequential in time, the image suffers due to image motion effects causing image artifacts, image blurring, image flickering, and improper reproduction of finer details. Any vertical image movement from field-to-field smears the image in a complex fashion. Horizontal movement serrates vertical lines. With excessive movement, the two fields may have images that are displaced one from the other.

Progressive images The progressive image transfer or progressive scanning exposes the entire CCD array at the same time and transfers the information sequentially to the processor or display. Hence, the progressive scan devices do not suffer from the image motion effects as envisaged with the interlaced scans. It scans all the lines in one pass from top to bottom (1, 2, 3, $\cdots$), so the entire frame is transmitted at once rather than as two separate, sequential fields. This method results in sharper images. There are no image artifacts and image blurring in this mode. For video applications, the array aspect ratio (ratio between the height and width of a sensor), number of elements, and output timing are consistent with standard video formats. The camera electronics converts progressively scanned data into a standard video format, facilitating the output to be viewed on any standard display. As the technology improves and the bandwidth increases, such systems are developed with higher frame rates, allowing smoother motion.

6.2.3 *CCD Camera System*

A CCD camera system consists of three parts, such as (i) CCD camera chip, (ii) the electronics that are the intermediate between the CCD chip and the host computer, what is termed as the controller, and (iii) camera control software, whose primary features are to (a) open and close the shutter of the CCD camera, (b) capture the image data into the chip, (c) read the data from parallel port, and (d) store the data. The shutter drive that generates the required exposure times and temperature control are the axillary functions of controllers. Figure 6.8 shows the block diagram of a typical CCD camera system.

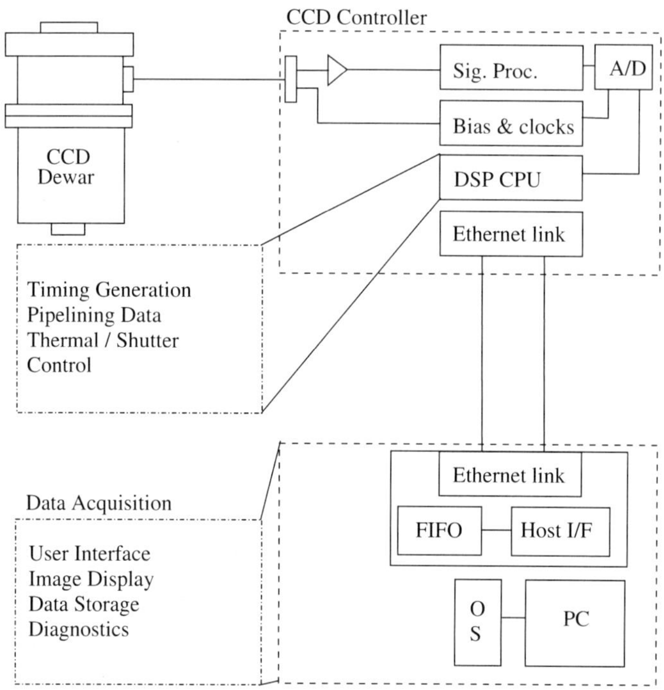

Figure 6.8 A block diagram of a typical CCD camera system.

6.2.3.1 Dewar

As stated in Section 4.1.1.8, the thermal generation of charges (dark current), as well as the dark current shot-noise (see Section 4.5.2.3) reduces the detected quantum efficiency (DQE) particularly in the red region. Also, the additive noise sources need to be reduced in order to obtain a precise measurement of light, particularly in the case of long-exposure imaging, which contributes to the accumulation of a large thermal charge.

The functions of a Dewar (see Section 4.3.1) are cooling the sensor to reduce the thermal signal, vacuum chamber, cooled by liquid nitrogen (LN_2), temperature control, etc. In order to obtain a stable dark current statistics (see equation 4.32), the CCD needs to be operated at a stabilized temperature of about 153 K to 173 K, which should be adequate for negligible dark contributions even in long exposures. Typically, at 153 K operating temperature, the dark

(a)

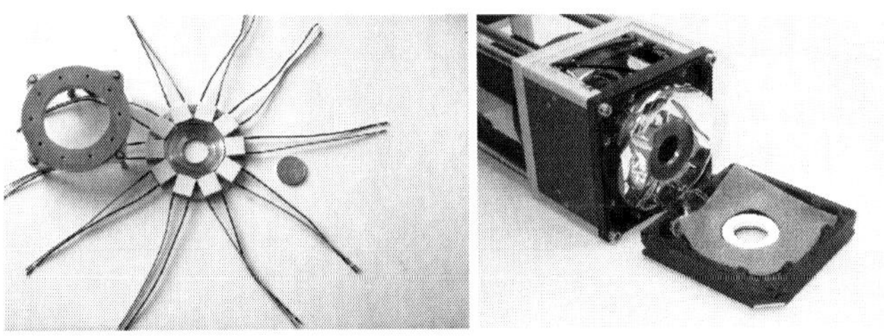

(b)

Figure 6.9 (a; left): An evacuated dewar with a pre-amplifier stage, (a; right): CCD dewar is filled with liquid nitrogen (LN_2) to cool the CCD, (b; left): Peltier cooling (copper ring), and (b; right) input window and insulation for the CPNG (comptage de photon nouvelle gènèration) camera (see Section 7.6.1); there is insulation on left and right parts (Courtesy: A. Blazit).

current would be just about a few electrons in hour-long integration. The CCD temperature is servo controlled using a temperature sensor and a heater resistor. The temperature sensor provides a current output. The signal is compared with the set temperature by a gain comparator that regulates the current into a heating resistor and, thereby the temperatures in the CCD. For slow scan mode (see Section 4.3.1) operation of large CCD, it is generally housed in a cryogenic environment that is filled with LN_2. Figure 6.9a shows a CCD dewar where a pre-amplifier board is kept inside, which coordinates output signal of the CCD to its controller that is connected to a PC. The analog output of the CCD chip is applied to the pre-amplifier. The output of such an amplifier is applied to the CDS circuit that processes the video waveform and converts to a digital number proportional to the size of the charge packet contained in the pixel (Holst, 1996). For smaller and faster acquisition systems, thermoelectric coolers (see Figure 6.9b) are employed.

6.2.3.2 CCD controller

The main functions of a CCD controller are to generate bias voltages, producing clock signals to transfer charge with high charge transfer efficiency, signal processing and digitization with low read noise, communicating with host computer, etc. Such an equipment is used for implementing essential operations, such as read-out of the CCD, interfacing with the host and a correlated double sampling (CDS; signal processing technique) for reset noise elimination. For example, it can be based on a digital signal processor (DSP) chip (Naidu and Srinivasan, 1997, 1998, 2001). The controller architecture has a host computer centered on a PC and a front-end electronics implemented with a DSP chip. The CCD camera system comprises the following subsystems.

DSP CPU board The main functions of DSP central processing unit (CPU) board are (i) to receive and execute the program, from the host, (ii) to load all the digital-to-analog converters (D/A converter or DAC) in the bias and clocks board, (iii) to generate the timing sequences for read-out of the CCD, (iv) to send the digitized data to the host, with a strobe signal, and (v) generate hand-shaking signals with the host computer.

Clock and bias board The key feature of any CCD control system is timing. Numerous events should occur in strict time order and at precise time intervals. The bias and clock board generates all the bias voltages required for a single CCD. The timing waveforms with required voltage levels for charge shifting and analog processes area also generated on this board.

Signal processing board The signal processing board implements various functions, such as DC offsets, CDS, bias offset, and digitization of the signal. The charge is converted to a voltage by a floating diode or floating diffusion. As stated in Section 6.2.1, the charge in every pixels is transferred through number of pixels (both in parallel and serial directions) to get to the output node where it is converted into a voltage proportional to the charge. The diode, which acts as a capacitor (see Section 2.2.7), which is pre-charged at a reference level. The capacitance or sense node, is partially discharged by the amount of charge transferred. The difference in voltage between the final status of the diode and its pre-charged value is linearly proportional to the number of electrons, N_e. If q is the charge and C_{n} is the node capacitance across the output gate, the voltage developed is $V = q/C_{\mathrm{n}}$. For one electron charge and 0.1 pf capacitance the voltage is 1.6 μV. In order to measure the charge collected under each pixel, the output node capacitor, C_{n}, is pre-charged to a reference voltage by turning on the reset transistor. Before the charge from a pixel is transferred to the node capacitor, it is recharged to a fixed potential by pulsing the reset transistor connected to it. This introduces noise in the measurement of the charge transferred. The thermal agitation of the electrons in the transistor causes a certain amount of uncertainty in the mean value of the final reset value on the capacitor. The reference level of the sense capacitor is, therefore, different from pixel to pixel, and the resulting noise is given by

$$n_{\mathrm{r}} = \sqrt{4k_{\mathrm{B}}T\,\Delta f R}, \tag{6.1}$$

where n_{r} designates the uncertainty in the amount of charge remaining on the capacitor following reset, k_{B} Boltzmann's constant,

T the absolute temperature, Δf the noise power bandwidth, and R the effective channel resistance.

Across the capacitance, C_{n}, the charge q has variance, that is

$$\langle \sigma_{\mathrm{q}}^2 \rangle = k_{\mathrm{B}} T C_{\mathrm{n}} \Delta f, \tag{6.2}$$

which is the fundamental limit of the noise across C_{n}.

In the absence of any noise, the dynamic range of this capacitor is written as

$$\text{Dynamic range} = V \sqrt{\frac{C}{k_{\mathrm{B}} T}}. \tag{6.3}$$

The reset noise can also be expressed in units of equivalent electrons at the detector output, by introducing the fundamental value for electron charge, q, that is

$$n_{\mathrm{r}} = \frac{\sqrt{k_{\mathrm{B}} T C_{\mathrm{n}}}}{q} \quad \mathrm{e}^{-}\ \mathrm{RMS}. \tag{6.4}$$

After shifting the accumulated charges serially along the output register, the charge is moved across the gate of the output field-effect transistor (FET; see Section 2.2.9) that functions as a switch and also a clock. The CCD containing a read-out circuit transfers the charges of a given column one after the other into an output series register and then moves on to the next column, until all the stored charges have been read. However, the output transistor has intrinsic noise, which increases as $1/f$-noise (see Section 4.5.2.4) at low frequencies.

Correlated double sampling (CDS): A correlated double sampling device is used in many analog circuits to reduce offsets and reset noise. It assumes that the same voltage fluctuation is present on both the video and reset levels. This may be performed in the analog domain by a clamping circuit with a delay-line processor. In the digital domain, a high-speed analog-to-digital (A/D) converter is required; it converts the electrons to a digital signal. However, the output signal is limited to the read-out noise (see Section 4.5.4), particularly at lower frequencies. The CDS technique removes an unwanted electrical signal, associated with resetting of the tiny on-chip CCD output amplifier, which would otherwise compromise the performance of the detector.

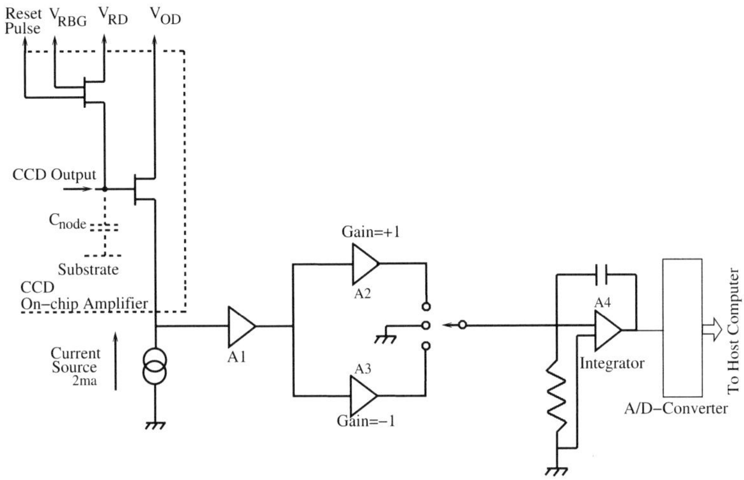

Figure 6.10 Concept of the signal processing.

Figure 6.10 describes the double correlated sampling, in which the signal from the output transistor is amplified by A1 and then fed to an integrator A4 through unity gain amplifiers of opposite polarity alternatively. First, the integrator is connected to A3 of gain $+1$ for a fixed period of time and then to the ground, while the charge is transferred to the output transistor. It is then connected to A2 of gain -1 for the same time period. Since A2 and A3 provide identical output signals of opposite sign for identical times, the integrator measures the difference of the signal present before and after the pixel charge is transferred. Hence, it effectively nullifies the noise factors due to the resetting process and the intrinsic property of the output transistor.

Analog-to-digital (A/D) converter: Also known as, digitization, the analog-to-digital conversion is the process by which charge from a detector is translated into a binary form used by the computer. The term binary refers to the base 2 number system used. The number of counts is dictated by $\Delta t(V_{\text{camera}}2^n/V_{\max})$, in which n is the number of bits provided by the A/D converter, $V_{\max}$ the maximum output, and Δt the exposure time. A 16-bit camera system would output $2^{16} = 65536$ as the maximum value. For applications requiring higher speed and less dynamic range that is limited by the number of digital bits of the output, 8 to 16-bit digitization is common. The higher the digital resolution, the slower the system throughout. The linearity of response of a CCD depends on the signal processing electronics, as well as on the analog-to-digital converter. The output of the integrator is connected to a fast A/D converter from which the signal is measured as a digital number by a computer. A buffer before the A/D converter presents a low impedance to the sample and hold of the A/D converter. An offset signal can be added to avoid clipping at its input. An anti-aliasing filter followed by the buffer, albeit before the A/D converter, ensures that this converter cannot envisage any high frequency noise caused by the buffer. A 16-bit A/D converter comprising a channel multiplexer (see Section 6.2.1) and a sampling device, may be used to digitize the processed signal.

Host interface card The host computer communicates with the DSP board in the CCD controller (for bootstrapping commands), through an asynchronous serial communication link. This link implemented with a serial link for sending commands to the CCD camera electronics and the transfer of image data from the camera electronics to the host using parallel data transfer with a strobe. The host interface handles

- acquire, which includes acquisition of bias, dark current, and expose,
- image display functions that allow to select a method to convert the image pixel values to screen pixels, zoom, and pan the image, and
- analysis functions; the plot menu provides the access to represent the image data graphically, for instance, a row or column plot displays pixel intensities along a straight line, while a histogram plot depicts the pixel intensity distribution in the image.

Data acquisition The primary function of the CCD data acquisition software is to send commands to the CCD controller (e.g., to time the exposures), acquire the digitized data from the controller and process the data to display and store the data.

FPGA-based Controller As an alternative to the DSP controller, approach of using a field programmable gate array (FPGA) offers significant improvement in terms of small size, low power consumption, enhanced reliability and image processing strategies. An FPGA is an integrated circuit (IC; see Section 2.3) comprising an array of logic blocks[a] together with an interconnect network; such blocks may include memory elements. Its configuration is, in general, specified by employing a hardware description language (HDL), similar to that used for an application-specific integrated circuit (ASIC). However, unlike the latter, where the device is custom built for the particular design, the former can be programmed to the desired application or functionality requirements. The FPGA offers the highest gate density (Hamblen and Furnan, 2000) and reduces hardware, reconfigurability, and multiplicity of operations as well. Applications of FPGAs include digital signal processing, medical imaging, bioinformatics, radio astronomy, telecommunications (Maxfield, 2004). The FPGA-based controller may be appreciable for building a space-borne application of a CCD camera.

A FPGA forms the front-end camera controller, which can be configured by the user at the point of application. Many of the functions currently performed by the Host can be incorporated into the FPGA. This would permit flexibility in changing the configuration as and when required. Also, the time taken for commissioning a camera can be reduced. The serial and parallel copper wire link that are generally used in DSP controller can be replaced by a high-speed fiber-optic link (see Section 3.4) in the FPGA architecture (Srinivasan et al., 2009). Apart from speeding up the communication, the noise interference to the data transmission is greatly reduced in this approach.

A VHSIC (very-high-speed integrated circuit) hardware description language, VHDL, is used in electronic design automation to describe digital and mixed-signal systems, such as FPGAs and IC (Creasey, 1985). It can be employed as a general-purpose parallel programming language as well. The VHDL has file input and output

---

[a] Logic block is a small circuit block that is replicated in a field programmable device (FPD), a type of IC used for implementing digital hardware. When a circuit is implemented in such an FPD, it is first decomposed into smaller subcircuits that can each be mapped into a logic block.

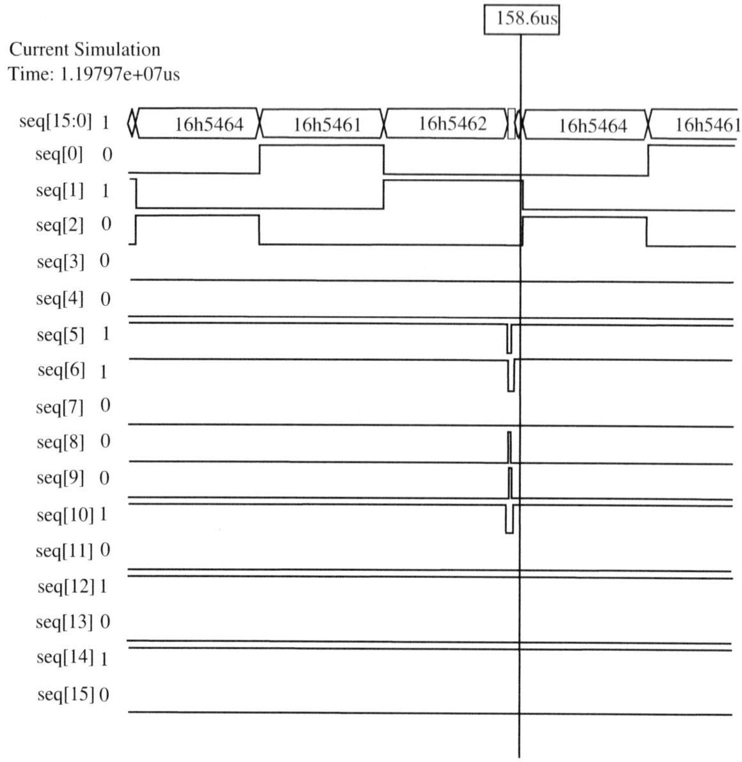

Figure 6.11 Parallel waveform generated with FPGA (Srinivasan et al., 2008).

capabilities and can be used as a general-purpose language for text processing. There are some VHDL compilers that build executable binaries. In this case, it may be possible to use VHDL to write a test-bench in order to verify the functionality of the design using files on the host computer to define stimuli, to interact with the user, and to compare results with those expected. Srinivasan et al. (2008) have written this code using the state machine and state transitions as required in the CCD controller. The generated waveform (see Figure 6.11) with FPGA contains the levels of 16 state bits at any given time. The patterns are time sliced and time field is then combined with state field to produce a combined word (24 bits).

Sequences of such words are used to represent the parallel transfer, serial transfer, or pre-flush waveforms.

6.2.4 *Read-Out Procedure*

As stated in Section 6.2.2.2, in CCD sensors, the signals of separate elements are transferred to a single output from which they are read out sequentially, one by one. All columns of data are read through the same set of electronics at end of the multiplexer (see Section 6.2.1) row. The charges are shifted horizontally from pixel to pixel onto an on-chip output amplifier, where the collected charge would be converted into a working voltage. These detectors perform close to an ideal detector, however the contribution of detector noise (see Section 4.5), which is limited by the electronic noise in the amplifiers, prevents them from being used in applications, such as very low-light level imaging. The transfer of photocharges, and also amplification, are accompanied by two sources of noise, whose quadratic sum is observed. There is a fluctuation in this number of charges, which has thermal origins and depends on the capacitance storing the charges. Furthermore, there exists a non-negligible fluctuation due to transfer and amplification of the signal, what is termed as the read-out noise (see Section 4.5.4) of the CCD. It is reiterated that the noise powers additive and equivalently, the noise sources are added in quadrature (Holst, 1996), that is

$$\langle n_{\mathrm{sys}}\rangle = \sqrt{\langle\sigma_Q^2\rangle + \langle\sigma_{\mathrm{patt}}^2\rangle + \langle\sigma_{\mathrm{r}}^2\rangle + \langle\sigma_{\mathrm{on-chip}}^2\rangle + \langle\sigma_{\mathrm{off-chip}}^2\rangle + \langle\sigma_{\mathrm{A/D}}^2\rangle}, \tag{6.5}$$

where $\langle\sigma_Q^2\rangle (= \langle\sigma_{\mathrm{ph}}^2\rangle + \langle\sigma_{\mathrm{D}}^2\rangle)$ the shot-noise (see Section 4.5.2.3) variance that include photon shot-noise and dark current shot-noise, which are caused by the random generation of electrons, $\langle\sigma_{\mathrm{patt}}^2\rangle$ the pattern noise variance, $\langle\sigma_{\mathrm{r}}^2\rangle$ the reset noise variance that can be reduced substantially by the correlated double sampling (see Section 6.2.3.1); the expression for the reset noise is given in equation (6.5), $\langle\sigma_{\mathrm{on-chip}}^2\rangle$ the on-chip amplifier noise variance, $\langle\sigma_{\mathrm{off-chip}}^2\rangle$ the off-chip amplifier noise variance, which is small compared to the on-chip amplifier noise, and $\langle\sigma_{\mathrm{A/D}}^2\rangle$ the A/D converter noise variance.

For system analysis, it may be prudent to treat the amplifier noise (see Section 4.5.3) and other noise present as noise floor, $n_{\rm fl}$. The simplified noise model is

$$\langle n_{\rm sys}\rangle = \sqrt{\langle\sigma_{\rm Q}^2\rangle + \langle\sigma_{\rm fl}^2\rangle + \langle\sigma_{\rm patt}^2\rangle}. \tag{6.6}$$

Ideally, the A/D converter noise is less than noise floor, which can be achieved by selecting a high-resolution A/D converter. The noise floor is a measure of the signal created from the sum of all the noise sources and unwanted signals within a measurement system. The read-out noise depends on the read rate; the noise level decreases as the CCD is read slowly. The read-out is white noise dominated, and thus the lowest read-out noise is achieved by closing down the system bandwidth, which implies low read-out speeds, typically about $3e^-$ RMS at a 20 kHz pixel rate. On modifying the electronic subsystems, the total system noise power is

$$\langle\sigma_{\rm sys}^2\rangle = \int_0^\infty S(f_{\rm e})|H_{\rm sys}(f_{\rm e})|^2 df_{\rm e}, \tag{6.7}$$

in which $S(f_{\rm e})$ is the total noise power spectral density from all sources and $H_{\rm sys}(f_{\rm e})$ the frequency response of the system electronics.

On assuming the noise is white over the spectral region of interest $[S(f_{\rm e}) = S_0]$, the noise bandwidth, $\Delta f_{\rm e}$ is expressed as

$$\Delta f_{\rm e} = \frac{\int_0^\infty S(f_{\rm e})|H_{\rm sys}(f_{\rm e})|^2 df_{\rm e}}{S_0}. \tag{6.8}$$

Here $\Delta f_{\rm e}(= 1/(2\tau_{\rm cl} = f_{\rm cl}/2)$, in which $\tau_{\rm cl}$ is the time between the pixels, applies to the white noise but not to the flicker noise (see Section 4.5.2.4). The array signal-to-noise (S/N) ratio is given by

$$S/N = \frac{N_{\rm pe}}{\sqrt{\langle\sigma_{\rm Q}^2\rangle + \langle\sigma_{\rm fl}^2\rangle + \langle\sigma_{\rm patt}^2\rangle}}, \tag{6.9}$$

with $N_{\rm pe}$ as the number of photoelectrons.

Conventional astronomy requires a slow frame read-out, typically tens of seconds for large CCDs, which are unsuitable for high time resolution imaging applications. High read-out rates ($\leq$1 megapixels.s^{-1}) are sought for the short-exposure imaging (see Section 1.5.5). The data rates are limited by the bandwidth of the amplifier, as well as the conversion capability of the A/D converter (see Section 6.2.3.2). It is reiterated that a CCD has one or two read amplifiers and all the pixel charges are to serially pass through them. In order to increase the frame rates from the CCD, different read modes like frame transfer and kinetic modes are normally used. Because of the architecture of CCD, even to read 10×10 pixels occupied by a single star, one has to read the entire device. This increases the reading time thus limiting the number of frames that can be read with this system.

In order to increase the effective read-out rate, without increasing the read-out noise, of course, the full frame may be divided into two, four, or eight subarrays, with a separate exit for each part; each subarray is read out simultaneously. Figure 6.12 depicts the basic principle of subarray read-out. The minimum integration time

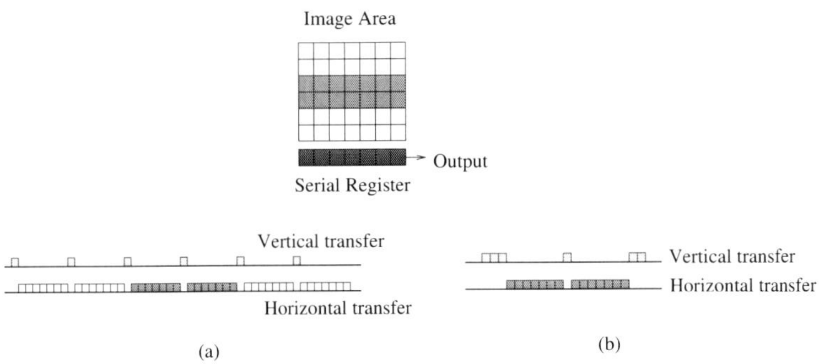

Figure 6.12 Principle of subarray; (a) Normal read-out and (b) Subarray read-out.

for a subarray depends of the location on the array. The necessary software is able to reconstruct the original image. The large array devices permit to select a particular subarray for read-out; thus one can trade off the frame rate for the image size. In this process, one is able to obtain high frame rates over the area of interest (subframe). The maximum frame rate, $F_{\max}$, is

$$F_{\max} = \frac{1}{t_{\text{array}}}, \tag{6.10}$$

where t_{array} is minimum time required to clock out the array.

In kinetic mode, region of interest, say 100×100 pixels can be read, which may be digitized by A/D converter after the charge is read from CCD, the remaining area charges are dumped out from

CCD after reading without being digitized. Let τ_r be the time taken to read the CCD; therefore,

$$\tau_r = N_x N_y(\tau_{sr} + \tau_v) + N_x \tau_i, \tag{6.11}$$

where N_x, and N_y are the number of pixels in x and y direction of CCD, respectively, τ_{sr} the time required to shift one pixel out to shift-register, τ_v the time taken to digitize one pixel, τ_i the time to shift one line into shift-register, and τ_s the time needed to discard a pixel charge.

The pixels of the output register are of bigger size to hold more electrons so that as many number of rows can be added (binned) to the output register. The pixel binning is a clocking scheme, which is used to combine the charge collected by several adjacent CCD pixels. The adjacent pixels are summed according to a pre-defined geometry (2×2, 4×4, etc.). It is designed to reduce noise and improve the S/N ratio, since the signal is increased for the same level of read-out noise (see Section 4.5.4) and frame rate of digital cameras. This mode, which reduces the volume of data collected and the read-out time. However, it reduces the spatial resolution of the final image. The hardware binning is the addition of electronic charges of several pixels generated directly on the CCD chip during the read-out, while the software binning is the addition of adjacent pixel intensities in the image memory after the image acquisition. The addition increases the number of electrons of one image element, and therefore the S/N ratio corresponding to the square root of the number of binned pixels. The advantage of the hardware binning is that of reducing the read-out time. However, combination of hardware and software binning factors results in a good compromise between the spatial and time resolution for acquisition of time-dependent image sequences.

6.2.5 *Characteristic Features*

The CCD is characterized by high system efficiency, large dynamical range (see Section 4.1.1.7), and linear response compared to other detectors. The other characterizations include quantifying the noise floor since it determines the minimum detectable signal, full well capacity, pixel non-uniformity (see Section 6.2.6.1), and signal-to-noise (S/N) ratio (see Section 4.1.1.4). The output of the CCD camera depends on both the system spectral response (see Section 4.4.1.3) and the color temperature of the illuminating source (Holst, 1996). The signal is modified by the spatial response of the optics and detector, while the noise originates in the detector and electronics. The noise characteristics may be different from the signal characteristics (see Section 4.3.2). The system design needs optimizing the S/N ratio, which depends on (i) the natural photon statistics of the incoming photons, (ii) the dark noise, and (iii) the read-out noise resulting from the CCD and the processing electronics. The sources of noise are a function of the detector temperature as well. The predicted performance of a CCD may deviate significantly from the actual performance in the presence of $1/f$-noise (see Section 4.5.2.4) or any other noise.

6.2.5.1 Charge transfer efficiency (CTE)

There are two interrelated quantities of interest in enumerating the transfer of charges from one electrode to the next. These are (i) time to transfer the charge and (ii) charge transfer efficiency (CTE) defined by the ratio of charge transferred from one electrode to the other and measured at the output to the charge collected in the pixels. The time constant, τ, for transfer of charge from one electrode to the next is of the order

$$\tau = \frac{L^2}{4D}, \tag{6.12}$$

where L is the linear dimension of the electrode and D the diffusion constant; for $L = 10^{-3}$ cm and $D = 10$ cm^2 s^{-1}; the time constant, τ, turns out to be 2.5×10^{-8} s.

It is assumed that the spacings and the applied voltages are such that no potential barrier exists between the electrodes. In addition, there is a field-enhanced component resulting from the change of surface potential with charge density. This effect results in a field parallel to the interface, which adds to the diffusion component and its effect is expected to be significant. In the surface channel CCD architecture (see Section 6.2.2), the electrons get trapped at the lattice irregularities near the surface. The result is very poor charge coupling and severe image smear. The CTE, which quantifies this, is sensor dependent, and an improper camera system design can cause less than optimum CTE. Let a discrete charge transfer device that is to a good approximation, discrete in both space and time, be considered. The perfect charge transfer implies

$$q(x, t) = q(x - 1, t - 1), \tag{6.13}$$

where $q(x, t)$ is the charge in station x at time t; x and t are assumed to be having integer values, in which the unit of the former is taken as the separation of the station (center-to-center) and the unit of the latter is taken as the stepping interval.

The real devices are typically characterized by the fraction, α, of the charge, which is successfully advanced per step. If a fraction, ϵ, is unable to move forward and remains in its original station, the process is described by

$$q(x, t) = \alpha q(x - 1, t - 1) = \epsilon q(x, t - 1). \tag{6.14}$$

A fraction $(1 - \alpha - \epsilon)$ of the charge is thus lost per step. With significant transfer inefficiency, the point images would get broadened and have their position shifted; the sinusoidal images would experience a shift in phase and the reduction of amplitude compared to the uniform background. The passage of the signal information through an imperfect CCD register can result in a degradation of the modulation transfer function (MTF) of the system.

The maximum charge that can be stored depends on the available density of state, which is related to the extent of band bending, i.e., on Ψ_s, which in turn, is dependent on the applied voltage, V. The surface density, N_{max}, is given by the following relationship:

$$N_{max} = \frac{\epsilon_{o'} V}{d}, \tag{6.15}$$

in which $\epsilon_{o'}$ stands for the dielectric constant of the insulating layer and d its thickness; for $SiO_2 (\epsilon_{o'}/\epsilon_o = 3.9)$, the maximum electric field for the oxide, V/d, is of the order of 10^6 V cm^{-1} and therefore, $N_{max} = 2 \times 10^{12}$ electrons per cm^2.

6.2.5.2 Quantum efficiency

The most appealing aspect of CCDs over other detectors is their great efficiency. Most of the CCDs that have been made are capable of registering above 50% across a broad spectral range from near-IR (1–5 μm), visible, UV, extreme UV to soft X-ray. Peak efficiency may exceed over 80% for some wavelength of light. In addition, the CCD is responsive to wavelengths in the region 380 nm to 1100 nm, where most of the other detectors have low quantum efficiency (QE; see Section 4.1.1.3). However, there seems to be a great reduction in sensitivity if the CCDs are used in the blue (<400 nm) since silicon absorbs photons at these wavelengths.

The quantum efficiency of the CCD is governed by the ability of the incident photons to be absorbed in the depletion region of the sensor. It is dependent on the gate voltage and the thickness of the material; low voltages produce small depletion regions. The long wavelength QE depends on the thickness of the substrate and the diffusion length. The light sensitivity is important for selecting a CCD camera system for applications in the low-light level region. Figure 6.13 depicts the quantum efficiency plot for Tectronix 1k×1k CCD device.

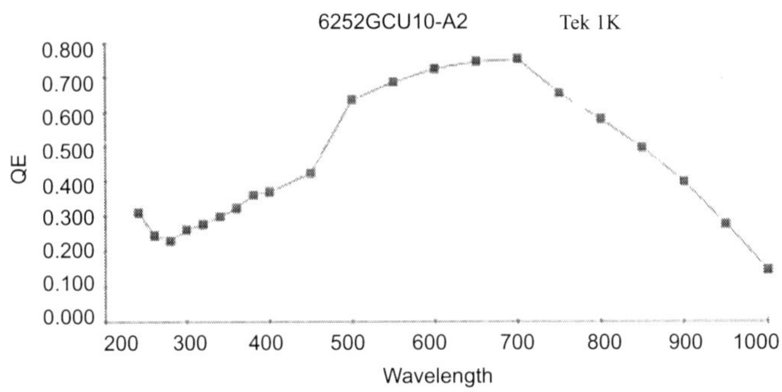

Figure 6.13 A typical quantum efficiency curve.

For the front-illuminated devices, the spectral responsivity (see Section 4.4.1.3) deviates from the ideal spectral response due to the polysilicon gate electrodes. These are more sensitive in the 500 to 800 nm range than they are to UV region. Since the electrodes and gates are in the path of the incident light, they absorb or reflect photons in the shorter wavelengths, and the spectral range becomes limited. There is absorption in the bulk substrate of 1–2 mm thickness, region below the front-sensitive area and photons absorbed in this area do not form part of the signal. In order to improve the UV response, the CCD can be given phosphor coatings, which absorb UV photons and emit visible photons. In the case of interline transfer devices (see Section 6.2.2.2), photodiodes do not have polysilicon gates; hence, they tend to have a response approaching the ideal silicon spectral response. There is no interference effect and the UV region response is restored.

On the contrary, a back-illuminated CCD has the quantum efficiency of ~85–90% around 600 nm. The CCD is thinned out from the back to ≈ 10 to 15 μm, and the illumination is from the back. Hence, there is an enhancement of QE since there is no loss either in the bulk substrate or in the electrode structure. Because of this thinning, the CCD starts showing interference effects from 700 nm up. A thinned CCD also requires anti reflection coatings with Hafnium oxide, aluminum oxide, lead fluoride, zinc sulfide or silicon monoxide to reduce the reflection losses, which are 60% in UV and 30% in visible. After thinning silicon oxidizes and a thin layer of SiO_2 is formed at the back. The presence of impurities in this oxide layer causes a net positive charge. This results in a backside potential well along with one at the electrode. The signal

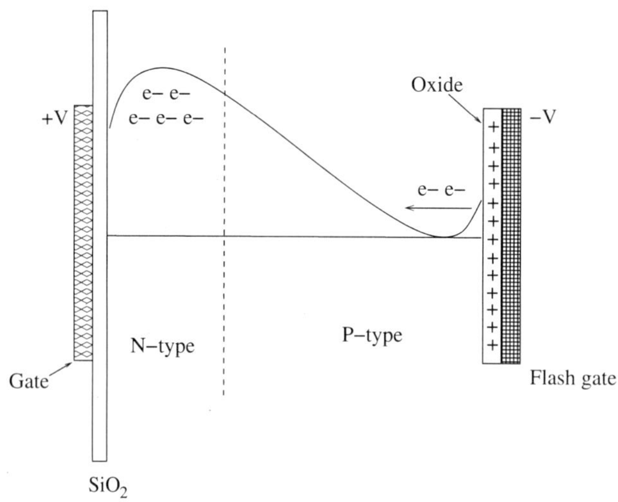

Figure 6.14 Flash-gate to counter backside potential well.

electrons diffuse into it and recombine resulting in low quantum efficiency. Hence, the back surface must be treated to compensate for the backside potential well. The abruptly broken bonds at the silicon lattice due to thinning need to be tied-up. An oxide layer is grown at the back. A backside discharge mechanism, such as UV flooding, corona charging, gas charging, flash-gate, biased flash-gate are then used to direct the signal electrons towards the electrodes. Figure 6.14 depicts the flash-gate to counter backside potential well.

6.2.5.3 Dynamic range

As stated in Section 4.1.1.7, the dynamic range is the ratio of a pixel's saturation level to the signal threshold, which provides an indication of the number of digitization levels that might be appropriate for a given sensor. It specifies the ability to measure very dim and very bright parts, and therefore the intensity range of a single image. For a CCD system, the dynamic range can be expressed as the ratio of the full well, N_{well} to the system noise, $\langle n_{\text{sys}} \rangle$, measured in electrons RMS. A system with a well depth of 85,000 electrons and a read-out noise (see Section 4.5.4) of 12 electrons would have a dynamic range of 20 log (85000/12) or 77 decibels (dB). The total camera noise is related to the array noise and the A/D converter noise, that is

$$\langle n_{\text{sys}} \rangle = \sqrt{\langle n_{\text{fl}}^2 \rangle + \langle n_{\text{A/D}}^2 \rangle}, \tag{6.16}$$

where n_{fl} is the noise floor and $N_{\text{A/D}}$ the A/D converter noise.

6.2.5.4 Gain

The CCD camera gain may be determined precisely by measuring signal and variance in a flat-fielded (pixel-to-pixel sensitivity difference) frame. The variance of the flat-fielded frame should be halved to account for the increase of the noise by square root of two in the difference frame. In spite of the negligible value of the read noise compared with the variance, an input guess is applied at the read noise. Different regions in the frame are selected randomly. A number of gain values are generated as well. These values are plotted in a histogram form. The value of the gain corresponding to the peak of the histogram is known as the system gain. The gain in the integrated amplifier (see Section 4.3.2.2), G is related to the variation in voltage between the reference level and the signal level (in volts), ΔV

$$G = \frac{\Delta V C_{\mathrm{d}}}{q N_{\mathrm{e}}}, \tag{6.17}$$

with q as the electronic charge, N_{e} the number of electrons per charge packet, and C_{d} the capacitance of the output diode.

The values of gain that are obtained from regions with defects give rise to erroneous values and fall outside the main histogram peak. The gain of the signal processing chain is selected so as to cover the range of the A/D converter used, as well as the full well capacity of the CCD pixel. To note, the full well capacity is the maximum number of electrons that one pixel can contain before saturation depending on the size of the pixel, the kind of CCD image sensor, and its operating characteristics. In CCD devices, the electrons over the saturation charge flow to the adjacent pixel with the resulting effect of blooming and influence the qualitative and quantitative image performance. The V_{max} corresponds to a full well, N_{well} (Holst, 1996)

$$V_{\mathrm{max}} = G' \frac{Gq}{C_{\mathrm{s}}} N_{\mathrm{well}}, \tag{6.18}$$

where G' is the off-chip amplifier gain.

6.2.5.5 Responsivity

The average responsivity (see Section 4.1.1.2), $\langle \mathcal{R} \rangle$, at the cut-off wavelength, λ_c, is expressed as

$$\langle \mathcal{R} \rangle \approx \frac{2^n}{N_{\mathrm{well}}} \frac{\lambda_c}{1.24} \mathtt{A}_{\mathrm{d}} \eta_{\mathrm{d}}(\lambda_c), \tag{6.19}$$

with η_{d} as the quantum efficiency of the CCD and $\mathtt{A}_{\mathrm{d}}$ the active area of the CCD.

For the arrays, the output may be normalized to photometric units (see Section 1.2.2). The array voltage output, V_s is given by

$$V_{\mathrm{s}} = \mathcal{R}_{\mathrm{ph}} \left(\frac{683}{\mathtt{A}_{\mathrm{d}}} \int_{0.36\ \mu\mathrm{m}}^{0.8\ \mu\mathrm{m}} \mathcal{E}_{\lambda} V(\lambda) d\lambda \right), \tag{6.20}$$

in which $\mathcal{E}_\lambda$ is the power in units of $W.\mu\text{m}^{-1}$ and $V(\lambda)$ the spectral luminous efficiency for photopic vision.

On combining equations (4.93) and (6.19), the photometric responsivity, $\mathcal{R}_{\text{ph}}$ is derived as

$$\mathcal{R}_{\text{ph}} = \frac{\dfrac{G}{C}\dfrac{\mathtt{A_i}}{\mathtt{A_d}} s_o^2 \displaystyle\int_{\lambda_1}^{\lambda_2} \mathcal{L}_e(\lambda)\mathcal{R}(\lambda)\Delta t d\lambda}{\dfrac{683}{\mathtt{A_d}} \displaystyle\int_{0.36\ \mu\text{m}}^{0.8\ \mu\text{m}} \mathcal{E}_\lambda V(\lambda) d\lambda} \quad \text{V.lx}^{-1}, \qquad (6.21)$$

where $\mathcal{L}_e(\lambda)$ is the spectral radiance (see Section 1.2.1), $\mathcal{R}(\lambda)$ the spectral responsivity (see Section 4.1.1.2), and Δt the integrating time.

6.2.5.6 Dark current

Performance of a CCD depends on the dark current (see Section 4.1.1.8) of the device, whose output signal in the dark is caused by thermally generated minority carriers, as well as by the background generation. These minority carriers partially fill up the well and cause pattern noise. The contribution of thermally generated electrons in the surface states at the Si–SiO_2 interface is large. Most of the electrons, which are generated thermally deep in the CCD (neutral bulk), are diffused and recombined before they come under influence of the nearest potential well. The thermal electrons generated in the deep depleted regions are further diffused, and some of them may be collected by the neighboring pixels. The dark current generation due to the surface states depends on the density of the surface states at the interface and the density of the free carriers that populate the interface. This contribution of dark current can be substantially reduced by the passivation technique, which makes a material passive in relation to another material prior to use the materials together, at the time of fabrication of the CCDs, or by operating the device in the inversion mode. In the case of CCDs, such a technique is used to reduce the number of the interface states by growing a thin layer of oxide to tie-up the dangling bonds at Si–SiO_2 interface. When a gate is biased such that the surface potential of the phase is equal to the substrate potential, the N-channel at the Si–SiO_2 interface gets inverted, i.e., the holes from the nearby channel stops are attracted and pinned at the surface. This pinning condition eliminates further hopping of the electrons from the valance band to the conduction band, and thereby, reduces the dark current. If the two barrier phases in a three phase CCD are biased into the inversion state (partial inversion operation), the rate of dark current generation decreases two-thirds of the non-inverted mode of operation.

Dark current builds up with time and the acquired frame would become saturated, if the device is not cooled even for a few seconds integration time. The dark current can be measured by taking exposures with the shutter closed, subtracting of which from the observed image provides the real number of electrons due to incident light. However, the intrinsic dark current is negligibly small for the short-exposure used in interferometric experiments (see Section 1.5.5), but the thermal background signal may pose a problem if the detector sees large areas at room temperature. The dark signal must be calibrated on average and subtracted from the signal, what is referred to as doing a dark frame subtraction.

6.2.6 *Calibration of CCD*

The linearity of a CCD can be measured by illuminating it with a very stable luminous source and by tracing the detector's response as a function of integration time. The limitation of CCDs, of course, comes from the heterogeneities of response, which is due to unidentical elements. In general, a scientific grade CCD has more than 99% operative pixels. The pixels have a Gaussian distribution (see equation 4.45) of read noises, but with a tail extending to high values.

6.2.6.1 Sources of non-uniformities

The following sources of non-uniformities are required to be removed in a series of processing steps:

Bias variations They have fixed amplitudes, independent of signal level, exposure time, and elapsed time. The electronic bias needs to be subtracted to eliminate signal registered by the detector in complete darkness. The required bias image, $B(\mathbf{x})$, is constructed by averaging a series of zero exposure images. Such exposures are averaged out in order to mitigate random noise. A pixel by pixel bias subtraction is done to obtain a bias subtracted data.

Pixel-to-pixel variations A large number of pixels in CCD arrays can contain defects (Holst, 1996). Every pixel cannot be used effectively due to the possible imperfections located within the pixel. The common encountered problems are: high dark current, rate saturating the pixel's potential well; non-linear dark current dependencies (Widenhorn et al., 2010; Popowicz, 2011; Dunlap et al., 2012); transient events of the dark current due to the irradiation (Hopkinson et al., 2008), pixel non-linear light response and CCD fabrication defects (Janesick, 2001). There are two types of defects, namely:

(1) **Point defects** (see Section 2.2.2.1): The point defects may arise during the manufacturing of the CCD (the impurities in technology processes) and also can be the result of the high-energy particles striking the matrix and causing a vacancy or an atomic displacement as a result of a collision. The output of such a pixel deviates more than 6% compared to the adjacent pixels. There exist hot pixels as well, which are abnormally receptive with high output voltages; typically the dark current in such a pixel is 10 times higher than the average dark current. The dead pixels either do not work or have low output voltage; they may have poor responsivity.
(2) **Spatial defects**: The spatial defects (dislocations; see Section 2.2.2.3) can occur during the fabrication process (Giannattasio et al., 2002) and can be envisaged in the form of column (or line) defects in which several, typically 10 or more point defects occur in a single column. The cluster defects where the cluster of pixels with point defects may also appear. The traps interfere with the charge transfer process and results in either a partial or entire bad column.

In order to identify such elements, calibration may be made by analyzing flat-fields with various intensities and spectral distributions. The impact of bad pixels (Hilbert, 2012) on the image quality can be reduced by taking many images with slight shifts. Such a set of dithered pictures is shifted back and averaged, ignoring the pixels from the mask (Fruchter and Hook, 2002); the bad pixel masks are created to help during the image reduction. However, it is improbable if there is a single image and in such cases an interpolation over bad pixels is necessary. A popular method is a linear interpolation, which is a standard procedure in the astronomical software package, called Image Reduction and Analysis Facility (IRAF; Massey, 1997). The large scale response variations

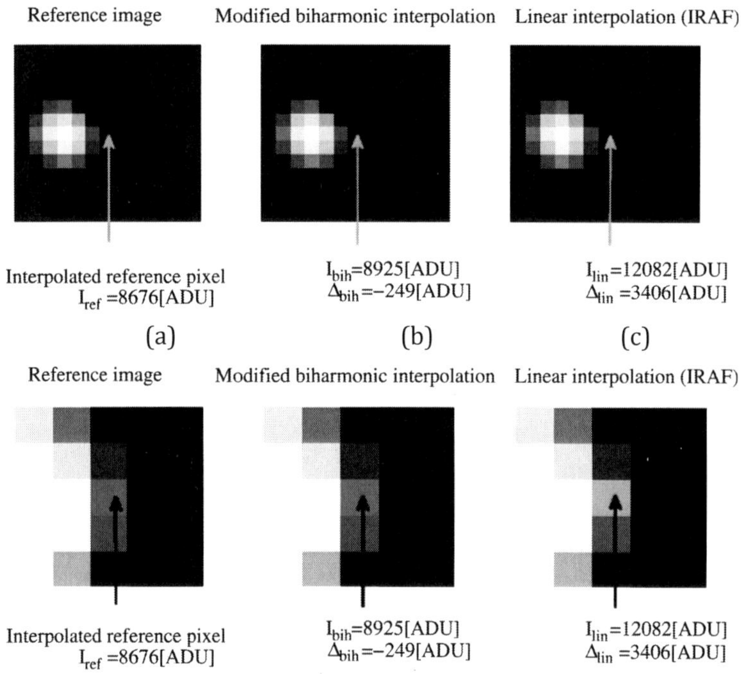

Figure 6.15 Exemplar comparison: (a) reference image with pointed central pixel (treated as bad pixel), (b) image with a central pixel corrected by Popowicz et al. (2013), and (c) image with a central pixel linearly interpolated by IRAF with corresponding values and errors labeled (Courtesy: A. Popowicz); the figures have different magnification (Top: with a slightly lower magnification to see whole star and bottom: higher magnification). The interpolated pixel is pointed by an arrow. Here ADU designates for analog-digital units and is a measure of signal from a CCD/CMOS pixel.

are removed through division by flat-field factor, F. A pair of identical such frames close to full well capacity ($\sim$80%) are required to be obtained. By subtracting these frames pixel by pixel, pixel non-uniformity is removed. Usually a small window of 20×20 pixels in the flat-fielded frame is selected to compute variance. The mean signal counts from the original two frames are obtained.

Recently, Popowicz et al. (2013) have proposed an improved version of interpolation for the pixel brightness estimation in astronomical images; it can also be applied to single image correction. They made the error absolute and relative value histograms for given interpolation methods, which can be seen in Figure 6.15. In this a mean error, $E_{\mathrm{r}}(= \sum_{i=1}^{N} |\Delta_i)$, in which Δ_i is the estimation error (absolute or relative) and N the total number of estimations, was computed to enable a comparison of the methods' effectiveness. The brightness estimation was compared with the true value to calculate an error.

Thermal agitation The thermal noise (see Section 4.5.2.1) is an important effect of presence of defects in CCD pixel causing the accumulation of additional charge not related to the photoelectric effect (Ong and Pierret, 1975). However, in the case of dislocations, the situation is complex; knowledge of some dislocation properties, for example, position and the arrangement in depletion area are required. The specific defects seating on the edge of the depletion area in a pixel may cause a gradual decrease in the rate of thermal generation due to the shrinking of the depleted area (Widenhorn et al., 2010; Dunlap et al., 2012). The thermal agitation in a CCD produces free electrons in a manner that varies from one pixel to another, as well as non-zero electronic noise. The calibration for thermal noise is also carried out by aforementioned manner, but with same integration time and at the same temperature as the actual astronomical observations. This offset map is used in the reduction of the observed data. In order to fulfill the condition of performing the observations at the same temperature it may be necessary to produce several offset maps for different observing conditions before and after the observing session.

Zero variations These are caused by electronic or temperature drifts, which depend on the elapsed time, the temperature stability of the array, and its read-out electronics, which can be mitigated by appropriate use of reference pixel outputs (Reike, 2007). A set of short-exposure images are generally taken in the dark, which can be averaged to obtain a high signal-to-noise image of the offset variations. Such an image can be subtracted from the image of the target source.

6.2.6.2 Flat-field corrections

The flat-field images are debiased individually and combined by weighted averaging. The main advantage of using several well exposed images is to construct the template image, which reduces statistical errors introduced during the division. The normalized flat-field map is used to reduce the astronomical observations (Chromey and Hasselbacher, 1996). The actual stellar counts, $D(\mathbf{x})$, is determined by

$$D(\mathbf{x}) = \frac{R(\mathbf{x}) - B(\mathbf{x})}{F(\mathbf{x}) - S(\mathbf{x})}, \tag{6.22}$$

in which $\mathbf{x}(= x, y)$ denotes the 2-D position vector, $R(\mathbf{x})$ the raw CCD image, $B(\mathbf{x})$ the bias image, $F(\mathbf{x})$ the flat-field image, and $S(\mathbf{x})$ the sky background.

The sky background, $S(\mathbf{x})$, interpolated for the position spectra is subtracted from the debiased and flat-fielded image to obtain the stellar counts, $D(\mathbf{x})$. In the case of high-resolution stellar spectra, a least square low order polynomial fit to the sky data at side of the object spectrum is to be obtained. The 1-D stellar spectrum is extracted out from the 2-D image by summing contributions from a range of spatial pixels containing object spectrum. Such a spectrum is calibrated to wavelength scale using the coefficient obtained by fitting a low order polynomial to the comparison spectrum (known wavelength). This 1-D wavelength calibrated data is used for fitting the continuum in order to determine equivalent widths of stellar lines.

The image factor, F, is constructed from several flat-field images obtained by exposing CCD to a specially uniform continuum source. There exit a few ways of obtaining field frames, for instance, (i) the twilight flat method, in which images of the twilight sky that should be a reasonably uniform light source, are taken, and (ii) use of a illuminated screen in the telescope dome. Also, scattered light that reaches the CCD without going through the complete optical system, needs to be corrected for. It is necessary to image a bright star near zenith on a clear night at a number of positions across the CCD. For each image, one may look at the brightness of the star on a field that has been bias- and dark- subtracted, but not flat-fielded, following which the flat-fielding can be carried out and look at the constancy of the star brightness across the chip.

The saturated image, which occurs due to large signal generated electrons filling the storage, should be avoided. The discharge is deferred so that the recognition of pixel is biased resulting in displacement of image. To add to this misery, a ghost image of the saturated field may remain for several hours on the CCD. In order to eliminate such an effect, a pre-flash that makes the effect uniform over the whole array, should be done. All calibrations may be repeated after a pre-flash.

6.3 CMOS Sensor

Like CCDs, the complementary metal-oxide semiconductor (CMOS; see Section 2.3.3) imaging devices, made from silicon, are formed from a grid of light-sensitive elements, each capable of producing an electrical signal. The latter devices are popular in the consumer market (Litwiller, 2001), although the former sensors have a firm position in scientific applications (Janesick, 2001). The operating principle of these sensors is based on the photoelectric effect (see Section 2.1). Each CMOS pixel employs a photosensitive diode, a capacitor, and a few transistors, which are integrated inside. Prior to the commencement of the integration period, the capacitor is charged to a known voltage, following which the charge is allowed to slowly drain away through the photodiode; the rate of drain is directly proportional to the level of incident light. At the end of the integration period, the charge remaining in the capacitor is read out and digitized (Hurwitz et al., 1997). There exist two forms of CMOS image sensor:

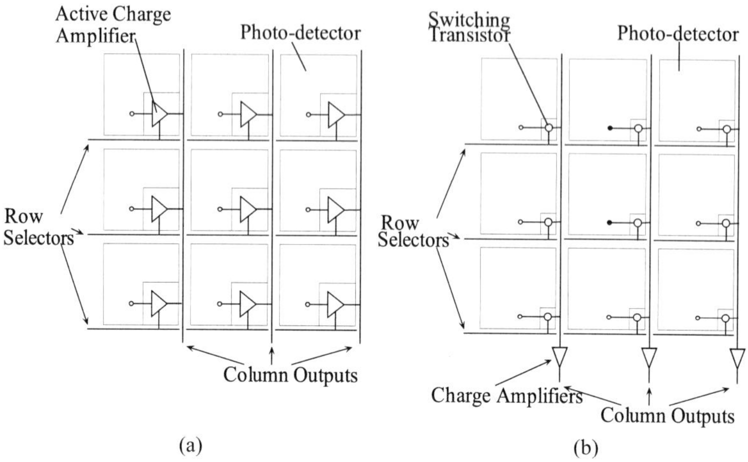

Figure 6.16 (a) Active pixel array and (b) passive pixel array.

Active pixel The active pixel contains an amplifier. The charge in each pixel can be read on an x, y-coordinate system that enables to detect a photon directly, converts it to a voltage and transfers the information directly to the output. Figure 6.16a depicts the active pixel array.

Passive pixel A passive pixel array (see Figure 6.16b) comprises the photodetector without the amplifier inside. These devices have charge amplifiers at the bottom of each column of pixels, with each pixel having a lone transistor in addition to the photodiode and capacitor. Such a transistor is used as a charge gate and switches the contents of each pixel's capacitor to the charge amplifier.

However, the addition of circuitry inside the chip may lead to more structured noise, such as stripes and other patterns, and also increases the complexity of the design. The area occupied by such a circuitry is insensitive to light. The effect of this greatly reduces the fill-factor (the ratio of the total light-sensitive area to the total area of the chip) of each pixel. This implies that the overall light sensitivity of CMOS sensors is smaller ($\sim$60%) than that of CCDs (close to 100%). In order to alleviate this problem, micro-lenses are incorporated directly in front of the pixel in an attempt to focus the incident light towards the active portion. Such an arrangement may increase the captured light by a factor of two at least. As stated earlier, the CMOS pixels are sensitive to photons, but not to color. So, they capture light creating a black and white image. In order to distinguish between colors, filters are put on top of a pixel permitting certain colors to pass, turning the rods (see Section 4.2.1.1) of the array into cones. The individual color schemes are deposited on top of the pixel array. After being read from the sensor, the software takes the values and recombines the colors to match the original picture.

The read-out procedure for both active and passive pixel arrays is same. Each of the row selectors are sequentially clocked, which in turn, causes the switching transistor (or charge amplifier) for that row of pixels to activate, and thus transfer each pixel's charge to the column outputs. Following this, a read-out register serially transfers the column output values to an A/D converter. It is pertinent to note that clocking the row selectors sequentially allows the full image to be progressively read out from the array. It is also possible to clock a limited number of row selectors, which has the effect of reading out a tiny horizontal strip from the array. By discarding pixel values from the beginning and end of each of these rows, a specific area of the image can be obtained. Sampling a small number of relevant pixels in such a method enhances the read-out speed (Graydon, 1997).

In the visible region, the CMOS technology couples a silicon photodiode with the MOS-transistors, and fabricates in the same Si-substrate. Each pixel on a CMOS sensor has several transistors located next to it, due to which the light sensitivity tends to be lower. Many of the photons hitting the chip hit the transistors instead of the photodiodes. In CMOS area- and line-scan imagers, the front-end of the interface between the imager and the off-chip circuitry (data path) is highly parallel allowing each amplifier to have low bandwidth. By the time the signal reaches this data path, CMOS data are in the digital domain.

The responsivity (see Section 4.1.1.2) of a CMOS is marginally superior to CCD sensor, but is susceptible to noise since each CCD amplifier has higher bandwidth and also requires better uniformity. Of course, the latter has a better dynamic range, while the former has the advantage in the case of the speed that is required for wavefront sensing, as well as the ability to read out a portion of the image sensor. The notable other characteristics of CMOS devices are high noise immunity and low power consumption. Intensified-CMOS camera (see Section 7.6.2) can also be used for recording images of the photon-starved astronomical sources (Hain et al., 2007).

The CMOS technology was first developed for the near-IR wavelengths, and was extended to the mid- and far-IR. Based on the photovoltaic (PV) effect (see Section 4.3.1.1), these detectors couple a photodiode (see Section 5.3.1) with a read-out circuit made from silicon, the whole thing being in a large format array (called hybrid array; see Section 5.5.2). The concept of a hybrid focal plane array (FPA; see Section 5.5.1) that is combining a CMOS-based read-out integrated circuit with a photodiode detector array has been applied to several image sensors. In order to obtain an image in such wavelengths, the imagers need to have a thicker photon absorption region since the photons in these wavelengths are absorbed deeper than visible photons in Si-substrate. Increasing the thickness of the epitaxial layer (substrate) to improve the IR sensitivity would degrade the ability of the imager to resolve spatial features. The thicker substrate needs to be coupled with higher pixel bias voltages. However, changing the voltage or epitaxial layer doping would affect the operation of the CMOS analog and digital circuits. In order to avoid dark current (see Section 4.1.1.8), very low energies involved (a fraction of an electron-volt) make it necessary to cool (see Section 4.3.1) the devices.

6.4 Intensified CCD

All modern image intensifier (see Section 5.2.6) tubes consist of three main components, a photocathode, a micro-channel plate, and a phosphor screen. The characteristics of each component would contribute to the overall quality of the resulting image. In short-exposure (see Section 1.5.5) imaging, usage of such an intensifier is a necessity. As stated in Section 5.2.7, in an MCP image intensifier, the photoelectron is accelerated into its channel releasing secondaries and producing an output charge cloud. However, it is incapable of storing information, so it is required to be optically linked (Talmi, 1994) to a CCD with either a fiber-optic bundle (see Section 3.4) or a relay lens; the combined unit is referred to as an intensified CCD (ICCD; Lemonier et al., 1988). Each fiber is in contact with the phosphor and transmits the photons corresponding to one pixel to the CCD. Although, the fiber-optic bundles reduces camera size and weight, fiber-coupling is complex since it requires critical alignment of the fibers to the CCD pixels. The fiber area and pitch should, in principle, match the CCD detector size and pitch for the maximum sensitivity and resolution. Typically, the fiber diameter is one-half of the pixel pitch, which reduces sensitivity. There exist a few spatial samplers in the ICCD, for example, (i) sampling by the discrete channels in the MCP, (ii) the fiber-optic output window within the intensifier, (iii) the input to the fiber-optic minifier, and (iv) the CCD array (Holst, 1996). The image is resampled at these interfaces. Depending on the location of the minifier, Moiré patterns can develop and the modulation transfer function (MTF; see Section 1.3.1) can degrade. On the contrary, the lens-coupled system, although relatively heavy, avoids the alignment problems and does not produce Moiré patterns.

The spectral response of an ICCD is determined by the intensifier photocathode responsivity. The responsivity depends on the window material, as well as the photocathode material; the former controls the short-wavelength cut-off, while the latter controls the long wavelength cut-off. In a frame transfer (see Section 6.2.2.2) ICCD operates at a standard CCIR video output of the recorded scene. The interface between the intensifier screen and the CCD chip is a fiber-optic bundle that reduces the image size by a factor of 1.69. The MCP gain is of 10^4 to 10^5 and the taper loses about a factor of 10 of those. The video frame grabber cards digitize the video signal and store the images in the memory buffer of the card. Depending on the buffer size, the number of interlaced frames (see Section 6.2.2.3) stored in the personal computer (PC) can vary from 2 to 32 (Saha et al., 1997a). By changing the voltage on the image intensifier, the gain can be varied. The electronic shutter action is produced by pulsing the MCP voltage.

A speckle camera system, developed by Saha et al. (1999) for the 2.34 m Vainu Bappu Telescope (VBT), VBO, India, is displayed in Figure 6.17a, while Figure 6.17b illustrates speckles (see Section

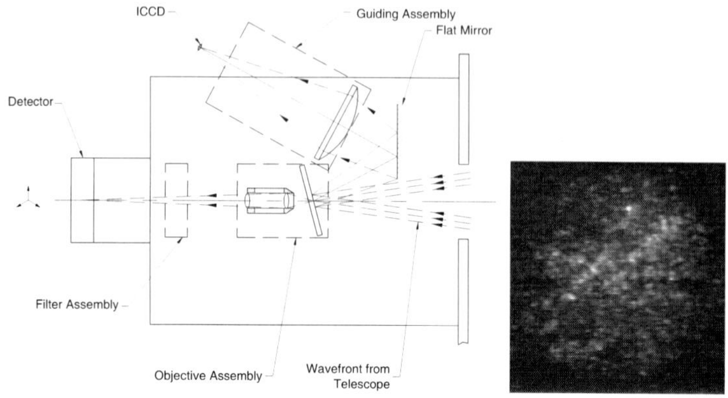

(a) (b)

Figure 6.17 (a) Optical layout of the speckle interferometer (Saha et al., 1999) and (b) instantaneous specklegram of a close binary star HR 4689 taken with (a).

1.5.5) of a binary star HR 4689, recorded on February 21, 1997, with the same instrument (Saha and Maitra, 2001).

Speckle interferometer A speckle interferometer is a high quality diffraction-limited camera where magnified (~f/100) short-exposure images can be recorded. Additional element for atmospheric dispersion corrections is necessary to be incorporated. At an increasing zenith distance speckles get elongated owing to this effect. Either a pair of Risley prism[a] or a narrow-band filter is used. In this instrument, the wavefront falls on an optical flat having an aperture of ~350 μm diameter. The image of the object passes on to the scientific camera via a microscope objective and a narrow-band filter. The surrounding star field of diameter 10 mm, gets reflected from the optical flat on to a plane mirror and is re-imaged onto an ICCD.

[a]Risley prism is a beam steering device comprising a pair of rotatable wedged elements that redirect the optical beam by refraction. The refracting angle of each wedge is typically less than 5°. By rotating each wedge independently, such a beam can be steered to any position within a solid angle defined by the maximum deflection.

Unlike bare CCDs, where the signal-to-noise (S/N) ratio (see Section 4.1.1.4) is strongly influenced by the read-out noise and is also a significant function of the pixel rate, the ICCD cameras are not affected by the CCD chip temperature and are shown to be by and large photon-noise (see Section 4.5.2.3) limited since the read-out (see Section 4.5.4) and dark current noises are negligible. However, the intensifier tube limits the ICCD resolution. Such a tube is circular, and since the CCD array is rectangular, it cannot sense the entire amplified scene. Also, the quantum efficiency of such an intensifier is low ($\sim$12% at 500 nm).

In order to obtain high quality images from the speckle image reconstruction, correction of the photon-noise bias should be performed with sufficient accuracy. In the case of high light level regime, in which a bare CCD is used, the photon-noise bias is very small; hence, no correction is required, while in the case of low-light level data sets, the signal sits on a bias, whose amplitude is proportional to the number of detected photons. Its shape is determined by the point response of the detector; the fainter the detected source, the larger the ratio of bias to signal, and therefore the greater the accuracy of the correction needed.

With an ICCD system, the number of photons should be estimated from the power spectrum (see equation 1.145) of the data, and the detected shape of the photon is determined from the combined point spread functions of the intensifier, optical relay system, and the CCD. The intensified data increases the bias level by the factor of several hundreds. Estimates of the shape of the bias are needed to be produced using low-light level flat-fields image (see Section 6.2.6.2), recorded at the time of the data recording, although inaccuracies in these estimates remain at large. The detector noise (see Section 4.5) may cause additional bias terms in the average power and bispectrum (Lohmann et al., 1983; see Section 1.5.5). Hofmann et al. (1995) recovered Fourier modulus from the bias compensated average power spectrum and Fourier phase from the bias compensated average bispectrum.

Guiding unit An ICCD can be used as detector for remote guiding unit of a telescope for a wide-field coverage (Chinnappan et al.,

1991), which helps in detecting the instrumental errors, such as improper tracking of telescope, obstruction in the light path of the main detector, and the necessary corrective action can be taken. A drift in star's position across the focal plane detector due to inaccurate tracking rate or an unbalanced telescope, as well as due to disturbances to the telescope like a gust of wind or movement of observers at the prime cage can also be corrected. An auto-guider generates error information proportional to the change in image position and feeds this error in the feedback loop of the telescope drive electronics. The ICCDs are also being used in a wide range of fields, namely, engineering research and physical or biological sciences.

Chapter 7

Photon-Counting Systems

7.1 Introduction

High-resolution imaging is an essential requirement in many scientific applications. However, detection of a celestial object in the presence of atmospheric turbulence (see Section 1.5.3) is a difficult task, more so, in the case of spatial interferometry (Labeyrie, 1975, 1986; Labeyrie et al., 2006; Saha, 2002, 2010, and references therein) in particular, since the interference fringes must be gathered within a few milliseconds, or the source would move from the field-of-view (FOV). Such an interferometry requires detection of very faint signals and reproduction of interferometric visibilities to high precision and, therefore, demands detectors and electronics with extremely low noise.

Also, an adaptive optics (AO) system (see Section 1.5.6) needs such a detection system; however, as stated in Section 4.5, the limitation of measurement arises from the detector noise. In a system that consists of a segmented mirror controlled by a Shack–Hartmann sensor, let θ'' be the width of a subimage. The mean square angular error, $\langle\sigma_{\theta'}\rangle^2$, on local slopes is of the order of θ'^2, and

High-Resolution Imaging: Detectors and Applications
Swapan K. Saha

ISBN 978-981-4613-27-9 (Hardcover), 978-981-4613-28-6 (eBook)
www.panstanford.com

therefore, the mean square angular error, $\langle \sigma_{\theta'} \rangle^2$, is expressed as

$$\langle \sigma_{\theta'} \rangle^2 = \frac{\theta''^2}{p_n}, \tag{7.1}$$

where $p_n (= N_{\text{ph}} d^2)$ are the independent photon events provided by the guide star, N_{ph} the number of photons, and d the size of the subaperture over which an error, θ', on the slope angle produces an error, $\delta = \theta' d$, on the optical path with variance

$$\langle \sigma_\delta \rangle^2 = \frac{\theta''^2}{N_{\text{ph}}}. \tag{7.2}$$

In the presence of atmospheric turbulence, the resolution of a large telescope is dictated by the atmospheric coherence diameter, r_0; hence, the minimum number of photons, N_{ph}.cm^{-2}, is given by

$$N_{\text{ph}} = \frac{4\pi^2}{0.134} d^{-5/3} r_0^{-1/3}, \tag{7.3}$$

in which d is the spot size.

It is reiterated that the flux of a zero magnitude star with spectral type, A0, at $\lambda = 0.63$ µm, the value at which silicon detectors have maximum efficiency, is 2.5×10^{-12} W.cm^{-2} per µm bandwidth (Johnson, 1966) and the photon energy, hc/λ, in which c the velocity of light, and λ the wavelength of light, is calculated as 3.5×10^{-19} joules. It is noted from the Section 1.1.3 that for a zero magnitude star, the flux, F_0, with spectral type A0 is written as

$$m - m_0 = -2.5 \log \frac{F}{F_0}, \tag{7.4}$$

where m and m_0 are the apparent magnitudes of two stars of fluxes F and F_0, respectively.

The term, F/F_0 in equation (7.4), is the ratio of the observed stellar flux over the flux given by a zero magnitude A0 star. By dividing F_0 by the photon energy, the corresponding photon flux turns out to be, $F_0 \approx 8 \times 10^6$ photons s^{-1} cm^{-2}. This value can be placed in this expression, in order to derive the photon flux of a given star, that is

$$F = 8 \times 10^6 \times 10^{(3-0.4m_v)}. \tag{7.5}$$

The number of detected photons, N_{ph} per cm^2, for a star of visual magnitude, m_v, striking the Earth's surface is

$$N_{\text{ph}} = 8 \times 10^{(3-0.4m_v)} \Delta t \eta_{\text{tr}} \int \eta_{\text{d}}(\lambda) d\lambda \qquad \text{cm}^{-2}, \tag{7.6}$$

with Δt as the integration time (seconds), η_{tr} the transmission coefficient of the system, $\eta_{\text{d}}(\lambda)$ the quantum efficiency of the detector and the integral is over the detector bandwidth expressed in nanometers; the number of stars brighter than 12 m_{v} is $1.45e^{0.96m_{\text{v}}}$ stars rad^{-2}, according to which there are 150,000 stars rad^{-2} brighter than 12 m_{v}.

With the help of equations (7.4) and (7.6), one derives

$$10^{-0.4m_{\text{v}}} = \frac{3.68 \times 10^{-2}\, d^{-5/3} r_0^{-1/3}}{\Delta t \eta_{\text{tr}} \int \eta_{\text{d}}(\lambda) d\lambda}. \tag{7.7}$$

An intensified CCD (see Section 6.4) can be used as a wavefront sensing device from which phase errors in the incoming wavefront are computed; the data is used to correct the image on-line. This would enable narrow-slit widths in high-resolution spectrograph and increase the signal-to-noise (S/N) ratio (see Section 4.1.1.4) considerably (Beckers, 1993). A number of various device architectures have been developed to optimize resolution, sensitivity, and other parameters, which run as follows.

7.2 Photon-Counting Methods

As stated in Section 1.1, a light signal consists of the emission of photons for a period of time that may be over nanoseconds (ns) or even hours. The rate of emission of photons over a certain time contains all the signal information, such as intensity, time structure, and a total light output. The perfect detector would be the one that is able to detect individual photons (or photoevent) in an image plane, thus giving the maximum possible S/N ratio. This ability is called photon-counting. It is a quantum process containing intrinsic amplification without noise (avalanche) and exhibits a certain spectral selectivity $\leq$100). The photon-counting system, in which flux levels as low as a few tens of photons per second can be measured, eliminates the disturbing factors from the measurement, namely, the DC component of the dark current, low-amplitude pulses originating in the electron multiplier, and high amplitude pulses of other than photoelectric origin. It can be used for determining the shape of fast, low-intensity light pulses as well.

All the output signals above a threshold are, in general, counted as photon events provided the incoming photon flux is of a sufficiently low intensity such that there is no more than one electron generated in any pixel during the integration period, and the dark noise (see Section 4.5.2.5) is zero, and gain (see Section 4.3.2.2) can be set at a suitable level with respect to the amplifier read-out noise (see Section 4.5.4).

A window discriminator is usually employed to eliminate all pulses lower than a distinct threshold, as well as those higher than a higher threshold. However, this procedure is not able to separate pulses due to the thermal emission of the photocathode of a device, such as photomultiplier tube (PMT; see Section 5.2.2) since they have the same pulse height distribution (see Section 7.2.1) as the photo pulses. The output signal of such a sensor changes if the light intensity is lowered. The output signal at higher levels is a superposition of analog pulses, which is inseparable per time. As the light intensity decreases, the output pulses gradually separate from each other, and when the light intensity becomes very low, one can envisage discrete pulses corresponding to single incident photons. Each of these pulses corresponds to individual photons and the number of pulses per unit time is proportional to the light intensity. In photon counting method, the number of pulses is digitally counted to measure the light intensity. The counting efficiency is determined by the quantum efficiency and any electronic losses that are present in the system.

The analog or pulse counting is, in general, used where the light source is intense and slowly varying, typically over milliseconds (ms) or even seconds. To note, in pulse-mode operation the integration time constant has to be short depending on the pulse characteristics that need to be measured (amplitude, frequency), as well as the requirement by the electronics (rise time, duration). However, in some cases, fast transients of the order of nanoseconds (ns) need to be encoded, for instance, applications, such as scintillator involve fast transients and a wide dynamic range. The resolution time is determined either by the decay time constant of the scintillator or the time characteristics of the electronics. If the dead time, τ_{d}, of the counter is comparable with the mean interval separating successive scintillations, the true count rate, n_{tr}, corrected for the dead time is

$$n_{\mathrm{tr}} = \frac{n_{\mathrm{obs}}}{1 - n_{\mathrm{obs}}\tau_{\mathrm{d}}}, \tag{7.8}$$

where n_{obs} is observed count rate.

The primary specification parameter is the bandwidth of the measuring electronics, which may be anywhere from 1 Hz to 100 MHz. These modules use negative high voltage and are DC-coupled. The measurement of these discrete pulses is referred to as the photon-counting mode of light measurement. It is probable to get photon-counting with sufficient accuracy approximately up to 10^7 photons per second (Heering, 1998). Such a limit arises from the finite width (which is due to the transit time spread and is of the order of 10 ns) of a pulse at the output of the detector, which is the response to a single photon.

7.2.1 *Detection of Photoelectrons*

Photocurrents are sensed by measuring the voltage across a series resistor when no current passes through it. However, this voltage is sensitive to changes in the external resistance or in the intrinsic resistance of the semiconductor. Two methods are generally employed to record light, such as

(1) The electronic signal from the photodetector is integrated for a time interval, Δt and record a photocurrent that is proportional to the power of the light, $P(t)$, on the detector. The rate at which photons reach the detector is given by

$$i(t) = \frac{N_{\mathrm{e}}(t)q}{\Delta t} = \frac{P(t)q}{h\nu\eta_{\mathrm{d}}}, \tag{7.9}$$

in which q is the charge of a single electron and $N_{\mathrm{e}}(t)$ the number of electrons generated.

(2) Existence of photons means that for a given collecting area, there exists a physical limit on the minimum light intensity for any observed phenomenon. The probability of obtaining a photon at a given location is proportional to the intensity at that location. Thus, the probability distribution of photon impacts is the intensity distribution in the image. One single electron can be turned into an electronic pulse using processes

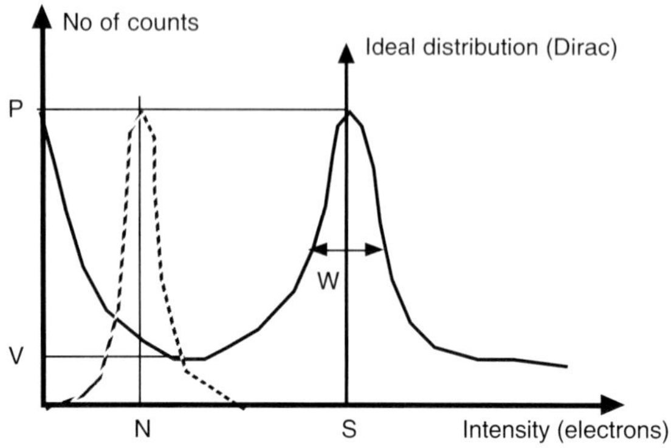

Figure 7.1 Example of a usual PHD, in which the dashed curve stands for the noise statistics of the detector read-out; the usual PHD is a Dirac's peak (after Morel and Saha, 2005).

for secondary electron emission. The pulses correspond to individual detection events (individual photons), which can be counted and processed. The count rate, $R(t)$ is proportional to optical power of the light incident on the detector, that is

$$R(t) = \frac{N_e(t)}{\Delta t} = \frac{P(t)}{h\nu\eta_d}. \tag{7.10}$$

In order to characterize a photon-counting device, the statistics of the intensity of the output signal (expressed in electrons) triggered by a photoevent, what is known as pulse height distribution (PHD), of the same is measured. Figure 7.1 represents PHD by a curve number-of-counts vs. output signal intensity. A PHD curve shows a peak, which may be characterized by its normalized peak-to-valley ratio, and its normalized FWHM. Ideally, peak-to-valley should tend to infinity and normalized FWHM to zero. Of course, the intensity corresponding to the peak has to be much larger than the maximum level of read-out noise of the sensor (anode, CCD chip, etc.) whose output signal is the photon information carrier and that terminates the chain in the photon-counting device. In these conditions, false detection, FD, tends to 0.

The marked advantage of photon-counting technique is that of reading the signal a posteriori to optimize the correlation time of short exposures in order to overcome the loss of fringe visibility due

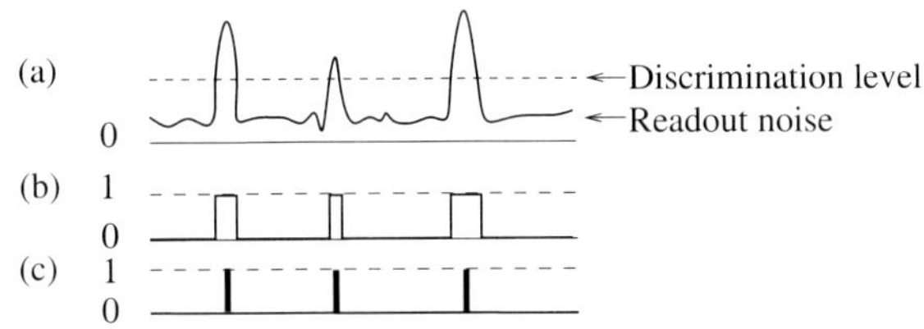

Figure 7.2 Processing of centroiding system: (a) output signal, (b) slice mode, and (c) center-of-gravity mode.

to the speckle lifetime; the typical values for an object of $m_v = 12$ over a field of 2.5″ are less than 50 photons/ms through a narrow-band filter. The other notable features are

- capability of determining the position of a detected photon, to 10 μm to 10 cm,
- providing spatial event information by means of the position sensitive read-out setup, the encoding systems identify each event's location; the centroiding of each photon provides enhanced spatial resolution (see Figure 7.2),
- ability to register individual photons with equal statistical weight and produces signal pulse (with dead time of a few ns), and
- possessing low dark noise, which is typically of the order of 0.2 counts $cm^{-2}\ s^{-1}$.

7.2.2 *Necessity of Photon-Counting Systems*

During the initial phases of the development in single aperture interferometry, speckle imaging in particular (see Section 1.5.5), a problem in the early 1970s was the data processing. Computers were not powerful enough for real-time processing and video recorders were expensive. One of the first cameras for such purpose, built by Gezari et al. (1972), was an intensified film movie[a] camera. Subsequently, a few observers used photographic films with an intensifier attached to it for recording speckles of astronomical objects (Breckinridge et al., 1979). Saha et al. (1987) used a bare

[a] Movie is known to be a film running at 16 or more frames per second (FPS).

movie camera, which could record the fringes and specklegrams of a few bright stars; a water-cooled bare CCD was also used for certain interferometric observations by Saha et al. (1997). However, the main drawback of a CCD is its serial read-out architecture, limiting the speed of operation. Hofmann et al. (1995) reported the speckle masking observations of the central object HD 97950 in the giant H II region[a] and NGC 3603, a luminous star cluster,[b] using an image intensifier with a gain of 500,000 coupled optically to a fast CCD camera (512×512 pixels) having 4 FPS and correlated double sampling (see Section 6.2.3.2) at the 2.2 m ESO telescope.

Modern telescopes fitted with new-generation photon-counting sensor have led to major advances in observational astronomy, high angular resolution imaging in particular. Such a detector is able to count how many photons it has received, but is unable to provide any information about the angle of incidence for individual photon. The contribution of a single aperture interferometry to the following important fields in astrophysics would increase sharply if an intensified photon-counting detector is incorporated in the speckle interferometer. Mapping of following interesting objects using image processing algorithms can reveal the structure and orientation of the same:

(1) Studies of binary stars play a fundamental role in measuring stellar masses, providing the benchmark for stellar evolution calculations. A long term benefit of speckle imaging is a better calibration of the main-sequence mass-luminosity relationship. Most of the available late type stars, within 5 pc radius from the sun, are red dwarfs (K- and M- dwarfs) with $m_{\rm v}$ 15; these dwarfs may often be close binaries.

---

[a]The H II region, where star formation is taking place, is composed primarily of hydrogen and have the temperatures around 10,000 K. Generally, less than 10% of the available gases of this region is converted into stars, with the remainder of the gases dispersed by radiation pressure, supernova explosions, and strong stellar winds from the most massive stars, leaving behind open clusters. Later, the cluster breaks out, the gas is blown away, and the stars evolve.

[b]Embedded in the H II region, NGC 3603 is a luminous star cluster several O-type and Wolf-Rayet stars. It is one of the most distant H II regions (7.0 ± 0.5 kpc; 1 pc = 3.08568×10^{16} m = 3.26156378 light years; ly.) which are optically visible.

(2) The physical processes, viz., temperature, density and velocity of gas in the active region of the active galactic nuclei[a] can be studied.
(3) The spatial distribution of circumstellar matter surrounding objects that eject mass, particularly young compact planetary nebulae, can be determined.
(4) Capability of resolving the gravitationally lensed quasars (QSO) in the range of 0.2″ to 0.6″ would allow the discovery of many more lensing events.

7.3 Genesis of Photon-Counting Detectors

The photon-counting cameras are engineered to address the challenges of ultra-low light level imaging applications. Such cameras detect every bias corrected image for a local maximum above a threshold. The photon-noise (see Section 4.5.2.3) bias can be corrected if the number of detected photons is known; each photon should be detected as a delta function. For an ideal photon-counting system with unity quantum efficiency of the detector, $\eta_{\rm d}$, vanishing dead times, no pulse piles up. In addition to its pulse height, individual event is given time tag as well as a position tag.

In spite of the high sophistication of modern photon counting cameras, they are unable to target the photon-starved astronomical objects. The sampling rate of these sensors scales with the turbulence in the atmosphere up to kHz and is limited by the number of photons received in a short exposure. For high-resolution imaging, photon-counting systems of very high temporal resolution

[a]Some galaxies are active, referred to as active galaxies, in which a significant portion of the total energy output from the galaxy is emitted by a source other than the stars, dust, and interstellar medium. They exhibit violent activity that is produced in the nucleus, which appear to be extremely bright at any given epoch. Their nuclei containing a large quantity of gas are called active galactic nuclei (AGN; Binney and Merrifield, 1998). AGNs were first discovered in the 1940s as point-like sources of powerful optical emission with spectra showing very broad and strong emission lines in their nuclei, indicating large internal velocities. These lines exhibit strong Doppler broadening (see Section 3.3.5.1), which may be due to either rotational velocities of the order of several thousand kilometers per second near a black hole or due to explosive events in the nucleus.

of the order of several MHz is necessary to tune the integration time according to the value of r_0. The photon-counting hole is also a problem of such systems connected with their limited dynamic range (see Section 4.1.1.7).

An important consideration for photon-counting is the level of clock induced charges, which do not arise from photons but are due either to spurious charges created by clocking charge over the surface of the CCD, or charges that are thermally generated. Since such a sensor and the processing electronics have a dead time, two or more photons arriving at the sensor at time intervals less than such a dead time can generate single electronic pulse. In order to overcome the shortcoming due to the former, the count rate should be larger than the dark noise. Cooling of the sensor can significantly reduce the dark noise, which is typically <100 counts s^{-1}, while the latter effect can be looked into by adjusting input intensity so that average count rate is smaller than the dead time, else yields saturation.

Classical charge-coupled device (CCD; see Section 6.2) is not considered to be a photon-counting detector, even if one photon produces one electron with $\eta_d \approx 0.9$, and with an A/D-converter (see Section 6.2.3.2) set that one step (ADU) corresponds to one electron, because their read-out noise (see Section 4.5.4) is too significant to determine whether a photon or no photon at all has been received in an image element when the light level is low (less than 1 photon/pixel/frame). Gen I intensifiers (see Section 5.2.6) did not allow photon-counting even if the phosphor was placed before a TV camera to quantify the signal since

- the pulse height distribution (PHD) of such an intensifier has a negative exponential shape with no peak, and
- the output energy that is generated by any photoevent is statistically weaker than the electronic noise of a TV camera.

Any TV camera equipped with a stack of cascaded Gen I intensifier and Gen II image intensifiers (see Section 5.2.7) is able to produce TV frames at the quantum limit, hence in the photon-counting mode. However, because of the sparse nature of photon-counting mode images, it was necessary, in order to optimize the data processing, to convert each frame into a set of (x, y) coordinates of the photoevents. Nevertheless, photon-counting cameras without photon coordinate extraction were mostly used as still cameras by integration of the TV frames. Such systems were employed in the 1970s. For example, Lowrance et al. (1976) designed an intensified TV camera system for photon-counting, in which the output signal was digitalized and added to an image memory with 16 bits per pixel.

7.3.1 *Initial Experiments*

A host of photon-counting systems have been developed inspired by the tubes that were used for commercial television in the 1950s. A vidicon tube is a television camera tube that uses the photoelectric effect (see Section 2.1) in light-sensitive material, in which incident light forms a charge-density pattern on a photoconductive (PC) surface. Scanning by a low velocity electron beam discharges the surface, producing a current in an adjacent conducting layer. Such a tube has three sections, such as (i) target section comprising glass plate and lens assembly, (ii) scanning section consisting of alignment coil, deflection coil (both horizontal and vertical deflections), and (iii) electron gun. The released photoelectrons are stored in a layer of micro-capacitors, which are periodically recharged by a scanning electron beam. The recharge current of the micro-condenser is modulated by the light intensity. Passing through the charge resistance of the target, it produces a potential difference, which is the video signal, at a frequency of a few tens of megahertz (MHz). The read-out noise is high and is observed on the TV screen.

Ever since the successful development of a photon-counting detector system, widely known as Image Photon-Counting System (IPCS; Boksenburg, 1975), detectors for visible light interferometry have made incredible advances and operating near their fundamental limit in most wavelength regions. Photon-counting cameras for low-energy photons are the result of parallel progress in different fields of research, for example, gamma-ray imaging, night vision, and photometry. However, these cameras did not directly attempt low-energy photon-counting imaging, and therefore separately brought the elements like micro-channel plates, image intensifiers, position sensitive anodes. Such elements are being employed in present-day photon-counting cameras. These cameras have high quantum efficiencies, high frame rates, and read noise of a few electrons. These devices are produced for real time applications like adaptive optics.

7.3.2 *Image Photon-Counting Device*

The first photon-counting camera using the coordinates of the photoevents for the visible spectrum was designed by Boksenberg and Burgess (1972). It was similar to the systems like the one built by Lowrance et al. (1976), except that each time a photoevent was detected in a frame (when the TV signal was above a threshold), its coordinates (given by line and column counters) were used to address a digital number in a RAM and to increase it. The advantage was a reduction of the number of additions to be performed in real time.

The Image Photon-Counting System (IPCS) is an imaging detector that counts individual photon events and is based on the principle of event centering. It combines the linearity and sensitivity of the electronographic camera with the convenience of the TV camera, and a capacity for digital storage of information. The individual photon events in a 2-D image are detected by means of an image intensifier (S 20 photocathode; Boksenberg, 1982), on the front of which a photocathode is mounted. The incoming photons on the light sensitive target are amplified by an image-tube intensifier, producing a signal of about 10^7 electrons at the output. This splash of electrons is detected by a Vidicon detector or CCD. That signal is analyzed by hard-wired logic to determine the positions of individual photon events. The event-center detection logic identifies the centroid of each photon event. The (x, y) coordinates of each photon event are passed on to the detector memory system, which counts the number of photons detected in each pixel by incrementing an appropriate memory location. During the integration, a 2-D image (or spectrum) is built up in the said memory system.

The sensitivity of the IPCS is determined by the characteristics of the image tube phosphorus, the S 20 photocathode. The IPCS is sensitive up to about 700 nm, with an overall quantum efficiency of about 20%. In reality, the IPCS is used for observations where the S/N ratio is limited by the detector. The first-generation IPCS was in use until about 1992. This detector made use of a Vidicon tube for event detection. It required elaborate setup procedures for alignment and minimizing of image distortion. The image size is restricted to 2048×256 pixels, but for spectroscopic use a format of 2048×32 and a pixel size of 16 μm in the dispersion direction and 74 μm in the cross-dispersion direction. Because of the large format, non-linearity effects become evident at counting rates as low as 0.2–0.4 Hz. Also, the calibration of such a system was cumbersome and, in particular for long-slit spectroscopy; special calibration exposures are needed. The second-generation IPCS with CCD read-out was in use since 1991 until about 1996 at WHT instruments, such as ISIS and UES.

7.4 PMT-Based Photon-Counting System

In order to detect the photons individually, it is essential to increase the gain of an intensifying device to reach the quantum limit. As far as measuring the photon positions in an $(\mathbf{x})$, in which $\mathbf{x}(= x, y)$ is the 2-D position vector, in the focal plane is concerned, the image resolution of a single element photon sensor (photomultiplier) should be increased by miniaturizing and multiplying the basis element. Actually, modern photon-counting cameras inherit both approaches that were used alternatively, for example, from photomultipliers to micro-channel plates (MCP; see Section 5.2.7), or from low gain MCP-equipped imaging devices to photon-counting cameras.

An important problem in designing a photon-counting cameras, which needs to be addressed, is to convert as fast as possible the position in the image plane of an incoming photon into a set of coordinates (x, y) that are digital signals. Iredale et al. (1969) addressed the problem of photon position encoding in the image intensifiers for one dimension case (x-coordinate to be estimated). They presented the results of three possible optical setups (see Figure 7.3) of which the last setup (see Figure 7.3c) was considered to be the most accurate, and proposed to extend it for 2-D imaging:

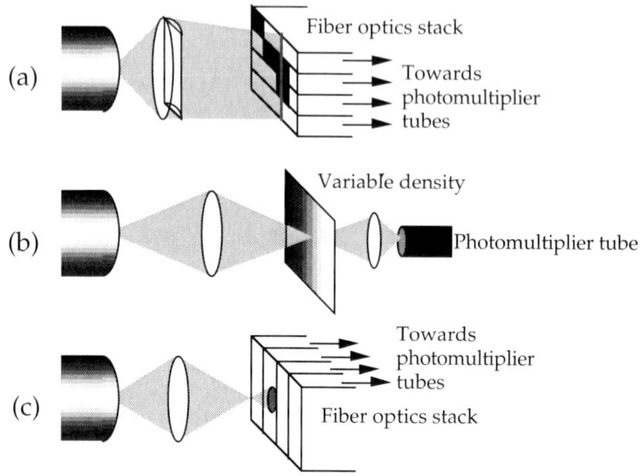

Figure 7.3 Three systems for measuring the photon position by Iredale et al. (1969).

(1) The spot on the intensifier output (corresponding to a photo-event) was re-imaged as a line along *y* (by means of a cylindrical lens) onto a binary code mask that was located at the entrance of a stack of fiber-optics, each fiber having a rectangular cross section (see Figure 7.3a). The output of each fiber fed a photomultiplier tube (PMT; see Section 5.2.2). Therefore, the PMT outputs gave the binary value of the photoevent coordinate.
(2) The spot was re-imaged onto a neutral density filter whose attenuation was varying along *x* (see Figure 7.3b). A PMT behind this filter gave a signal with an amplitude proportional to the photoevent *x*-position.
(3) The spot was re-imaged with a certain defocus on a stack of fibers connected to PMTs (see Figure 7.3c) as in Figure 7.3a. By combination of the analog signals at the PMT outputs, *x* was found out.

7.4.1 *Digicon*

The Digicon (Beaver and McIlwain, 1971) was one of the first alternatives to Lallemand tubes (see Section 5.2.6) for astrophysics and provided photon-counting ability. This device was not a real imaging device, since it measured the photons only in the one dimension. Its main application was, therefore, spectrometry. Based on the principle of a Gen I intensifier (see Section 5.2.6), its originality comes due to the fact that instead of bombarding a photographic plate (see Section 1.4.2), accelerated electrons in the Digicon collide with an array of photocathodes. The signal provided by each diode, the binarized signal resulting from the collision incremented 16-bit registers. A Digicon, with a larger number of photodiodes (see Section 5.3.1) providing an improved resolution, has been employed in the Hubble Space Telescope (HST). To note, Herrmann and Kunze (1969) introduced a photon-counting spectrometer working in the UV and featuring an array of 40 miniaturized photomultipliers.

From the principle of Digicon, Cuby et al. (1988) investigated the electron-bombarded CCD concept. It consists of a CCD array placed in a vacuum tube with photocathode. Electrons are accelerated by a 25 kV voltage to make them strike the CCD pixels. Each accelerated electron liberates a charge of around 7500 electrons in the CCD. With a unique diode, the characteristics of the PHD were PV = 0.33 and NFWHM = 0.22. Since these devices were not used for high-resolution astronomy, they were replaced by high performance CCDs.

7.4.2 *Precision Analog Photon Address*

The marked advantage of the detection scheme for Precision Analog Photon Address (PAPA; Papaliolios et al. 1985) is the capability to record the photon arrival time, which allows one to tune the integration time according to the value of Fried's parameter, r_0, during the off line data treatment. This camera, a 2-D photon-counting camera, is based on a high gain image intensifier and a set of photomultiplier tubes (PMT). It finds the centroid (see Section 7.2.1) of each photon detected by the front photocathode and outputs the x and y position of that photon with 10 or 11 bit precision, as well as the time of arrival of each detected photon (Papaliolios and Mertz, 1982). However, the PMT used for photon-counting should have a high responsivity (see Section 4.1.1.2) to the wavelengths of light signal, high gain, at the first dynode in particular, fast rise time followed by a rapid and smooth return to baseline, low dark counting rate, and low transit time spread (Heering, 1998). Also, in PMT, a relatively high dark signal is obtained by thermal emission of electrons from dynodes producing charge pulses with low height, and also some pulse rate by glass emission, etc., giving enhanced pulse heights. The thermal emission can be reduced by PMT cooling.

The front-end of the camera is a high gain image intensifier that produces a bright spot (brighter than those caused by a photoevent) on its output phosphor for events detected by the photocathode. The back face (phosphor) of the intensifier is then re-imaged by an optical system that is made up of a large collimating lens and an array of smaller lenses. Each of the small lenses produces a separate image of the phosphor on a mask to provide position information of the detected photon. Behind each mask is a field lens that relays the pupil of the small lens onto a small photomultiplier.

A camera developed by Papaliolios et al. (1985) has a set of 19 PMTs, out of which 9 + 9 PMTs provide a format of 512×512 pixels optical configuration. The 19th tube acts as an event strobe, registering a digital pulse if the spot in the phosphor is detected by the instrument. Nine tubes are used to obtain positional information for an event in one direction, while the other nine are used for that

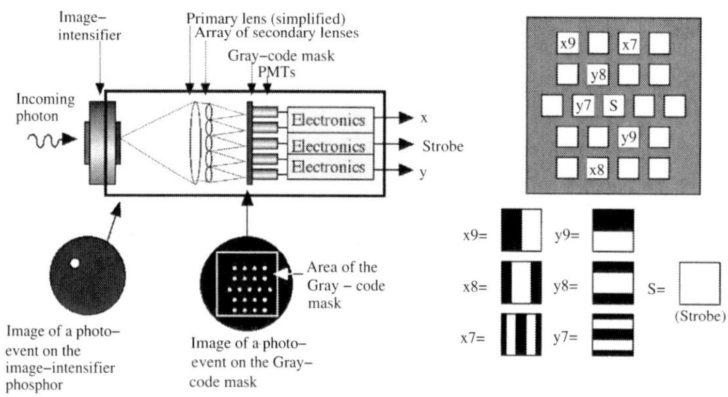

Figure 7.4 The PAPA camera. Coding mask elements are shown on the right (after Morel and Saha, 2005).

in the orthogonal direction. If the photon image falls on a clear area, an event is registered by the phototubes. The masks use gray code (see Figure 7.4), which ensures that mask stripes do not have edges located in the same place in the field.

Each mask provides a Gray-code bit, either of the x or the y photoevent coordinate. The re-imaged spot may either be blocked or not by a mask, the PMT thus giving a signal that is binarized to yield a value, either 0 or 1. This value corresponds to a Gray-code bit of x or y. One of the secondary lens + PMT system has no mask and is used to detect the presence of a photoevent by sampling the outputs of the other PMTs. For a $2^N \times 2^N$ pixel resolution, $2N + 1$ secondary lens + mask + PMT sets are required. With the PAPA detector, the time of arrival of each event is recorded, so photons may be grouped into frames in a way that maximizes the S/N ratio in the integrated power spectrum. This camera provides 1k$\times$1k pixels (22 μ) with 3–5$\times 10^6$ photons per second count rate. It can supply the arrival time of detected photon in 100 ns as well.

The first-generation PAPA detector was used to record the speck-legrams of the stellar objects by the group at Center for Astrophysics,

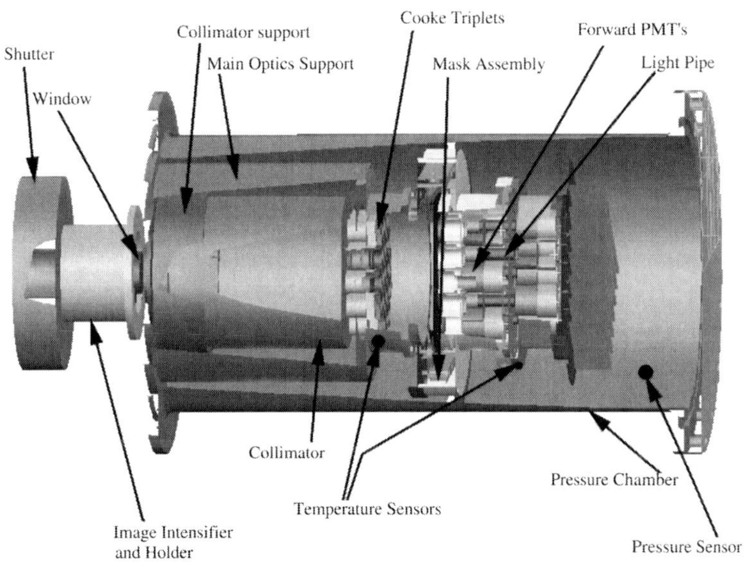

Figure 7.5 An overview of the PAPA camera (Courtesy: P. Nisenson).

Harvard-Smithsonian Observatory, Harvard (Nisenson, 1988, and references therein), as well as to record the fringes of stars using long baseline interferometers by groups at Mt. Wilson (Kaplan et al., 1988) and at Sydney (Lawson, 1994). A modified PAPA camera has been used to observe close binary stars (Aristidi et al., 1999). This version has a new binary mask and refurbished image intensifier. Another modified version of the system with 2k×2k pixels has been developed by using a CeTe photocathode intensifier for the NASA UV applications (Nisenson, 1997). The photocathode is sensitive down to 100 nm and operates in vacuum. The system runs at a maximum count rate of 2 million per second. Figure 7.5 depicts the overview of PAPA camera made for NASA UV applications.

Diamicon The Diamicon (Diffraction-Addressed Mask Iconographer) is an attempt to modernize the PAPA camera, by using a simpler optics (Morel, 2004). The concept features only one lens, coupled to a special diffraction filter (see Figure 7.6). This optics re-images each spot from the image intensifier phosphor as a cross

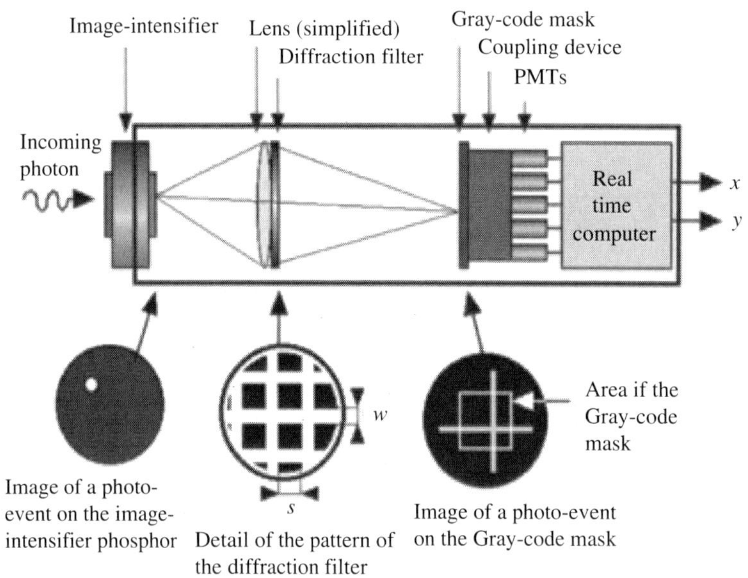

Figure 7.6 The Diamicon camera (after Morel and Saha, 2005).

pattern. A binary mask is located in front of fibers feeding an array of PMTs. According to the position of the cross-pattern on the mask, some PMTs are illuminated, yielding a signal that codes the (x, y) photoevent coordinates. For an array of $N \times N$ PMTs, the resolution of a Diamicon camera would be $N(2^{N-1} - 1) \times N(2^{N-1} - 1)$ pixels. Hence, if $N = 8$, the resolution would be 1016×1016 pixels. Since for an equivalent resolution, more PMTs are required for a Diamicon than for a PAPA camera, PMTs could be replaced by an array of avalanche photodiodes (APD; see Section 5.3.5).

7.4.3 *Position-Sensing Detector*

Photodiodes with at least two terminals are able to detect the positions of incident light spots on their active surfaces, which are referred to as the position sensitive devices (PSD). There are 1-D PSDs having three connectors, one at the center and one on either side. It detects the light spot between the two outer terminals (Haus, 2010). A 2-D PSD has five terminals indicating the 2-D position of the said spot. Similar to these devices, segmented photodiodes have several independent active areas on a single chip. The light spot can be determined with a suitable wiring.

The PSDs do not have a mechanically defined center and are uniformly accurate in measuring beam position across much of the detector area. They are used to characterize lasers and align optical systems during the manufacturing process. These devices provide good resolution and linearity, and fast response for a wide range of light intensities. In order to measure the X and Y positions from the PSD, four electrodes are attached to the detector and an algorithm then processes the four currents generated by photo absorption.

Position-sensitive photomultipliers The position-sensitive photomultipliers technology (PSPMT) uses dynodes, like classical PMTs (see Section 5.2.2). Such a photomultiplier tube consists of an array of dynode chains, packed into a vacuum tube. Like the Photicon (see Section 7.5.1.2), currents measured on electrodes at the output of the last dynodes are interpolated to find out the position of the photoevent. A photon-counting camera, based on PSPMT, installed at the exit of an image intensifier, has been built by Sinclair and

Kasevich (1997). The count rate is satisfactory (500,000 photons per second), but the resolution is poor (360 μm FWHM with a 16 mm image field). Another camera based on the same concept has been built by Pruksch and Fleischmann (2000). The resolution of their camera only reaches 128×128 pixels and presents some distortion in the image.

Quadrant detector The quadrant detector is a uniform disk of silicon with two gaps across its surface as shown in Figure 7.7a. For optimum performance and resolution, the spot size should be as small as possible, while being bigger than the gap between the cells. Typically, the gap is of 10–30 μm and the active sensing area is 77 mm^2 or 100 mm^2 (depending on the exact model). When illuminated, the cells generate an output signal proportional to the magnitude of the illumination. The local areas of A and B vary with Y and the local areas of C and D vary with X. The intensity on each electrode is proportional to the number of electrons received by the electrode, and therefore the local area. The intensity difference between A and B yields Y, while the intensity difference between C and D yields X.

Let A, B, C, D be the four quadrants, respectively, and R is the radius of the incident beam illuminating the detector. The beam

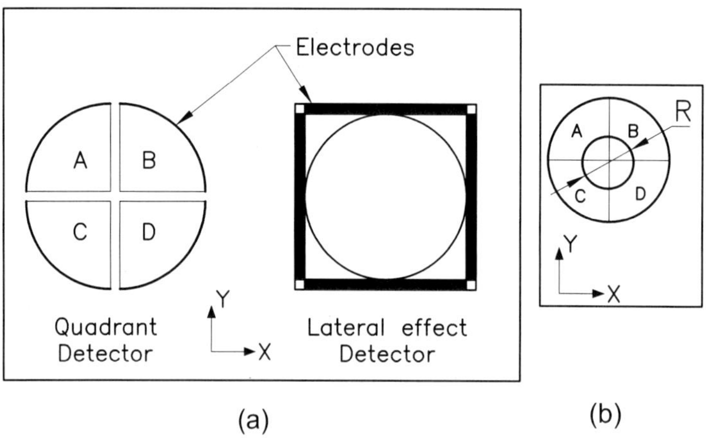

Figure 7.7 (a) Quadrant detector. (b) Beam movement relative to the X or Y direction.

position is calculated using the following formulae:

$$X = \frac{(\mathrm{B}+\mathrm{D})-(\mathrm{A}+\mathrm{C})}{P}; \qquad Y = \frac{(\mathrm{A}+\mathrm{B})-(\mathrm{C}+\mathrm{D})}{P}, \qquad (7.11)$$

with P (total power) $= \mathrm{A}+\mathrm{B}+\mathrm{C}+\mathrm{D}$.

It is the electronic card that digitizes the output signal, and the host computer that then processes the signal. The computer and software perform basic calculations of the position and power of the monitored beam. The output position is displayed as a fractional number or as a percentage figure, where the percentage represents the fraction of beam movement relative to the X or Y direction as shown in Figure 7.7b. The minimum detectable displacement of a spot of light on the detector active area, referred to as the position resolution, is limited by the S/N ratio of the system. It depends on light intensity, detector noise, and electronics bandwidth.

The present-day gamma camera, for instance, the Mammary Gland Specific Gamma Imaging System (MSGIS; Pal et al., 2008) that can aid to the early detection of breast cancer uses a PSPMT device for the conversion of light photons to electrical charge.

MSGIS Unlike the gamma camera developed by Anger (1952; see Section 5.2.5.1), which suffers from high resolution and hence inhibits early detection, the MSGIS is more effective diagnostic tool especially for patients having dense tissues, suspicious areas on a mammogram, lumps that can be felt but not seen with mammography or ultrasound or having implants or scaring from surgeries. Unlike the ultrasound or magnetic resonance imaging (MRI) equipment, which offers additional diagnostic information that is sensitive to changes, the MSGIS is specific to determine metabolic changes within the breast indicating suspicious cells.

The MSGIS camera uses the same photo peak detection technique as used in the camera developed by Anger (1952) for an intensity-based image generation where the spatial and energy resolution of the overall system is enhanced by the advanced detector and simultaneous data acquisition of thousands of detectors with the support of application specific integrated circuits. It allows imaging with no compression of the tissues. The MSGIS adopts molecular breast imaging technique, which depends on the localization of radio-tracer. As stated in Section 5.2.5.1, the radio-tracer, technetium-99m (^{99m}Tc) is the most common radioisotope used for diagnostic purposes in medicine. A small amount of this tracing element is injected, which is absorbed by all the cells of the body. Because cancerous cells have higher rate of metabolic activity, the tracing usually concentrates in these cells, if present. The tracing agent emits gamma rays, which are detected by the imaging system. ^{99m}Tc has half-life of 6 hours, which allows sufficient measurement and receive less radiation by body organs. The 2-D pixilated camera captures these gamma rays from different positions of the target

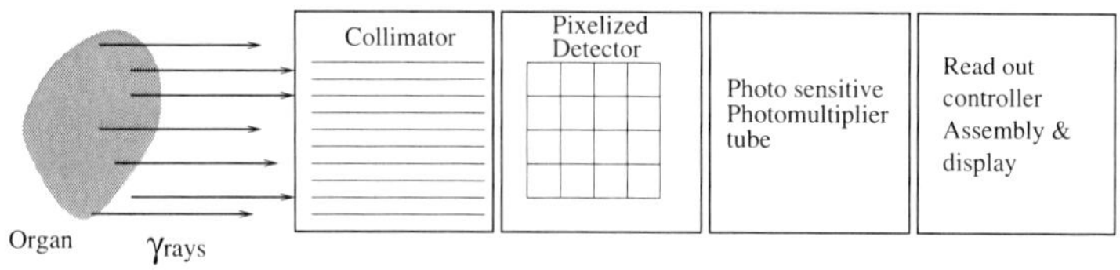

(a)

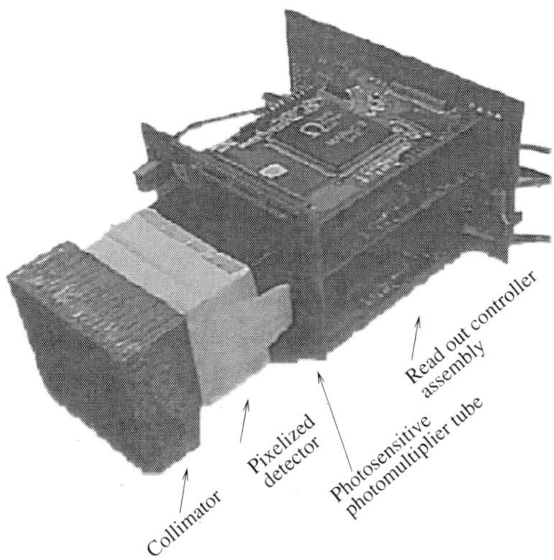

(b)

Figure 7.8 (Top) Schematic block diagram of the MSGIS camera system and (b) the prototype gamma imaging system employing a PSPMT (Courtesy: S. Pal).

and could form a high-resolution 2-D image based on the spatial intensity distribution of the gamma-ray. Unlike standard devices, such as X-ray mammography, this imaging detector is capable of capturing close views of the tumor, and increasing accuracy in detection and localization of small lesions.

The basic design of the camera consists of three main sections depicted in Figure 7.8a. The first section is a focusing device, which allows the camera to pass only the parallel gamma-rays from the lesion by which the scattering problem is avoided in the image. Gamma-ray photon energy to electrical signal conversion could be achieved by the two-stage conversion like scintillator in first stage and a detector in the second stage (Pal et al., 2009). A PSPMT device is used for the conversion of light photons to electrical charge (see Figure 7.8b; Pal, 2014); the digitized signal of all channels are collected by a field programmable gate array (FPGA; see Section 6.2.3.2)-based controller board, which is the master controller of the system. The energy resolution and rise-time of the first-stage detector, the S/N ratio and speed of the second-stage detector are important for the design of the gamma camera. The radio medicine, used for the purpose, emits the gamma-ray of this energy 140 keV. The desired objective is to extract such gamma-ray information from the electrical pulse output of the second stage.

7.5 MCP-Based Photon-Counting systems

Several electronic read-out techniques have been developed to detect the charge cloud from a high gain MCP. Some of these systems provide 2-D imaging capabilities (Lampton and Carlson, 1979; Martin et al., 1981; Timothy, 1986; Siegmund et al., 1994). The main advantage of such systems is to obtain spatial event information by means of the position sensitive read-out systems; the encoding systems identify each event's location. The shortcomings of the MCPs are notably due to its local dead-time, which essentially restricts the conditions for use of these detectors for high spatial resolution applications (Rodriguez et al., 1992). These constraints are also related with the luminous intensity and the pixel size.

7.5.1 *Special Anode Cameras*

A variety of photon-counting cameras consisting of a tube with a photocathode, one or several MCPs, and a special anode to determine the photoevent (x, y) coordinates are developed. However, these sensors detect the charge cloud from a high gain MCP. They provide spatial event information by means of the position sensitive read-out setup; the encoding systems identify each event's location. In what follows, a few such systems are enumerated in brief.

7.5.1.1 Wedge-and-strip anodes detector

Anger (1966) developed the concept of wedge-and-strip camera system that uses conductive array structure, in which the geometrical image distortions might be eliminated. The target for the cloud of electrons coming out from the MCP is a four-electrode (A, B, C, and D) anode (see Figure 7.9) as discussed in Section 7.4.3 with special shapes. It comprises multiple terminals with the x, y co-ordinates of a charge cloud determined through ratios of charge deposited onto the various terminals. The amplitudes of the signals detected on the wedge, strip, and electrodes are linearly proportional to the x, y co-ordinates of the detected photon event. For example, the local areas of A and B vary with y and the local areas of C and D vary with x. The intensity on each electrode is proportional to the number of electrons received by the electrode,

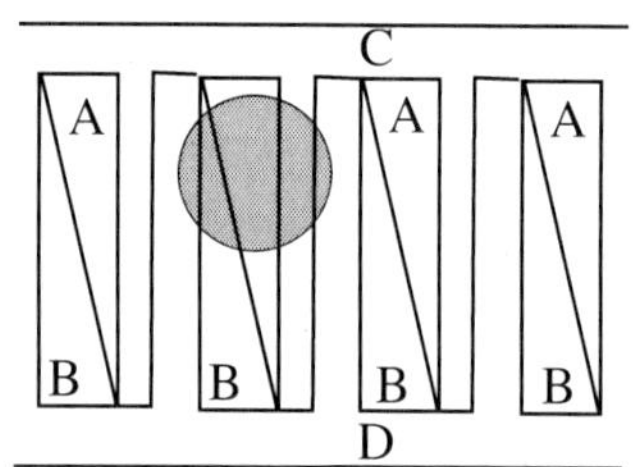

Figure 7.9 Wedge-and-strip anode. The gray disk corresponds to the cloud of electrons that is spread onto a local area of the anode (after Morel and Saha, 2005).

and therefore to the local area. The difference between A and B yields y, while the intensity difference between C and D yields x. Cameras based on this concept have been developed by Martin et al. (1981) at University of California, Berkeley, as well as by Siegmund et al. (1983) at the University of London. In wedge and strip system, the spatial resolutions of the order of 40–70 μm FWHM and position sensitivities of 10 μm are obtained at high MCP gains. High-resolution wedge-and-strip detectors can operate at event rates up to about 5×10^4 photons per second. The problems of the wedge-and-strip technique comes from the limitations on the anode capacitance, which restricts the maximum count rate 40,000 ph/s and from the defocusing (required to spread the cloud of electrons onto the anode) which is sensitive to the ambient magnetic field.

7.5.1.2 Resistive-anode position sensing detector

Another solution for photon position encoding is the resistive anode, used in the Ranicon camera, developed by Lampton and Paresce (1974). It consists of a single resistive plate located at the exit of an MCP. By measuring the voltages at four edges of the plate anode, the position of the photoevent can be estimated. Rees et al. (1980) developed a prototype of Ranicon obtained a 150×150 pixel resolution. However, in order to enhance the resolution, it was necessary to increase the electron gain by using a stack of MCPs. With three MCPs instead of two, the resolution of a Ranicon camera could be multiplied by two (Rees et al., 1981).

In a system developed by Clamping et al. (1988), a continuous uniform resistive sheet with appropriately shaped electrodes provides the means for encoding the simultaneous location and arrival time of each detected photon. This is coupled to a cascaded stack of MCPs acting as the position sensitive signal amplifier. A net potential drop of about 5 kV is maintained from the cathode to the anode. Each primary photoelectron results in an avalanche of 10^7 to 10^8 secondary electrons onto the resistive anode. The signals resulting from the charge redistribution on the plate are amplified and fed into a high-speed signal processing electronics system that produces 12 bit x, y addresses for each event. A PC-based data acquisition system builds up a $1k \times 1k$ image from this asynchronous stream of x, y values that can be stored in a PC (Paresce et al., 1988). The drawback of this system is the large pixel response function; the nominal resolution of the system is about 60 μm. For certain applications, this system was used on the GI2T (Mourard et al., 1994).

In order to improve the special anode cameras enumerated above, the induced-charge technique is used (Battistoni et al., 1982). In this, a resistive plate inserted between the MCP exit and the anode is used to induce electrons in the anode. The electrons are then received from the MCP. This method has been tested with delay-line and wedge-and-strip anodes (Jagutzki et al., 2002), and reduces the read-out noise and the image distortion.

Photicon camera Kellog et al. (1977) developed another concept of camera, called Photicon, which used as an anode, two orthogonal networks of 17 wire electrodes each (see Figure 7.10). By interpolating the intensities at the electrode exit, a 128×128 pixel resolution could be obtained. As performance was not so good compared to the Ranicon system, this concept was discarded.

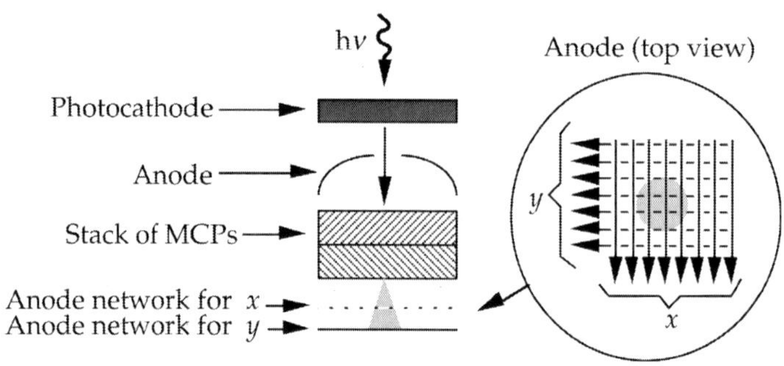

Figure 7.10 Schematic diagram of a Photicon camera system (after Morel and Saha, 2005).

7.5.2 *Multi-Anode Micro-Channel Array*

Nevertheless, the Multi-Anode Micro-channel Array (MAMA; Timothy and Bybee, 1975) camera concept was based on the Photicon,

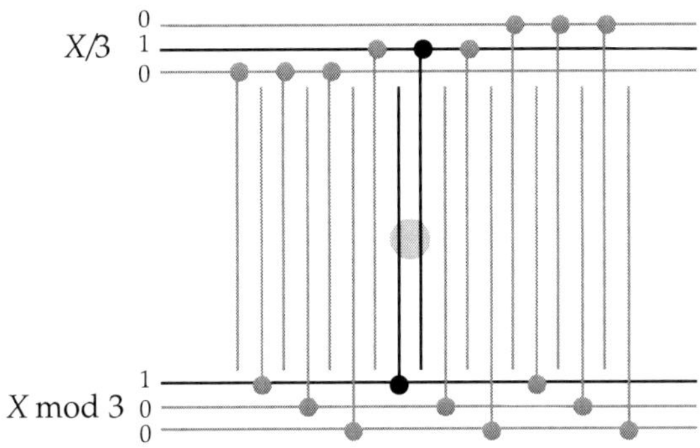

Figure 7.11 Principle of coordinate encoding of a MAMA camera (imaginary case with $N = 3$). The gray zone represents the impact of the cloud of photons on the anode. The activated electrodes are represented in black (after Morel and Saha, 2005).

but later evaluated by using a different wiring of the electrodes to digitally encode the photon position (Timothy, 1983). This detector allows high speed, discrete encoding of photon positions and makes use of numerous anode electrodes that identify each event's location. The electron amplification is obtained by an MCP and the charge is collected on a crossed grid coincidence array. The idea is to slightly defocus the electron cloud at the exit of the MCP, so it falls into two wire electrodes for each coordinate (x or y). The position of the event is determined by coincidence discrimination. The resulting electron cloud hits two sets of anode arrays beneath the MCP, where one set is perpendicular in orientation to the other; the charge collected on each anode is amplified. One electrode is used to encode the coordinate divided by an integer N, and the other encodes the coordinate modulo N. Figure 7.11 shows an example with $N = 3$. Hence, the number of electrodes to encode X possible coordinate values is $N + (X \bmod N)$. To reach $X = 1024$, if $N = 32$, then 64 wire electrodes are needed (128 for a 2-D imager). The MAMA camera presented in 1985 had a 256×1024 pixel resolution (Timothy, 1985). If the analog signal at the output of the electrodes is processed, the resolution can reach 4k$\times$4k pixels. MAMA detector is used for speckle interferometric observations by the group at Stanford University (Horch et al., 1996).

7.5.3 *Delay-Line Anode*

Another system for position encoding is the delay-line anode (Rindi et al., 1970). One of the first prototypes of this system was designed by Sobottka and Williams (1988). Such a system offers high-resolution pixels of the order of 15 μm FWHM, as well as high event rate of the order of more than 4×10^5 photons per second (Siegmund et al., 1994). Such a system has a zigzag micro-strip transmission line etched onto a low loss, high dielectric substrate. The position of the charge cloud event is encoded as the difference in arrival times of the charge pulse at both ends of the transmission line. Figure 7.12a shows a ceramic plate (152 mm × 152 mm), on which two

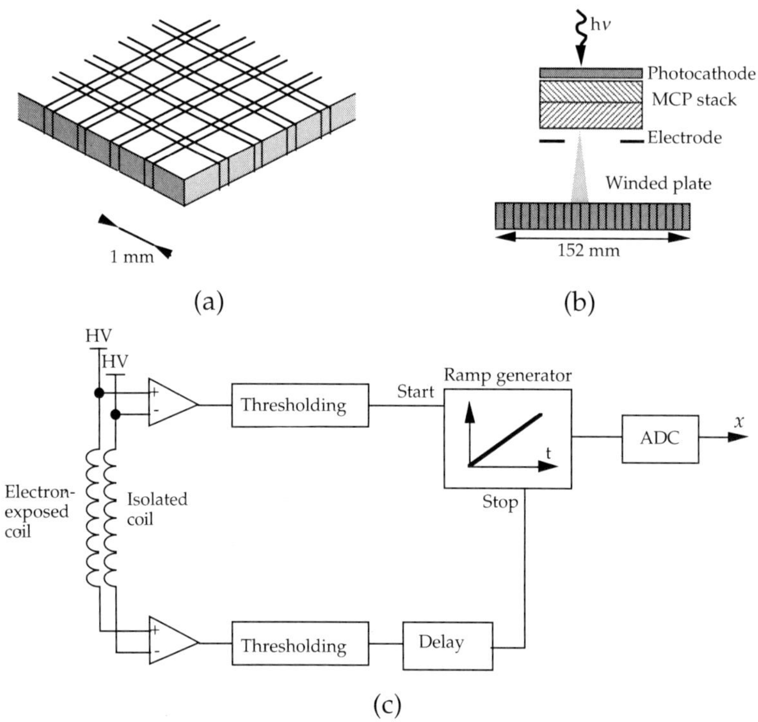

Figure 7.12 (a) Sobottka and Williams' delay-line anode, (b) camera using this anode, and (c) read-out electronics of this camera (after Morel and Saha, 2005).

orthogonal pairs of coils are winded. Each pair is used to encode a coordinate x or y. Within a pair, a coil is exposed to the electrons from a MCP, while the other is isolated and is used as a reference (see Figure 7.12c). The difference of current at an end of the coil pair is used to trigger a ramp generator, and at the other end to stop this generator. The voltage at the output of the generator, when it is stopped, depends on the delay between the pulses received at both ends, and therefore to the position of the electron cloud on the exposed coil. This system (see Figure 7.12b) allows a high count rate (10^6 ph s^{-1}). The problem of the system is the size of the anode target, larger than any MCP and requiring a distortion-free electronic lens.

Similar detector systems have also been used in far-ultraviolet spectroscopic explorer mission, namely, FUSE and GALEX space missions. These missions use double delay-line and cross delay-line anode systems for read-out (Jelinsky et al., 2003; Sahnow, 2003; Morrissey et al., 2005, 2007).

FUSE The Far-Ultraviolet Spectroscopic Explorer (FUSE), a NASA-supported astronomy mission, was launched into orbit in 1999 and continued to detect light in the far-UV portion of the spectrum between 90.5 and 119.5 nm until 2007. It became inoperable when the satellite lost its ability to point accurately. Its primary mission was to characterize universal deuterium to learn about the stellar processing times of deuterium left over from the Big Bang. The science instrument collected the light of distant objects and contains the equipment necessary to disperse and record the light, for instance, the telescope mirrors, the spectrograph and its electronic detectors, as well as an electronic guide camera. FUSE observed more than 3000 different astronomical objects.

GALEX The Galaxy Evolution Explorer (GALEX; Martin et al., 2005) is a NASA small explorer, launched in 2003. It has imaged about 75% of the sky in two spectral bands (far-UV; 153.1 nm and near-UV; 232.1 nm) with a spatial resolution of about 5″ over a field-of-view of 0.6°. Catalogs of several millions UV sources enable advances in a variety of fields, from hot stellar objects in the Galaxy to quasi-stellar objects (Blanchi et al., 2009) and star-forming regions (Thilker et al., 2009; Zhang et al., 2010).

Hybrid systems From the aforementioned concept, Friedman et al. (1996) built a camera using two serpentine delay lines (see Figure 7.13a). The Fourier-width half maximum (FWHM) resolution was 32 μm (the size of the anode was 80 mm × 80 mm), and the count rate was 110,000 photons per second. A hybrid system based on wedge-and-strip and delay-line has been experimented by Lampton et al. (1987). In this case, *x* was determined by a zigzag delay-line electrode, while *y* was determined by measuring the charge partition on two electrodes (see Figure 7.13b). Therefore, flaws of the wedge-and-strip system appeared. To correct them in order to have a better *y* resolution, a low-capacitance double delay-line was imagined by Raffanti et al. (1993). They also designed in 1994 a more accurate system for measuring the time delay, reaching a 4 ps RMS resolution (Lampton and Raffanti, 1994).

One of the latest types of special anode is the "vernier" anode (Lapington, 1995), which is a hybrid of the wedge-and-strip technique and the MAMA anode. A set of three superimposed strip electrode anodes, with different orientations, provides 3 analog values A, B, and C, from which can be computed *x* mod *N*, *y* mod *N*, *x*/4 and *y*/4. Despite of a 10 μm announced resolution, the count

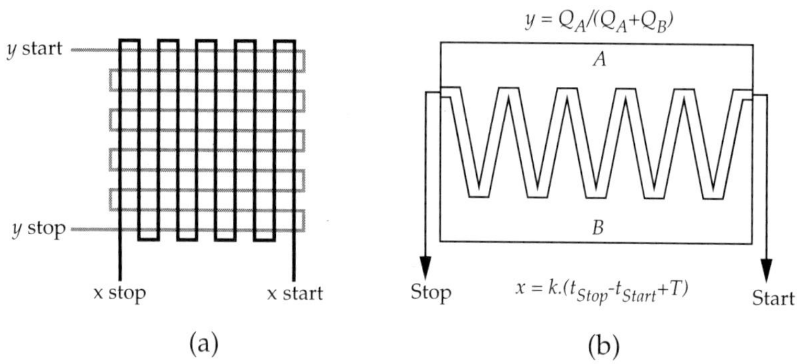

Figure 7.13 (a) Friedman's serpentine delay-line anode and (b) Lampton's hybrid wedge-and-strip and delay-line anode.

rate of a vernier anode camera is limited, like other special-anode cameras, by the anode remanence.

7.6 Charge Transfer Device–Based Photon-Counting Systems

As stated in Section 6.4, the intensified charge-coupled devices (ICCD) have provided the sensitive detection capabilities. However, the drawback of the frame-integrated ICCD system is the poor gain statistics resulting in the introduction of a noise factor between 2 and 3.5. Also, such a system is subjected to limitations in detecting fast photon-event pairs. Non-detectability of a pair of photons closer than a minimum separation by the detector yields a loss in high frequency information; this, in turn, produces a hole in the center of the autocorrelation–Centreur hole–resulting in the degradation of the power spectra or bispectra of speckle images. Also, the pixel-based read-outs introduce systematic effects in the centroiding (see Figure 7.2) and counting of photon events, which require frequent calibration and testing (Kuin and Rosen, 2008; Breeveld et al., 2010). The spatial resolution and photometric properties of the reconstructed images depend on several factors, such as (i) detector hardware design, (ii) choice of centroid algorithms, (iii) the effects of background, and (iv) frame acquisition rate. The non-linearity (see Section 4.4.1.2), in particular, is expected to be present in reconstructed images due to finite exposure time of photon-counting detectors.

7.6.1 *CP40*

Blazit et al. (1977) developed a photon-counting system that is coupled with a micro-channel image intensifier to a commercial television camera. This camera operates at the fixed scan line (312) with 20 ms exposure. A digital correlator discriminates the photon events and computes their positions in the digital window. It calculates the vector differences between the photon positions in the frame and integrates in memory a histogram of these difference vectors.

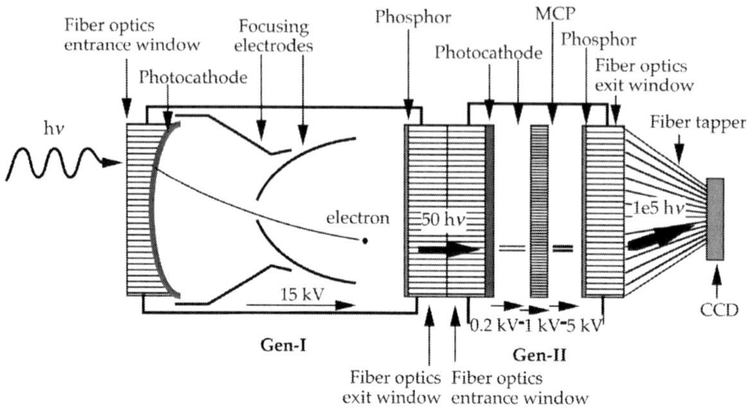

Figure 7.14 Classical design of an intensified-CCD photon-counting camera. The represented Gen-II intensifier features proximity focusing (which reduces the image distortion).

Subsequently, Blazit (1986) has developed another version of the photon counting camera system (CP40), consisting of a mosaic of four 288×384 pixel CCD chips (Thomson TH 7861) with a common stack of a 40 mm diameter Gen I (Varo) image intensifier (see Section 5.2.6) and a Gen II (MCP; see Section 5.2.7) that intensifies the spot at the exit of the latter (see Figure 7.14).

Coordinates are extracted in real time from the CCD frames by a dedicated electronics and sent to a computer system, either for speckle imaging or for dispersed fringe imaging for long-baseline interferometry. The maximum count rate for artifact-free images was about 50, 000 photons/s. The read-out of this system is standard, 20 ms. The amplified image is split into four quadrants through a fiber-optics (see Section 3.4) reducer and four fiber-optics cylinders. Each of these quadrants is read out with a CCD device at the video rate (50 Hz). This camera is associated with the CP40 processor—a hardware photon centroiding processor to compute the photocenter of each event with an accuracy of 0.25 pixel. The major shortcomings of such system arise from

- calculations of the coordinates which are hardware-limited with an accuracy of 0.25 pixel and
- limited dynamic range of the detector.

This system was used by the groups at Observatoire de la Cote d'Azur, France, for the speckle interferometric programs, as well as to record the fringes of stars using Grand Interféromètre à Deux Télescope (GI2T; Labeyrie et al., 1986) and multi-slit spectroscopy. It was also used by other teams for frame selection imagery.

Algol-CPNG A lighter version of this camera called, CP20 (Abe et al., 1999), featuring a single CCD chip and 20 mm photocathode intensifier was built from the CP40 design. Later, Thiébaut et al. (2003) have built a new-generation camera, called comptage de photon nouvelle gènèration (CPNG. featuring a Gen-III (GaAs photocathode) image intensifier (see Section 5.2.7), coupled to a Gen-II image intensifier, and various CCD or CMOS cameras, from a standard TV camera to a 262 FPS CCD camera (Dalsa) with a 532×516 pixel resolution. The processing electronics consists of a real-time computer that extracts the photoevent positions. The software can extract these positions at subpixel resolution and offers a $2k \times 2k$ pixel resolution. The maximum count rate of the CPNG is around 10^6 photons per second. The histogram of the intensities of the peaks detected in the CPNG read-out CCD images is shown in Figure 7.15.

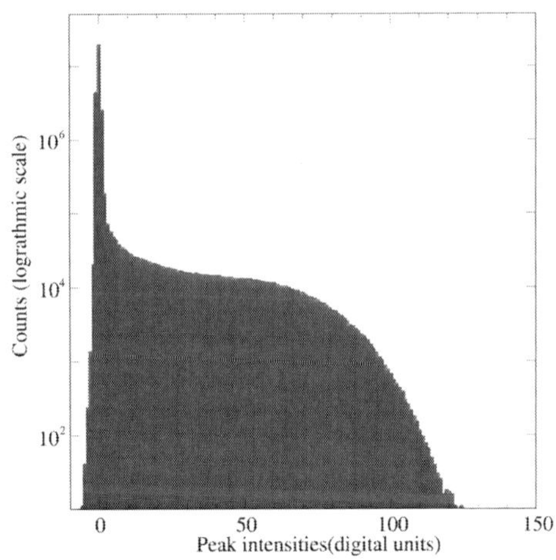

Figure 7.15 (a) Identification of photon events; threshold, set at 5, in order to separate photon events from amplifier noise and the remnants can be removed by comparison with preceding image (Courtesy: A. Blazit).

Blazit et al. (2008) reported the evolution of Algol-CPNG, in which new micro-channel intensifiers and image processing reach up to 36% and 26% quantum efficiencies in the blue (GaAsP) and in the red (GaAs), respectively; this is 65% of the QE of these photocathodes, which is marked increase in the sensitivity compared to CP40. The number of pixels (768×640) is read at a frame rate of 50 Hz. The dark current is found to be 5×10^{-4} e^- per pixel per second. The Algol-CPNG camera is used on VEGA-CHARA interferometer (Mourard et al., 2009). Recently, this system is upgraded to a frame rate of 100 Hz (Blazit, 2014). It is coupled to a CHARA-Vega instrument enabling to reach a visual magnitude, m_V (see Section 1.1.3), of 8. Figure 7.16 displays a snapshot of a K5-type

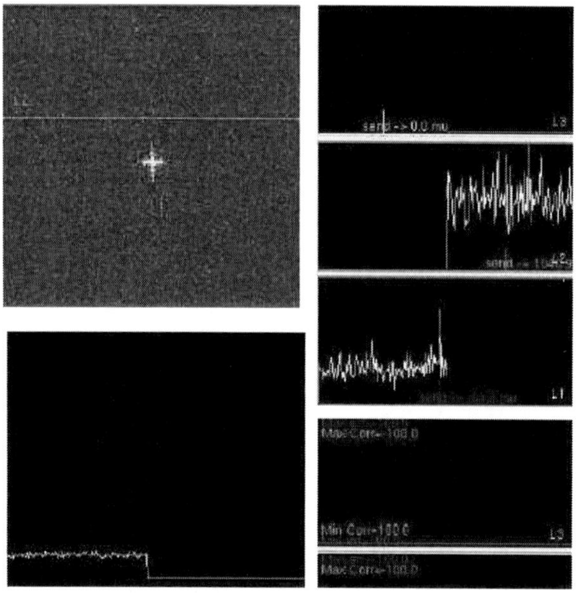

Figure 7.16 A recent image of HD 11969 taken on September 21, 2012, with Algol-CPNG camera at the VEGA-CHARA interferometer (Courtesy: D. Mourard). The upper left picture shows the 2-D spectral density computed in real time with the presence of fringe signal. Other information displayed on the panel are photometry (bottom left), 1-D spectral densities in the upper right panel, and optical path difference offsets as a function of time in the lower right panel.

star, HD 11969 ($m_V = 7.99$) in the constellation Andromeda, taken with this camera (Mourard, 2014).

GI2T The GI2T is an optical interferometer with a pair of 1.52 m spherical telescopes (independent Cassegrain-coudé telescopes) movable on a North-South baseline (variable between 12 and 65 m) and an optical recombining system feeding with two afocal beams housed in a central laboratory at Plateau de Calern, France. It uses a spectrograph as fringe detector to achieve simultaneously high spatial and spectral resolution. The boule telescope is composed of a reinforced concrete spherical structure having 3.5 m diameter, a steel rotating support, and an electronic control unit. Each telescope has three mirrors, kept inside the cylindrical tube like structure, directing the horizontal afocal coudé beam (see Figure 7.17a) to the recombiner optics. The driving system of this sphere consists

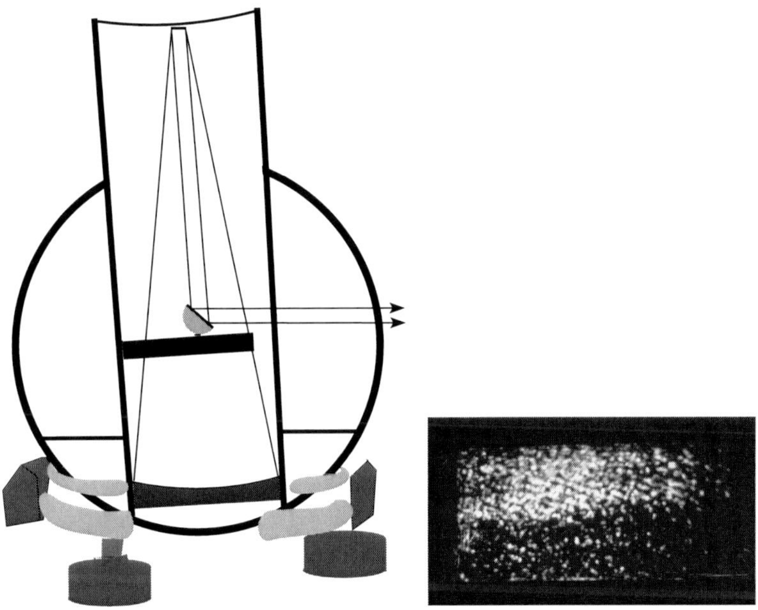

(a) (b)

Figure 7.17 (a) Schematic diagram of the GI2T, France and (b) a dispersed image of a fringe pattern across a star, γ Cassiopeiae with CP40 at GI2T (Mourard et al., 1989; Courtesy: D. Mourard).

of a pair of rings; each ring is motorized by three actuators acting in three orthogonal directions within two different tangential planes (Mourard et al., 1994). The two rings alternately carry the sphere, which in turn produces continuous motion. Each telescope is equipped with two sets of three small DC-motors and three linear encoders. The resolution obtained here is of the order of 1 μm-equivalent to 0.1″ on the sky. The clutching of the rings is performed by computer-controlled electro-valves and hydraulic pistons, which carry each of the 60 mm balls that raise the rings by 2 mm to carry the sphere.

The primary mirror sits on a static cell at the bottom of the sphere. The secondary mirror has a mechanical focusing system, while the tertiary mirror is in a fixed position in order to obtain a focal point inside the sphere; the tertiary is motorized to have a fixed beam toward the laboratory. The configuration of the boule telescopes, the baseline, the choice of a Michelson pupils reconfiguration, and the main science drivers have defined the necessary functions of the interferometric system, such as (i) beam combiner and pupils remapping, (ii) pupil stabilization, (iii) image stabilization, (iv) optical path difference (OPD) compensation, and (v) atmospheric dispersion compensation. The fringes (see Figure 7.17b) are recorded at short exposure using a photon-counting detector.

CHARA The Center for High Angular Resolution Astronomy (CHARA) array is an interferometric facility operated by the Georgia State University and is situated at Mt. Wilson, California, USA (ten Brummelaar et al., 2005). Equipped with six telescopes of 1 m in diameter, this interferometer is arranged on a Y-shaped configuration with 15 baselines ranging from 31 to 331 m. It operates at optical (470–800 nm, 0.15 mas limiting resolution) and IR (2.0–2.5 μm, 0.6 mas limiting resolution) wavebands (McAlister et al., 1998). Light from the individual telescopes, each of them is an afocal beam reducer with a 12.5 cm output beam, is sent through vacuum transport tubes to the centrally located beam synthesis facility, a L-shaped long building (McAlister et al., 1998) that houses the optical path length equalizer and the beam combination laboratory.

VEGA VEGA (Visible spEctroGraph and polArimeter; Mourard et al., 2006) is a beam combiner developed for operating at visible wavelengths (0.45–0.85 μm) in a spectrally dispersed fringe mode, is used to record three-telescope fringes on two different triplets of baselines ranging from 34 to 211 m (Mourard et al., 2009). This allows it to combine high spectral resolution (up to 30,000) and high angular resolution (better than 1 mas with the longest baselines). VEGA spectrograph, similar to a classical spectrograph with a collimator, a slit, a grating, and a camera for reimaging the spectrum on the detector, is fed by four CHARA parallel beams vertically aligned so that fringes appear as horizontal lines at the coherent focus. Equipped with two photon-counting detectors, it allows simultaneous recording of data, in medium spectral resolution, of the spectral regions around Hα and Hβ.

DELTA camera The DELTA camera (Morel and Koechlin, 1998) was a project attempting to use the sparse nature of the frames at the output of an intensifier phosphor (if the frame time is short enough). Spots at the output of the phosphor (corresponding to photoevents) can be projected onto three axes. The coordinates of these projections allow to retrieve the (x, y) coordinates, even if there are more than one photoevent in a frame. Concretely, the image of the phosphor is re-imaged, by means of a cylindrical optics (Iredale et al., 1969) onto three 1024 pixel linear CCD arrays (Thomson TH7809A) noticeable by a very short frame time (2.6 μs). One problem came from the coupling efficiency between the phosphor and the linear CCDs. These CCDs had to be coupled to compact Gen-II image intensifiers. Because of the complexity of the system and the lack of support (especially when Thomson decided to stop the production of the TH7809A), the DELTA camera project had to be abandoned.

7.6.2 *CMOS-Based Photon-Counting Detector*

In a photon-counting system, it may be advantageous to use a complementary metal-oxide semiconductor (CMOS) instead of a charge-coupled device (CCD). As stated in Section 6.3, the pixel addressing and the read-out circuits, as well as the analogue-to-digital (A/D) converters (see Section 6.2.3.2) are integrated into the CMOS chip. Although the CMOS has higher read-out noise (see Section 4.5.4) than the CCD read-out, the aforementioned characteristic makes the CMOS active pixel sensor a compact, low-power-consumption photon-counting system. The read-out noise in a photon-counting system can be overcome by setting a higher gain in the micro-channel plate (MCP; see Section 5.2.7).

A photon-counting system can be constituted by a MCP image intensifier with a phosphor screen coupled to a CMOS through a relay optics (Bonanno et al., 2004). The electrons generated in a photocathode are multiplied by the MCP, and the emerging electron cloud illuminates a phosphor (with a fast decay time), whose light passes through a fiber-optic taper that is bonded to the surface of a CMOS-chip. The photon event can be detected in the form of a light splash, which follows roughly a Gaussian distribution on the CMOS sensor. Such an event can be centroided to a fraction of a CMOS pixel; the centroid position can be used to accumulate the final science images. The active pixel sensor and a field programmable gate array (FPGA; see Section 6.2.3.2) can be located on the same board connected to a small computer through a parallel interface and a serial connection. The intensified CMOS detectors can either be used in a photon-counting mode with a high gain or in integration mode with a low gain.

A CMOS-based photon-counting system (Hutchings et al., 2007; Postma et al., 2011) is developed for Ultra-Violet Imaging Telescope (UVIT) aboard ASTROSAT (see Section 5.3.7.2) The UVIT detectors are based on two MCPs, which are designed with the gap of 0.1 to 0.15 mm, between the photocathode and the MCP, to obtain a resolution of $\sim 1''$ (~ 23 μm on the photocathode). The photocathode, deposited on the inside of a 40 mm window, lies at the focal plane of the optical system. The released photoelectrons are accelerated across a gap (~ 100–200 μm) to these MCPs. As stated in this section, the resulting electron shower impinges on a phosphor screen providing a luminous spot that hits the CMOS having 512×512 pixels. Each pixel, square in size with 25 μm a side, extends $\sim 3 \times 3''$ on the sky.

The photon-counting mode is normal mode for UV channels while the visible channel works with integrating mode; the window mode for observing partial fields can get faster frame rates (>29 FPS). However, the integrating mode lowers the resolution, albeit permits centroiding to sufficient precision to track and compensate for spacecraft drift. In order to correct for such a drift, as well as to achieve the required spatial resolution, short exposures are to be taken, which would be integrated through a shift-and-add algorithm[a] on the ground.

7.7 Solid-State Technologies

All the devices that are enumerated thus far are based on the photoelectric effect (see Section 2.1), i.e., the emission in the vacuum of an electron from a photocathode by a photon. These devices suffer several problems (see Section 4.1). Research in solid-state electronic imaging detector began at the end of the 1960s (Boyle and Smith, 1970; see Section 6.2). Soon, the charge-coupled devices (CCD) replaced TV camera tubes for the photon-counting cameras using image intensifiers (see Section 7.6.1). However, the possibility to have a fully solid-state photon-counting camera is quite recent. In what follows, a few such devices are enumerated.

7.7.1 *Low-Light Level CCD*

Even though CCDs perform close to an ideal detector, they still suffer from non-negligible read-out noise (see Section 4.5.4), which prevent them from being used in applications, such as very low light level imaging where the read-out noise tends to dominate. The read-out is white noise (see Section 4.5.2) dominated and thus the lowest read-out noise is achieved by closing down the system bandwidth which implies low read-out speeds, typically about $3e^-$ RMS at a 20 kHz pixel rate. However, the slow read-out time results in

[a]Shift-and-add method (Lynds et al., 1976; Worden et al., 1976) involves calculation of the differential shifts of the short-exposure images. The image data frame is shifted so that the pixel with maximum S/N ratio in each frame can be co-added linearly at the same location in the resulting accumulated image. This provides an image with higher resolution (higher signal-to-noise at high spatial frequencies) than a long-exposure image; a Dirac impulse at the center of gravity of each speckle is put to estimate the same.

long frame read-out times, typically tens of seconds for large CCDs, which makes them unsuitable for high time resolution imaging applications. A new development in this regard came from Marconi, who developed a kind of CCD, called low light level CCD (L3CCD; Jerram et al., 2001) that addresses these issues. It is a solid-state–based non-intensified device, which shows promise for quantitative measurement of diffraction-limited stellar images. In the following, a few salient features of such a system are described in brief.

Architecture of an L3CCD An L3CCD is a standard CCD with an extended serial register where the charge is multiplied as it is shifted through this register. The electron multiplication takes place in the extended register due to the impact ionization caused by the high electric fields set by a high clock voltage (~40 V) in this serial register. This extended register is called the multiplication register or the avalanche register. The architecture of a frame transfer CCD with an avalanche gain register is shown in Figure 7.18.

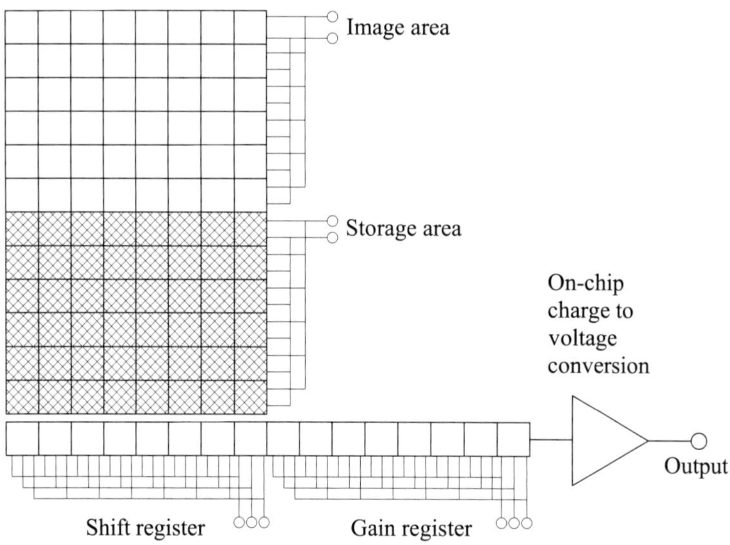

Figure 7.18 Architecture of a frame-transfer L3CCD.

The electron multiplication is produced by transferring electrons from a small potential well into large potential well under the influence of a large electric field (see Figure 7.19). The latter

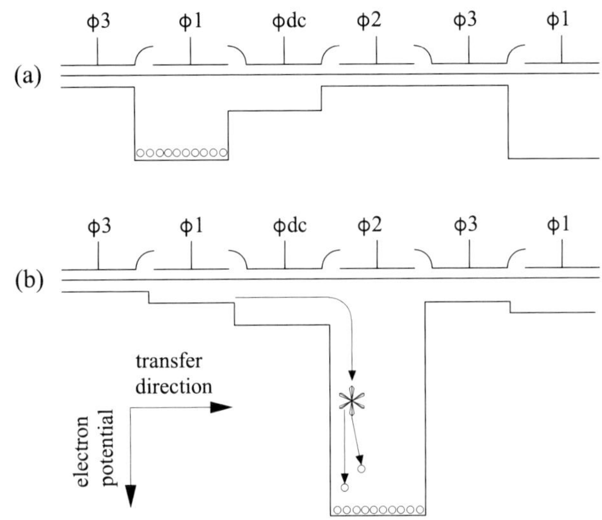

Figure 7.19 Probability of generating secondary electrons, which is dependent on the voltage levels of the serial clock and the temperature of the CCD.

causes some electrons to collide with the crystal lattice, liberating new electrons. This multiplication process is stochastic where the probability of generating a new electron is low, typically 1–2%. However, since there are a large number of multiplication stages in the multiplication register, the overall gain can be very high. The overall gain, G, for a multiplication register with n stages is given by

$$G = (1 + \mathcal{P})^N, \tag{7.12}$$

in which N is the number of pixels in the multiplication register, $\mathcal{P}(x)(= e^{-x/G}/G)$ the probability of generating a secondary electron; for example, a 600-stage avalanche register with the probability of 0.015 per gain stage can provide a gain of approximately 7500.

The overall gain is also dependent on the level of the high voltage clock and the operating temperature of the device. The gain increases at lower temperatures because the probability of the impact ionization increases as the lattice structure vibrates less at lower temperature, and hence, there is an increased chance of an electron collision with the crystal lattice. Since the avalanche gain is governed by a stochastic process (a statistical process involving a number of random variables depending on a variable parameter, which is, in general, time), the overall gain is not fixed but takes the form of a distribution. The standard deviation, σ, of the output distribution of the signal is increased by two times compared to the standard Poissonian distribution (see Section 4.1) seen in an astronomical image. Following equation (4.97), the S/N ratio (see Section 4.1.1.4) of an avalanche output signal is recast as

$$\mathrm{S/N} = \frac{i_\mathrm{s}}{\sqrt{2i_\mathrm{s} + (n_\mathrm{ro}/G)^2}}. \tag{7.13}$$

where i_s is the signal and n_ro the read-out noise; the total noise is dictated by equation (4.101).

Monte-Carlo simulations Figure 7.20a shows a Monte-Carlo simulation of the avalanche gain process in which an input signal of 30e is shifted through avalanche gain register to produce an overall gain of 100. The σ in the avalanche distribution is $\sqrt{2}$ times higher than the σ in the Poissonian distribution. This increase in the σ effectively halves the detected quantum efficiency (DQE). This makes the L3CCDs inefficient in photon-noise (see Section 4.5.2.3) dominated applications where a standard CCD performs better. This loss in the DQE can be regained if the device is used in photon-counting mode.

The L3CCD performs better in the read noise regime of operation, where the input signal is low and the read noise is the dominant noise source. The advantage of the L3CCDs is that these can be operated at very high speeds (few tens of MHz) and yet have negligible read-out noise by suitable adjustment of the avalanche gain; this gain can be adjusted by the high level of the high voltage clock.

The Monte-Carlo simulations for a single electron entering the multiplication register with 464 stages gives an overall avalanche gain of 1000 with probability of 0.015 is shown in Figure 7.20b. The output distribution shows a peak in the first histogram bin rather than corresponding to the 1000 e^- and has the following exponential form:

$$\mathcal{P}(x) = \frac{e^{-x/G}}{G}. \tag{7.14}$$

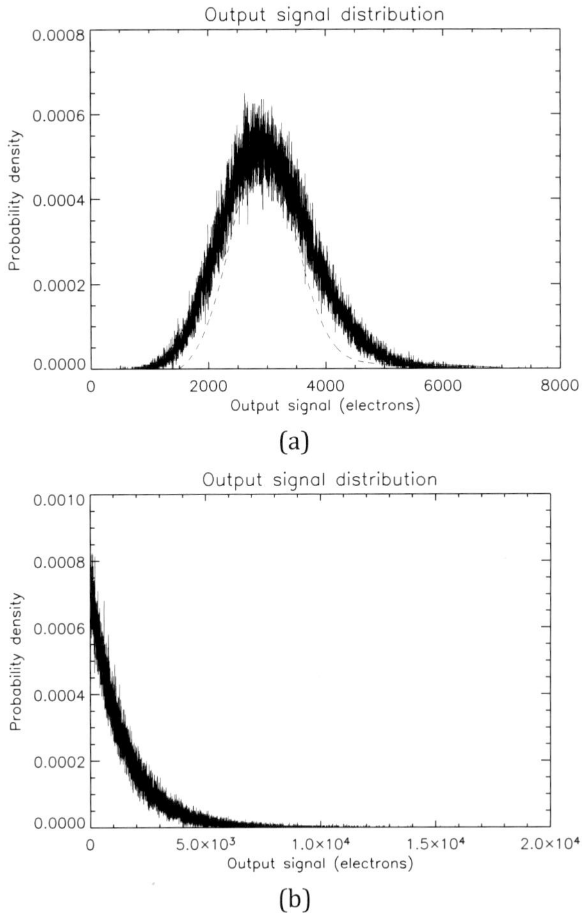

Figure 7.20 Monte-Carlo simulations of the avalanche gain process: (a) input = 30 e^-, mean gain = 100; solid line: modeled distribution with avalanche gain and dashed line: expected Poissonian distribution (distribution widens because of electron gain distribution); (b) input = 1 e^-, gain = 1000. Courtesy: D. Ives and N. Bezawada.

When plotted on natural log scale, this takes the form of a straight line where the inverse slope of the line gives the gain in the avalanche register.

Clock-induced charge The gain process works equally well for the thermally generated electrons and also for the electrons liberated during clocking (charge transfer process) due to the switching electric fields within the silicon lattice. The later form of charge generation is known as the clock-induced charge (CIC) or spurious charge. The thermal generation can be reduced to negligible levels by cooling the device sufficiently low while the CIC can be controlled by the actual clock levels, the shape of the clocks and the slope of the clocks. CIC is generated in all CCDs, but is not noticed in the standard CCD because the read-out noise tends to dominate. CIC is clearly noticeable in the L3CCDs because of the gain process. The CIC is indistinguishable from the real signal, it is effectively a noise source in the L3CCDs and therefore the S/N becomes

$$\mathrm{S/N} = \frac{i_\mathrm{s}}{\sqrt{2i_\mathrm{s} + (n_\mathrm{ro}/G)^2 + C_{CIC}}}, \tag{7.15}$$

where, C_CIC is the CIC expressed in electrons per pixel per frame read and is typically around 0.02 $\mathrm{e^-}$/pix/frame.

The E2V Technologies company and the Texas Instruments company have both simultaneously developed the electron multiplying CCD (EMCCD) technology where the charge in a CCD pixel is multiplied before read-out by a gain factor, thereby reducing the effect of the read-out noise by this gain factor. Figure 7.21 depicts the S/N ratios of an ideal case, a CCD and an EMCCD devices. The

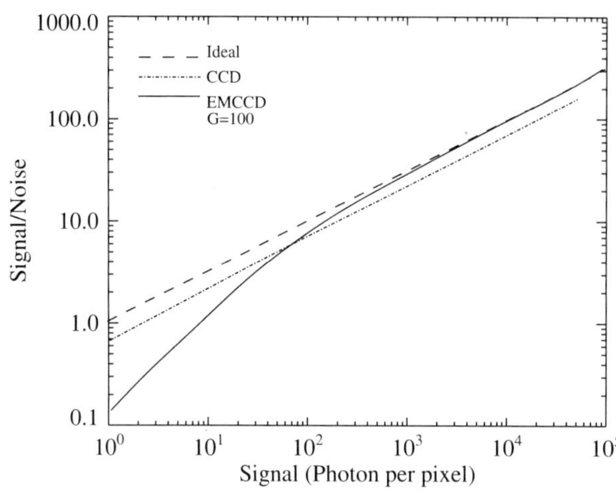

Figure 7.21 Comparison of an ideal case, a CCD and an EMCCD devices in terms of signal-to-noise ratios.

initial system from E2V had 576×288 pixels of size 20×30 μm in the image area, and was provided with Peltier cooling system (see Section 4.3.1) that operates to −65°C with air-cooling and with further additional water circulation, reaches to −80°C; each pixel data could be digitized to 16 bit resolution. An important factor influencing electron multiplication gain is that of cooling the CCD; the more cooler the CCD, a primary electron generates a secondary electron in the silicon, which gives rise to higher on-chip multiplication gain (see Figure 7.22). However, the spurious charge that is generated due to the movement of electrons through the multiplication register's pixels (the sharp inflection in the clock waveform produce a secondary electron in absence of the primary electron) occurs in numbers. The exposure has no effect on such a spurious charge.

There are a number of groups using the L3CCDs in astronomy doing adaptive optics (Tulloch, 2005), high time resolution imaging (Dhillon et al., 2007), lucky imaging (Mackay et al., 2005), in photon-counting mode (Daigle et al., 2004), and for speckle imaging (Saha and Chinnappan, 2002). In the Sydney University stellar interferometer (SUSI) instrument, the fringe detection system uses a cascaded EMCCD system that operates with a frame time of 3.5 ms and a duty cycle over 95%, due to the rapid frame transfer (Robertson et al., 2010).

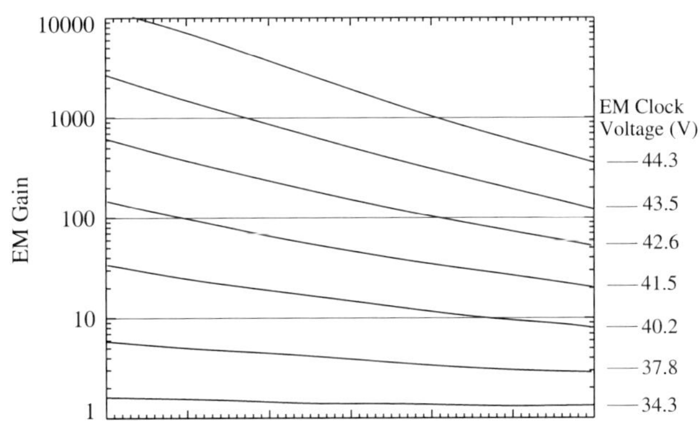

Figure 7.22 Temperature vs. gain.

Sydney University Stellar Interferometer (SUSI) After the successful venture of Narrabri intensity interferometer (see Section 5.2.4.2), the Sydney University had built a 12.4 m prototype stellar interferometer (Davis and Tango, 1986) and subsequently, Davis et al. (1999a) built the existing Sydney University Stellar Interferometer (SUSI) with a baseline ranging from 5 m to 640 m (North-South). It employs small apertures (14 cm; single r_0), adaptive optics (see Section 1.5.6) for wavefront tilt correction, and rapid signal sampling to minimize the effects of atmospheric turbulence. Two beam-combiners, such as (i) blue beam-combination system having spectral range between 430 and 530 nm and (ii) red beam-combination system having spectral range between 530 and 950 nm are used for the scientific program. Recent development of detector technology pave the way to decommission these combiners and replaced by the Precision Astronomical Visible Observations (PAVO; Ireland et al., 2007) that has several parallel spectral channels and spatial filtering.

7.7.2 *Superconducting Tunnel Junction*

Superconducting tunnel junction (STJ), also known as superconductor insulator superconductor (SIS) tunnel junction, has the property to get a charge proportional to the energy of an incoming photon. Such a device is a photon-counting sensor and can have promising applications for direct photon-counting from UV to the near-IR (Philips and Woody, 1982; Perryman et al., 1993; Delaët al., 2001). The properties of STJs, for example, high quantum efficiency, good temporal and spectral resolution, high-speed read-out (the STJs are read in parallel), and also photon-counting capabilities, which make them suitable for low light level photon detection in astronomy, particularly in the field of high angular resolution imaging.

A STJ consists of two superconducting electrodes separated by a thin ($\approx$10 nm) insulating tunneling barrier. Born from research in X-ray detectors, it is based on a stack (see Figure 7.23) of different materials ($Nb/Al/Al_2O_3/Al/Nb$). When there is no light flux and at temperature zero, a quasi-particle current may pass through the insulating layer when the applied potential difference exceeds the sum of the gaps of these two superconductors. At finite

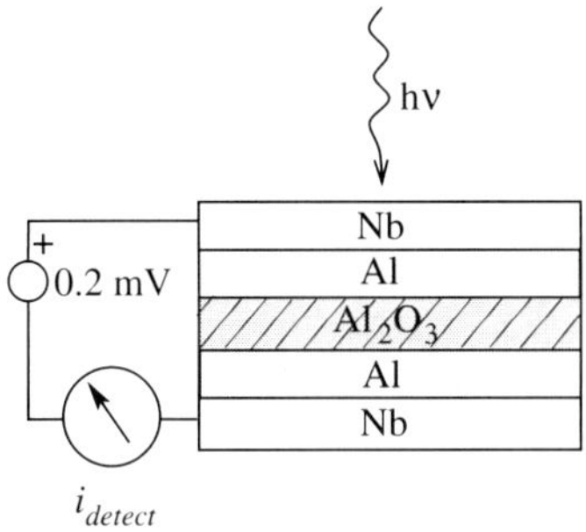

Figure 7.23 Principles of an STJ device.

temperature, i.e., $T > 0$, a dark current, i_d, circulates whenever $\delta V \neq 0$. Absorption of a photon in the superconductor breaks a number of Cooper pairs, which is directly proportional to the energy of the photon. This generates excess free charge carriers (quasi-particles), which can couple with a Cooper pair after crossing over the insulating barrier, cross it again and be counted once more. Theoretically, the STJ is capable of detecting single photons with a temporal resolution of about 10 μs and energy resolution of about 10 eV or better.

First prototypes of STJ detectors (Peacock et al., 1996) had QE $= 0.5$ and a count rate of the order of 2500 ph.s^{-1}. The photon-counting performances of the STJs have been improved by using niobium instead tantalum. In this case, PV$\rightarrow \infty$ and NFWHM $= 0.05$ (for $\lambda = 250$ nm). The spectral resolution is 8 nm at $\lambda = 200$ nm, and 80 nm at $\lambda = 1000$ nm. The main problem of the STJs is the very low temperature that they required (370 mK). Moreover, making STJ array detectors for imaging is a challenge. A 6×6 pixel STJ array has nevertheless been made and used in astronomy (Rando et al., 2000).

7.7.3 *APD-Based Photon-Counting System*

Avalanche photodiodes (APD; see Section 5.3.5) are the most common solid-state photomultipliers, which can work in counting mode, offering a combination of high speed and high sensitivity compared to P-I-N devices (see Section 5.3.2). These devices can sustain an avalanche multiplication process when reverse biased.

A standard APD is operated below the breakdown voltage (V_{br}; see Section 2.2.7), in the analog multiplication regime. When it is biased above its breakdown voltage, the electric field turns out to be so high (higher than about 3×10^5 V.cm^{-1}) that when an impact ionization chain is initiated, the charge carriers multiply faster than they can be extracted, leading to an exponential growth of current on a time scale of tens of picoseconds. The voltage has to be more than around 200 V to allow a chain reaction (avalanche) of electron liberation.

The bias voltage, V_B, is set above the breakdown voltage, V_{br}, by an amount called the excess bias voltage, $V_{ex}(= V_B - V_{br})$; V_{ex} is around 10% of V_B. This enables the avalanche to be maintained and an internal gain of up to 10^8 to be achieved. Such APDs are referred to as single-photon avalanche diodes (SPAD; Eisaman et al., 2011). The excess bias voltage has some effects on performance, for instance, it improves (i) the detection efficiency as V_{ex} increases, (ii) the resolution at higher electric field, and (iii) the dark count at higher bias voltages since the probability of a thermally generated carrier producing an avalanche increases. A SPAD is able to detect a single photon event with the limit imposed by the Poisson noise (see Section 4.1) of the signal. The photoimpulses can be counted individually (Fishburn, 2012); however, while counting number of photons, the SPADs should be used in Geiger-mode (Dautet et al., 1993). Since the current level quickly saturates due to the device resistance and space-charge effects, the APD acts as a digital device, switching from an off-state to a conducting-state in response to the detection of a single photon. This kind of operation is called Geiger-mode; hence, it is also referred to as Geiger-mode APD (G-APD).

The Geiger-mode operation at high gain (typically 10^5 to 10^6) leads to a very high current level within the APD. This should be kept under control by using a circuit to stop the avalanche. If the bias voltage, V_B, is kept constantly held above the breakdown voltage, V_{br}, the system cannot detect any further photons; the dissipation may damage the device. The pixel circuit should be able to detect the turn-on of the APD and quickly quench the SPAD by lowering its bias below breakdown. A pause of several hundred nanoseconds being required to restore the charges. The lower electric field is no longer able to accelerate carriers to impact-ionize with lattice atoms; therefore current ceases. In order to detect another photon, the bias voltage should be increased again above breakdown voltage. The current continues until the avalanche is quenched by reducing the bias voltage to a quenching voltage, V_q. This operation requires a circuit, known as quenching circuit, which is able to (i) sense the leading edge of the avalanche current, (ii) generate a standard output pulse synchronous with the avalanche build-up, (iii) quench the avalanche by lowering the bias down to the breakdown voltage, and (iv) restore the photodiode to the operative level. There exist two kinds of techniques to stop avalanche, for instance,

Passive quenching circuit In this design (Brown et al., 1986), a current-limiting resistor is placed in series with the photodiode, which would decrease the voltage, which would quench the avalanche if the resistor value is sufficiently large (about 100 Ω or more; see Figure 5.22b). The increase in current at avalanche results in a high voltage drop to bring the bias voltage below breakdown. The problem of passive quenching comes from the capacitance of the system that limits the bandwidth. However, such circuits typically have a long recovery time, which limits the effective maximum count rate (Stipcevic et al., 2010).

Active quenching circuit In this design (Brown et al., 1987), a fast discriminator is used to sense the steep onset of the avalanche current across a small resistor (of the order of 50 Ω) and also to provide a digital output pulse, synchronous with the photon arrival time. This circuit is able to detect the onset of the avalanche and then drops the voltage to below the breakdown voltage within a few nanoseconds. The result is a relatively short dead time (typically around 50 ns), following which quickly returns bias to above the breakdown voltage to sense the next photon. In this, the maximum count rates of 10 MHz and more can be achieved.

The silicon APDs have low bulk dark currents to be suitable for use in single photon-counting modules, although photon-counting has been demonstrated with germanium and InGaAs APDs (Cova et al., 1996). Recently, CMOS-based multi-pixel Geiger-mode silicon APDs have also been developed. This has advantages, such as (i) low production cost, (ii) low operating voltage, (iii) compact construction with a large overall active area, and (iv) good timing resolution. However, making integrated arrays of APDs working in photon-counting mode is a challenging task because of the photon-emission phenomenon that is caused by the electron avalanche. These photons may trigger avalanche in the pixels in their vicinity. Trakalo et al. (1987) built an integrated linear APD array with 32 pixels, but working in sub-Geiger mode (gain = 60). Also, it is difficult to assemble tens of diodes into arrays.

Applications for APDs include low light level detection, particularly in the field of diffraction and imaging where fast photon-counting are essential, instrumentation and aerospace applications, laser range finders and photon correlation studies, long distance optical communications, spatial light transmission, and bio-medical devices. They are also being used in high energy physics experiments. In the field of high energy astrophysics, APDs are being considered for Atmospheric Cerenkov Telescope (ACT; see Section 5.3.7.1) high energy imaging cameras (Otte et al., 2007). The previous fringe detection system in SUSI instrument consisted of two Perkin Elmer photon-counting module APD detectors (Davis et al., 2007).

Chapter 8

Radiation Detectors for Infrared Wavelengths

8.1 Atmospheric Transmission Windows

William Herschel (1738–1822) in 1800, the discoverer of the planet Uranus, got to know the heating effect in sunlight that occurred beyond the red end of the visible spectrum and understood its importance (Herschel, 1800). He called the region the thermometric spectrum, the term infrared (IR) being coined later (Krus et al., 1962; Liu et al., 2007). The IR-radiation extends from the red edge of the visible spectrum at ~0.70 nm to 1 mm (beyond which microwave region commences). The IR is, in general, is divided into five subregions (Byrnes, 2009), such as (i) near-IR (0.70–1.4 μm), (ii) short-wavelength IR (SWIR; 1.4–3.0 μm), (iii) mid-wavelength IR (MWIR; 3.0–8.0 μm), (iv) long-wavelength IR (LWIR; 8.0–14.0 μm), which provides a better penetration through smoke, smog, dust, water vapor, etc., and (v) very long-wavelength IR (VLWIR; 20.0–1000.0 μm).

The wavelength at which an object radiates most intensely depends on its temperature. The IR-radiation possesses low energy. As the temperature of an object cools, it shows up more prominently

High-Resolution Imaging: Detectors and Applications
Swapan K. Saha

ISBN 978-981-4613-27-9 (Hardcover), 978-981-4613-28-6 (eBook)
www.panstanford.com

at farther IR wavelengths, for example, a human body radiates IR energy with a peak wavelength of close to 10 μm at a temperature of 310 K. The IR-radiations are reflected, refracted, absorbed, and transmitted just like visible light. The basic laws of such radiations are discussed in Section 1.1.

The IR-radiation is invisible to human eyes (see Section 4.2.1) and has small energy, which is equal to the vibrational or rotational energy of molecules (see Section 1.5.1), the phenomenon of which makes it possible to identify molecules. The motion of atoms and molecules in an object produces IR-radiation. The emission- and absorption-lines of virtually all molecules, as well as atoms and ions lie in this part of the spectrum, whose primary source is heat (or thermal) radiation. The other sources include blackbody radiators (see Section 1.1.1), tungsten lamps, silicon carbide, etc.

Having longer wavelengths, the IR-radiations have better transmission through various mediums, such as the atmosphere and the gaseous medium. However, certain parts of the electromagnetic spectrum (see Section 1.1.6) pass through the Earth's surface. The atmospheric transmission windows occur in the visible and radio regions of the spectrum; gamma rays, X-rays, and UV rays are getting absorbed in the atmosphere. Also, most of the incoming IR-spectra are not useful for ground-based imaging since some of the atmospheric constituents, namely, water vapor (H_2O), ozone (O_3), and carbon dioxide (CO_2), are radiation absorbers at IR; the rest of the portions of the spectrum are referred to as the atmospheric transmission window in IR wavelengths (see Figure 8.1), which define the IR-bands that are usable on Earth (Hudson, 1969). It begins from 1.1 μm and is opened up approximately at wavelengths between 8 and 14 μm in the mid-IR region. The CO_2 is the dominant absorber (4.2 μm) in MWIR region, while H_2O is dominant absorber in LWIR region. The water vapor content in the atmosphere determines upper and lower wavelength limits for both MWIR and LWIR regions. The three spectral regions that are referred to as near-IR, mid-IR, and far-IR of the infrared spectrum (NASA, 2007) are described in Table 8.1.

The mid- and far-IR wavebands are important for the imaging of objects that emit thermal radiation, while the near-IR bands

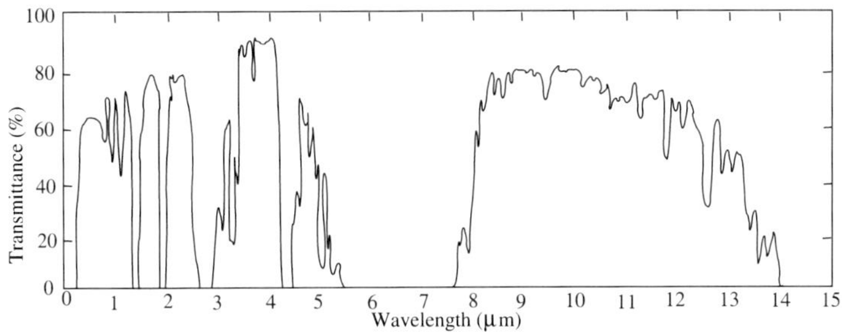

Figure 8.1 Electromagnetic spectral transmission windows in the IR-region.

Table 8.1 Spectral region and their temperature range

Spectral region	Wavelength range (μm)	Temperature range (kelvin, K)	Objects
I	0.7–1.0 to 5	740–5200	Cool red stars, red giants, interstellar dust
II	5 to 25–40	90–740	Planets, comets, asteroids, proto-planetary disks
III	25–40 to 200–350	10–90	Emission from cold dust, central regions of galaxies, very cold molecular clouds

are good for imaging scenes that reflect light. The bandwidths, 3–5 μm and 8–12 μm, where the rate of absorption is low, are used for remote sensing and satellite imaging applications. The other applications for the IR-radiation are in the fields of industry, agriculture, medicine, and communications (Jhabvala et al., 2007), in addition to the astronomical applications described in Section 8.2, for example, (i) forecasting weather and measurement of humidity, (ii) monitoring ozone and greenhouse effect, (iii) detecting volcanic eruptions, (iv) navigating through thick fog and clouds, (v) estimating the underground water, and (vi) detecting the mine. These wavelengths can also be applied for the purposes of (i) military and surveillance, (ii) non-destructive testing and thermal analysis, and (iii) night vision.

8.2 Infrared Astronomy

Many regions in space are hidden from optical telescopes since they are embedded in dense regions of gas and dust. Materials such as dust and gas absorb the light at visible wavelengths and re-emit them at mid-IR and far-IR wavelengths. Most of the optical observations of star-forming clouds of gas and dust in the interstellar medium (ISM) envisage the dark contours of the cloud. Dust and gas around a central hot object are heated by its radiation. The cloud lies primarily within the plane of the Galaxy (Milky Way), whose density, mass and temperature allow the formation of molecules. Having much longer wavelengths than optical light, IR-radiation can pass through such dusty regions of space. In the near-IR, the dust also becomes transparent, allowing one to envisage regions hidden by dust in the visible image. The obscuring effect of the dust is often entirely insignificant at the mid-IR wavelengths around 10 μm, allowing direct studies of the physical processes what takes place inside.

Following the discovery of Herschel, the IR-astronomy came into existence and provided a wealth of new information about the universe. These include cool stars whose temperatures are relatively lower, clouds of particles around stars, nebulae (the cloud of dust and gas molecules), stellar molecules, brown dwarfs, and exo-solar planets. The infrared emission beyond about 1.1 μm is primarily due to heat or thermal radiation. The temperature range for the near-IR spectral region falls between 740 and 3000–5000 K, while the temperature range for the mid-IR falls between 92.5–140 and 740 K. The IR photometric band system covering a particular wavelength region is given in Table 8.2; the various IR filter systems follow this pattern.

In the near-IR wavelength (J, H, K, L-bands) range, the objects like cool red stars, red giants, and interstellar dust can be seen. The objects such as planets, comets, asteroids, cosmic dust clouds, and star-forming clouds can be observed in the mid-IR wavelength range, particularly with N-band. A star is a massive luminous ball of plasma in space that generates energy through nuclear fusion in its core. The evolution of each star is controlled by their total

Table 8.2 Infrared photometric bands

Band	Wavelength range in μm	Centered at (μm)	Sky transparency	Sky brightness
J	1.1–1.4	1.25	High	Low at night
H	1.5–1.8	1.65	-do-	Very low
K	2.0–2.4	2.2	-do-	-do-
L	3.0–4.0	3.5	3.0–3.5 μm; Fair 3.5–4.0 μm; High	Low
M	4.6–5.0	4.75	Low	High
N	7.5–14.5	10.5	8–9 μm: Fair 10–12 μm; Fair Others: Low	Very high -do- -do-
Q	17–40		Very low	-do-
Z	330–370		-do-	-do-

mass, composition, and age. The stars form out of clouds of dust and hydrogen gas, which are composed primarily of H and He. The other notable molecules are H_2, H_2O, OH, CO, H_2CO, dust of silicates, iron, etc. These clouds are approximately in a state of virial equilibrium, which occurs when gas pressure equals gravity (Saha, 2007, and references therein). As the clouds become more massive, they begin to collapse into a smaller volume, causing the center to be under tremendous pressure, and to heat up. If the pressure is small one gets contraction by free fall of material toward the core, which increases the density. The release of gravitational energy heats up the material, which in turn, produce thermal pressure and stop the in-fall. A part of the gravitational energy lost in this collapse is radiated in the infrared, with the rest increasing the temperature of the core of the object. At sufficiently high pressure, a proto-star is formed as the cloud condenses by the gravitational accretion of gas and dust. As it grows, the contraction of the particles causes it to heat and begin to glow. Its center heats up to several millions of degrees, facilitating nuclear fusion and forming the star. The red giants are large in size and are running out of their nuclear fuel. They can swell up to 100 times their original size and have temperatures ranging between 2000 and 3500 K. These stars radiate most intensely in the near-IR region.

Stars are classified based on the spectral characteristics providing information about their temperature, luminosity, composition, velocity and rotation rate in space, and mass. The absorption spectra are used to classify stars, which can be observed with regard to appearance or disappearance of certain discrete features. The range of temperature involved shows atomic energy levels (see Section 1.1.6), which have been populated.

The stars are divided into mainly O, B, A, F, G, K, M classes (see Saha, 2007, and references therein). Within each type, these are further categorized by subdividing each class into 10 subdivisions, denoted by the numbers from 0 to 9, for example, an O2 star is hotter than O9 star. The O-type stars are relatively rare having very high surface temperature in the range of 25,000 K and above. They are very luminous, being strongly violet-blue in color. These stars have prominent ionized and neutral helium lines and emit most of their radiation in ultraviolet; they have weak hydrogen lines. The B-type stars are extremely luminous blue-white having surface temperature lying between 11,000 and 25,000 K, namely, β Orionis (Rigel), a blue super-giant. Their spectra have neutral helium and moderate hydrogen lines. These two types of stars live for a very short time. The A-types are white in color having surface temperature in the range of 7500–11,000 K. Some of the brightest stars in the sky belong to this classification. The stars of this class have strong hydrogen lines and also ionized metals. The spectra of the F-type stars is characterized by the weaker hydrogen lines and ionized metals. They have surface temperature between 6000 and 7500 K and are yellow-white in color. The G-type stars are yellow in color with surface temperature ranging between 5000 and 6000 K. The Sun, classified as a G2 star, falls under this category having effective temperature 5760 K. Stars of this class have weaker hydrogen lines than F-type stars, but along with the ionized metals, they have neutral metals. The K-type stars are orange in color, which have surface temperatures are between 3500 and 5000 K, which is a temperature low enough for simple molecules to form. Some of these stars are giants and supergiants (extremely massive and luminous stars, nearing the end of their lifespan), such as Arcturus, while others like α Centauri B are main sequence stars. They possess extremely weak hydrogen lines and have mostly neutral metals.

The red dwarfs are smaller in size and radiate strongly in the near-IR, many of which have been discovered in this wave band. They fall under the category of class M, for instance, Proxima Centauri. This type also hosts most giants and some supergiants, namely, Antares and Betelgeuse (α Orionis, an M2 star). These stars have very cool surface temperature, below 3500 K, which allows more complex molecules to form. The titanium oxide (TiO) can be strong in this group of stars. The spectrum of an M-type star shows lines belonging to molecules and neutral metals but hydrogen is absent, in general.

There are R, N, S type stars as well. The R, N-type stars, also known as carbon (C) stars. The former type represents the carbon star equivalent of the K-type stars, while the latter type represents the carbon star equivalent of the M-type stars. The S-type stars is similar to M, but with zirconium oxide (ZrO) replacing the regular TiO. The L-type stars are the brown dwarfs, which are substellar objects with masses insufficient to run the hydrogen nuclear fusion reactions in their cores. Also, they contain lithium, which is rapidly destroyed in hotter stars. These dwarfs may have masses between 10 and 84 Jupiter mass. Their surface temperatures are expected to be around 1000 K and the luminosities are of the order of 10^{-5} $L_{\odot}$, which indicates the absolute magnitude are going to be around $+17$.

The mid-IR sensor allows the spectroscopy of interstellar gas clouds, allowing the identification of their composition. The hydrogen gas can be identified by observing the vibrational states of molecular hydrogen and also by observing the hydrogen gas that has been stripped of one electron, designated as H II. The spectral shift of these emission lines allows one to deduce the temperature of the gas. One may observe using mid-IR sensors the temperature distribution of gas and dust clouds since the peak of the blackbody emission from many astronomical objects lies in this region. The detection of low temperature cosmic dust is important in the investigation of star formation as well, for example, the formation of stellar clusters in the blue compact dwarf starburst galaxy, NGC 5253, observed in the N-band (Martin-Hernandez et al., 2004).

In the far-IR region, where the temperature range is between 10 K and 90 K, the emission from cold dust, central regions of galaxies, and very cold molecular clouds can be seen. Near 100 μm, the transmission is low. The photons in this region of the spectrum are heavily absorbed by the atmosphere. This is primarily due to excitation of vibrational and rotational modes of various molecules in the atmosphere (see Section 1.5.1). The molecules exhibit this property are, H_2O, O_2, O_3, CO_2, N_2O, and CH_4. Observation in this range may be carried out through satellites. Spectrometers on-board infrared missions, such as the Infrared Space Observatory (ISO) and Kuiper airborne observatory, as well as near-IR spectra from ground-based observatories, have led to the discovery of many atoms and molecules in many different regions in space.

8.2.1 *Ground-Based IR Observations*

Most ground-based telescopes that observe in the near-IR wavelengths rely on the same telescopes as for optical astronomy. An IR-detector is attached to the focal plane of a telescope for detecting IR light. The atmospheric infrared radiation peaks at a wavelength of about 10 μm. The atmospheric transmission at these wavelengths is not 100%, but improves with height. Therefore, the IR telescopes are required to be placed near the top of high, dry mountains. A cluster of IR telescopes can be witnessed on top of Mauna Kea (a dormant volcano), Hawaii, at an elevation of 4145 m, which is much above of the IR absorbing water vapor. However, the complication in observing the cosmic sources arises from the reflection of external sources, for instance, Earth, sky, telescope and its enclosure, other objects, and atmospheric path radiance, which emit IR-radiation. The discrimination between IR emissions from the target and the surrounding background leads to target detection. In order to so, one measures both the emissions from the target, as well as from the Earth's atmosphere. In order to determine an accurate value, the subtraction of the latter from the former is carried out. The IR-telescopes are specially designed to limit the amount of the thermal emission from reaching the detectors. These telescopes need to be cooled; the detectors must be cooled as well to extremely low temperatures.

As stated in Section 1.5, when an electromagnetic radiation travels through the atmosphere, a part of it is absorbed by gaseous molecules and a part is scattered by suspended aerosol particulates. Absorption and scattering of scene radiation out field-of-view (FOV) reduces the target signature. Scattering into FOV and path radiance increases noise level and turbulence and aerosol forward scattering cause image distortion. The total reduction in radiation when an electromagnetic wave is propagated from a cosmic source through the Earth's atmosphere along the line-of-sight is called extinction (see Section 1.1.5).

Atmospheric turbulence and the concomitant development of thermal convection in the atmosphere distort the phase and amplitude of an incoming wavefront of starlight. Light reaching the entrance pupil of an imaging system is coherent only within patches of diameters of order $r_\circ$, Fried's parameter (Fried, 1966). This limited coherence causes blurring of the image, blurring that is modeled by a convolution with the point-spread function (PSF). Both the sharpness of astronomical images and the signal-to-noise (S/N) ratio (hence faintness of the objects that can be studied) depend on angular resolution, the latter because noise comes from as much of the sky as is in the resolution element. Thus, reducing the beam width from, say, 1 arcsec (″) to 0.5″ reduces sky noise by a factor of 4. Two physical phenomena limit the minimum resolvable angle at optical and infrared (IR) wavelengths—diameter of the collecting area and turbulence in the atmosphere. The crossover between domination by aperture size (λ/aperture diameter) and domination by atmospheric turbulence ('seeing') occurs when the aperture becomes larger than the size of a characteristic turbulent element.

At present, many large and moderate size ground-based telescopes are employing adaptive optics (AO) system to obtain near diffraction-limited information (Saha, 2007, and references therein). Based on the hardware-oriented approach, such a system compensates atmospheric effects on the image in real time. Most of the results that obtained from the ground-based telescopes equipped with AO systems are in the near-IR band, while results at visible wavelengths continue to be sparse. The advantage of such a system at IR wavelengths is that the optical path variations caused by atmosphere are small; the air refractive index has a larger dispersion in the visible wavelength. The atmospheric coherence diameter, r_0 is proportional to the 6/5 power of the wavelength (see equation 1.128) implying the width of seeing-limited image, $1.22\lambda/r_0 \propto \lambda^{-1/5}$, varies with λ. At 10 μm wavelength, the diameter of the cell should increase by a factor of 38. In such a situation, the size of seeing disk decreases as $\lambda^{-1/6}$. Therefore, the angular size of an image produced at 10 μm wavelength should be 1.8 times smaller than that in the visible wavelength, under similar atmospheric conditions. The number of degrees of freedom, i.e., the number of actuators on the deformable mirror (DM; see Section 1.5.6) and the number of subaperture in the wavefront sensor, in an AO system is determined by the following equation:

$$\left(\frac{D}{r_0}\right)^2 \propto \lambda^{-12/5}, \tag{8.1}$$

where D denotes the diameter of the telescope.

In the early 1900s, the IR-radiation was successfully detected from a few planets, such as Jupiter and Saturn, as well as from a few bright stars, namely, Vega and Arcturus. However, the insensitivity of the early IR instruments prevented the detection of other near-IR sources. Work in IR astronomy remained at a low level until breakthroughs in the development of sensitive IR detectors were achieved in the 1960s; however, their regular use in astronomy began a couple of decades ago. The near-IR observations that are carried out from the ground-based observatories are made in the same fashion as visible light observations for wavelengths less than 1 μm but need special IR detectors beyond 1 μm. The mid- and far-IR observations require the use of cooled detectors containing crystals whose electrical resistance is very sensitive to heat. The first IR survey of the sky was made at the Mt. Wilson Observatory employing a liquid nitrogen cooled lead-sulfide (PbS) detectors, which revealed as many as 20,000 IR sources. The PbS detectors operate in the 1 to 4 μm range but use materials that are more sensitive to IR. When IR-radiation falls on a PbS cell, it changes the resistance of the cell. This change in resistance is related to the amount of radiation that falls upon the cell.

8.2.2 *Space-Based IR Observations*

Infrared array detectors have been used on several IR satellite missions. The first cooled telescopes were placed on rockets, which could observe the sky for several minutes before reentry. A series of rocket flights by Air-force Cambridge Research Laboratory could survey the cosmos at wavelengths of 4, 10, and 20 μm. Although the total observation time by these flights was short (about 30 min), this project, called Hi Star, enabled to detect 2363 stars. Helium-filled mylar balloons have carried IR telescopes up to an altitude of about 40 km. A germanium bolometer was attached to a balloon to make observations of Mars in 1963. At the fall of 1966, the Goddard Institute of Space Sciences used balloons to survey the sky at 100 μm. Their program led to the discovery of about 120 bright IR sources near the plane of our galaxy.

IRAS The InfraRed Astronomical Satellite (IRAS) was launched in 1983, in which a telescope was housed in a dewar filled with liquid helium equipped with an array of 62 detectors. The entire telescope, in principle, should be cooler than the objects. IRAS provided the high sensitivity all sky map at wavelengths of 12, 25, 60, and 100 μm and detected about 500,000 IR sources. Among the discoveries made by this satellite included a disk of dust grains around the star Vega (α Lyrae), several comets, and strong IR emission from the interacting galaxies, as well as wisps of warm dust called IR-cirrus. The IRAS revealed the core of the Galaxy.

Airborne telescopes Infrared telescopes on-board aircraft, such as the Kuiper Airborne Observatory (KAO) were used to discover the rings of Uranus in 1977. The KAO has been used to gather IR astronomical data for over 20 years and can fly at an altitude of 41,000 ft, which is above 99% of the Earth's water vapor.

COBE satellite The National Aeronautics and Space Administration (NASA) launched the COBE satellite to study both IR and microwave characteristics of the cosmic background radiation (the remains of the extreme heat that was created by the Big Bang), which was able to map the brightness of the entire sky at several IR wavelengths and discovered that the cosmic background radiation showed tiny variations in temperature. These variations may have led to the formation of galaxies.

IRTS In the mid-1990s, the InfraRed Telescope in Space (IRTS), a satellite mission by Japan, was launched with a near and mid-IR spectrometer covering wavelengths of 1.4 to 4 μm and 4.5 to 11 μm, respectively. It also carried a far-IR line mapper for observing oxygen and carbon spectral lines at 63 and 158 μm, and a far-IR photometer for studying the sky at four bands centered at 150, 250, 400, and 700 μm.

ISO The European Space Agency (ESA) launched the Infrared Space Observatory (ISO) in the mid-1990s, which observed at wavelengths between 2.5 and 240 μm. It contained instruments that measured details of both the shorter and longer wavelength regions of the infrared spectrum, an infrared camera having two IR arrays, and a photometer. ISO has detected dry ice in interstellar dust and hydrocarbons in some nebulae.

8.3 Thermal Detectors

The thermal imaging systems exploit MWIR (3–5 μm) and LWIR (8–14 μm) atmospheric windows (see Section 8.1) where the atmosphere is transparent enough to permit sufficient self-emitted radiation from near ambient temperature objects to reach the sensors for passive imaging. One of the requirements for thermal imaging system is the optics module that captures radiations emanating from the objects in view (target and background) and focuses it onto the detector place at the focal plane of the optics. The focused light is scanned by the IR-detector elements creating electric impulses. The impulses are sent to a signal-processing unit that translates and sends the information from the elements into data for the display, where it appears as various colors depending on the intensity of the infrared emission.

The IR optics can be refractive, reflective or catadioptric (combination of both lens and mirrors) to make IR optics free of aberrations and to provide diffraction-limited performance. The materials used in IR optics include Ge, Si, zinc selenide (ZnSe), zinc sulfide (ZnS), and certain types of chalcogenide (a chemical compound, commonly reserved for sulfides, selenides, and tellurides; Greenwood and Earnshaw, 1997) glasses. In general, a combination of Ge with ZnS or ZnSe is used in LWIR, while a combination of Ge, Si, ZnS, and ZnSe is used in MWIR. Multi-layered anti-reflection coatings are done on these lenses to enhance their transmission in the preferred spectral band. A diamond-like carbon coating is done on front optical surface of IR lens assembly to provide resistance to abrasion, salts, acids, alkalis, and oil and protection from high velocity airborne particles. An important consideration in optics design is to achieve 100% cold shield efficiency. The exit pupil is placed at the cold stop (filter) of the detector, so that optics pupil matches with its stop, which ensures it views the radiation coming from the scene.

With the discovery of Seebeck effect (Seebeck, 1822), a group of detectors sprung up like thermocouples and thermopiles (see Section 8.3.4), in which thermoelectric effect was applied. In the mid-1850s, astronomers used thermocouples to detect IR-radiation from the Moon. A major breakthrough came in the early 1960s with the development of the germanium (Ge) bolometer (see Section 8.3.3). Such an instrument was more sensitive than the earlier detectors and was capable of detecting all IR wavelengths.

8.3.1 *Thermal Effects*

The physical mechanism of thermal detectors are the absorption of radiation in a near black surface and precise conversion into heat that flows through a sensing element through a weak thermally conducting path to a heat sink (see Figure 8.2). A thermal equilibrium temperature is achieved between the sensing element and heat sink, which is determined by the product of the heat current in the sensing element and the thermal resistance between the heat sink and sensing element. The time constant is the product of the heat capacity of the sensing element and the thermal resistance, which increases the thermal conductance to the heat sink decreases the sensitivity and shortens the response time. Two approaches have found the utility in IR systems, for example:

Bolometer effect In bolometers, a change in resistivity of a material in response to the heating effect of incident radiation is noticed. The temperature dependence is specified in terms of the temperature coefficient of resistance, α, defined as

$$\alpha = \frac{1}{R}\frac{dR}{dT}, \tag{8.2}$$

where R is the resistance and T the absolute temperature of the subject.

There exist two types of bolometers:

(1) **Thermistor**: A thermistor consists of a thin film or of semiconductor materials. As stated in Section 2.2.4.4, the conductivity, σ of a semiconductor is proportional to the density of free carriers (see equation 2.24), which may be enhanced by increasing either electrons or holes. From equation (2.29), the conductivity of germanium (Ge) is found to increase $\approx$6% per degree increase in temperature. Such a large change in conductivity with temperature places limits on the use of semiconductor devices in some applications. However, in some cases, this property of semiconductors is used. A semiconductor used in this manner, called thermistor whose resistance varies significantly with temperature, which finds application in thermometry, in the measurement of microwave frequency power, in control devices actuated by changes in temperature.

The thermistor detector, in general, contains two closely spaced thermistor flakes (or thin slabs); one is an active detector while the other acts as a reference detector. The active flake is coated with black material to increase its IR absorption, whereas the compensating flake is optically shielded from incident IR radiation. They are separately mounted on an insulating (sapphire) substrate that is placed on a heat sink. The response time depends on the rate of heat transfer between the active flake and heat sink. The detector sensitivity decreases for good thermal contact. The material can be an oxide of manganese (Mn), cobalt (Co), or nickel (Ni), which exhibits negative temperature coefficient. Since it is operated under electrical bias, the self-heating effect due to the internal power dissipation from the bias-current should be taken into account.

With increasing bias, the temperature rises rapidly and the dynamic resistance, dV/di, in which V denotes the voltage and i the current, decreases. Above a critical bias, the dV/di turns out to be negative. The thermistor bolometers that are operated below this critical bias, sense the heating effect of IR-radiation by virtue of the temperature dependent electrical resistance.

(2) **Cryogenic bolometer**: The cryogenic bolometer consisting of a flat slab cut from the core of a carbon resistor is mounted on a copper heat sink maintained at liquid helium temperature (4.2 K; see Section 4.3.1). The carbon cryogenic bolometers are highly sensitive, which are useful for astronomical applications.

Pyroelectric effect Pyroelectric effect depends on the rate of change of temperature. So detector is only sensitive to modulated radiation; a change in the internal spontaneous polarization is envisaged. However, at constant temperature, the internal charge distribution gets neutralized by free electrons and surface charges, so that no voltage is detectable. With the rapid change of temperature, the internal dipolement would change, producing a transient voltage. The pyroelectric effect can be exploited as a sensitive detector of modulated radiation, operating at ambient temperature.

8.3.2 *General Characteristics of Thermal Detector*

The detectors based on thermal effects produce an electrical response based on a change in temperature of the sensor. The simplest representation of the thermal circuit (Rieke, 2003) is shown in Figure 8.2. An ideal thermal detector has a noise (see Section 4.5) that is associated with the thermal fluctuations of the heat loss to the heat sink, and this coupling is purely radiative (Smith et al., 1968; Putley, 1980; Budde, 1983). The detector is a thermal mass coupled via a thermal link (heat conductance), $\mathcal{G}$, to a heat sink at a constant temperature. When a radiation input is received by the detector, the rise in temperature at thermal equilibrium is

$$\mathcal{W}\delta t - \mathcal{G}\Delta T\,\delta t = \mathcal{H}\delta(\Delta T), \tag{8.3}$$

where $\mathcal{W}$ is the radiation power incident upon the detector, of which a fraction, η, is actually absorbed, ΔT the rise in temperature of the detector after absorption of the incident radiation, $\mathcal{H}$ the heat

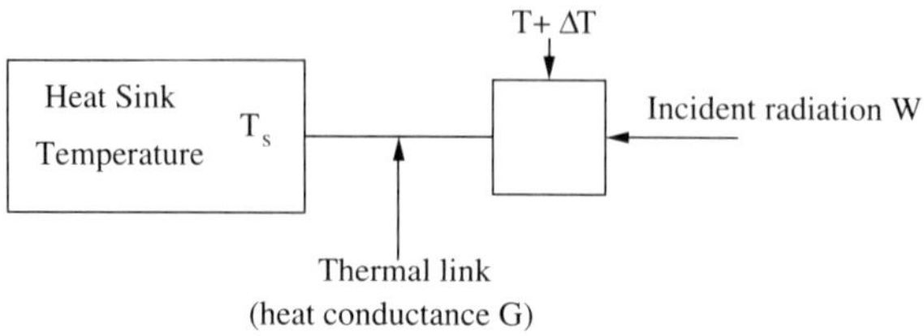

Figure 8.2 Schematic diagram of a thermal circuit.

capacity of the sensing element in which the temperature is $T + T_S$; T_S the heat sink temperature, and the amount of heat-loss through the thermal link is $\mathcal{G}\Delta T\,\delta t$.

At the limit, $\delta t \to 0$

$$\mathcal{W} = \mathcal{H}\left[\frac{d(\Delta T)}{dt}\right] + \mathcal{G}\Delta T. \tag{8.4}$$

Let $\mathcal{W}$ has a component in the form of $\mathcal{W}_0$, independent of time, and a component modulated at an angular frequency, ω; therefore

$$\mathcal{W} = \mathcal{W}_0 + \mathcal{W}_\omega e^{i\omega t} \qquad \mathcal{W}_\omega \not> \mathcal{W}_0. \tag{8.5}$$

Expressing the time dependent part of ΔT as

$$\Delta T = \Delta T_\omega e^{i(\omega t - \psi)}. \tag{8.6}$$

Equation (8.4) may be solved to obtain (Kruse et al., 1962; Smith et al., 1968)

$$\Delta T_\omega = \frac{\eta \mathcal{W}_\omega}{\sqrt{\mathcal{G}^2 + \omega^2 \mathcal{H}^2}}, \tag{8.7}$$

and

$$\psi = \tan^{-1}\left[\frac{\omega \mathcal{H}}{\mathcal{G}}\right]. \tag{8.8}$$

For good sensitivity, ΔT_ω should be as large as possible. This implies small values for both $\mathcal{H}$ and $\mathcal{G}$. As ω is increased, $\omega^2\mathcal{H}^2$ exceeds $\mathcal{G}^2$ and ΔT_ω falls inversely as ω

$$\Delta T_\omega = \frac{\eta \mathcal{W}_\omega}{\mathcal{G}\sqrt{1 + \omega^2 \tau_T^2}}. \tag{8.9}$$

A characteristic thermal response time can be defined as

$$\tau_T = \frac{\mathcal{H}}{\mathcal{G}}. \tag{8.10}$$

For good response at a frequency ω one requires $\tau \ll 1/\omega$. For typical detector design, τ_T falls within the range of milliseconds to seconds, which is longer than the typical response time of photon detector. For a given $\mathcal{H}$, $\mathcal{G}$ cannot be made arbitrarily small since it enhances τ_T that is undesirable. The limiting value of $\mathcal{G}$ can be estimated from Stefan–Boltzmann's radiation law (see Section 1.1.2). If the thermal detector has a receiving area, A_d of emissivity, η, under thermal equilibrium with the surroundings, it radiates a total flux $A_d\eta\sigma T^4$, in which σ designates Stefan–Boltzmann constant. If the temperature of the detector is increased by a small amount, $\Delta T = dT$, the flux radiated is enhanced by

$$4A_d\eta\sigma^3 dT = 4A_d\eta\sigma^3\Delta T. \tag{8.11}$$

Hence, the radiative component of thermal conductance is

$$\mathcal{G}_R = 4A_d\eta\sigma T^3. \tag{8.12}$$

When the detector is in thermal equilibrium with the heat sink, the root-mean-square (RMS) fluctuation in the power flowing through the thermal impedance into the detector (Smith et al., 1968) is $\sqrt{4k_B T^2\mathcal{G}}$, with k_B as Boltzmann's constant and T the absolute temperature, which is the smallest if $\mathcal{G} = \mathcal{G}_R$. The noise equivalent power (NEP; see Section 4.1.1.4) is given by

$$NEP = \sqrt{16A_d\sigma k_B T^5/\eta}, \tag{8.13}$$

for a unity measurement bandwidth. Equation (8.13) assumes that the detecting element is irradiated by a hemisphere of blackbody radiation (see Section 1.1.1) at the same temperature, T as the detector. The corresponding specific detectivity (D^*; see Section 4.1.1.4), assuming that the signal varies as the area and the noise as its square root, is

$$D^* = \frac{\sqrt{\eta}}{4\sqrt{\sigma k_B T^5}}, \tag{8.14}$$

where the detector and background are at the same temperature.

The detector must be supported in vacuum by providing zero thermal conductance. If the electrical contacts are required for the output, they should also possess zero thermal conductance. Since such conditions cannot be achieved, the actual value of $\mathcal{G}$ is greater than $\mathcal{G}_R$ and hence the NEP becomes poorer than the value determined by equation (8.13). Every detector is limited by the signal loss due to incomplete absorption at the surface and any transmission losses by the optical system that puts the radiation on the detector. In detectors employing an electrical read-out mechanism, electrical noise fluctuations (Johnson noise; see Section 4.5.2.1) and also along with other noise sources, such as low frequency noise (in some cases) must be taken into account. There would be Johnson noise associated with the output impedance of the detector. The final source of noise is that associated with the amplifier; a low-noise amplifier (see Section 4.5.3) is required. Having evaluated all the sources of noise the NEP can be found by referring all the noise sources to the input terminals of the amplifier. If an input signal power, P is incident on a detector with responsivity (see Section 4.1.1.2), $\mathcal{R}$, the output signal V_s becomes $V_s = \mathcal{R}P$. If ΔV_i represents the individual noise source voltages, the equivalent noise voltage, V_n, is related by $V_n^2 = \sum(\Delta V_i)^2$, and hence the NEP is derived as

$$NEP = \frac{\sqrt{\sum(\Delta V_i)^2}}{\mathcal{R}}. \tag{8.15}$$

8.3.3 *Bolometers*

A bolometer is a resistive element fabricated from a material having a large thermal coefficient and a low thermal conductance between the absorbing area and the read-out circuit. The absorbed IR-radiation is converted into heat, which produces a large change in resistance such that measurable electrical signals can be detected. If the incident radiation produces a minor change δR in resistance R of the bolometer through which a bias-current, i, flows, the output signal voltage, V_s, is given by

$$V_s = i\delta R. \tag{8.16}$$

If the input radiation causes a change ΔT in the temperature of the element and the relative temperature coefficient of resistance $(\frac{1}{R}\frac{dR}{dT})$ is α, the open-circuit voltage is

$$V_s = i\alpha R\Delta T. \tag{8.17}$$

Invoking equation (8.7), one gets

$$V_s = \frac{\eta \mathcal{W}_\omega i\alpha R}{\sqrt{\mathcal{G}^2 + \omega^2\mathcal{H}^2}}, \tag{8.18}$$

which yields the following expression for the responsivity, $\mathcal{R}$

$$\mathcal{R} = \frac{\eta i \alpha R}{\sqrt{\mathcal{G}^2 + \omega^2 \mathcal{H}^2}}. \tag{8.19}$$

For high responsivity, $\mathcal{R}$ (see Section 4.1.1.2), one should have large values of α, R, and i, as well as small values of $\mathcal{G}$ and $\mathcal{H}$. Equation (8.19) implies that the amplifier input impedance is large compared with R. This sets a limit on R so that the resistance cannot be increased independently. Also, with very high resistance elements, the input capacity of the connecting leads and the amplifier may produce a time-constant larger than the thermal time-constant. The bias-current, i, cannot be increased independently as well. The high such current raises the temperature of the element and with negative value of α, thermal runaway ensues, albeit an increase in noise with increasing i would set an optimum value. However, the problem of maximizing comes to choosing a suitable material. A bolometer focal plane array (FPA; see Section 5.5.1) consists of thin-film semiconductor photoresistors micro-machined on a silicon substrate, which is used for the far-IR through millimeter wave spectral range.

Metal bolometers The first bolometer used a platinum resistance element and later other metals, such as nickel (Ni), were employed. The thermal coefficient, α, decreases with temperature, and burnout does not occur (Bass et al., 1995). The coefficient is approximately equal to the inverse of the temperature, and is therefore never very high. The metal films have traditionally been used to make the bolometer detectors because of their low flicker noise (see Section 4.5.2.4). These devices use semiconductor films of 50 nm thickness having thermal coefficient of resistance of 2% per degree centigrade. One important element is the Pt100 resistor for the measurement of temperature between -100 °C and 400 °C. The effect is, in general, exploited by integrating the resistor into a Wheatstone bridge. The spacing between the micro-bridge and the substrate is selected to maximize the pixel absorption in the 8–14 μm wavelength range. The photolithographic techniques (see Section 2.3.1.1) pattern the thin film to form detectors for individual pixels.

Semiconductor bolometers For IR detectors, the metal film bolometers are replaced by semiconducting elements, which have a much larger temperature coefficient. The first semiconductor bolometer was the thermistor (see Section 8.3.1). The spectral response of the bolometer is flat across the IR-spectrum, since the sensing mechanism is independent of the photoexcited carriers jumping across an energy bandgap. The superconductor devices do not require critical temperature control and with the selection of correctly doped samples are better absorbing. The first cooled semiconductor bolometer used flakes of the carbon composition material; however, it was found that superior performance could be obtained with precisely doped germanium (Ge) or silicon (Si) elements. The detectors based on these elements are used in IR astronomy.

Superconducting bolometers At very low temperatures, it is probable to obtain much larger relative changes in resistance than near-room temperature in which the specific heat is much smaller. The cooled bolometer used a superconducting element with a transition in the helium temperature region. It is reiterated that the best performance of such a device can be envisaged at an extremely low temperature (at liquid helium temperatures, 4.2 K; see Section 4.3.1), which is, in general, immersed in metal dewar (insulated thermal flask). Under such a low temperature, the sensitivity of the detector would be orders higher than that for a room temperature device. At liquid helium with proper concentration of impurities, the resistance, R, follows an exponential temperature variation as (Bass et al., 1995)

$$R = R_0 e^{\beta/T}, \tag{8.20}$$

so that the temperature coefficient

$$\alpha = \frac{1}{R}\frac{dR}{dT} = -\beta/T^2. \tag{8.21}$$

The value of β depends upon the particular material; these detectors can burn out. Equation (8.21) shows that α increases as T is reduced, thus cooling is desirable and operation below 4 K is, in general, advantageous. The most used low-temperature bolometer is germanium in a bath of liquid helium (Low, 1961).

8.3.4 *Thermopiles*

The thermopile is one of the oldest IR detectors and is still used in its latest form (the thin-film thermopile) in space instrumentation; the thin-film thermopile is a series connection of thermocouples. A thermocouple that converts heat into electric current is made by joining two dissimilar conductors having a large Seebeck coefficient, Θ (the difference in the voltage change with temperature between the two materials), which gives rise to a potential difference (Felgett, 1949).

The thermopiles are arrays of thermocouples connected in series. However, for efficient operation, the materials should have a high electrical conductivity, σ, to minimize the heat loss and a low thermal conductivity, K, to minimize heat conduction loss between hot and cold junctions of the thermopile. These requirements are not compatible and a good choice of thermoelectric material that for which $\sigma^{\Theta^2/2K}$ is a maximum. Certain heavily doped semiconductors, such as bismuth telluride (Bi_2Te_3) and related compounds are used. The active or hot junctions of a thermocouple deposit on a freestanding film and are blackened. The reference or cold junctions are deposited on the same film but are located over a heat sink. For DC to low frequency measurements, the thermopile is superior to the other two because of its long-term stability and simple processing circuitry. Such a detector has a sensitivity of 20 $V.W^{-1}$ between 5 and 14 μm wavelength (PerkinElmer, 2003).

8.3.4.1 Golay cells

The Golay cell is a gas thermometer where changes in gas pressure at constant volume distorts a thin membrane and this provides a measure of absorbed power (Golay, 1947). The response time of the detector is long (between 2 and 30 ms) and the incoherent radiation should be modulated at about 10–20 Hz. In such a cell, the radiation absorbed by a receiver inside a closed capsule of gas (xenon, in general, for its low thermal conductivity) heats the gas causing its pressure to rise, which distorts a flexible membrane on which a mirror is kept. The movement of such a mirror is used to deflect a beam of light shining on a photocell, and therefore producing a change in the photocell current as the output. In present-day Golay cells, the beam of light is provided by a light-emitting diode (LED; see Section 3.3.8.1) and a solid-state photodiode (see Section 5.3.5) is used to detect it. Another method of obtaining an electrical output from the gas cell is to place a fixed conductor in close proximity of the distorting membrane forming a variable condenser, which can be measured with a suitable circuit.

8.3.4.2 Pyroelectric detectors

The pyroelectric detectors are dielectric materials with a temperature dependent spontaneous electrical polarization. The direction of this polarization is known as polar axis of the material. Although the external field produced by their inherent internal dipole moment would, in general, be neutralized by an extrinsic charge distribution near the surface of the material. In good pyroelectric material such charge distribution is relatively stable, so that slow changes in the sample's temperature, which produce changes in the internal dipole moment, causes charges to appear on the surfaces perpendicular to the polar axis of the material. When the pyroelectric detector absorbs radiation, its temperature rises, which in turn produces an observable current when the electrodes are connected to external circuitry. The electric field is directly proportional to crystal temperature because the degree of alignment (or polarization) changes. When parallel electrode are attached to the crystal, the electric charge will be attracted and stored on the electrode plates. Assuming that the element receives radiation over an area, A_d, normal to the polar axis of the material, the polarization change in temperature, ΔT, is described by $P = p\Delta T$, in which p is the component of the pyroelectric coefficient normal to the electrodes. The pyroelectric charge, Q, generated is $Q = pA_d\Delta T$, so that the effect of a temperature change on a pyroelectric material is to cause a current $i_{ph} = dQ/dt$ to flow in an external circuit (Rogalski, 2011) such that

$$i_{ph} = A_d p \frac{dT}{dt}, \tag{8.22}$$

with dT/dt as the rate of change of temperature with time.

In order to make a pyroelectric device work, it is necessary to modulate the source of energy. This is performed, in general, by mechanical chopping or by moving the detector relative to the source of radiation. A pyroelectric detector has fast response and high detectivity. Such a device can be employed over a wide range of spectrum, e.g., from near-IR to sub-mm region. The principal materials used for pyroelectric detectors are the members of triglycine sulfate (TgS) group, lithium niobium oxide ($LiNbO_3$), lithium tantalate ($LiTaO_3$), strontium barium niobate (SrBaNb), ceramics members of the lead zirconate titanate (PZT) group, and films of polyvinyl fluoride (PVF).

8.3.5 *Thermal Imagers*

As stated in Section 4.1, the thermal detectors absorb thermal energy raising the temperature of the detecting element, which in turn changes their temperature-dependent properties providing a resultant electrical signal. In these, the detector measures the total amount of heat that is absorbed, regardless of the number of photons. The thermal detectors are characterized by lower sensitivity and slow response time of the order of a few milliseconds (Rogalski, 2012, and references therein). A thermal imager is a non-contact temperature measurement device. It measures very small relative temperature differences and converts otherwise invisible heat patterns into distinct visible images that are seen through either a viewfinder or a monitor. It cannot see through walls, glass, or other solid objects, but can detect the IR energy emitted, transmitted, or reflected by all materials at temperatures above absolute zero K and converts the energy factor into a temperature reading or thermogram.

John Herschel (1792–1871) made the first thermal image in 1840 from sunlight using a suspension of carbon in alcohol and a lens placed in a beam of natural sunlight (Herschel, 1842). This method is known as the differential evaporagraph technique. He used the term thermogram to describe the image, which is still in common use today (Ring, 2000). Thermography is the acquisition and analysis using electronic optical devices to detect and measure radiation and determine surface temperature of an object. Later, Czerny (1929) developed thermal images using a film of volatile oil on a thin absorbing membrane. His evaporagraph was later developed into a laboratory instrument. The thermal picture appears as a colored pattern by optical interference of light reflected from oil film. Early infrared imaging systems were developed during the 1940s and became available to industry and medicine first in the late 1950s. During the 1960s and 1970s a new generation of thermal imaging systems were developed (Ring, 2004). Subsequent developments in thermal imaging devices are briefly highlighted below.

Absorption edge imager This system devised, by Harding et al. (1958), was the absorption edge converter, in which the optical absorption edge of a selenium film changes visibly. Hilsum and Harding (1961) discussed the theory and application of such a system, where they showed that bodies 10 K above the ambient can be imaged. Operation of the device was based on utilizing the temperature dependence of the absorption edge of semiconductor.

Pyroelectric vidicon The pyroelectric vidicon tube, an analogous to the visible television camera tube, in which the photoconductive target is replaced by a pyroelectric detector and germanium faceplate (Goss, 1987). Although it is a compact, rugged imager, it has suffered the disadvantage of low tube life and fragility, particularly the reticulated vidicon tubes required for enhanced spatial resolution. In this, the thermal scene is imaged on to the Piezo-electric target where the temperature differences induce a corresponding charge pattern, through the pyroelectric effect (see Section 8.3.1), and this charge is read by the electric beam. The vidicon tube can thus form the basis of a thermal imaging camera; in general the faceplate is made of germanium. Its main limitation is the resolution, which is limited by thermal spreading of the image on the target. The target consists of a thin photosensitive layer whose front face (which receives incident radiation) is covered with a transparent film that is a good conductor of electricity. The target material commonly used in such vidicons is triglycine sulfate (TgS) or a deuterated version DTgS. A signal charge is produced when the target temperature changes, therefore there is no video signal arises from a steady incident flux. In order to discern a stationary scene, the incident radiation should, in principle, be modulated by either a chopper or scanning (see Section 5.5.1.3). The spectral response of such a tube is determined by the transparency of the window. Usually it is germanium, which has a few absorption bands between 1.8 μm and the microwave region.

Liquid crystal imager This thermal imaging device was developed by Ennulat et al. (1974). In this device, the optical scattering of liquid crystals changes with temperature. Such a system was employed in medical imaging.

These imagers require very little power and possess high sensitivity but suffer from disadvantages of a spatial noise due to random temperature fluctuations across the film and limited dynamic range.

8.4 IR Detectors

It is to reiterate that there are two kinds of infrared detectors: (i) thermal detectors that are based on physical property change due to temperature change of detector element due to incident IR radiation, namely, bolometer, pyroelectric, Golay cell, etc. (see Section 8.3), and (ii) photo-detectors that are based on various internal photoelectric effects, for instance, photoconductive, photovoltaic, etc. The new-generation IR detectors are equipped with focal plane array (FPA) in which the optical section and the signal processing are performed at the focal plane, which facilitates high-speed images at high thermal and spatial resolution; the image quality has improved considerably as well. As stated in Section 5.5.1, an FPA refers to an assemblage of individual pixels located at the focal plane of an imaging system. The FPA arrays have found applications in Fourier IR spectroscopy, IR astronomy, IR imaging in medical imaging (Rogalski, 2003, and references therein). The IR FPAs are, in general, made up of two components: (i) IR detector arrays (linear or staring type; see Section 5.5.1.3) on the focal plane and (ii) silicon (Si) read-out integrated circuit (ROIC). Both these components are physically and electrically mated to each other through indium (In) bump. These FPAs are of two types: monolithic and hybrid (see Section 5.5.2).

8.4.1 *Evolution of IR Detectors*

The concept of generation of thermal imagers is mainly dictated by the evolution of IR detector technology.

First-generation IR detectors The first-generation detectors are based on single element, serial detector, serial-parallel (2×15, 4×11 or higher), parallel (60, 100, 120, 180, 240×1) or SPRITE detector configurations. The indium antimonide (InSb, a binary alloy) and mercury cadmium telluride (HgCdTe, a ternary alloy) based detectors are cooled to 77 K, which were encapsulated in vacuum dewar. Each detector element output is individually available from the detector. For each element, pre-amplifiers and post-amplifiers are required to be used on off focal plane electronics before digitization. These imagers required scanning (see Section 5.5.1.3) in one or two directions.

Second-generation IR detectors The second-generation systems are based on linear-FPA detectors (288×4, 480×4, 960×4, etc.). These thermal imagers required scanning in one direction and read-out electronics that provide on focal plane processing, such as, time-delay and integration (TDI; see Section 5.5.1.3), multiplexing (see Section 6.2.1), amplification etc., to improve the S/N ratio (see Section 4.1.1.4). Like the first-generation imagers, these are also cooled to 77 K; however, they require less electronics. The two-dimensional (2-D) scanner was simplified in the vertical direction, to include only the interlace motion.

Third-generation IR detectors The third-generation systems contain 2-D arrays with several columns of elements, which are based on third-generation IR detectors having staring (see Section 5.5.1.3) configuration (320×256, 640×512, 1280×1024, 1024×768 pixels). The images do not require any scanning as full matrix of pixel array in the image plane of an optical system covers the total FOV in the object plane. These detectors have functionalities like multi-color and multi-band operability, provide higher spatial resolution, higher frame rates, and offer better thermal resolution. There are cooled (HgCdTe, InSb, QWIPs, and antimonide-based type-II SLs) and uncooled A-Si- and VO_x-based microbolometer detectors available. Although scanning is not required in these imagers, microscanning can be done to avoid aliasing. The focal plane electronics contains signal amplification, multiplexing, control signal generation, and digitization. The wavelength tunability features, such as dual-band, dual-color, multi-spectral, and hyper-spectral, are also available.

8.4.2 *Emerging Trends in IR Detectors*

With the advancement in device fabrication technology, high-resolution (large formats, smaller pitch) detectors with advanced digital and low-noise integrated read-out electronics, higher spatial uniformity, higher temporal stability, and multi-spectral capabilities are becoming feasible. The emerging trends in IR detectors are enumerated briefly.

Dual-band and dual-color detectors The requirements for simultaneous detection of IR radiation in different wavelength bands (dual bands) or spectral ranges within single band (dual-color) are driven by the increase in reconnaissance performance. The suitable materials for such detectors are HgCdTe, quantum well (see Section 5.5.1.1), and super-lattice (SL; see Section 5.5.1.2) devices.

SNIR detector This type of IR detector consists of a read-out IC (ROIC; Shkedy et al., 2011) with advanced on-chip signal processing (Langgof et al., 2011). It can be used in both passive and active imaging in MWIR, as well as in SWIR spectral bands. The 0.18 μm CMOS technology allows for in-pixel advanced functionality, and at relatively low power consumption. The ROIC is a flip-chip bonded to a 640×512 InSb detector array with 15 μm pitch. Modern electro-optical systems that require integration of multiple sensors or functionalities, such as thermal imager, spot detection for laser designator, laser range finder (LRF), into single multi-function system can benefit from the SNIR detector.

Higher operating temperature (HOT) These detectors are high spatial resolution IR FPAs operating at 160 to 250 K, which serve to reduce cost, enhance reliability, and provide a sensor package where multiple arrays can be distributed on the system platform (Ashley et al., 2004). For a very high operating temperature, Klipstein et al. (2013) have developed an InAsSb detector with a 4.2 μm cut-off wavelength, which exhibits a background-limited performance (BLIP) at temperature of ~160 K at $F_{\#}/3$ and a reduction in cooling power by ~60%. These HOT detectors enable an improved range of solutions, including faster cool-down time and mission readiness, longer mission times, and higher cooler reliability.

IR retina detectors These systems are the fourth generation of IR detectors, incorporating enhanced functionality' in the imagers (preferably at the pixel level). The basic idea is to pack as much information in the pixel to develop an infrared retina (Krishna, 2009). The IR retina detector uses the concept of a bio-inspired infrared retina. The key idea is to engineer the pixel, which would have the ability to sense multi-modal data, such as color, polarization, dynamic range, and phase. Also, it has the intelligence to transmit a reduced data set to the central processing unit. The individual pixels in the FPA would act like the cones (see Section 4.2.1.1) in the human eye (see Section 4.2.1) and monitor the emissivity (see Section 1.2.1) of a given scene. If spectral or polarization information could be coded at the spectral level, one may obtain an IR retina; the similarity in the spectral shift as a function of voltage from a quantum dot detector and the spectral response of the cones (see Figure 4.4) in the human eye. In this approach, two material systems, namely, (i) InAs/InGaAs self-assembled quantum dots in well hetero-structure and (ii) InAs/InGaSb strain layer super-lattice detectors are used.

8.5 IR Photon Detectors

The photon detectors use direct detection principle that translates photons into electrons. The charge accumulated, the current flow, or change in conductivity are proportional to radiance of objects in the scene. On the contrary, the response of a thermal detector is proportional to the energy absorbed; they have photon-phonon type of interaction. To note, the energy of a phonon can be derived by considering the lattice (see Section 2.2.1.1) as a collection of oscillators. The phonon energies are typically comparable with IR-radiation. The coupling between lattice vibration and photons involves two phonons in crystals, such as germanium and silicon. Lax and Burstein (1955) have proposed two mechanisms for this coupling, for instance, (i) a transverse optical phonon created by the photon decays into two phonons and (ii) direct coupling between photon (see Section 1.1) and phonon.

The IR-radiation contains spatial and spectral details, both of which are dynamic in nature. The useful spectral range of the photon detectors is limited, while the thermal detectors exhibit a broad spectral coverage, and the responsivity (see Section 4.1.1.2) is independent of wavelength. Their wavelength of detection can be derived by Planck's law of radiation (see Section 1.1.1). As stated in Section 1.2.2, the photometric units are reasonable if the spectral response of the detector is within the human eye's response (see equation 4.38). If the response of such a detector is outside the eye's response curve, the photons contribute to the signal, but the photometry remains constant. An infrared filter may be added in order to restrict the wavelength response. In such a situation, the average responsivity becomes

$$\mathcal{R}_\nu = \frac{\int_{\lambda_1}^{\lambda_2} T_{\text{IR-Filter}}\mathcal{R}(\lambda)\mathcal{M}_e(\lambda, T)d\lambda}{683\int_{0.36\ \mu\text{m}}^{0.8\ \mu\text{m}} V(\lambda)\mathcal{M}_e(\lambda, T)d\lambda} \qquad \text{A.lm}^{-1}, \tag{8.23}$$

with h as Planck's constant, $\mathcal{R}(\lambda)[= q\eta_d/h\nu = \lambda\eta_d/1.24]$ the spectral responsivity (see Section 4.1.1.2), η_d the quantum efficiency of the detector, q the electron charge, $\mathcal{M}_e(\lambda, T)$ the spectral radiant emittance (see equation 1.78), and $V(\lambda)$ the spectral luminous efficiency for photopic vision (see Section 4.2.1).

8.5.1 *Quantum Well IR Photoconductor*

As stated in Section 5.3, the operation and working of a photovoltaic devices, such as P-N and P-I-N junctions rely on inter-band transitions. The devices based on hetero-structures formed by quantum wells (see Section 5.5.1.1) utilizing the inter sub-band or sub-band to continuum transitions caused by the absorption of incident IR photons are called quantum well IR photodetector (QWIP; Schneider and Liu, 2007). The absorption of these photons photoexcite electrons from the ground state of the quantum well into the continuum, causing a photocurrent. The absorption peak wavelength of inter sub-band transitions in quantum wells is determined by the ground (E_0) and excited (E_1) state energy difference. The energy eigen values of quantized states inside the wells may be derived by solving Schrödinger's equation (see Section 1.1).

The QWIPs are built from alternating thin layers (type-I superlattices) of $GaAs/Al_xGa_{1-x}As$ (gallium arsenide/aluminum gallium arsenide). The operation of such devices is similar to that of extrinsic detectors (see Section 4.3.1.1) where dopants are used to alter the band structure. However, in the case of the QWIPs, the dopants are concentrated in microscopic regions, creating quantum wells where band structure has shifted. The detection occurs when a photon knocks an electron or a hole out of the quantum well into the neighboring band. The transition takes place in the direction perpendicular to the growth of layer. In such a case, the electron and hole include a quasi-discrete energy spectra consisting of several sub-bands and form a part where the electron motion is three-dimensional one.

Since QWIP is photoconductive in nature, a potential difference is, in general, applied across the layers to collect the electrons from the well into the continuum. Use of an external electric field perpendicular to the well helps in changing both eigen states of quantum wells and wave function of the ground state of the quantum well. Variation in the form of electrical conductivity due to absorption of IR photons is accompanied by the transitions of electrons or holes between energy state in the conduction band. The change in concentration and mobility (see Section 2.2.4.4) of electron can be caused by the transition between the sub-bands within the conduction band. Similarly, the change in the mobility and concentration of holes can be caused by the transition between the sub-bands within the valence band. This phenomenon takes place whenever there is absorption of incident photons causing transfer of electrons or holes from bound states into quasi-bound or continuum states. Such intra-band detectors of this type are called

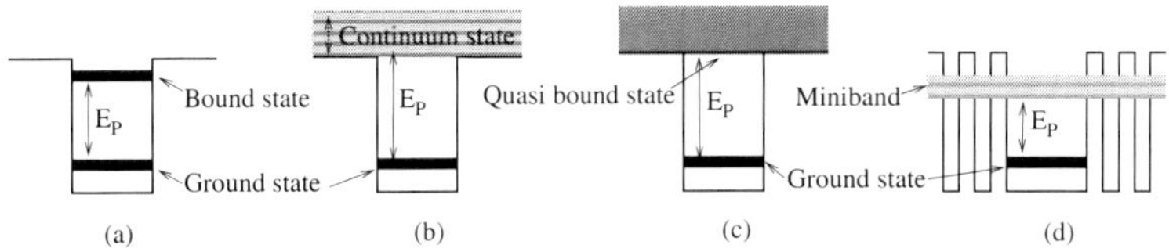

Figure 8.3 Energy band diagrams depicting (a) bound-to-Bound (B-B), (b) bound-to-continuum (B-C), (c) bound-to-quasi-bound (B-QB), and (d) bound-to-miniband (B-MB) transitions.

inter sub-band IR detectors. Depending on the carrier transition (see Figure 8.3), the operation of QWIPs is classified into following categories:

(1) Bound-to-bound (B-B) transition: In this transition, the electrons in the lowest state are excited by incoming photons into the excited state and then tunnel through the barrier under as external bias to form a photocurrent (Levine et al., 1987).
(2) Bound-to-continuum (B-C) transition: By reducing the quantum well thickness, it is probable to push the strong B-B inter sub-band absorption into the continuum, which results in a strong B-C inter sub-band absorption. The major advantage of B-C QWIP is that the electrons in the ground state are excited into the continuum where they contribute to the photocurrent without tunneling, and therefore, the barrier thickness has less bearing on the photocurrent.
(3) Bound-to-quasi-bound (B-QB) transition: In a B-QB QWIP, the wave function of the upper excited state is only half confined and the interaction between the initial and the final states is relatively smaller at the resonant photon energy. The absorption spectral width is wider than the B-B QWIP.
(4) Bound-to-miniband (B-MB) transition: In the B-MB states transition, the electron wave functions are well confined in both the initial and final states. The well width in a B-MB QWIP is much wider than that of the B-C and B-QB QWIPs. There is a large overlap of wave functions between the initial and the final states, and the interaction becomes strong. The density of states are delta functions and the absorption spectrum is sharp with a narrow absorption bandwidth. The absorption bandwidth can

be changed by adjusting the relative positions of the miniband and the excited state.

The integrated absorption is nearly identical for the bound-to-bound, bound-to-quasi-bound, and bound-to-miniband QWIPs. However, a B-MB QWIP has the smallest responsivity due to the lower electron mobility in the miniband. Due to the continuum of states and the multiple transitions that can occur, the photo response spectrum is much wider than in the B-B state, but the absorption coefficient is much smaller due to the low density of states above well. The quasi-bound and bound states belonging to the conduction/valence energy bands arises as a result of the electron and hole confinement in the wells of detectors with super-lattice (SL; see Section 5.5.1.2) hetero-structures form active region, consisting of parts with a narrow energy gap surrounded by a wide-gap material (Gunapala et al., 2001).

The QWIPs offer excellent thermal and spatial resolutions, good homogeneity, low fixed pattern noise, good reproducibility, low flicker noise (see Section 4.5.2.4) and high yield. They are supported by the materials, such as GaAs/AlGaAs and InGaAs/AlGaAs and processing technologies highly uniform and well controlled molecular beam epitaxy (MBE; see Section 5.5.1.1). In GaAs/AlGaAs QWIPs, the FPAs can be realized to detect light of wavelengths 6 to 25 μm (Gunapala et al., 1997, 1998). At normal bias, the QWIP has high impedance, in which case, the Johnson noise (see Section 4.5.2.1) is negligible compared to the shot-noise (see Section 4.5.2.3) of thermally generated dark current. As with the responsivity (see Section 4.1.1.2), the noise is a function of the applied bias. The noise in a QWIP (Beck, 1993) is

$$i_{\mathrm{n}} = \sqrt{4qG(V)\left[i_{\mathrm{ph}}(V) + i_{\mathrm{D}}(V)\right]\frac{1-\mathcal{P}_{\mathrm{c}}/2}{2\Delta t}} \qquad \mathrm{A}, \qquad (8.24)$$

where G is the photoconductive gain, q the electron charge, i_{D} the dark current (see equation 4.31), $\mathcal{P}_{\mathrm{c}}$ the capture probabilities of the carriers, and

$$i_{\mathrm{ph}} = \frac{q\mathrm{A}_{\mathrm{d}}}{1+4F_{\#}^{2}}\int \eta_{\mathrm{d}}(\lambda)t_{\mathrm{o}}(\lambda)\mathcal{M}_{\mathrm{e}}(\lambda,\, T_{\mathrm{b}})d\lambda \qquad \mathrm{A}, \qquad (8.25)$$

the photocurrent, T_{b} is the background temperature, $\mathcal{M}_{\mathrm{e}}(\lambda,\, T_{\mathrm{b}})$ the spectral emittance (see equation 8.32) of the scene, A_{d} the area of the detector, $F_{\#}(= f/D)$ the aperture speed of the optics, f its focal length, D the diameter of the aperture, t_o the optics (filter) transmission spectrum, and λ the wavelength.

Detectivities (see Section 4.1.1.4) of QWIP have improved considerably and are high enough so that large FPAs with LWIR imaging performance comparable to state-of-the art of HgCdTe are being fabricated (see Section 8.6.2). stringent uniformity, large tunneling currents, could be eliminated by using the super-lattice. In HgTe/CdTe super-lattice detector, it would be possible to control the wavelength, tunneling current can be reduced and operating temperatures may be reasonable. In comparison to other QWIP's, this material has the advantage of polarization-independent absorption and large absorption coefficient. QWIP using HgTe/CdTe super-lattice has been demonstrated for the 4.5 μm wavelength (Harris et al., 1991).

Work has also been carried out on the double-barrier QWIP, in which a barrier layer is introduced within the well layer to divide it into a broad and a narrow well. There are two energy levels below the barrier in the broad well that is doped and one level in the narrow well that is undoped, close to the top level of the broad well (see Figure 8.4a). When a quantum well structure is subjected to an applied bias, the potential well would tilt in the direction opposite to the electric field (see Figure 8.4b). In the absence of light electrons tunnel sequentially through the narrow well, while under illumination the excited electrons tunnel coherently through the narrow well and reach the continuum and flow in the direction of the growth of the super-lattice. Further, on applying a suitable bias voltage the carriers can be collected. The depth of potential well can be varied by the amount of dopants. To note, addition of dopant to form a N-type or P-type semiconductor material as impurity is called modulation doping. If the doping density in the quantum wells remains constant, the absorption quantum efficiency is determined by the oscillator strength and the density of states. The oscillator strength is at its maximum when the incoming photon energy, $h\nu$, is in resonance with the transition energy, in which h is Planck's constant and ν the frequency of the wave in question.

The double-barrier QWIPs have good responsivity with low dark current (see Section 4.1.1.8). In such detectors, the dark

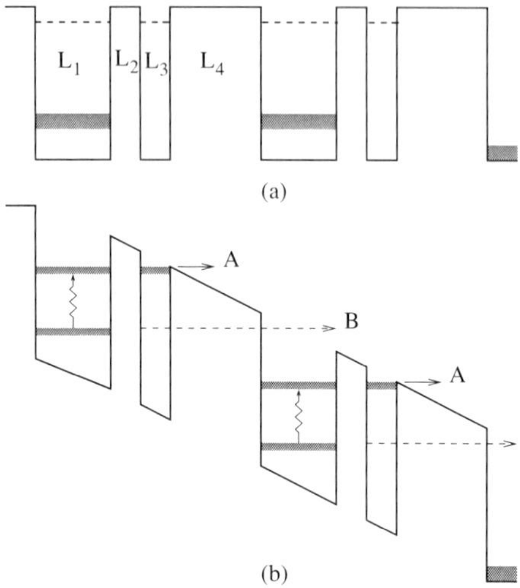

Figure 8.4 Energy band diagram of a double-barrier QWIP: (a) potential distribution without any bias, (b) potential distribution under forward bias depicting the tilt in the direction opposite to the electric field; L_1 is GaAs well (doped), L_2 the $Al_{0.33}Ga_{0.67}As$ undoped barrier, L_3 the undoped GaAs well, L_4 the undoped $Al_{0.433}Ga_{0.67}As$ barrier, A the photo-excited electrons, and B the tunneling electrons (after Bandara et al., 1998).

current is the main source of noise at high temperature. These devices have been constructed with GaAs/AlAs/AlGaAs (Harris, 1993) and AlGaAs/AlAs/InGaAs (Park et al., 1990; Harris et al., 1991; Kiledjian et al., 1991). The other characteristics, such as high sensitivity, high selectivity, and multi-spectral capability for detection and identification of space targets have been found in QWIP that is capable of detecting long wavelength IR (LWIR; Levine et al., 1988). Also, the current density of such a device is lower at low temperature over mercury cadmium telluride (MCT or HgCdTe), which helps to lower tunneling current effect in QWIP (Rogalski, 2003). However, the QWIP detectors have relatively low quantum efficiencies, typically less than 10%. The spectral response band is narrow as well for such a detector, with a FWHM of about 15%. The QWIPs having 1k×1k system (18 μm pixel size) hybrid LWIR and MWIR have been developed (Jhabvala et al., 2004; Gunapala et al., 2005).

The QWIP device has a narrow bandgap, whose gain (see Section 4.3.2.2) is inversely proportional to the transit time of the photo-excited carriers across the active region of the device. In other words, the gain is the ratio of the electron lifetime, τ_e and the transit time, τ_d, across the active quantum well region to the contact electrode

$$G = \frac{\tau_d + \Delta t}{N_w \tau_d}, \tag{8.26}$$

with v_d as the drift velocity given in equation (2.19) (see Section 2.2.4.2), Δt the capture time, and N_w the number of wells.

The photoconductive gain increases with increase in the bias voltage, which can be changed by varying the number of quantum wells. A QWIP device, in general, consists of 20 to 25 periods of quantum wells. When the number of wells is reduced the thickness of the active region decreases and reduces the electron transit time. Reducing the number of wells would enhance the photoconductive gain but would reduce the absorption quantum efficiency as well if the light paths and the doping density are fixed. If the absorption is kept constant by using effective optical coupling schemes, then the responsivity (see Section 4.1.1.2) may increase with decreasing number of quantum wells. The total photoconductive gain, defined as the ratio of the photoconductive gain for a single-quantum well to the total number of quantum wells, shows the transport property of the QWIP device, that is

$$G = \frac{v_d \tau_{ec}}{L}, \tag{8.27}$$

where τ_{ec} is the lifetime in the upper state or the energy continuum and L the active length of the device.

The probability with which the carriers can escape from the well and get collected can be expressed as

$$\mathcal{P} = \frac{1}{1 + \tau_d/\Delta t}. \tag{8.28}$$

A single-quantum well photoconductive gain is related to the capture probability, $\mathcal{P}_c$, of the carriers as

$$\mathcal{P}_c = \frac{\tau_d}{\tau_d + \Delta t}.$$

The total photoconductive gain, G, can be expressed in terms of the capture probability as (Goldberg, 2002)

$$G = \frac{1}{N_{\mathrm{w}}[\tau_{\mathrm{d}}/(\tau_{\mathrm{d}} + \Delta t)]} = \frac{1}{N_{\mathrm{w}}\mathcal{P}_{\mathrm{c}}}. \tag{8.29}$$

The S/N ratio for a charge well-limited QWIP (well kept half full; Goldberg, 2002) is given by

$$\mathrm{S/N} = \frac{qN_{\mathrm{e}}/2\Delta t}{\sqrt{4qG(qN_{\mathrm{e}}/2\Delta t)/(2\Delta t)}} = \frac{1}{2}\sqrt{\frac{N_{\mathrm{e}}}{2}}, \tag{8.30}$$

with N_{e} as the number of electrons the well can hold.

The expression for the minimum detectable temperature difference (see Section 8.6), $NE\Delta T$, under these conditions is

$$NE\Delta T = \frac{T_{\mathrm{b}}}{2}\sqrt{\frac{G}{N_{\mathrm{e}}}} \quad \mathrm{K}. \tag{8.31}$$

The QWIPs are used in radars, night vision cameras in missile defense systems, and metrology weather monitoring systems from the satellite vision. In astronomical observations, these detectors are attractive since they offer good sensitivity. They are stable, and also, large-size devices can be manufactured (Rogalski, 2003).

8.5.2 *Strained-Layer Super-Lattice Detectors*

The strained-layer super-lattice (SLS; also called type-II super-lattice) is a material that can be used to make high-quality cooled IR photon detectors with a cut-off wavelength ranging from 2 to 30 μm covering from SWIR to VLWIR wavelength bands. In a super-lattice (SL; see Section 5.5.1.2) structure, a repeating sequence of thin layers of different materials are employed. The semiconductors, namely, InAs, GaSb, and AlSb, form an approximately lattice-matched set around 6.1 Å, with (room temperature) energy gaps ranging from 0.36 eV (InAs) to 1.61 eV (AlSb). The hetero-structure combining InAs with the two antimonides offers band lineups that are different from those of AlGaAs system (Rogalski, 2012). The type III–V materials offer much stronger chemical bonds and hence higher chemical stability compared to HgCdTe. The 6.1 Å materials can be epitaxially grown on GaSb and InAs substrates.

It is reiterated that gallium arsenide (GaAs) and aluminum gallium arsenide ($Al_xGa_{1-x}As$) are transparent to the IR-region. The mismatch in the lattice constant of these compounds being less (~0.12%), it was probable to grow layers of GaAs and mixed compounds of GaAs and AlAs on each other with crystalline perfection. A transistor was realized by using a $Al_{0.7}Ga_{0.3}As$/GaAs hetero-structure by Mimura et al. (1980). The $Al_{0.7}Ga_{0.3}As$ layer was doped N-type and the GaAs layer was undoped. The electron concentration in GaAs and the in-plane current could be controlled by using a conventional MOSFET (see Section 2.3.3) like structure with source, drain, and gate contacts. The multi-quantum well structures are used in lasers and photodetectors, and as non-linear optical elements. The multi-quantum well detectors based on the aforementioned compounds are useful for very long wavelength applications.

There are alternative IR materials for HgCdTe and QWIPs comprising InAs/$Ga_{1-x}In_xSb$ (InAs/GaInSb) SLSs, which have shown great potential for the very long wave IR (VLWIR) region at a relatively higher operating temperature as compared to HgCdTe with the same cut-off wavelength (Rogalski, 2007). However, the problems exist in material growth, processing, substrate preparation, and device passivation. The optimization of super-lattice growth depends on the trade-off between interface roughness, with smoother interfaces at higher temperature, and residual background carrier concentrations, which are minimized on the low end of this range. Smith and Mailhiot (1987) demonstrated the potential of SLSs using InSb/InAsSb. The GaInSb/InAs has a large absorption coefficient and potential for high detectivity, longer wavelength and multi-spectral capability.

In the SL, the electrons are located in the InAs-layers, while holes are confined to the GaInSb-layers, which suppresses non-radiative recombination (Auger, 1923) mechanisms (see Section 2.2.4.1) and increases carrier lifetime. However, at present time, the measured carrier lifetime is below 100 ns and is limited by Shockley–Read–Hall recombination mechanism (see Section 2.2.4.1; Shockley and Read, 1952; Hall, 1952) in both MWIR and LWIR compositions. The type-II SL structures can provide high quantum efficiency and also the responsivity (see Section 4.1.1.2), of type-II SL detectors has proved to be similar to those made of HgCdTe (Rogalski et al., 2009). The uncooled strained-layer super-lattice photon detector arrays, as well as multi-color IR detectors using type-II InAs/GaSb nanopillar array, which has a theoretically predicted higher performance than HgCdTe, have been demonstrated by Razeghi et al. (2003).

8.5.3 *Quantum Dot IR Photodetectors*

Like QWIP devices, the quantum dot IR photodetectors (QDIP) utilize quantum confinement in the wide bandgap materials for their operation. The success of quantum well structures for IR detection applications has stimulated the development of QDIPs. In general, QDIPs are similar to QWIPs but with the planner shaped quantum wells replaced by the quantum dots, which have size confinement in all spatial directions (Martyniuk and Rogalski, 2008). The detection mechanism is based on the intra-band photoexcitation of electrons from confined states in the conduction band dots into the continuum. The emitted electrons drift toward the collector in the electric field provided by the applied bias, and photocurrent is created. The wavelength response can be varied by changing the physical parameters of the dots.

For deposition amounts greater than 1.6 monolayers (ML), self-assembled QDs form spontaneously. Typically, coherent InAs islands grow laterally up to a base length of $\sim$20 nm, beyond which the dominant increase in size occurs through an increase in the steepness of the pyramidal side walls. The heights may vary with deposition amounts, for instance, for 2 MLs, it may be $\sim$3.6 nm and for 3 MLs, it may be $\sim$7.5 nm. These InAs/GaAs quantum dots are referred to as shallow and steep QDs, respectively. In order to create QDIP structures, either type is capped with any one of the capping materials, for example, GaAs, AlGaAs, InGaAs, InGaAlAs. The fabrication of such a device requires photolithography (see Section 2.3.1.1), wet-chemical etching, metal deposition, lift-off and rapid thermal annealing. In the recent past, FPAs based on inter sub-band transitions have been demonstrated. Gunapala et al. (2004) highlighted 640×512 pixels LWIR quantum dot IR photodetector imaging FPA by exploiting the properties of epitaxially grown, self-assembled QDs operating at higher temperatures.

Two types, such as vertical and lateral structures have been proposed for QDIPs. In the former case, the photocurrent is collected through the vertical transport of carriers between top and bottom contacts. These devices are grown employing solid source molecular beam epitaxy (MBE; see Section 5.5.1.1) technique. The substrate used is semi-insulating GaAs over which binary InAs is deposited. The quantum dots are directly doped (usually with silicon) in order to provide free carriers during photoexcitation, and an AlGaAs barrier can be included in the vertical device hetero-structure in order to block dark current (see Section 4.1.1.8) created by thermionic emission (Wang et al., 2001). Due to the intricate dependence of the operating wavelengths on the size and shape of the dot, which in turn depends on the inherently random and stochastic self-assembly process, it is difficult to obtain a desired spectral response. Attempts have been made to develop detectors based on inter sub-band transitions in a quantum dots-in-a-well (DWELL; Krishna et al., 2001) hetero-structures. The lateral QDIP collects photocurrent through transport of carriers across a high-mobility channel between two top contacts, operating much like a field-effect transistor (see Section 2.2.9). Instead of blocking the dark current, these barriers are used to both modulation-dope the quantum dots and to provide the high-mobility channel. Lateral QDIPs have demonstrated lower dark currents and higher operating temperatures than vertical QDIPs since the major components of the dark current arise from inter-dot tunneling and hopping conduction (Lee et al., 1999).

8.5.4 *Cooled and Uncooled Detectors*

According to the requirements of sensitivity, detectors can be Cooled (generally cryogenic) and Un-cooled. A direct detection (photon counting) technique translates the photons directly into electrons. The charge accumulated, the current flow, or the change in conductivity are proportional to the radiance of objects in the scenery viewed. This category detectors consists of materials like, lead selenide (PbSe), HgCdTe, InSb, platinum silicide (PtSi), etc. Except for thermal imagers, working in the SWIR range, all IR cameras based on this technology are detectors cooled to cryogenic temperatures (see Section 4.3.1). The new-generation photon type IR detectors operating at elevated temperatures allowed solid-state thermal electric coolers and closed cycle Stirling coolers. The Stirling coolers have a continuous operating life around 10,000 hours. These coolers are available in split or integrated configurations; the thermoelectric coolers are, in general, used to stabilize uncooled detectors at room temperatures. Vuillermet (2013) reported to the development of a product based on HgCdTe technology, which is a hybrid of a matrix of photodiodes made of MCT material onto a silicon read-out circuit. This matrix converts the photons into small electrical signals, which are amplified and multiplexed by the read-out circuit. The interconnection between the detection module and the read-out circuit uses indium bumps. Due to the intrinsic dark current of photodiode, this detector can operate at cryogenic temperature. The hybrid is packaged into a vacuum dewar and cooled by Stirling cycle cryo-cooler. He presented some examples of developments of MCT infrared devices: (i) 640×512 with 15 μm pitch that is sensitive in the long wavelength, (ii) 1280×1024 with 15 μm pitch sensitive in MWIR wavelength, (iii) HOT detectors operating at 150 K, and (iv) the 10 μm pitch detectors.

On the contrary, a thermal detector uses secondary effects, such as the relation between conductivity, capacitance, expansion and detector temperature. Bolometers, thermocouples, thermopiles, pyroelectric detectors, etc., fall in this category. They do not require cryogenic temperatures, but usually stabilized to temperatures between -30°C and $+30^\circ$C. The uncooled detectors used in thermal imagers are of two types, for instance, microbolometer FPA, which is a 2-D array of small temperature-sensitive electrical resistors, and ferroelectric. The microbolometer resistor is a tiny vanadium oxide (VO_x) or amorphous silicon resistor with a large temperature coefficient on a silicon element. The microbolometer resistance distribution stems essentially from resistance geometry distribution, while remaining stable regarding focal plane temperature variation. However, the main drawback of microbolometer detectors is that they require low $F_\#(= f/D)$ optics, in which $F_\#$ is the aperture speed of the optics, f its focal length, and D its diameter. Recently, Tissot et al. (2013) reported to have developed an uncooled micro-bolometers made from amorphous silicon facilitating to develop uncooled IR focal plane array (IRFPA) with 17 μm pixel-pitch.

However, InSb FPAs are generation-recombination (GR; see Section 4.5.2.2) limited. The dark current in InSb FPAs is significantly higher than in HgCdTe FPAs. Both InSb and MCT detectors must be cooled cryogenically, typically with a miniature Joule Thomson or Stirling cycle refrigerator. Klipstein et al. (2013) discussed a technology involving molecular beam epitaxy (MBE; see Section 5.5.1.1) growth, where in the case of InSb diodes, a high quality homo-structure is grown on an InSb substrate and diodes are isolated by etching a mesa structure through the P-N junction. The dark current is reduced according to the ratio of concentrations of GR centers in both the standard and MBE grown structures.

8.6 IR Imaging Detectors for Astronomy

The present-day imagers with higher performance use a detector or detector array at the focus of an optical scanning system. Their performance is determined by two parameters: thermal sensitivity, which is defined in terms of noise equivalent temperature difference (NETD), and spatial resolution, which is determined by the modulation transfer function (see Section 1.3.1). For the analysis of the design and performances of an IR imaging system, the radiation emitted from the scene needs to be taken into account. The relationship between spectral emittance, $\mathcal{M}_e(\lambda)$, temperature, and radiant energy for a blackbody is given by Planck's law (see Section 1.1.1)

$$\mathcal{M}_e(\lambda, T) = \frac{c_1}{\lambda^5}\frac{1}{e^{c_2/(\lambda T)} - 1} \qquad \text{W.cm}^{-2}.\mu\text{m}^{-1}, \qquad (8.32)$$

with λ as the wavelength, $c_1 (= 2\pi hc^2 = 3.7418 \times 10^4$ W μm^4.cm^{-2}) the first radiation constant, $c_2 (= hc/k_B = 1.4388 \times 10^4$ μmK) the second radiation constant, h Planck's constant, c the velocity of light, and k_B Boltzmann's constant.

As stated in Section 1.1.4, the blackbody curves exhibit a peak of spectral emittance at a wavelength, λ_{max} with absolute temperature, T of the body and is determined by Wien's law, i.e., $\lambda_{max} = 2897.8\ T^{-1}\mu$m. The total radiation from a body is obtained by integrating equation (8.32) over all wavelengths from zero to infinity and is given by Stefan–Boltzmann's law (see Section 1.1.2)

$$\mathcal{M}_\mathrm{e} = 5.67 \times 10^{-12} T^4 \qquad \mathrm{W.cm^{-2}}, \tag{8.33}$$

indicating the total power radiated from a blackbody varies as the fourth power of the absolute temperature. In thermal imaging, one concerns with the objects and scenes around 300 K. These have a broad peak of emission at 10 μm, which is bracketed by the atmospheric window (see Section 8.1) at 8–13 μm. The radiation emitted in this window by a blackbody at 300 K is 150 $\mathrm{W.m^{-2}}$. The incident solar radiation in the spectral band is 1.5 $\mathrm{W.m^{-2}}$; hence, the reflected sunlight plays little part in imaging in the band. However, the heating effect of the Sun's radiation in outdoor environment can be significant. Although the emission at 3–5 μm band is small, which is 6 $\mathrm{W.m^{-2}}$, this window is also used for imaging. This can be compensated by the better detectivity of detectors in that band. In this band, the reflected solar radiation plays a significant role in imaging since the incident solar radiation amounts to 20 $\mathrm{W.m^{-2}}$.

In thermal scenes, the radiation is, in general, incident on an object from all other bodies around and some of this is reflected toward the imager. Since, according to Kirchhoff's Law, the sum total of the emissivity, ϵ and the reflectivity, ρ, amounts to 1, i.e., $\epsilon + \rho = 1$, one may expect the contrast in the scene to be zero, which is rare; generally, sufficient temperature does exist to provide adequate contrast for a picture. The temperature differences in the object of interest is often small fractions of 1°C.

For IR imaging systems, the figure of merit (see Section 4.1.1) for determining the performance is dictated by the minimum detectable temperature difference (Ciupa and Rogalski, 1997), $NE\Delta T$, i.e., the difference of temperature of the object required to produce an electric signal equal to the RMS noise voltage, which is given by (Lloyd, 1975)

$$NE\Delta T = \frac{4F_\#^2\sqrt{\Delta f}}{t_\mathrm{o}\sqrt{A_\mathrm{d}}}\left[\int_0^\infty \left(\frac{d\mathcal{M}_\mathrm{e}}{dT}\right)_\lambda t_{a\lambda} D_\lambda^* d\lambda\right]^{-1}, \tag{8.34}$$

where $F_\#$ is the aperture speed of the optics, $t_{a\lambda}$ the atmospheric transmission at wavelength, λ, t_o the optics transmission factor, ΔT the temperature difference in the scene, $d\mathcal{M}_\mathrm{e}/dT$ the differential spectral radiance from a blackbody.

It is apparent from equation (8.34) that the $NE\Delta T$ value depends on the bandwidth that is determined for the value of the detector noise. The maximum gain in sensitivity can be obtained by minimizing $F_{\#}$. The $NE\Delta T$ is a single measurable quantity that includes all the major imaging characteristics of the system including NETD, MTF, eye integration time, frame rate, target spatial frequency, and S/N ratio.

In imaging applications, for detection of weak radiation, the photon detectors (see Section 4.3) have better resolution, as well as have larger noise equivalent power (NEP; see Section 4.1.1.4) than the thermal detectors. However, in the mid- and far-IR region, the photon detectors require expansive cooling arrangement but a thermal detector needs no expensive cooling arrangement. As stated in Section 8.1, the atmospheric transmission is highest in (i) 3–5 μm and (ii) 8–14 μm bands (see Figure 8.1) and the emissivity maximum of the objects at $T \approx 300$ K is at $\lambda \approx 10$ μm; hence, interest in the ground-based IR astronomy has centered mainly on these bands. For the space applications, there has been increasing interest in longer wavelengths as well.

The choice of an IR detector for a particular application depends on the conditions, such as (i) spectral range, (ii) operating temperature, (iii) background temperature range, (iv) effective FOV (see Section 4.4.1.4), and (v) operating frequency and amplifier band-pass. Its performance depends on the (i) current-voltage characteristics, (ii) effect of background radiation on various detector parameters, (iii) effective time constant, (iv) noise characteristics, and (v) minimum detectable power.

Infrared detectors used in astronomy today are basically two-dimensional arrays of P-N junctions (see Section 5.3.1). The junction devices (see Section 5.3) fabricated from silicon (Si) can also be used in the near-IR band. However, in general, the near-IR devices are made from Ge, gallium arsenide (GaAs), and indium gallium arsenide (InGaAs) and with other semiconductor materials, for instance, indium antimonide (InSb), indium arsenic (InAs), mercury cadmium telluride (HgCdTe), Ge doped with copper (Cu) in the mid-IR, and silicon arsenic (SiAs) in the far-IR (5–28 μm). Case (1920) used thallous sulfide to detect IR-radiation for wavelengths up to 1.3 μm. This material was sensitive at wavelengths up to 3 μm. Later for very long wavelength IR-detectors, HgCdTe, the Gr II-Gr VI material system was used especially in medical applications (Gunapala et al., 2001).

Unlike CCDs (see Section 6.2), the IR detectors permit the accumulated signal in each pixel to be measured non-destructively multiple times. The capability to sample the signal multiple times during the integration can be exploited to reduce the effective read-out noise (see Section 4.5.4) significantly. The IR detectors are capable of very high dynamic-range observations; however, they are intrinsically non-linear. Since the current interferometry projects are mostly for IR observations, it is worth mentioning the detectors for this band. There have been recent progress in IR photon-counting that may also be applied to interferometry.

The development of IR array detectors caused a giant leap in the sensitivity of IR observations. Such detectors allow one to obtain images containing hundreds of pixels at the same time. Today, a variety of approaches are now employed for detector arrays operating from 1 μm to 1 mm and beyond (Rieke, 2007, and references therein). The chips for detecting IR-radiation that are made of direct hybrid arrays (see Section 5.5.2) of indium antimonide (InSb) or mercury cadmium telluride (HgCdTe) are preferred by the astronomers. These detectors operate in a similar fashion to the lead sulfide (PbS) detectors (see Section 8.2.1).

8.6.1 *Indium Antimonide Detectors*

The bimetallic compound of indium and antimony forms a semiconductor with an energy gap of 0.19 eV at 300 K, 0.22 eV at 90 K, and 0.23 eV at 4.2 K. Unlike PbS, InSb has a higher value of energy gap at the lower temperature. The long wavelength limit to intrinsic absorption is approximately 6.5 μm at liquid helium temperatures. InSb can be prepared in a highly purified condition in the form of single crystals. The InSb detector collects the light in the 3–5 μm spectral band providing a better spatial resolution. The InSb detector elements are usually smaller in size compared to the microbolometer detector elements. For the required spatial resolution, InSb detectors require lenses with shorter focal lengths. In order to detect IR-radiation, InSb may be used in three forms:

Photovoltaic The technique, in general, used for developing photovoltaic (PV; see Section 4.3.1.1) InSb consists of preparing a P-N junction by diffusing P-type dopant into N-type single crystal with plane face. Contacts are made to the thin P-type surface layer and the photovoltage is measured between these contacts when P-type layer is illuminated. The latter must be thin so that the excess carriers may diffuse to the P-N junction before recombining. The production of a photovoltage at a P-N junction has the relationship between the signal current and the voltage (see equation 5.32). For the photoelectromagnetic (PEM) detectors, the long wave cut-off comes at a slightly shorter wavelength. This is due to the fact that the incident radiation has to be absorbed in close vicinity of the front surface to produce the full effect. In order to realize the full detection capability of the photovoltaic InSb detector, it is required to use care in the proper design of the input circuit to which the detector is connected. Since the impedance of many of the detectors at 77 K is the neighborhood of 100 Ω (ohms), transformer input coupling has been used with vacuum tube and transistor pre-amplifiers. When the cooled detector is exposed to a room background radiation (see Section 4.5.1.2) causing BLIP noise, the response of the detector is such that an undesirable biasing of the junction is produced. The operation with an applied back bias to cancel the biasing due to background would result in minimum noise output from the detector. The characteristics of this noise is of the order of the $1/f$-type (see Section 4.5.2.4) at 1 KHz modulation frequency.

Photoelectromagnetic (PEM) Characteristic of a high carrier mobility exhibited by InSb has made possible the construction of a detector making use of the PEM effect (see Section 4.3.1.3) in the material for IR detection. A notable feature of such a detection method is that it functions at room temperature. Crystals of about 1 mm square in area are mounted between the poles of a small magnet (about 25 mm diameter). There is no requirement of any additional bias. The output current is the result of photon-generated carriers separated according to charge by the fixed magnetic field. The effective time constant of the detector is less than 1 μsec. Since the noise level of the PEM detector is set by the Johnson noise (see Section 4.5.2.1) of the element, the detectivity (see Section 4.1.1.4) at ambient temperature is considerably less than the theoretical limit. However, effective magnetic shielding should be employed in order to eliminate stray magnetic fields from the vicinity of the detector. For such detectors, the long wave cut-off arises at a shorter wavelength since the incident radiation is absorbed very close to the front surface to produce the full effect.

Photoconductive InSb, as a photoconductor (see Section 4.3.1.1), shows appreciable response at room ambient. Since the operation of the semiconductor in the intrinsic state is required, room temperature InSb photoconductors can be easily developed. The residual impurity levels are ionized at the higher temperatures, so that the purity of the material is not critical for application to the uncooled detectors. However, it is necessary to use very high purity material for high detectivity characteristics at 77 K. The effective photoconductive (PC) InSb devices are made in the form of a very thin ($< 10\mu$m) wafer mounted on a metal block. The maximum spectral response of such a detector is at 5.5 μm at 77 K; the response falls off rapidly at longer wavelengths. Crystals with parallel faces about 100 μm in thickness are prepared, which are treated with an oxidizing effect (see Section 2.3.1.1). This produces uniform layers of oxide, which are striped off until a layer thickness of the order of 20 μm in thickness is obtained. In order to take the advantage of the high minority carrier mobility (see Section 2.2.4.4), P-type material is, in general, preferred. Detectors having an area of 0.1 cm^2 and a resistance of the order of 20 kΩ may be made having a responsivity (see Section 4.1.1.2) of the order of 10^4 $V.W^{-1}$ and are capable of detecting powers of the order of 10^{-11} W with 1 Hz bandwidth. The minimum detectable power is of the order of 10^{-9} W, while the time constant is of the order of 10^{-8}s.

In the case of the large-area detectors, the PV-type has some advantages, but it is difficult to make thin crystals of large area. The optimum dimensions of InSb detector to obtain maximum responsivity for a given power, P per unit area incident on a detector. Let the detector be of unit width and breadth and of thickness, w. If the incident radiation causes a change in resistance, ΔR, a current, i, flowing through the strip would give rise to a signal, ΔV, given by

$$\Delta V = \Delta R \equiv \Delta R \sqrt{\frac{P}{R}}. \tag{8.35}$$

Since the resistance, R, is equal to $1/(\sigma w)$, in which σ is the conductivity (ohm.cm), the thickness of the strip should be as small as possible; the purest available material is suited for this purpose.

InSb photovoltaic detectors are used for ground-based IR astronomy and for applications aboard the space IR telescope facility. Its bandgap at 77 K provides response out to about 5.6 μm, which matches the J, H, K, L, M atmospheric windows. Linear array formats of 64, 128, 256, and an array size of 1k×1k are developed. InSb detector arrays have found applications in mid-IR due to their spatial uniformity, low dark current (see Section 4.1.1.8), and image quality. For missile warning systems (Hirsch et al., 2012) applications, the IR imagers require to exhibit high dynamic range, fast frame rates, high-resolution, very wide-FOV, and high sensitivity.

The first InSb array to exceed one million pixels was the ALADDIN array, which was demonstrated on a telescope by National Optical Astronomy Observations (NOAO), Tucson, Arizona in 1994 (Fowler et al., 1994). An InSb detector array (1k×1k) has been employed at the focal plane of the vacuum cryogenic spectrometer operating at about 60 K, called NIRSPEC, at the 10 m Keck II telescope, Mauna Kea, Hawaii. The quantum efficiency of this detector is about 80%, with full well capacity of 105,000 electrons. Figure 8.5 displays a spectral image obtained with said instrument.

Figure 8.5 A spectral image from the near-IR echelle spectrometer (NIRSPEC; Courtesy: I. S. McLead).

Hoffman et al. (2004) reported the development of InSb FPAs having 2k×2k ORION SCA for astronomical observations. Four such systems were deployed as a 4k×4k focal plane in the NOAO near-IR camera, at the Mayall 4 m telescope on Kitt Peak.

NIRSPEC A spectrometer for the Near-IR echelle Spectrometer (NIRSPEC) has been developed by McLean et al. (1998). It is a single-slit cross-dispersed echelle spectrometer operating at the Nasmyth focus on Keck II telescope. The NIRSPEC features the range between 0.96 and 5.5 μm at resolutions of $R = 2000$ or $R = 25{,}000$. The slit-viewing camera provides additional capability to image fields in a 46″-square field from 1 to 2.5 μm.

Keck interferometer The Keck interferometer, built at Mauna Kea, Hawaii, USA, combines light from the twin 10 m apertures main telescopes with a fixed baseline of 85 m. These large telescopes are being used for visible and near-IR astronomy. Each telescope sits on alt-azimuth mount, whose focal length is 17.5 m (f/1.75) and the collecting area is 76 m^2. The telescopes are equipped with a host of instruments, such as cameras and spectrometers. The interferometer has a spatial resolution of 5 mas at 2.2 μm, and 24 mas at 10 μm. The system transports the light to a laboratory located between the main telescopes, where a beam combiner and infrared camera combine and process the light. In order to equalize the path length between the arms of this interferometer, the delay-lines consisting of movable mirrors, are used to enable interference. Both Keck telescopes are equipped with laser guide star (LGS; see Section 3.5.4) adaptive optics (AO; see Section 1.5.6) system (Wizinowich et al., 2000).

8.6.2 *HgCdTe-Based Detectors*

The semiconductor crystal (see Section 2.2.1) mercury cadmium telluride (HgCdTe or MCT) is the most widely used mid-IR detector material due to the high density of dislocations present in the material (Brill et al., 2009). The bandgap of $Hg_{1-x}Cd_xTe$ is a function of the alloy composition ratio, x, of cadmium telluride (CdTe) ($E_g = 1.6$ eV, $\lambda_c = 0.725$ μm) to a semi-metallic compound mercury telluride (HgTe) with a negative energy gap ($E_g = 0.3$ eV,

$\lambda_c = 4.13$ μm). By adjusting the alloy composition and mixing HgTe and CdTe, suitable IR detector materials are obtained. The original interest in this alloy material is to obtain a semiconductor with a narrow energy gap needed for an intrinsic IR detector, which could be sensitive in the 8–14 μm atmospheric window. It was found that $x = 0.2$ was a promising material, which yielded both photovoltaic (PV; see Section 5.3) and photoconductive (PC) detectors (see Section 5.4) operable at 77 K with high performance. With higher values of x the spectral response can be tailored to 0.7 μm corresponding to the bandgap of CdTe. Since the bandgap of HgTe is negative, the alloy can be grown to achieve arbitrarily small bandgaps (Norton, 2002). However, HgCdTe has not been used beyond 25 μm. As stated in Section 2.2.7, a small bandgap allows a small contact voltage in order to maintain the diode depletion region. Thus, long wavelength detectors tend to have high dark current (see Section 4.1.1.8). However, Bacon et al. (2004) found promising results for MCT detectors fabricated to have cut-off wavelength, λ_c, of 9.6 μm at 30 K.

To note, the near zero-bias requirement of PV detectors make them free from power dissipation problems as compared to PC detectors that generate i^2R heat due to their low resistance. Also the high impedance of PV HgCdTe matches well with the impedance of the read-out circuit. However, HgCdTe has several drawbacks. These compounds formed by lead tin telluride ($Pb_{1-x}Sn_xTe$) alloys, whose energy gap, E_g, is a function of composition (Hatto and Crocker, 1982) and temperature, suffer from high thermal expansion coefficient and a large non uniformity. Besides, due to their high dielectric constant, diode capacitance is relatively high and limits the high frequency response of these devices. There is surface instability because of large lattice mismatch. As stated in Section 4.3.1, the thermal conductivity and the coefficient of thermal expansion are the two important factors to decide the method of mounting an IR detector in a cooler. Typically, the thermal conductivity of pure single crystal CdTe is 0.4 $W.cm^{-1}$ at 77 K and varies approximately as $T^{-1.3}$ between 30 and 300 K, while the polycrystalline ($\sim 10^{18}.cm^{-3}$ acceptor concentration) HgTe has a thermal conductivity of 0.15 $W.cm^{-1}$ at 77 K and varies approximately as T^{-1} between 77 and 300 K. The HgCdTe device has a moderate thermal coefficient of expansion, α, which is of the order of 2×10^{-6} at 77 K; for CdTe it is about 5.5×10^{-6}°C between 25°C and 100°C and for HgTe, it is about 4×10^{-6}°C in the range 25–50°C.

The fabrication of such a photosensitive material is based on the double layer hetero-junction structure (see Section 5.3.3). It is made of HgCdTe grown with a molecular beam epitaxial process (see Section 2.3.1.1) on a ternary crystal, cadmium zinc telluride (CdZnTe) substrate that is very rugged and provides a good thermal contraction match to silicon multiplexer. This sapphire substrate interconnects through indium (In) bumps to a silicon (Si) substrate, where circuitry for addressing the pixel and analog electronics can be built. The fraction of Hg and Cd controls the long wavelength cut-off of the material, whereas the doping material (arsenic; As, for the P-type, indium; In, for the N-type) creates the P-N junction. The properties of CdZnTe alloy depend on the concentration of zinc in the bulk, as well as on the surface of the material. The base P-type (or N-type) layers are sandwiched between CdZnTe substrate and high doped or wider-gap regions (Henini and Razeghi, 2002).

An intrinsic PC detector is a parallel-piped of $Hg_{1-x}Cd_xTe$ with electrical contacts on its ends and it is used in a constant current circuit because of its low electrical resistance. Excitation of current carriers in the crystal arises, in general, from three sources:

(1) The thermal excitation provides the thermal equilibrium carriers that give the crystal its dark conductivity, σ_D (see Section 5.4). This gives rise to Johnson noise (see Section 4.5.2.1).

(2) There would also be electromagnetic radiation (largely IR) falling on the detector from a warm background to which it exposed, and this radiation may excite excess carriers above the thermal equilibrium concentration, thereby enhancing the conductivity. This is a source for the BLIP noise (see Section 4.5.1.2). In general, a detector is mounted on a flat base holding it at the operating temperature and to be exposed to a 2π sr (steradian) FOV (see Section 4.4.1.4) of a background with a temperature of ~300 K, although a narrow FOV reduces the amount of radiation received may be used.

(3) The signal radiation from the IR-emitting object to be detected excites additional excess carriers, which cause generation-recombination (GR) noise (see Section 4.5.2.2), and it is chopped to give a periodically fluctuating component of the conductivity, which can be distinguished from the dark and background radiation induced DC conductivities by AC method. The photoconductor detectivity is limited by whatever noise voltage appears with the signal voltage, V_c at the output.

The noise, i_n, in HgCdTe (MCT) detectors operating at 77 K is due primarily to two sources, such as Johnson noise from the resistance and the shot-noise (see Section 4.5.2.3) from the photocurrent, i_{ph}. The noise current may be expressed as

$$i_{\mathrm{n}} = \sqrt{\left[2qi_{\mathrm{ph}} + \frac{4k_{\mathrm{B}}T_{\mathrm{d}}}{R_{\mathrm{D}}}\right]\frac{1}{2\Delta t}} \qquad \mathrm{A}, \tag{8.36}$$

where k_B is Boltzmann's constant, T_d the detector temperature, R_D the dynamic resistance of the detector, q the electron charge, and Δt the integration time.

It has been found that for both MWIR and LWIR HgCdTe detectors, the Johnson noise is much smaller than the shot-noise from the photocurrent at tactical background flux levels and the detector is said to have BLIP (Goldberg, 2002), in which case the second term of this equation (8.36) may be neglected. On a single element basis, such detectors operate at or near theoretical levels of S/N performance in tactical environments (Kinch and Yariv, 1989). In the case where the number of electrons collected in a frame is limited by the capacity of the read-out IC (ROIC) charge well, the S/N ratio is given by

$$\mathrm{S/N} = \frac{qN_{\mathrm{e}}/2\Delta t}{\sqrt{2q(qN_{\mathrm{e}}/2\Delta t)\frac{1}{2\Delta t}}} = \sqrt{\frac{N_{\mathrm{e}}}{2}}, \tag{8.37}$$

with N_e as the number of electrons the well (half full) can hold.

The expression for the minimum detectable temperature difference, $NE\Delta T$, under these conditions is

$$NE\Delta T = \frac{T_b}{2\sqrt{2N_e}} \quad \text{K}. \tag{8.38}$$

in which T_b is the background temperature.

The spectral range of MCT detectors is suitable for a wide range of thermal imaging applications. Its other properties are (i) high quantum efficiency (QE), (ii) direct bandgap with high absorption coefficient, (iii) moderate dielectric constant, (iv) requirement of minimal cooling because the detection mechanism relies photoexcitation across an intrinsic bandgap, and (v) availability of wide bandgap lattice-matched substrates for epitaxial growth (Norton, 2002). The SWIR, MWIR, and LWIR electronically scanned HgCdTe arrays with CMOS multiplexer are commercially available from several manufactures. An example of a large mosaic, developed by Teledyne imaging sensors, is a 147 megapixel FPA comprising 35 arrays, each with 2k×2k pixels (Beletic et al., 2008). In what follows, a few such detectors that are used for astronomical observations are highlighted.

8.6.2.1 NICMOS

An IR focal-plane array called Near-IR Camera and Multi-Object Spectrometer (NICMOS) is a scientific instrument for infrared astronomy. It provides imaging capabilities in broad, medium, and narrow-band filters, broadband imaging polarimetry, a method to determine the degree of alignment of wave vibrations in a beam, coronagraphic imaging (Saha, 2010, and references therein), and slit-less grism spectroscopy, in the wavelength range 0.8–2.5 microns. NICMOS has three adjacent but not continuous cameras, designed to operate independently, each with a dedicated array at a different magnification scale. Built by Ball Aerospace, a similar system in the range of 1 to 5 μm has been installed for use in the HST in 1997. It is a hybrid direct read-out integrating detector with a cut-off wavelength of 2.4 μm, which consists of 256×256 of 40 μm square pixel size, organized in four independent 128×128 quadrants that can be addressed simultaneously and independently (Cooper et al., 1993). Each pixel has a detector diode, which is electrically connected to its unit cell in the silicon layer. This unit cell circuits contain a field-effect transistor (FET; see Section 2.2.9) with high gate resistance, so as to enable to read the voltage across the diode following integration without loss of any charge. The near-IR (0.8–2.5 μm) imager around a 512×512 HgCdTe array (with 18 μm pixel) is also built. The image scale of such an instrument would help in subarcsec imaging and reaching out to fainter limit.

Unlike a CCD, the individual pixels of the NICMOS arrays are strictly independent and can be read non-destructively. Since the array elements are independently addressed, such a sensor does not suffer from some of the artifacts that afflict CCD. However, the dark current associated with NICMOS is quite substantial compared to that produced by the new generation CCDs. The typical NICMOS 3 FPAs; see Section 5.5.1) have read noise less than 35 e- with less than 1 e-/sec detector dark current at 77 K and broadband quantum efficiency is better than 50% in the range of 0.8 to 2.5 μm. The Infrared Optical Telescope Array (IOTA; see Section 3.4.4.2) employed such a camera capable of reading ten fringe frames per second. Each fringe frame, containing 256 samples, is made by scanning the optical path difference between the two beams with a mirror mounted on a 60-micron stroke PZT. Four pixels are read (two for the interferometric fiber outputs, two for the photometric fiber outputs).

NICMOS 3 near-IR camera was employed at the 6 m telescope of the Special Astrophysical Observatory (SAO), Russia, to record specklegrams of an evolved object, Red Rectangle (AFGL 915; see Figure 8.6a), a reflection nebula[a] around the star, HD 44179, near the Monocerotis constellation and a carbon star IRC+10216 (CW Leo) around a resolved central peak surrounded by patchy circumstellar matter (Osterbart et al., 1996). The speckle masking reconstruction of the former object exhibits two lobes with the separation of ~0.15″. The dark lane between the lobes is due to an obscuring dust disc containing icy dust grains and hydrocarbon molecules formed in the cool outflow and the central star is a close binary system. Since it is viewed edge-on, the boundary edges of the cone shapes appear to form a rectangle. The reconstruction of the object

[a]The reflection nebula displays a bipolar flow carrying a significant amount of mass away from the central stars. These stars may create a pair of jets, which might throw gas into a thick disk.

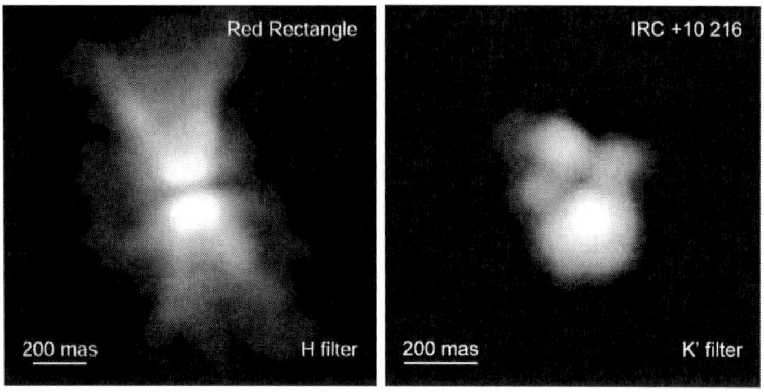

(a) (b)

Figure 8.6 Speckle masking reconstructed images of (a) an evolved object, Red Rectangle and (b) a carbon star IRC+10216; the respective resolutions of these objects were 75 mas for the H-band and 76 mas for the K′ band. Courtesy: R. Osterbart.

revealed the separation between bright clouds was 0.13–0.21″, implying a stochastic behavior of the mass outflow in pulsating carbon stars. Five individual clouds (see Figure 8.6b) were resolved within a 0.21″ radius of the central object in high resolution K′ band observation (Weigelt et al., 1998); the structures may be produced by circumstellar dust formation.

8.6.2.2 PICNIC

A PICNIC focal plane array (Cabelli et al., 2000; Kozlowski et al., 2000; Pedretti et al., 2004), an evolved version of the NICMOS 3, is composed of four 128×128 pixel quadrants of a HgCdTe-based detector and sensitive to the 0.8–2.5 μm wavelength region; the active area of each pixel is 40 μm^2. Each quadrant contains two digital shift-registers (horizontal and vertical) for addressing pixels in the array. Each register requires two clocks, with one being a dual-edge triggered clock and one a level triggered clock. The vertical register selects the row to read and/or reset depending on the reset and read clock inputs. The quantum efficiency of these detectors is more than 50% for wavelengths in the range 0.8–2.5 μm. This system is being used at the MIRC-CHARA interferometer (see Figure 8.7; Pedretti et al., 2009); it was also employed at the IOTA

Figure 8.7 Dispersed fringe pattern recorded on the PICNIC detector at the MIRC-CHARA interferometer (Pedretti et al., 2009; Courtesy: E. Pedretti).

interferometer. The PICNIC camera is equipped with standard near-IR broadband filters, such as J(1.11–1.39 μm), H(1.50–1.80 μm), and K′(2.00–2.32 μm), as well as a variety of narrow-band filters within the those bands.

MIRC The Michigan Infra-Red Combiner is an image plane combiner, which was developed to combine light beams from four telescopes of CHARA array (see Section 7.6.1) simultaneously (Monnier et al., 2006). These beams are filtered by single-mode fibers (see Section 3.4.2), which are rearranged into a one-dimensional non-redundant pattern and brought to a focus. These overlapping beams create six interference fringes, each with a unique spatial frequency. The pattern is focused by a cylindrical lens into a line of fringes, which are subsequently dispersed by a spectrograph with spectral resolution $\Delta\lambda \sim 0.035$ μm. MIRC can measure six visibilities, four closure phases, and four triple amplitudes simultaneously over eight spectral channels spanning the astronomical H-band.

8.6.2.3 HST Wide-field camera

A Wide-Field Camera (WFC 3) system for the Hubble Space Telescope (HST; see Section 6.2.2.2) is configured as a two-channel (ultraviolet-visible and near-IR) instrument. Its wide wavelength coverage is made using two detectors: (i) a charge-coupled device

(CCD; see Section 6.2) for the UV-visible channel and (ii) a 1k×1k pixel HgCdTe array detector for the near-IR channel. In the second case, a cooling device, based on the thermoelectric cooler (see Section 4.3.1), keeps the detector at an operating temperature of ~145 K. The read-out circuitry is made on a separate complementary metal-oxide semiconductor (CMOS; see Section 6.3) chip (multiplexer; see Section 6.2.1), which is hybridized to the detector with an indium (In) contact for each pixel. Following hybridization of two chips, the CdZnTe substrate is removed to reduce the susceptibility of the device to cosmic-ray events, as well as to enhance the sensitivity at short wavelengths. The final chip is, thus, a CMOS device connected through In columns to a thin layer of HgCdTe substrate.

8.6.2.4 HAWAII

A device called HgCdTe Astronomical Wide Area Infrared Imager (HAWAII) has been developed for scientific applications particularly for the ground- and space-based astronomy. The first prototype of a hybrid (see Section 5.5.2) focal plane array (FPA; see Section 5.5.1) with 1k×1k pixels was built in the 1990s and tested at the ground-based telescopes where the images of high resolution were obtained using tip-tilt correction by Hodapp et al. (1995). Depending on the desired wavelength sensitivity, HAWAII devices can be hybridized to a variety of detector materials, including HgCdTe, InGaAs, and Si. For instance, HAWAII 2k×2k-based FPAs with MCT detector arrays have low dark current of $< 0.01\ e^{-}.s^{-1}$ (Loose et al., 2007). This detector is employed in the following various types of equipment for recording high-resolution spectra or images of celestial objects.

Using VLTI/AMBER instrument, Weigelt et al. (2007) carried out near-IR interferometry of a LBV, η Carinae.[a] A HAWAII array detector having 512×512 pixels was used to obtain the spectrally dispersed

[a]The enigmatic object, η Carinae, is one of the most luminous and most massive ($M_\star \sim 100\ M_\odot$, in which $M_\star$ is the mass of the star and $M_\odot$ the solar mass) unstable luminous blue variables (LBV) suffering from a high mass-loss. Its distance is $\approx 2300 \pm 100\ pc$ (Davidson and Humphreys, 1997; Davidson et al., 2001; Smith, 2006). The η Carinae is a highly variable southern object located in the constellation Carina (α 10 h 45.1 m, δ $-59^\circ 41'$) surrounded by the expanding bipolar Homunculus nebula ejected during the Great Eruption in 1843.

fringes of this object. The interferograms are spectrally dispersed with low resolution; $R = \lambda/(\Delta\lambda) = 75$, medium resolution; $R = 1500$, and high resolution $R = 12{,}000$. The fibers (see Section 3.4) in AMBER limit the FOV to the diameter of the fibers on the sky, which is of the order of $\sim$60 mas. Figure 8.8a shows the spectrally

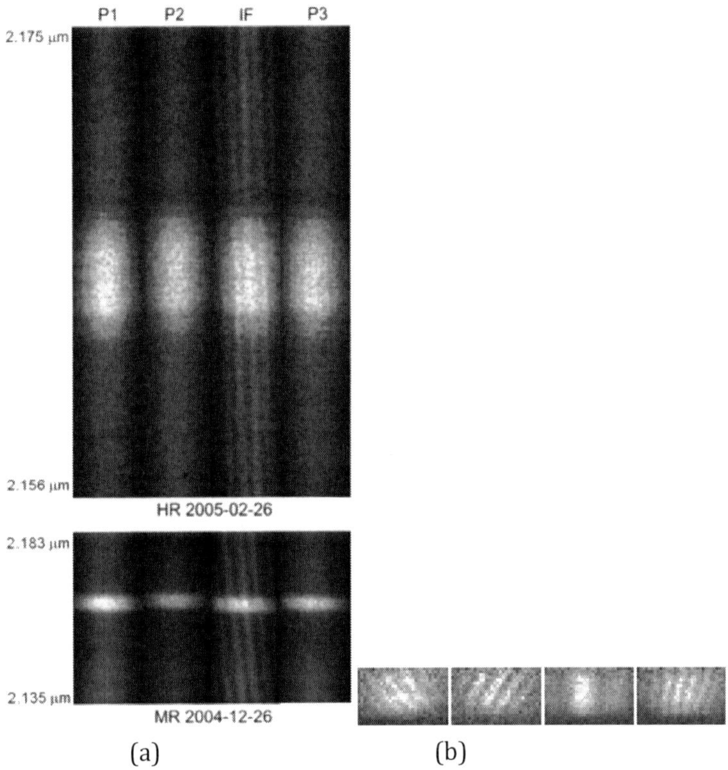

(a) (b)

Figure 8.8 (a) Spectrally dispersed AMBER/VLTI Michelson interferograms of η Carinae (Weigelt et al., 2007) depicting the spectrally dispersed fringe signal (IF), as well as the photometric calibration signals from the three 8.2 m unit telescopes UT2, UT3, and UT4 (P1, P2, and P3) in high (HR, upper panel) and medium spectral resolution mode (MR, lower panel) and (b) two-telescope interferograms of NGC 3783 (Weigelt et al., 2012); from left the first and second interferogram are recorded with UT2–UT3 (46.6 m projected baseline), the third one with the universal telescopes (3 and 4; UT3-UT4, baseline 62.5 m), and the fourth one with UT2-UT4 (baseline 89.4 m) of VLTI. Courtesy: G. P. Weigelt.

dispersed AMBER/VLTI Michelson interferograms of η Carinae; in both panels, the bright regions are associated with the Doppler-broadened (see Section 3.3.5.1), Brγ, emission line. Weigelt et al. (2012) have also recorded interferograms (see Figure 8.8b) of NGC 3783 with the VLTI/AMBER instrument in the K-band by using long detector integration times of 400 and 800 ms to be able to recognize faint fringes during data recording and correct drifts of the optical path differences (OPDs) between the telescope beams.

HAWAII 1 (1k$\times$1k) array is employed at the TIFR near-IR Spectrometer and Imager (TIRSPEC; Ojha et al., 2012) that is mounted on 2 m Himalayan Chandra Telescope (HCT), Indian Astronomical Observatory (IAO), India. Figures (8.9a and 8.9b) depict the near-IR (*JHK*) images of the NGC 7538 star-forming region and the region around a supernova, SN 2014 J (see Figure 6.5a), in the galaxy M 82, respectively.

(a) (b)

Figure 8.9 (a) Image of the NGC 7538 star-forming region and (b) region around SN 2014 J in the galaxy M 82. Courtesy: D. Ojha.

Another detector, HAWAII 1 array, is being used at the PISCES near-IR camera (*JHK*-imager), operating at the focal plane of Large Binocular Telescope (LBT). With the narrow-band Brγ (2.16 μm) and [Fe II] (1.644 μm) filters, which minimize saturation on the array, Close et al. (2012) obtained high-resolution images of θ 1 Ori B and θ 1 Ori C fields in the Orion Trapezium and opined that all the components of the θ 1 Ori B are likely to be gravitationally bound system. These images are taken with LBT adaptive optics (AO)

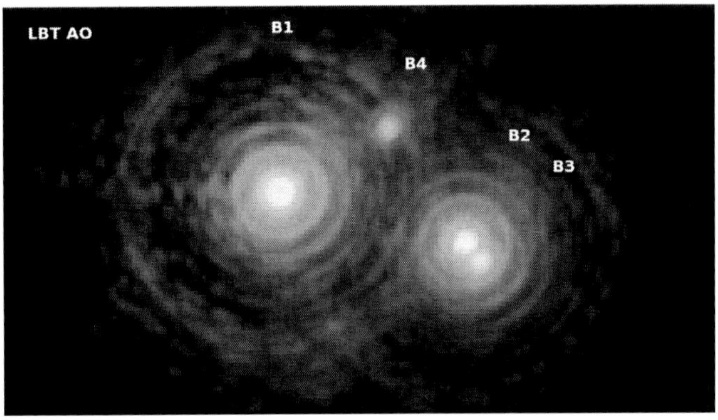

Figure 8.10 A view of four young stars, θ 1 Ori B1, θ 1 Ori B2, θ 1 Ori B3, and θ 1 Ori B4, in the Orion Trapezium cluster 1350 light-years away, as seen through the LBT's adaptive optics (AO; Courtesy: L. Close).

system. Figure 8.10 displays an image taken of four components of these stars, which are tightly located within 1 arcsec of each other. By comparing this infrared image to past images of this group (see Figure 6.7b), astronomers can envisage the motion of each star with respect to the others.

A Teledyne HAWAII 2RG HgCdTe-based detector with 2k$\times$2k pixels is being used as a focal plane camera for the Multi-Object Spectrometer for InfraRed Exploration (MOSFIRE) instrument (Kulas et al., 2012) at the Keck I telescope, Mauna Kea, Hawaii.

Another instrument, called OH-suppressing infrared imaging spectrograph (OSIRIS; Larkin et al., 2006), also employs a Rockwell HAWAII 2 MCT detector (2k$\times$2k pixels) as camera. The diffraction-limited spectroscopic observations of the Sagittarius A* region[a] using this instrument was carried by Krabbe et al. (2006). The faint

[a]Sgr A* is a point-like radio source within the core (about 5″ in size) of the galactic center of the Galaxy. Studies of radio properties of this source reveal that it is a compact object approximately one astronomical unit in size. However, with the speckle interferometry and adaptive optics system at the 10 m Keck-I telescope, Ghez et al. (1998) have observed a 6″$\times$6″ (0.23$\times$0.23 pc) region to monitor the proper motion of 90 stars. The shift-and-add algorithm (Lynds et al., 1976; Worden et al., 1976) was employed to process the data, which showed that the stellar surface density and the velocity dispersion were consistent with the position of Sgr A*. Their measurements allow an independent estimate of the central density, which is at least 10^{12} $M_\odot$ pc^{-3}. Subsequent AO observations of Sgr A* covering an area of 76″ $\times$ 76″

close-in substellar companions in halos of bright stars have also been studied (McElwain et al., 2007). Fu et al. (2012) identified two binary active galactic nuclei (AGN) where the line splitting is driven by the orbital motion of the merging nuclei.

LBT Large Binocular Telescope interferometer (LBT), commissioned on Mt. Graham (elevation 3200 m), Arizona, USA, in which two 8.4 m primary mirrors, each with f/14.2 (Hill, 2000), are mounted in a single mounting, spanning 22.8 m baseline; the equivalent circular aperture of LBT is 11.7 m. The primary mirrors are co-mounted on a fully steerable alt-az mounting, where variable delay-lines for the path equalization are not needed. Such an interferometer provides information in the u, v-plane, which can be continuously combined or co-added. From this, the complex visibility can be recovered. The configuration allows essentially complete sampling of all spatial frequencies in the image within a variable baseline of 0 to 22 m using interferometric imaging between the two pupils. The LBT uses an Adaptive Secondary Mirror (ASM) consisting of 672 voice coil actuators that push (or pull) on 672 small magnets glued to the back surface of a thin (1.6 mm), 0.911 m aspheric ellipsoidal glass shell (Esposito et al., 2010, 2011). Combination of long-baseline interferometry (Saha, 2010, and references therein), mimicking a wide aperture, and AO techniques to improve the images offers the best of both approaches and shows great promise for applications, such as the search for extra-solar planets. At this point, it seems clear that interferometry and AO are complementary, and neither can reach its full potential without the other. The wavefront slopes are measured with the pyramid wavefront sensor. A pyramid sensor, consists of a four-faces optical glass pyramidal prism, which is placed with its vortex on the nominal focus of the optical system (Ragazzoni, 1996). The four faces deflect the beam in four different directions, depending on which the face of the prism gets the incoming ray. Using a relay lens located behind the pyramid, these four beams are then re-imaged onto a high-resolution detector, obtaining four images of the telescope pupil.

---

by Ghez et al. (2005) at the 10 m Keck-II telescope found significant IR intensity variations in L′ (3.8 μm); the magnitudes varied from 12.6 to 14.5 magnitude.

AMBER The AMBER (Astronomical Multiple BEam Recombiner) is a near-IR (J, H, K-band) beam-combiner instrument of the Very Large Telescope (VLTI; see Section 3.4.4), which allows the measurement of visibilities, differential visibilities, differential phases, and closure phases (Millour et al., 2006; Petrov et al., 2005, 2007; Tatulli et al., 2007).

MOSFIRE The MOSFIRE is a vacuum cryogenic instrument mounted in a rotating module at the Cassegrain focus of 10 m Keck I telescope (McLean et al., 2010, 2012). This instrument provides near-IR (0.97 to 2.45 μm) multi-object spectroscopy over a $6.14' \times 6.14'$ FOV (see Section 4.4.1.4) with a resolving power of R ~3270 for a $0.7''$ slit-width and imaging over a FOV of $6.8'$ diameter with $0.18''$ per pixel sampling. The instrument employs a single, fixed diffraction grating hat can be set in multiple orders (3, 4, 5, and 6) for dispersion in the K, H, J and Y bands, respectively.

OSIRIS The OSIRIS is a medium resolution ($R = 3700$) IR integral field spectrograph for the 10 m Keck II operated with the laser guide star AO system. It takes spectra in a small FOV to provide a series of images at different wavelengths. This instrument uses a lenslet array to sample a rectangular patch of the sky and produces spectra at up to 3000 locations simultaneously. Each lenslet focuses the incident light to a pupil plane located behind the lenslet array. These pupils are dispersed by a diffraction grating and focused onto a detector subsequently. Both the camera and spectrograph can operate at wavelengths between 1 and 2.4 μm. The center of the imaging camera's field is about $20''$ offset from the center of the spectrograph field and both can be used simultaneously with the same or different filters.

8.6.2.5 SAPHIRA

The SAPHIRA detector is designed for high-speed infrared applications, adaptive optics in particular. The key drivers are to achieve photon sensitivity (<10 photons RMS) and high-speed, non-destructive read-out (>10 K frames per second). SAPHIRA is an HgCdTe avalanche photodiode (APD) array incorporating a full custom read-out integrated circuit. The unique properties of HgCdTe enable almost noise-free avalanche gain, which is ideal for very low photon flux applications (Baker et al., 2008). Most applications of SAPHIRA would be in the 1 to 2.5 μm range or the 3 to 4.2 μm range. A key aspect of the array is the ability to perform multiple non-destructive read-outs, which can allow down the slope sampling to reduce the noise significantly and increase the sensitivity. The architecture allows multiple, independently resettable windows and a selectable number of parallel output. In this procedure, HgCdTe is grown into a complex stack of layers on a gallium arsenide (GaAs) substrate, by a technique, called metal organic vapor phase epitaxy (MOVPE; Pillans et al., 2012). The SAPHIRA detector has been selected for use in the wavefront sensors and fringe trackers of the European Southern Observatory's GRAVITY program (Kendrew et al., 2012). The SELEX/ESO SAPHIRA arrays have been able to reach read noise levels of <3 e^- and quantum efficiency (QE) of 40% at 80 K (Finger et al., 2012).

GRAVITY GRAVITY (general relativity analysis via very large telescope interferometry; Eisenhauer et al., 2008) is an interferometric imager with 10 μas (micro-arcsecond) astrometry and interferometric imaging in the K-band (2.2 μm) with 4 milliarcsec resolution at the VLTI, Mt Paranal, Chile. It combines the light from four UTs or ATs of VLT, measuring simultaneously the stellar interferograms from six baselines. The fringes of both science and fringe tracking targets are spectrally dispersed by the spectrometers inside the beam combiner cryostat, with the science spectrometer providing spectral resolutions of R $\sim$20, $\sim$500 or $\sim$4000 (Fisher et al., 2012). This instrument has the beam combiner instrument using fiber-fed integrated optics (IONIC; see Section 3.4.4.2), infrared wavefront sensors (see Section 1.5.6.2) mounted to each of the four UTs, fringe tracker, beam stabilization, and a laser metrology system. GRAVITY will have the ability perform extremely accurate astrometric measurements on very faint sources; it may enable to measure interaction of celestial sources through gravity. The GRAVITY instrument now at the testing phase. It will be shifted to the Paranal in late 2014, and it is expected to obtain first light in the middle of 2015.

8.7 Heterodyne Interferometry

An optical heterodyne system consists of a conventional PV detector or photoconductor employed with a laser (see Section 3.2), termed as the local oscillator that is set at a close-by frequency, which irradiates the detector at a relatively intense level. The desired outcome is the difference frequency (beat note). The signal-to-noise (S/N) ratio (see Section 4.1.1.4) of an ideal optical heterodyne detector is independent of all material parameters except for quantum efficiency. The limit to the minimum detectable signal power arises from the fluctuations in the stream of signal photons (signal fluctuation limit; see Section 4.5.1.1).

However, at optical wavelengths, heterodyne is considered to be inefficient and successive exposures made with different baselines are instead combined incoherently. Such a technique has the following advantages in the case of beam recombination in IR interferometry: a larger coherence length, a simplification of the transport of the signal from the collector to the recombiner. Let $U_s(t)$ and $U_l(t)$ be, respectively, the signals of a wave coming from a star and that of an artificial source (laser), which are expressed as

$$U_s(t) = a_{s0}.e^{-i(\omega_s.t-\psi)}, \tag{8.39}$$

$$U_l(t) = a_{l0}.e^{-i\omega_l.t}. \tag{8.40}$$

The laser is the phase reference. A detector like a photodiode, illuminated by the sources (star + laser) yields an electrical signal corresponding to the light intensity:

$$\begin{aligned} I(t) &= |U_s(t) + U_l(t)|^2 \\ &= \left[a_{s0}.e^{-i(\omega_s.t-\psi)} + a_{l0}.e^{-i\omega_l.t}\right]\left[a_{s0}.e^{i(\omega_s.t+\psi)} + a_{l0}.e^{-i\omega_l.t}\right] \\ &= a_{s0}^2 + a_{l0}^2 + 2.a_{l0}.a_{s0}.\cos[(\omega_l - \omega_s) + \psi]. \end{aligned} \tag{8.41}$$

If ω_l and ω_s are close, the frequency of I is low enough to fit in the bandwidth (see Section 4.1.1.1) of the detector and its electronics (a few GHz) and I carries the phase information from the radiation of the star. By correlating (multiplying) the signals I_1 and I_2 yielded by two apertures with heterodyne systems, one can extract a visibility term. However, the lasers must have the same phases for the two apertures.

Heterodyne interferometry at 8–11.5 μm spectral range was employed on the SOIRDÉTÉ, Observatoire de Calern, France, an IR interferometer that consisted of a pair of 1 m telescopes with a 15 m East-West horizontal baseline (Gay and Mekarnia, 1988). In this project, beams were received in the central laboratory on a double cat's eye delay-line on a step by step movable carriage; natural OPD drift due to the Earth-rotation was used for acquiring fringes. This technique was also used on the IR Spatial Interferometer (ISI), Mt. Wilson, California (see Figure 8.11; Townes et al., 1998), which features three movable telescopes. Each telescope possesses a stable CO_2 laser (see Section 3.3.7.1) acting as local oscillator. Starlight collected by the telescope comes into a Schwarzschild optical system consisting of a pair of spherical front-surface mirrors that provide a less divergent beam to optical components and allow matching of the incident stellar wavefront with that of the local oscillator. The star light from each aperture is first mixed with the CO_2 laser local oscillator, converting the signal to microwave frequencies, followed by path length matching and fringe detection in a correlator. On ISI,

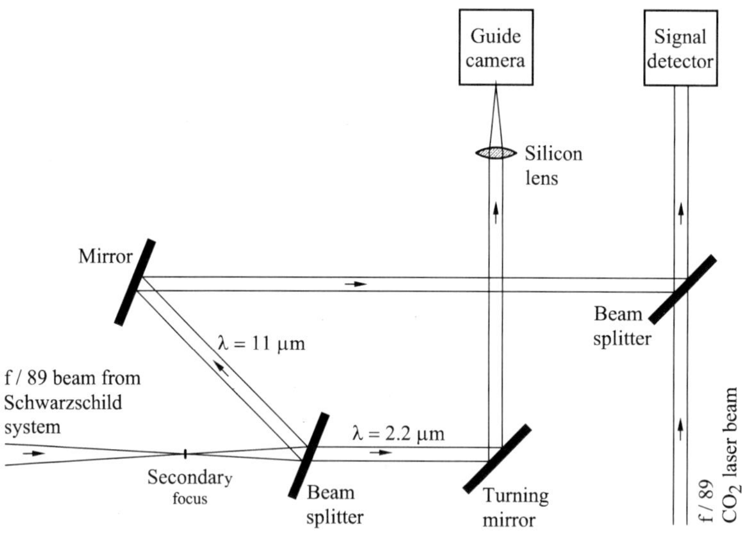

Figure 8.11 Schematic of the Infrared Stellar Interferometer (ISI) instrument; an infrared sensitive photodiode (liquid nitrogen cooled HgCdTe) signal detector is employed to record the intensity.

two CO_2 lasers are used, the phase of one being controlled by the other. Here the interferometer noise is dominated by shot-noise (see Section 4.5.2.3) of the laser and thermal background is negligible for setting the sensitivity limit (Hale et al., 2000). ISI utilizes Earth rotation and periodic discrete changes of the baseline to obtain a wide range of effective baselines and map the visibility functions of the stellar objects.

8.7.1 *Conventional Heterodyne Detection*

Conventional heterodyne detection (see Section 4.3.1.4) has some advantages in the form of (i) high sensitivity, (ii) frequency selectivity, and (iii) strong detectivity. In the optical/IR, it is a useful technique for applications, such as optical communications, spectroscopy, and radiometry. For radar applications, it provides a major method of recovering desired signals and removing clutter. The significant improvement in sensitivity that it provides over direct detection arises from the knowledge of the Doppler frequency (see Section 3.3.5.1), also called heterodyne frequency or intermediate frequency (IF), which permits narrow receiver bandwidth centered around IF. This technique, generally used in radio-astronomy, has the following advantages in the case of beam recombination in IR interferometry: (i) a larger coherence length, and (ii) a simplification of the transport of the signal from the collector to the recombiner (coaxial cables instead of mirrors). However, in addition to a tunable stable local oscillator, it requires to have (i) a good knowledge of the velocity of the source, (ii) a target presenting a minimum of frequency broadening, and (iii) at least some photons per measurement interval.

8.7.1.1 Two-frequency single-photon heterodyne detection

On using two Zeeman components of a visible (incoherent) spectral line, Forrester et al. (1955) observed this effect. The photomixing became considerably easier to observe with the development of laser and was studied by Javan et al. (1962) at 1.15 μm using a He-Ne laser (see Section 3.3.7.1) and by Siegman et al. (1963) at 6943 Å with a ruby laser (see Section 3.3.7.3). Subsequently,

Teich et al. (1966) used CO_2 laser at 10.6 μm in conjunction with a copper doped germanium photoconductive detector operated at 4 K. Later, Teich (1968) observed with a lead-tin selenide (PbSnSe) photovoltaic detector operated at 77 K. The spectral density of noise referred to the amplifier input is

$$\Psi(\nu) = \frac{h\nu}{e^{h\nu/(k_B T)} - 1} + h\nu, \tag{8.42}$$

with k_B as Boltzmann's constant, $h\nu$ the photon energy, and T the absolute temperature.

The first term of equation (8.42) is due to the thermal noise (see Section 4.5.2.1) and the second term is contributed by the quantum noise (see Section 4.5.2.3). If $\nu \gg k_B T / h$, the second term turns out be significant, and hence

$$\Psi(\nu) \cong h\nu. \tag{8.43}$$

If the bandwidth, $\Delta\nu$, of the system is centered around, the Signal-to-noise power ratio is given by

$$\begin{aligned} \text{S/N} &= \frac{P_s}{\int_{\nu-\Delta\nu/2}^{\nu+\Delta\nu/2} \Psi(\nu) d\nu} \\ &= \frac{\eta_d P_1}{h\nu\Delta\nu}, \end{aligned} \tag{8.44}$$

where P_1 is the received signal radiation power and η_d the detector quantum efficiency.

Heterodyne detectors in IR-region include InSb photoconductive detector (see Section 8.6.1), bolometer, Golay cell, and pyroelectric detector (see Section 8.3). In general, for such type of detectors, Johnson noise predominates ($h\nu \ll k_B T$), and the input S/N ratio is

$$(\text{S/N})_{\text{in}} = \frac{P_1}{k_B T_{\text{eff}} \Delta\nu}. \tag{8.45}$$

For simplicity, a number of detector parameters and operating conditions are lumped into the receiver effective temperature, T_{eff}.

8.7.1.2 Two-frequency multi-photon heterodyne detection

The single-photon photoelectric effect (see Section 2.1) requires that the incident photon energy, $h\nu$, be greater than the work function of the material, ϕ_m. When $h\nu$ is less than the ϕ_m, two photons must be absorbed in order to observe a photoelectric effect. Therefore, for the two-quantum case, it becomes prudent when

$$\frac{\phi_m}{2} < h\nu < \phi_m. \tag{8.46}$$

Since the emission current from a cathode is given by equation (5.1), the probability of exciting an electron initially in the ground state per unit time is given by $\sum_n \frac{d}{dt} \mid a_{0,n}(t) \mid^2$. Therefore, the total number of excited electrons per unit time, N' is given by the integral of the probability of exciting an electron over all momentum states κ. Thus,

$$N' = \frac{2\Omega}{8\pi^3} \sum_n \frac{d}{dt} \int\int\int \mid a_{0,n}(t) \mid^2 d\kappa_x d\kappa_y d\kappa_z, \tag{8.47}$$

in which Ω is the volume of the photocathode involved in the emission.

In order to calculate $|a_{0,n}(t)|^2$, the probability of exciting an electron, it is assumed that the electron in the ground state can reach the final state via two transitions: (i) by absorbing two photons simultaneously and reaching the final excited state, n, or (ii) by first absorbing one photon to reach the intermediate state and then absorbing another photon to reach the final state. A two-quantum DC photocurrent, $I_{DC}^{(2)}$, is proportional to the square of the incident radiation power, P, and inversely proportional to the irradiated area, A_d (Teich and Wolga, 1968), which is given by

$$I_{DC}^{(2)} = \Lambda^{(2)}(\lambda, T)P \propto IP, \tag{8.48}$$

with $\Lambda^{(2)}$ is the two-quantum yield expressed in ampere/watt, λ the wavelength of radiation, T the sample temperature, and I the irradiance at the detector expressed in watt per cm^2.

The two-quantum efficiency, denoted by $\eta_d^{(2)}$, is related to the two-quantum yield by the relationship

$$\Lambda^{(2)} = (q/h\nu)\eta_d^{(2)} \propto I. \tag{8.49}$$

Here the quantity $h\nu/q$ is the incident photon energy expressed in eV and is of the order unity. For the k- photon process, one obtains

$$I_{DC}^{(k)} = \Lambda^{(k)}(\lambda, T)P \propto I^{k-1}P, \tag{8.50}$$

and

$$\Lambda^{(k)} = (q/h\nu)\eta_{\rm d}^{(k)} \propto I^{k-1}. \tag{8.51}$$

The typical values for the two-quantum yield are $\Lambda_{Na}^{(2)}$ (8450 Å, 300 K)$\sim 8 \times 10^{-16} I$ and $\Lambda_{Cs_3Sb}^{(2)}$ (10600 Å, 300 K)$\sim 5 \times 10^{-11} I$ (Teich and Wolga, 1968). Using the relationship between the two-quantum yield, $\Lambda^{(2)}$ and the two-QE, $\eta_{\rm d}^{(k)}$, the signal-to-noise (S/N) ratio is expressed as

$$(\mathrm{S/N})^{(2)} \simeq \frac{4\eta_{\rm d}^{(k)} P_1}{h\nu(\Delta\nu)}. \tag{8.52}$$

The two-quantum minimum detectable power (MDP) is given by (Teich and Keyes, 1966; Teich, 1968)

$$\mathrm{MDP}^{(2)} \simeq \frac{h\nu\Delta\nu}{4\eta_{\rm d}^{(2)}}. \tag{8.53}$$

This corresponds to a minimum number of photons, $N_{\rm ph}^{(2)}$, detectable in the resolution time of the receiver ($t_r \sim (\Delta\nu)^{-1}$), given by

$$N_{\rm ph}^{(2)} \simeq (4\eta_{\rm d}^{(2)})^{-1}. \tag{8.54}$$

From equation (8.49), it is evident that the two-quantum efficiency, $\eta_{\rm d}^{(2)}$, is proportional to the irradiance, I at the detector. This I can be enhanced by increasing the local oscillator (LO) irradiance, and hence $\eta_{\rm d}^{(2)}$ may be enhanced by increasing LO irradiance. Thus, in contra-distinction to the single-quantum case, performance is not limited by the detector QE, $\eta_{\rm d}^{(2)}$. This technique may, therefore, be useful in regions of the electromagnetic spectrum where detectors with high QE are not available.

The S/N ratio at the fundamental difference frequency, $(\omega_1 - \omega_2)$, may also be obtained for coherent beam mixing in the k-photon absorption heterodyne detector

$$(\mathrm{S/N})^{(k)} \simeq \frac{C_k\eta_{\rm d}^{(k)} P_1}{h\nu(\Delta\nu)}, \tag{8.55}$$

$$\mathrm{MDP}^{(k)} \simeq \frac{h\nu\Delta\nu}{C_k\eta_{\rm d}^{(2)}}, \tag{8.56}$$

$$N_{\rm ph}^{(k)} \simeq (C_k\eta_{\rm d}^{(k)})^{-1}. \tag{8.57}$$

where C_k is a constant and $\eta_{\rm d}^{(k)}$ is proportional to I^{k-1}.

8.7.2 *Non-Linear Heterodyne Detection*

An alternative detection technique is the non-linear heterodyning, which eliminates the difficulties faced by the conventional heterodyne technique. The technique is useful in the region of the spectrum in the absence of high quantum efficiency (QE) detectors as well since the performance similar to that of conventional unity QE heterodyne receiver can be achieved theoretically.

Three-frequency non-linear heterodyne detection scheme is important in the case of the radial velocity of the target being tracked or the frequency difference between the transmitter and LO in a communication system has not been established. Under these circumstances, since the heterodyne frequency is not known, it is improbable to acquire a weak signal using conventional heterodyne technique. A transmitter, in general, emits two waves of frequencies, ν_1 and ν_2, whose difference, $\nu_c = \mid \nu_1 - \nu_2 \mid = \Delta\nu$, is known to high accuracy. The waves are Doppler-shifted (see Section 3.3.5.1) by the moving target, the nature of which is inconsequential. Thus, a wave of frequency, ν would return with a frequency, ν' given by the relativistic Doppler shift

$$\nu' = \nu\left(1 + \frac{2v_{\parallel}}{c}\right), \tag{8.58}$$

where $v_{\parallel}$ is the radial velocity of the target and c the velocity of light.

After scattering from the target, and choosing $\nu_1 > \nu_2$, the new frequency difference between two waves, ν_c' is given by

$$\nu_c' = \nu_1' - \nu_2' = \nu_c \pm \left(\frac{2v_{\parallel}}{c}\right)\nu_c \cong \nu_c. \tag{8.59}$$

Thus, the radiated and received difference frequencies are independent of Doppler shift to a good approximation when $v_{\parallel}/c \ll 1$. The two radiation fields, ν_1 and ν_2 are mixed in a heterodyne detector with strong, coherent, and polarized LO signal at frequency, ν_L, yielding two beat signals at $\mid \nu_1' - f_L \mid$ and $\mid \nu_2' - f_L \mid$, along with a DC, which is prevented. The third signal at $\mid \nu_1' - f_2' \mid$ is weak and may be ignored. The AC output of the mixer should be, in principle, broadband coupled, through a filter of bandwidth, $\Delta\nu$ to a non-linear device. The $\Delta\nu$ should be small in order to maximize the S/N ratio but must encompass the difference frequencies generated in the mixer. The non-linear device that should have a response over $\Delta\nu$, generates a component at the frequency (Teich, 1977)

$$\nu_c' = \mid \nu_1' - f_2' \mid . \tag{8.60}$$

Since the output of the non-linear device is essentially independent of the Doppler shift, as well as the LO frequencies, variation in these quantities have little effect on the system output. A narrow-band filter centered at $\nu_c \cong \nu_c'$ and of bandwidth, $\Delta\nu$, placed after the non-linear device, achieves a low-noise bandwidth. To note, the noise response, i_Q, of the detector is arising from shot-noise (see Section 4.5.2.3). The S/N power ratio, $(S/N)_p \equiv (S/N)^1$ is given by

$$(S/N)_p = \frac{\eta_d(P_1 + P_2)}{h\nu_L\Delta\nu}, \tag{8.61}$$

which is independent of P_L. On defining $P_T \equiv P_1 + P_2$, and let $\nu = \nu_L \approx \nu_1 \approx \nu_2$, and $(S/N)_i = (S/N)_p$. one gets

$$(S/N)_i = \frac{\eta_d P_T}{h\nu\Delta\nu}, \tag{8.62}$$

in which P_T is the total optical power.

Intensity interferometry (see Section 5.2.4.2; Hanbury Brown, 1974) has found use in gathering detailed information about objects far removed from the observation capability of human eye e.g., telescopic objects and submicroscopic objects in the range of nanometer (nm; 10^{-9} m). In the few-nanometer range, the matter (comprising from 10 to couple of thousand atoms say, atomic size being 0.1 nm) is arranged/distributed over a tiny (submicroscopic) volume and experiences energy transfer from one region to another region at picoseconds (10^{-12} s), and electron transfers at femtoseconds (10^{-15} s) time scale. Multi-point correlation parameters are required and challenge to be measured under these conditions utilizing multi-photon, multi-frequency, sum and difference, transition signals. Much in the manner of two-photon transition heterodyne technique illustrated for astronomical interest in this chapter technological development can be extrapolated for the fast developing fields of submicroscopic researches (Tewari, 2014). The use of interferometry is indeed growing fast (Labeyrie et al., 2006; Saha, 2010, and references therein) in revealing atomic scale processes. Its future is awaiting ideas and technological growth and offers an opportunity for the progress of novel sources and detectors that would help measuring parameters of interest for biological applications (Palmer III and Thompson, 1987; Shtengel et al., 2009), for example, control over cancerous growth and/or development of materials of unusual properties, for instance, extremely strong but very light in weight, or storage of energy without degradation in storage efficiency or in release of the energy at a desired fast or slow rate.

Appendix

Typical Tables

Table A.1 Radiometric units

Quantity	Symbol	SI unit	Physical description
Radiant energy	$\mathcal{Q}_e$	joule (J)	Energy
Radiant power/flux	Φ_e	watt (W)	Power/flux
Radiant flux density	$\mathcal{E}_e$	$W.m^{-2}$	Irradiance
Radiance	$\mathcal{L}_e$	$W.m^{-2}.sr^{-1}$	Angular flux density
Radiant intensity	$\mathcal{I}_e$	$W.sr^{-1}$	Intensity
Radiant emittance	$\mathcal{M}_e$	$W.m^{-1}$	Power emitted from a surface

Table A.2 Photometric units

Quantity	Symbol	SI unit	Conversions
Luminous power/flux	Φ_ν	lumen (lm)	$1\ lm = 1\ cd.sr$
Illuminance	$\mathcal{E}_\nu$	lux (lx)	$1\ lx = 1\ lm.m^{-2}$
Luminous intensity	$\mathcal{I}_\nu$	Candela (cd)	$1\ cd = 1\ lm.sr^{-1}$
Luminance (brightness)	$\mathcal{L}_\nu$	$1\ cd.m^{-2}$	
Luminous energy	$\mathcal{Q}_\nu$	1 lm.s	
Luminous emittance	$\mathcal{M}_\nu$	lx	

Table A.3 Zernike polynomials, Z_j for $j = 1, 2, \cdots, 15$ and the common optical aberrations

j	n	m	Z_j	Optical aberrations
1	0	0	1	Piston or Bias
2	1	1	$2\rho\cos\theta$	Tilt-x (Lateral position)
3	1	1	$2\rho\sin\theta$	Tilt-y (Longitudinal position)
4	2	0	$\sqrt{3}(2\rho^2 - 1)$	Defocus
5	2	2	$\sqrt{6}\rho^2 \sin 2\theta$	Astigmatism-y (3rd order)
6	2	2	$\sqrt{6}\rho^2 \cos 2\theta$	Astigmatism-x (3rd order)
7	3	1	$\sqrt{8}(3\rho^3 - 2\rho)\sin\theta$	Coma-y (3rd order)
8	3	1	$\sqrt{8}(3\rho^3 - 2\rho)\cos\theta$	Coma-x (3rd order)
9	3	3	$\sqrt{8}\rho^3 \sin 3\theta$	Trefoil-y
10	3	3	$\sqrt{8}\rho^3 \cos 3\theta$	Trefoil-x
11	4	0	$\sqrt{5}(6\rho^4 - 6\rho^2 + 1)$	Spherical aberration
12	4	2	$\sqrt{10}(4\rho^4 - 3\rho^2)\cos 2\theta$	Secondary astigmatism-x
13	4	2	$\sqrt{10}(4\rho^4 - 3\rho^2)\sin 2\theta$	Secondary astigmatism-y
14	4	4	$\sqrt{10}\rho^4 \cos 4\theta$	Tetrafoil-x
15	4	4	$\sqrt{10}\rho^4 \sin 4\theta$	Tetrafoil-y

Note: n is the radial order and m the azimuthal order. Even values of j represent the symmetric modes defined by $\cos(m\theta)$, while j values correspond to the antisymmetric modes given by $\sin(m\theta)$.

Bibliography

Abe L., Vakili F., Percheron I., Hamma S., Ragey J.-P., Blazit A., 1999, Catching the perfect wave. *Symp. Proc. (110th ASP Conf.)*, 21.

Agrawal P. C., 2006, *Adv. Sp. Res.*, **38**, 2989.

Ahmed S. N., 2007, *Physics and Engineering of Radiation Detection*, Elsevier, Amsterdam.

Alexander S. B., 1997, *Optical Communication Receiver Design*, SPIE, Washington DC.

Anger H. O., 1952, *Nature*, **170**, 220.

Anger H. O., 1958, *Rev. Sci. Instrum.*, **29**, 27.

Anger H. O., 1966, *Trans. Instrum. Soc. Am.*, **5**, 311.

Aristidi É., Prieur J.-L., Scardin M., Koechlin L., Avila R., Carbillet M., Lopez B., Rabbia Y., Nisenson P., Gezari D., 1999, *Astron. Astrophys. Suppl.*, **134**, 545.

Ashley T., Qineti Q., Gordon N. T., 2004, *SPIE*, **5359**, 89.

Auger P., 1923, *C.R.A.S.*, **177**, 169.

Ayers G. R., Dainty J. C., 1988, *Opt. Lett.*, **13**, 457.

Babcock H. W., 1953, *Pub. Astron. Soc. Pac.*, **65**, 229.

Bacon C. M., et al., 2004, *SPIE*, **5167**, 118.

Bahcall J. N., et al., 1991, *The Decade of Discovery in Astronomy and Astrophysics*, Natl. Acad. Press. Washington DC.

Baker I. M., Owton D., Trundle K., Thorne P., Storie K., Oakley P., Copley J., 2008, *SPIE*, **6940**, 69402L.

Baldwin J. E., et al. 1996, *Astron. Astrophys.*, **306**, L13.

Bandara S. V., Gunapala S. D., Liu J. K., Luong E. M., Mumolo J. M., Hong W., Sengupta D. K., McKelvey M. J., 1998, *Appl. Phys. Lett.*, **72**, 2427.

Bass M., Van Stryland E. W., Williams D. R., Wolfe W. L., 1995, *Handbook of Optics*, McGraw-Hill Inc.

Bates R. J., 2001, *Optical Switching and Networking Handbook*, McGraw-Hill, New York.

Battistoni G., Campana P., Chiarella V., Denni U., Iarocci E., Nicoletti G., 1982, *Nucl. Instrum. Methods*, **202**, 459.

Baum W. A., 1954, Astron. J., **59**, 314.

Beaver E., McIlwain C., 1971, *Rev. Sci. Instrum.*, **42**, 1321.

Beck W. A., 1993, *Appl. Phys. Lett.*, **63**, 3589.

Beckers J. M., 1993, *Ann. Rev. Astron. Astrophys.*, **31**.

Becquerel E., 1839, *Comptes Rendus*, **9**, 561.

Beiser A., 1973, *Concepts of Modern Physics*, McGraw-Hill Kogakusha Ltd.

Beletic J. W., Blank R., Gulbransen D., Lee D., Loose M., Piquette E. C., Sprafke T., Tennant W. E., Zandian M., Zino J., 2008, *SPIE*, **7021**, 70210H.

Berns R. S., 2000, *Principles of Color Technology*, John Wiley & Sons, Inc., New York.

Bezerra H. G., Costa M. A., Guagliumi G., Rollins A. M., Simon D. I., 2009, *JACC: Cardiovascular Interventions*, **2**, 1035.

Bhandari P., Prina M., Bowman R. C., Paine C., Pearson D., Nash A., 2004, *Cryogenics*, **44**, 395.

Binney J., Merrifield M., 1998, *Galactic Astronomy*, Princeton Series in Astrophysics, Princeton, New Jersey.

Blanchi L., et al., 2009, *Astron. J.*, **137**, 3761.

Blazit A., 1986, *SPIE*, **702**, 259.

Blazit A., 2014, Private communication.

Blazit A., Bonneau D., Koechlin L., Labeyrie A., 1977, *Astrophys. J.*, **214**, L79.

Blazit A., et al., 2008, *Appl. Opt.*, **47**, 1141.

Bloch F., 1928, *Z. Physik*, **52**, 555.

Bohr N., 1913, *Phil. Mag.*, **26**, 151.

Boksenberg A., 1975, *Proc., 'Image Processing Techniques in Astronomy'*, Eds., C. de Jager, H. Nieuwenhuizen.

Boksenberg A., 1982, *Phil. Trans. R. Soc. London A*, **307**, 531.

Boksenberg A, Burgess D, *Advances in Electronics and Electron Physics*, Academic Press, London, 33B (1972), 835.

Boltzmann L., 1884, *Ann. der Physik und Chemie*, **22**, 291.

Bonanno G., Belluso M., Cali A., Carbone A., Cosentino R., Modica A., Scuderi S., Timpanaro C., Uslenghi M., 2004, Scientific detectors for astronomy, In *The Beginning of a New Era*, Eds., P. Amico, et al., 21.

Born M., 1926, *Zeitschrift für Phys.*, **37**, 863.

Born M., Wolf E., 1984, *Principles of Optics*, Pergamon Press.

Bouguer P., 1729, *L'Essai d'Optique*.

Boyle W. S. Smith G. E., 1970, *Bell System Tech. J.*, **49**, 587.

Bravais A., 1850, *J. Ecole Polytech.*, **19**, 1.

Breckinridge J. B., McAlister H. A., Robinson W. A., 1979, *Appl. Opt.*, **18**, 1034.

Breeveld A. A., et al., 2010, *Mon. Not. R. Astron. Soc.*, **406**, 1687.

Brennan K. F., 2003, *The Physics of Semiconductors: With Applications to Optoelectronic Devices*, Cambridge University Press, Cambridge.

Brenner D. J., Hall E. J., 2007, *N. Engl. J. Med.*, **357**, 2277.

Brill G., Chen Y., Wijewarnasuriya P., Dhar, N. K., 2009, *SPIE*, **7419**, 74190L.

Brillouin L., 1953, *Wave Propagation in Periodic Structures*, Dover, New York.

Brown R. G. W., Ridley K. D., Rarity J. G., 1986, *Appl. Opt.*, **25**, 4122.

Brown R. G. W., Ridley K. D., Rarity J. G., 1987, *Appl. Opt.*, **26**, 2383.

Bruns D., Barnett T., Sandler D., 1997, *SPIE*, **2871**, 890.

Budde W., 1983, *Physical Detectors of Optical Radiation*, Academic Press, New York.

Burger J. F., ter Brake H. J. M., Rogalla H., Linder M., 2002, *Cryogenics*, **42**, 97.

Burke B. F., Tonry J. L., 1997, *SPIE*, **3019**, 233.

Burns D., et al., 1997, *Mon. Not. R. Astron. Soc.*, **290**, L11.

Byrnes J., 2009, *Unexploded Ordnance Detection and Mitigation*, Springer, New York.

Cabelli C. A., Cooper D. E., Haas A. K., Kozlowski L. J., Bostrup G. L., Chen A. C., Blackwell J. D., Montroy J. T., Vural K., Kleinhans W. E., Hodapp K., Hall D. N., 2000, *SPIE*, **4028**, 331.

Carleton N., et al., 1994, *SPIE*, **2200**, 152.

Carroll J. A., Moss E. B., 1930, *Mon. Not. R. Astron. Soc.*, **91**, 191.

Case T. W., 1920, *Phys. Rev.*, **15**, 289.

Chapman S., 1931, *Proc. Roy. Soc.*, **A132**, 353.

Chauhan D. S., Marshall J., 1999, *Invest Ophthalmol Vis Sci.*, **40**, 2332.

Chinnappan V., 2006, PhD thesis, Bangalore University.

Chinnappan V., Saha S. K., Faseehana, 1991, *Kod. Obs. Bull.* **11**, 87.

Chromey F. R., Hasselbacher D. A., 1996, *Pub. Astron. Soc. Pac.*, **108**, 944.

CIE Pub., 15.2, 1986, *Colorimetry*, Central Bureau of the CIE, Kegelstrasse 27, Vienna, Austria.

Ciupa R., Rogalski A., 1997, *Opto-Electr. Rev.*, **5**, 257.

Clampin M., Croker J., Paresce F., Rafal M., 1988, *Rev. Sci. Instrum.*, **59**, 1269.

Close L. M., et al., 2012, *Astrophys. J.*, **749**, 180.

Close L. M., et al., 2013, *Astrophys. J.*, **774**, 94.

Compton A. H., 1923, *Phys. Rev.*, **21**, 483.

Cooper D., Bui D., Bailey R., Kozlowski L., Vural K., 1993, *SPIE*, **1946**, 170.

Coudé du Foresto V., Ridgway S. T., 1992, *Proc. ESO-NOAO, Conf.*, Eds., J. M. Beckers and F. Merkle, 731.

Coutancier J., 1940, *Revue Générale de l'Électricité*, **48**, 31.

Cova S., Ghioni M., Lacaita A., Samori C., Zappa F., 1996, *Appl. Opt.*, **35**, 1956.

Crawford D. L., 1958, *Astrophys. J.*, **128**, 185.

Creasey D. J., 1985, Advanced Signal Processing, IEE Telecom. Ser.

Crouzier A., et al., 2013, arXiv:1309.3448v3[astro-ph.IM]18Sep2013.

Csorba I., 1985, *Image Tubes*, Howard Sams, Indianapolis, Indiana.

Cuby J.-G., Baudrand J., Chevreton M., 1988, *Astron. Astrophys.*, **203**, 203.

Czerny M., 1929, *Z. Physik*, **53**, 1.

Daigle O., Gach J.-E., Guillaume C., Carignan C., Balard P., Boissin O., 2004, *SPIE*, **5499**, 219.

Dautet H., Deschamps P., Dion B., MacGregor A., MacSween D., McIntyre R., Trottier C., Webb P., 1993, *Appl. Opt.*, **32**, 3894.

Davidson K., Humphreys R. M., 1997, *Ann. Rev. Astron. Astrophys.*, **35**, 1.

Davidson K., Smith N., Gull T. R., Ishibashi K., Hillier D. J., 2001, *Astron. J.*, **121**, 1569.

Davis J., et al., 2007, *Pub. Astron. Soc. Austr.*, **24**, 138.

Davis J., Tango W. J., 1986, *Nature*, **323**, 234.

Davis J., Tango W. J., Booth A. J., ten Brummelaar T. A., Minard R. A., Owens S. M., 1999a, Mon. Not. R. Astron. Soc., **303**, 773.

de Broglie L., 1925, *Ann. Phys.*, **3**, 22.

Decker R., 1969, *Advances in Electronics and Electron Physics*, Academic Press, London, **28**A, 357.

Dekker A. J., 1981, *Solid State Physics*, Prentice-Hall.

Delaët B., Feautrier, P., Villégier, J.-C., Benoît, A., 2001, *IEEE Trans.*, **11**, 824.

Dember H., 1931, *Phys. Zeits.*, **32**, 554.

Dereniak E. L., Boreman, G. D., 1996, *Infrared Detectors and Systems*, Wiley, New York.

Dhar N. K., Dat R., Sood A. K., 2013, http://dx.doi.org/10.5772/51665.

Dhillon V., Marsh T., Copperwheat C., Naidu B., Ives D., Vick A., O'Brien K., 2007, *The Messenger*, **127**, 41.

Diament P., Teich M. C., 1992, *IEEE J. Quantum Electron.* **28**, 1325.

Dirac P. A., 1926, *Proc. Roy. Soc. Series*, **A112**, 661.

Dunlap J. C., Blouke M. M., Bodegom E., Widenhorn R., 2012, *IEEE Trans.*, **59**, 1114.

Einstein A., 1905a, Ann. der Physik, **17**, 132.

Einstein A., 1905b, Ann. der Physik, **18**, 639.

Einstein A., 1905c, Ann. der Physik, **17**, 891.

Einstein A., 1905d, Ann. der Physik, **17**, 549.

Einstein A., 1916, *Sitzungsber. Preuss. Akad. Wiss.*, 688.

Einstein A., 1917, *Physikalische Zeitschrift*, **18**, 121.

Eisaman M. D., Fan J., Migdall A., Polyakov S. V., 2011, *Rev. Sci. Instrum.*, **82**, 071101.

Elster J., Geitel H., 1893, *Ann. der Physik*, **48**, 338.

Elster J., Geitel H., 1916, *Zeitschrift Phys.*, **17**, 268.

Ennulat R. D., Garin L. E., White J. D., 1974, *Mol. Cryst. Liq. Cryst.*, **26**.

Esaki L., Tsu R., 1970, *IBMJ. Res. Dev.*, **14**, 61.

Esposito S., et al., 2010, *SPIE*, **7736**, 773609.

Esposito S., et al., 2011, *SPIE*, **8149**, 814902.

Estrera J. P., et al., 2000, *SPIE*, **4128**, 46.

Estrera J. P., Saldana M. R., 2003, *SPIE*, **4796**, 60.

Felgett P. R., 1949, *Proc. Phys. Soc.*, **B62**, 351.

Fermi E., 1926, *Z. Physik*, **36**, 902.

Finger G., et al., 2012, *SPIE*, **8453**, 84530T.

Fischer S., et al., 2012, *SPIE*, **8445**, 84451W.

Fishburn M., 2012, *Fundamentals of CMOS Single-Photon Avalanche Diodes*, Delft, the Netherlands.

Floryan R., Devoe N., Peck T., 2003, *SPIE*, **5071**, 397.

Forrester A. T., Gudmundsen R. A., Johnson P. O., 1955, *Phys. Rev.*, **99**, 1691.

Fowler A. M., Bass D., Heynssens J., Gatley I., Vrba F. J., Ables H. D., Hoffman A., Smith M., Woolaway J., 1994, *SPIE*, **2268**, 340.

Foy R., Labeyrie A., 1985, *Astron. Astrophys.*, **152**, L29.

Fried D. C., 1966, *J. Opt. Soc. Am.*, **56**, 1972.

Friedman P. G., Cuza R. A., Fleischman J. R., Martin C., Schiminovich D., Doyle D. J., 1996, *Rev. Sci. Instrum.*, **67**, 596.

Fruchter A. S., Hook R. N., 2002, *Pub. Astron. Soc. Pac.*, **114**, 144.

Fu H., Yan L., Myers A. D., Stockton A., Diorgovski S. G., Aldering G., Rich J. A., 2012, *Astrophys. J.*, **745**, 67.

Gabor D., 1948, *Nature*, **161**, 777.

Gay J., D. Mekarnia, 1988, *Proc. ESO-NOAO conf.*, Ed., F. Merkle, 811.

Geiger W, 1955, *Zeitschrift Phys.*, **140**, 608.

Geiger H., Müller W., 1928, *Zeitschrift Phys.*, **29**, 389.

Gezari D. Y., Labeyrie A., Stachnik R., 1972, *Astrophys. J.*, **173**, L1.

Ghez A. M., Klein B. L., Morris M., Becklin E. E., 1998, *Astrophys. J.*, **509**, 678.

Ghez A. M., et al., 2005, *Astrophys. J.*, **635**, 1087.

Giannattasio A., Senkader S., Falster R. J., Wilshaw P. R., 2002, *J. Phys. (Condensed Matter)*, **14**, 12981.

Gibbons M. D., Wang S. C., 1983, *SPIE*, **443**, 151.

Glindemann A., 1997, *Pub. Astron. Soc. Pac.*, **109**, 68.

Glindemann A., 2011, *Principles of Stellar interferometry*, Springer, New York.

Glover A. M., 1941, *Proc. IRE*, **29**, 413.

Golay M. J. E., 1947, *Rev. Sci. Instrum.*, **18**, 357.

Goldberg A., J., 2002, *Int. J. High Speed Electron. Syst.*, **12**, 659.

Goodman J. W., 1968, *Introduction to Fourier Optics*, McGraw Hill book Co., New York.

Goodman J. W., 1985, *Statistical Optics*, Wiley & Sons Inc., New York.

Goodrich G. W., Wiley W. C., 1961, *Rev. Sci. Instrum.*, **32**, 846.

Goodrich G. W., Wiley W. C., 1962, *Rev. Sci. Instrum.*, **33**, 761.

Goss A. J., 1987, *SPIE*, **807**, 25.

Graydon O., 1997, *Opto & Laser Europe*, June issue.

Greenwood D. P., 1977, *J. Opt. Soc. Am.*, **67**, 390.

Greenwood N. N., Earnshaw A., 1997, *Chemistry of the Elements*, Oxford.

Grindlay J. E., Helmken H. F., Hanbury Brown R., Davis J., Allen L. R., 1975, *Astrophys. J.*, **197**, L9.

Grundmann M., 2010, *The Physics of the Semiconductors*, Springer, N.Y.

Gunapala S. D., Bandara S. V., 1995, *Physics of Thin Films*, Eds. M. H. Francombe and J. L. Vossen, **21**, Academic Press, New York, 113.

Gunapala S. D., et al., 1997, *IEEE Trans. Electron. Dev.*, **44**, 51.

Gunapala S. D., et al., 1998, *IEEE Trans. Electron. Dev.*, **45**, 1890.

Gunapala S. D., et al., 2001, *Infrared Phys. Technol.*, **42**, 267.

Gunapala S. D., et al., 2004, *IEEE Trans. Electron. Dev.*, **50**, 2353.

Gunapala S. D., et al., 2005, *Semicond. Sci. Technol.*, **20**, 473.

Hain R., Kahler C. J., Tropea C., 2007, *Expt. Fluids*, **42**, 403.

Hale D., Bester M., Danchi W., Fitelson W., Hoss S., Lipman E., Monnier J., Tuthill P., Townes C. H., 2000, *Astrophys. J.*, **537**, 998.

Hall E. H., 1879, *Am. J. Math.*, **2**, 287.

Hall R. N., 1952, *Phys. Rev.*, **87**, 387.

Hamblen J. O., Furnan M. D., 2000, *Rapid prototyping of digital system*, Kluwer Academic Publishers, Boston.

Hanbury Brown R., 1974, *The Intensity Interferometry: Its Applications to Astronomy*, Taylor & Francis, London.

Hanbury Brown R., Twiss R., 1956a, Nature, **177**, 27.

Hanbury Brown R., Twiss R., 1956b, Nature, **178**, 1046.

Harding W. R., Hilsum C., Northrop D. C., 1958, *Nature*, **181**, 691.

Harris K. A., Myers T. H., Yanka R. W., Mohnkern L. M., Otsuka N., 1991, *J. Vac. Sci. Technol. B*, **9**, 1752.

Hatto P., Crocker A. J., 1982, *J. Crystal Growth*, **57**, 507.

Haus J., 2010, *Optical Sensors: Basics and Applications*, Wiley-VCH Verlag GmbH & Co. KGaA.

Heering W., 1998, *Biophotons*, Eds., J. J. Chang, et al., Kluwer Academic Publishers, 143.

Heisenberg W., 1927, *Zeitschrift für Physik.*, **43**, 172.

Henini M., Razeghi M., 2002, *Handbook of Infrared Detection Technologies*, Elsevier.

Heroux L., Hinterreger H. E., 1960, *Rev. Sci. Instrum.*, **31**, 280.

Herrmann H., Kunze C., 1969, *Advances in Electronics and Electron Physics*, Academic Press, London, 28B, 955.

Herschel J., 1842, *Phil. Trans. R. Soc.*, London, **182**, 214.

Herschel W., 1800, *Phil. Trans. R. Soc., London*, **90**, 284.

Hertz H., 1887, *Ann. der Phys.*, **267**, 983.

Heyes G. J., Mill A. J., Charles M. W., 2006, *Br. J. Radiol.*, **79**, 195.

Hilbert B., 2012, *Instrument Science Report WFC3*, Space Telescope Science Institute.

Hill J. M., 2000, *SPIE*, **4004**, 36.

Hilsum C., Harding W. R., 1961, *Infrared Phys.*, **1**, 67.

Hiltner W. A., 1962, *Astronomical Techniques*, University of Chicago Press.

Hirsch L., Shkedy L., Chen D., Fishler N., Hagbi Y., Koifman A., Openhaim Y., Vaserman I., Singer M., Shtrichman I., 2012, *SPIE*, **8353**, 83530H.

Hobson G. S., 1978, *Charge-Transfer Devices*, John Wiley & Sons Inc., New York.

Hodapp K. W., et al., 1995, *SPIE*, **2475**, 8.

Hoffman A. W., Corrales E., Love P. J., Rosbeck J. P., Merrill M., Fowler A., McMurtry C., 2004, *SPIE*, **5499**, 59.

Hoffman A. W., Love P. J., Rosbeck J. P., 2004, *SPIE*, **5167**, 194.

Hofmann K.-H., Seggewiss W., Weigelt G., 1995, *Astron. Astrophys.*, **300**, 403.

Holst G. C., 1996, *CCD Arrays, Cameras, and Displays*, SPIE Opt. Eng. Press, Washington, USA.

Holst G., DeBoer J., Teves M., Veenemans C., 1934, *Physica*, **1**, 297.

Hopkins H. H., Kapany N. S., 1954, *Nature*, **173**, 39.

Hopkinson G. R., Goiffon V., Mohammedzadeh A., 2008, *IEEE Trans.*, **55**, 2197.

Horch E. P., Dinescu D. L., Girard T. M., van Altena W. F., López C. E., Franz O. G., 1996, *Astron. J.*, **111**, 1681.

Hudson R. D., 1969, *Infrared Systems Engineering*, Wiley Interscience.

Hull A. W., 1918, *Proc. Inst. Radio Electron. Eng. Austr.*, **6**, 5.

Hulst H. C., 1981, *Light Scattering by Small Particles*, Dover Publications.

Hurwitz J. E. D., Denyer P. B., Baxter D. J., 1997, *SPIE*, **3019**, 115.

Hussain M. M., Pant T. K., Vineeth C., Sumod S. G., Sridharan R., 2010, *Ann. Geophys.*, **28**, 2071.

Hutchings J. B., Postma J., Asquin D., Leahy D., 2007, *Pub. Astron. Soc. Pac.*, **119**, 1152.

Ingerson T. E., Kearney R. J., Coulter R. L., 1983, *Appl. Opt.*, **22**, 2013.

Iredale P., Hinder G., Smout D., 1969, *Advances in Electronics and Electron Physics*, Academic Press, London, **28**B, 965.

Ireland M. J., Mérand A., ten Brummelaar T. A., Tuthill P. G., Schaefer G. H., Turner N. H., Sturmann J., Sturmann L., McAlister H. A., 2008, *SPIE*, **7013**, 701324.

Ishimaru A., 1978, *Wave Propagation and Scattering in Random Media*, Academic Press, New York.

Jagutzki O., Lapington J. S., Worth L. B. C., Spillman U., Mergela V., Schmidt-Böcking H., 2002, *Nucl. Instrum. Methods Phys. Res. A*, **477**, 256.

Janesick J., 2001, *Scientific Charge Coupled Devices*, SPIE Press Monograph, **PM83**.

Janesick J., Elliot T., Fraschetti G., Collins S., Blouke M., Freeman J., 1987, *Opt. Eng.*, **26**, 692.

Javan A., Ballik E. A., Bond W. A., 1962, *J. Opt. Soc. Am.*, **52**, 96.

Jeans J. H., 1905, *Phil. Mag.*, **10**, 91.

Jelinsky P. N., et al., 2003, *SPIE*, **4854**, 233.

Jennison R. C., 1958, *Mon. Not. R. Astron. Soc.*, **118**, 276.

Jerram P., Pool P., Bell R., Burt D., Bowring S., Spencer S., Hazelwood M., Moody L., Carlett N., Heyes P., 2001, *Marconi Appl. Technol.*

Jhabvala M., et al., 2004, *SPIE*, **5167**, 175.

Jhabvala M., Choi K. K., Monroy C., La A., 2007, *IR Phys. Techn.*, **50**, 234.

Johnson H. L., 1966, *Ann. Rev. Astron. Astrophys.*, **4**, 201.

Johnson H. L., Morgan W. W., 1953, *Astrophys. J.*, **117**, 313.

Johnson J. B., 1925, *Phys. Rev.*, **26**, 71.

Kao C. K., Hockham G. A., 1966, *Proc. IEEE*, **113**, 1151.

Kaplan G. H., Hershey J. L., Hughes J. A., Hutter D. J., Johnston K. J., Mozurkewich D., Simon R. S., Colavita M. M., Shao M., Hines B. E., Staelin D. H., 1988, *Proc. ESO-NOAO conf. High Resolution Imaging Interferometry*, Ed., F. Merkle, 841.

Kapron F. P., Keck D. B., Maurer R. D., 1970, *Appl. Phys. Lett.*, **17**, 423.

Kasap S. O., 1999, *Optoelectronics*, Prentice Hall.

Kellog E., Murray S., Briel U., Bardas D., 1977, *Rev. Sci. Instrum.*, **48**, 550.

Kendrew S., et al., 2012, *SPIE*, **8446**, arXiv:1207.2945 [astro-ph.IM].

Kervella P., Coudé du Foresto V., Glindemann A., 2000, *SPIE*, **4006**, 31.

Kiledjian M. S., Schulman J. N., Wang K. L., 1991, *Phys. Rev. B*, **44**, 5616.

Kinch M. A., Yariv A., 1989, *Appl. Phys. Lett.*, **55**, 2093.

Kittel C., 1996, *Introduction to Solid State Physics*, John Wiley & Sons, New York.

Klipstein P., Mizrahi U., Fraenkel R., Shtrichman I., 2013, *Defence Sci. J.*, **63**, 555.

Kneubühl F. K., Sigrist M. W., 1999, *Laser*, Teubner, Stuttgart.

Knoll G. F., 2000, *Radiation Detection and Measurement*, Wiley

Kolmogorov A., 1941a, in *Turbulence*, Eds., S. K. Friedlander, L. Topper, 1961, Wiley-Interscience, New York, 151.

Kolmogorov A., 1941b, in *Turbulence*, Eds., S. K. Friedlander, L. Topper, 1961, Wiley-Interscience, New York, 156.

Koozekanani D., Roberts C., Katz S. E., Herderick E. D., 2000, *OCT Invest. Ophthalmol. Vis. Sci.*, **41**, 1486.

Kothare A., Mirza I., Singh K. P., Abbey A. F., 2009, *Nucl. Instrum. Methods Phys. Res. A*, **604**, 747.

Kozlowski L. J., Kosonocky W. F., 1995, Infrared detector arrays, in *Handbook of Optics*, Eds., M. Bass, et al., McGraw-Hill, New York.

Kozlowski L. J., Montroy J. T., Cabelli C. A., Cooper D. E., Chen A. C., Bostrup G. L., Bai Y., Vural K., Hodapp K., Hall D. N., 2000, *SPIE*, **4008**, 1240.

Krabbe A., Iserlohe C., Larkin J. E., Barczys M., McElwain M., Weiss J., Wright S. A., Quirrenbach A., 2006, *Astrophys. J.*, **642**, L145.

Krishna S., 2009, *J. Phys. D: Appl. Phys.*, **42**, 234005.

Krishna S., et al., 2001, *IEEE*, **95**, 1838.

Kronig R. de L., Penney W. G., 1930, *Proc. Roy. Soc.*, **A130**, 499.

Kruse P. W., McGlauchlin L. D., McQuistan R. B., 1962, *Elements of Infrared Technology*, Wiley, New York.

Kuin N. P. M., Rosen S. R., 2008, *Mon. Not. R. Astron. Soc.*, **383**, 383.

Kulas K. R., McLean I. S., Steidel C. C., 2012, *SPIE*, **8453**, 63.

Kulov S. K., et al., 2005, *SPIE*, **5834**, 203.

Kweon G., 2003, *K. Kor. Phys. Soc.*, **43**, 714.

Labeyrie A., 1970, *Astron. Astrophys.*, **6**, 85.

Labeyrie A., 1975, *Astrophys. J.*, **196**, L71.

Labeyrie A., 1996, *Astron. Astrophys. Suppl.*, **118**, 517.

Labeyrie A., 2013, *EAS Publ. Ser.*, **59**, 5.

Labeyrie A., Lipson S. G., Nisenson P., 2006, *An Introduction to Optical Stellar Interferometry*, Cambridge University Press, UK.

Labeyrie A., Schumacher G., Dugué M., Thom C., Bourlon P., Foy F., Bonneau D., Foy R., 1986, *Astron. Astrophys.*, **162**, 359.

Laforce F., 2009, *SPIE*, **7212**, 721210.

Lallemand A., 1936, *C. R. Acad. Sci. Paris*, **203**, 243.

Lallemand A., Duchesne M., 1951, *C. R. Acad. Sci. Paris*, **233**, 305.

Lampton M., Carlson C. W., 1979, *Rev. Sci. Instrum.*, **50**, 1093.

Lampton M., Paresce F., 1974, *Rev. Sci. Instrum.*, **45**, 1098.

Lampton M., Raffanti R., 1994, *Rev. Sci. Instrum.*, **65**, 3577.

Lampton M., Siegmund O., Raffanti R., 1987, *Rev. Sci. Instrum.*, **58**, 2298.

Langof L. et al. 2011, *SPIE*, **8012**, 80120F-13

Lapington J. S., 1995, *SPIE*, **2518**, 336.

Larkin J., et al., 2006, *SPIE*, **6269**, 62691A.

Laval J., 1958, *Rev. Mod. Phys.*, **30**, 222.

Lawson P. R., 1994, *App. Opt.*, **33**, 1146.

Lax M., Burstein E., 1955, *Phys. Rev.*, **97**, 39.

LeBohec S., Barbieri C., de Witt W., Dravins D., Feautrier P., Foellmi C., Glindemann A., Hall J., Holder J., Holmes R., Kervella P., Kieda1 D.,

Le Coarer E., Lipson S., Morel S., Nunez P., Ofir A., Ribak E., Saha S., Schoeller M., Zhilyaev B., Zinnecker H., 2008, *SPIE*, **7013**, 7013-2E.

LeBohec S., Holder J., 2005, *Proc. 29th Int. Cosmic Ray Conf.*, Pune, **5**, 411.

LeBouquin J.-B., et al., 2004, *Astron. Astrophys.*, **424**, 719.

LeBouquin J.-B., et al., 2006, *Astron. Astrophys.*, **450**, 1259.

Lee S. W., Hirakawa K., Shimada Y., 1999, *Appl. Phys. Lett.*, **75**, 1428.

Lemonier M., Richard J.-C., Riou D., Fouassier M., 1988, *SPIE*, **980**, 27.

Lena P., Rouan D., Lebrun F., Mignard F., Pelat D., 2012, *Observational Astrophysics*, Springer, New York.

Lenard P., 1902, *Ann. der Physik*, **313**, 149.

Leo W. R., 1994, *Techniques for Nuclear and Particle Physics Experiments*, Springer-Verlag, Berlin.

Levenduski R., Scarlotti R., 1996, *Cryogenics*, **36**, 859.

Levine B. F., 1993, *J. Appl. Phys.*, **74**, R1.

Levine B. F., Choi K. K., Bethea C. G., Walker J., Malik R. J., 1987, *Appl. Phys. Lett.*, **50**, 1092.

Levine B. F., et al. 1988, *Electron. Lett.*, **24**, 747.

Lewis G. N., 1926, *Nature*, **118**, 874.

Liang J., Williams D. R., Millar D. T., 1997, *J. Opt. Soc. Am. A.*, **14**, 2884.

Litwiller D., 2001, *Photonics Spectra*, **1**, 154.

Liu H. C., Luo H., Song C. Y., Wasilewski Z. R., Thorpe A. J. S., Cao J. C., 2007, *Infrared Phys. Techn.*, **50**, 191.

Lloyd J. M., 1975, *Thermal Imaging Systems*, Plenum Press, New York.

Locher G. L., 1932, *Phys. Rev.*, **42**, 525.

Loferski J. J., 1963, *IEEE*, **51**, 667.

Loferski J. J., Wysocki J. J., 1961, *RCA Rev.*, **22**, 38.

Logeeswaran V. J., et al., 2011, *IEEE*, **17**, 1002.

Lohmann A. W., Weigelt G. P., Wirnitzer B., 1983, *Appl. Opt.*, **22**, 4028.

Loose M., Beletic J., Garnett J., Xu M., 2007, *SPIE*, **6690**, 6690DDC-1.

Low F. J., 1961, *J. Opt. Soc. Am.*, **51**, 1300.

Lowrance J., Renda G., Zucchino P., 1976, *Advances in Electronics and Electron Physics*, Academic Press, London, 40B, 711.

Lynds C., Worden S., Harvey J., 1976, *Astrophys. J.*, **207**, 174.

Mackay C., 2005, ING Scientific Detectors for Astronomy 2005, *Astrophys. Space Sci. Lib.*, **336**, 93.

Maiman T., 1960, *Nature*, **187**, 493.

Malbet F., et al., 2011, *Expt. Astron.*, September, 109.

Malbet F., et al., 2013, *SPIE*, **8864**.

Mandel L., 1963, *Progress in Opt.*, **II**, 181.

Mandel L., Wolf E., 1995, *Optical Coherence and Quantum Optics*, Cambridge University Press, Cambridge.

Manley B., Guest A. J., Holmshaw R., 1969, *Advances in Electronics and Electron Physics*, Academic Press, London, **28**A, 471.

Marcuse D., 1980, *Principles of Quantum Electronics*, Academic, New York.

Martin C., Jelinsky P., Lampton M., Malina R. F., Anger H. O., 1981, *Rev. Sci. Instrum.*, **52**, 1067.

Martin D. C., et al., 2005, *Astrophys. J.*, **619**, L1.

Martin-Hernandez N. L., Schaerer D., Saugave M., 2004, arXiv: astro-ph/0408582.

Martyniuk P., Rogalski A., 2008, *Prog. Quant. Electron.*, **32**, 89.

Massey P., 1997, *A User's Guide to CCD Reductions with IRAF, Instruction Manual for IRAF Software Package.*

Maxfield C., 2004, *The Design Warrior's Guide to FPGAs: Devices, Tools and Flows*, Elsevier.

Maxwell J. C., 1865, *Phil. Trans. R. Soc. London*, **155**, 459.

McAlister H., Bagnuolo W., ten Brummelaar T., Hartkopf W. I., Shure M., Sturmann L., Turner N., Ridgway S., 1998, *SPIE*, **3350**, 947.

McElawin M. W., et al., 2007, *Astrophys. J.*, **656**, 505.

McLean I. S., 2008, *Electronic Imaging in Astronomy: Detectors and Instrumentation*, Springer, New York.

McLean I. S., et al., 1998, *SPIE*, **3354**, 566.

McLean I. S., et al., 2010, *SPIE*, **7735**, 77351E.

McLean I. S., et al., 2012, *SPIE*, **8846**, 84460J.

Meinel A. B., 1950, *Astrophys. J.*, **111**, 555.

Menn N., 2004, *Practical Optics*, Elsevier Inc.

Michelson A. A., Morley E., 1887, *Am. J. Sci.*, **34**, 333.

Michelson A. A., Pease, F. G., 1921, *Astrophys. J.*, **53**, 249.

Millikan R., 1914, *Phys. Rev.*, **4**, 73.

Millikan R., 1916, *Phys. Rev.*, **7**, 355.

Millman J., Halkias C. C., 1972, McGraw-Hill, New York.

Millour, F., Vannier, M., Petrov, R. G., et al. 2006, in *EAS Publ. Ser.*, **22**, 379.

Mimura T., Hiyamizu S., Fuzi T., Nanbu K., 1980, *J. Appl. Phys. (Jpn)*, **19**, L225.

Monet D. G., 1988, *Ann. Rev. Astron. Astrophys.*, **26**, 413.

Monnier J. D., et al., 2006, *SPIE*, **62681P**.

Morel S., 2004, *SPIE*, **5501**.

Morel S., Koechlin L., 1998, *Astron. Astrophys. Suppl. Ser.*, **130**, 395.

Morel S., Saha S. K., 2005, *21st Century Astrophysics*, Eds., S. K. Saha, V. K. Rastogi, Anita Publications, New Delhi, 237.

Morrissey P., et al., 2005, *Astrophys. J.*, **619**, L7.

Morrissey P., et al., 2007, *Astrophys. J. Suppl.*, **173**, 682.

Mott N. F., 1939, *Proc. Roy. Soc.*, **A171**, 27.

Mourard D., 2014, Private communication.

Mourard D., Bosc I., Labeyrie A., Koechlin L., Saha S., 1989, *Nature*, **342**, 520.

Mourard D., Tallon-Bosc I., Blazit A., Bonneau D., Merlin G., Morand F., Vakili F., Labeyrie A., 1994, *Astron. Astrophys.*, **283**, 705.

Mourard D., et al., 2006, *SPIE*, **6268**, 62683R.

Mourard D., et al., 2009, *Astron. Astrophys.*, **508**, 1073.

Nag B. R., 2002, *Physics of Quantum Well Devices*, Kluwer Academic Publishers, New York.

Naidu B. N., Srinivasan R., 1997, *IETE Techn. Rev.*, **14**, 331.

Naidu B. N., Srinivasan R., 1998, *Bull. Astron. Soc. Ind.*, **26**, 387.

Naidu B. N., Srinivasan R., 2001, *IETE Electro/Information Techn. Conf.*, Chicago, 7.

NASA IPAC, 2007.

Nevshupa R., et al., 2008, *J. Phys., (Conf. Ser.)*, **100**, 042012.

Ng K. K., 2009, Negative-electron-affinity photocathode, in *Complete Guide to Semiconductor Devices*, John Wiley & Sons Inc, New Jersey, USA.

Nisenson P., 1988, *Proc. NATO-ASI, 'Diffraction Limited Imaging with Very Large Telescopes'*, Eds., D. M. Alloin, J.-M. Mariotti, Cargése, France, 157.

Nisenson P., 1997, Private communication.

Nishizawa J., Suto K., 2004, Terahertz wave generation and light amplification using Raman effect, in *Physics of Semiconductor Devices*, Eds., K. N. Bhat and A. DasGupta, Narosa Publishing House, New Delhi, 27.

Norton P., 2002, *Opto-electron. Rev.*, **10**, 159.

Norton P., 2003, in *Encyclopedia of Optical Engineering*, Ed., R. Driggers, 320, Marcel Dekker Inc., New York.

Nunez P. D., Labeyrie A., Riaud P., 2014, *Mon. Not. R. Astron. Soc.*, **439**, 1787.

Nyquist H., 1928, *Phys. Rev.*, **32**, 110.

Ohta N., Robertson A. R., 2005, *Colorimetry Fundamentals and Applications*, Wiley, Chichester, England.

Ojha D. K., Ghosh S. K., D'Costa S. L. A., Naik M. B., Sandimani P. R., Poojary S. S., Bhagat S. B., Jadhav R. B., Meshram G. S., Bakalkar C. B., 2012, *Bull. Astron. Soc. Ind. (ASI Conf. Ser.)*, **4**, 191.

Ong D. G., Pierret R. F., 1975, *IEEE Trans.*, **22**, 593.

Osterbart R., Balega Y., Weigelt G., Langer N., 1996, *Proc. IAU symp. 180*, Eds., H. Habing, Lamers, 362.

Otte A. N., et al., 2007, *Proc. 30th ICRC, OG.2.4*, Merida, Mexico.

Pal B. P. (Ed.), 1992, *Fundamentals of Fiber Optics in Telecommunication and Sensor Systems*, John Wiley, New York.

Pal S., 2014, Private communication.

Pal S., Bhattacharjee T., Saha S. K., 2009, *Emerging Trends in Electronic & Photonic Devices & Systems*, Eds., P. Chakrabarti, et al., MacMillan Publishers Ind. Ltd., New Delhi, 282.

Palmer III A. G., Thompson N. L., 1987, *J. Biological Soc.*, **52**, 257.

Papaliolios C., Mertz L., 1982, *SPIE*, **331**, 360.

Papaliolios C., Nisenson P., Ebstein S., 1985, *Appl. Opt.*, **24**, 287.

Paresce F., Clampin M., Cox C., Crocker J., Rafal M., Sen A., Hiltner W. A., 1988, *Proc., 'Instrumentation for Ground Based Optical Astronomy: Present and Future'*, Ed., L. Robinson, Springer Verlag, New York, 542.

Park J. S., Karunasiri R. P. J., Wang K. L., 1990, *J. Vac. Sci. Technol. B*, **8**, 217.

Paul H., 1989, *The Art of Electronics*, Cambridge University Press, UK.

Peacock T., Verhoeve P., Rando N., van Dordrecht A., Taylor B. G., Erd C., Perryman M. A. C., Venn R., Howlett J., Goldie D. J., Lumley J., Wallis M., 1996, *Nature*, **381**, 135.

Pedretti E., Millan-Gabet R., Monnier J. D., Traub W. A., Carleton N. P., Berger J.-P., Lacasse M. G., Schloerb F. P., Brewer M. K., 2004, *Pub. Astron. Soc. Pac.*, **116**, 377.

Pedretti E., Monnier J. D., ten Brummelaar T., Thureau N. D., 2009, *New Astron. Rev.*, **53**, 353.

PerkinElmer, 2003, *TPS 534 Thermopile Detector*, Product Datasheet.

Perryman M. A. C., Foden C. L., Peacock A., 1993, *Nucl. Instrum. Methods Phys. Res. A.*, **325**, 319.

Petrov R., et al., 2005, in *The Power of Optical/IR Interferometry: Recent Scientific Results & 2nd Generation VLTI Instrumentation*, Eds., F. Delplancke, A. Richichi.

Petrov R., et al., 2007, *Astron. Astrophys.*, **464**, 1.

Philips T. G., Woody D. P., 1982, *Ann. Rev. Astron Astrophys.*, **20**, 285.

Pickering E. C., 1910, *Harvard Coll. Obs. Circ.*, **155**, 1.

Pillans L., Ash R. M., Hipwood L., Knowles P., 2012, *SPIE*, **8353**, 83532W.

Planck M., 1900, *Ann. d. Physik*, **1**, 99.

Planck M., 1901, *Ann. d. Physik*, **4**, 553.

Pogson N. R., 1857, *Mon. Not. R. Astron. Soc.*, **17**, 12.

Popowicz A., 2011, *Przeglad Elektrotechniczny* (Electrical Review), **87**, 260.

Popowicz A., Kurek A. R., Filus Z., 2013, *Pub. Astron. Soc. Pac.*, **125**, 1119.

Postma J., Hutchings J. B., Leahy D., 2011, *Pub. Astron. Soc. Pac.*, **123**, 833.

Prabhakar R. R., Mathews N., Jinesh K. B., Karthik K. R. G., Pramana S. S., Varghese B., Sow C. H., Mhaisalkar S., 2012, *J. Mater. Chem.*, **22**, 9678.

Pratt T., 1947, *J. Sci. Instrum.*, **24**, 312.

Pruksch M., Fleischmann F., 2000, *Appl. Opt.*, **39**, 2443.

Putley E. H., 1980, in *Optical and Infrared Detectors*, Ed., R. J. Keyes, Springer-Verlag.

Raffanti R., Lampton M., 1993, *Rev. Sci. Instrum.*, **64**, 1506.

Ragazzoni R., 1996, *J. Mod. Opt.*, **43**, 289.

Rando N., Peacock T., Favata, Perryman M. A. C., 2000, *Expt. Astron.*, **10**, 499.

Rao A. R., Naik S., Patil M., Malkar J. P., Kalyan Kumar R. P. S., 2010, *Nucl. Instrum. Methods Phys. Res. A*, **616**, 55.

Rao N. K., et al. 2004, *Asian J. Phys.*, **13**, 367.

Rappaport R., 1959, *RCA Rev.*, **21**. 373.

Rayleigh J. W. S., 1900, *Phil. Mag.*, **49**, 539.

Rayleigh J. W. S., 1905, *Nature*, **72**, 54.

Razeghi M., et al., 2003, *Microelectron. J.*, **34**, 5.

Rees D., McWhirter I., Rounce P. A., Barlow F. E., 1981, *J. Phys. E*, **14**, 229.

Rees D., McWhirter I., Rounce P. A., Barlow F. E., Kellock S. J., 1980, *J. Phys. E*, **13**, 763.

Rieke G. H., 2003, *Detection of Light from the Ultraviolet to the Submillimeter*, Cambridge University Press, UK.

Rieke G. H., 2007, *Ann. Rev. Astron. Astrophys.*, **45**, 77.

Rindi A., Perez-Mendez V., Wallace R. I., 1970, *Nucl. Instrum. Methods*, **77**, 325.

Ring E. F. J., 2000, *Imaging Sci. J.*, **48**, 1.

Ring E. F. J., 2004, *Rheumatology*, **43**, 800.

Roberts C. G., 1983, *SPIE*, **443**, 131.

Robertson J. G., Ireland M. J., Tango W. J., Davis J., Tuthill P. G., Jacob A. P., Kok Y., ten Brummelaar T. A., 2010, *SPIE*, **7734**, 773405.

Roddier F., 1981, *Progress in Optics*, **XIX**, 281.

Roddier F., 1999, in *Adaptive Optics in Astronomy*, Ed., F. Roddier, Cambridge University Press, UK.

Rodriguez L. F., Sosa N., Rosa F., Fuensalida J. J., 1992, *Proc. ESO-NOAO Conf. 'High Resolution Imaging Interferometry'*, Eds., J. M. Beckers, F. Merkle, 621.

Rogalski A., 2003, *Prog. Quant. Electron.*, **27**, 59.

Rogalski A., 2007, *Infrared Phys. Technol.*, **50**, 240.

Rogalski A., 2011, *Infrared Detectors*, CRC Press, Boca Raton, FL.

Rogalski A., 2012, *Opto-Electron. Rev.*, **20**, 279.

Rogalski A., Antoszewski J., Faraone L., 2009, *J. Appl. Phys.*, **105**, Article ID: 91101.

Rothman A. J., Bickel P. J., Levina E., Zhu J., 2008, *Electron. J. Statistics*, **2**, 494.

Rouaux E., Richard J.-C., Piaget C., 1985, *Advances in Electronics and Electron Physics*, Academic Press, London, **64**A, 71.

Rousset G., 1999, in *Adaptive Optics in Astronomy*, Ed., F. Roddier, Cambridge University Press, 91.

Rutherford E., Geiger H., 1908, *Proc. Roy. Soc. London A*, **81**, 141.

Saha S. K., 1999, *Bull. Astron. Soc. Ind.*, **27**, 443.

Saha S. K., 2001, *Ind. J. Phys.*, **75**B, 285.

Saha S. K., 2002, *Rev. Mod. Phys.*, **74**, 551.

Saha S. K., 2007, *Diffraction-limited Imaging with Large and Moderate Telescopes*, World-Scientific, New Jersey.

Saha S. K., 2009, Detectors for the astronomical applications, in *International Conference on 'Emerging Trends in Electronic and Photonic Devices & Systems'*, Eds., P. Chakrabarti, et al., MacMillan Publishers Ind. Ltd., New Delhi, 452.

Saha S. K., 2010, *Aperture Synthesis: Methods and Applications to Optical Astronomy*, Springer, New York.

Saha S. K., Chinnappan V., 2002, *Bull. Astron. Soc. Ind.*, **30**, 819.

Saha S. K., Jayarajan A. P., Rangarajan K. E., Chatterjee S., 1988, *Proc. 'High Resolution Imaging Interferometry'*, Ed., F. Merkle, 661.

Saha S. K., Maitra D., 2001, *Ind. J. Phys.*, **75B**, 391.

Saha S. K., Morel S., 2000, *Bull. Astron. Soc. Ind.*, **28**, 175.

Saha S. K., Rajamohan, Rao P. V., Som Sundar G., Swaminathan R., Lokanadham B., 1997, *Bull. Astron. Soc. Ind.*, **25**, 563.

Saha S. K., Sudheendra, G., Umesh Chandra, A., Chinnappan, V., 1999, *Exp. Astr.*, **9**, 39.

Saha S. K., Venkatakrishnan P., 1997, *Bull. Astron. Soc. Ind.*, **25**, 329.

Saha S. K., Venkatakrishnan P., Jayarajan A. P., Jayavel N., 1987, *Curr. Sci.*, **56**, 985.

Saha S. K., Yeswanth L., 2004, *Asian J. Phys.*, **13**, 227.

Sahnow D. J., 2003, *SPIE*, **4854**, 610.

Saleh B. E. A., Teich M. C., 1991, *Fundamentals of Photonics*, John Wiley & Sons. Inc., New York.

Sandel B. R., et al., 1977, *Appl. Opt.*, **16**, 1435.

Schartmann M., Burkert A., Alig C., et al., 2012, *Astrophys. J.*, **755**, 155.

Scheer J. J., van Laar J., 1965, *Solid State Commun.*, **3**, 189.

Schneider H., Liu H. C., 2007, *Quantum Well Infrared Photodetectors*, Springer, Berlin.

Schottky W., 1918, *Ann. Phys.*, **57**, 541.

Schottky W., 1942, *Z. Physik*, **118**, 539.

Schroder D. K., 2006, *Semiconductor Device and Material Characterization*, Wiley-Interscience Publication, New Jersey.

Schrödinger E., 1926, *Phys. Rev.*, **28**, 1049.

Schwabel F., 2008, *Advanced Quantum Mechanics*, Springer.

Scribner D. A., Kruer M. R., Killiany J. M., 1991, *IEEE*, **79**, 66.

Seebeck T. J., 1822, *Abh. Deutsch. Akad. Wiss. Berlin*, 265.

Seeger K., 2004, *Semiconductor Physics*, Springer, Berlin.

Seetha S., et al., 2006, *Adv. Space Res.*, **38**, 2995.

Shao M., Colavita M. M., Hines B. E., Staelin D. H., Hutter D. J., 1988, *Astron. Astrophys.*, **193**, 357.

Shkedy L., Markovitz T., Calahorra Z., Hirsh I., Shtrichman I., 2011, *Opt. Eng.*, **50**, 061008.

Shtengel G., Galbrath J. A., Galbrath C. G., Scwartz J. L., Gillette, Manley S., Sougrat S., Waterman C. M., Kanchanawong P., Davidson M. W., Fetter R. D., Hess H. F., 2009, *Proc. Nat. Acad. Sci.*, **106**, 3125.

Shockley W., Read W. T., 1952, *Phys. Rev.*, **87**, 835.

Siegel R., Howell J. R., 1981, *Thermal Radiation Heat Transfer*, Hemisphere Publishing, Washington DC.

Siegman A. E., Harris S. E., McMurtry B. J., 1963, in *Optical Masers*, Ed., J. Fox, Wiley-Interscience, New York, 511.

Siegmund O. H. W., Clossier S., Thornton J., Lemen J., Harper R., Mason I. M., Culhane J. L., 1983, *IEEE Trans. Nucl. Sci.*, NS-30(1), 503.

Siegmund O. H. W., et al., 1994, *SPIE*, **2280**, 89.

Sigg D., 2006, *Class. Quant. Grav.*, **23**, S51.

Sigrist M. W., 1994, *Air Monitoring by Spectroscopic Techniques*, John Wiley & Sons, New York.

Sihota R., Tandon R., 2004, *Parsons' Diseases of the Eye*, Elsevier.

Silfvast W. T., 2004, *Laser Fundamentals*, Cambridge University Press, UK.

Sinclair A. G., Kasevich M. A., 1997, *Rev. Sci. Instrum.*, **68**, 1657.

Smith D. L., Mailhiot C., 1987, *J. Appl. Phys.*, **62**, 2545.

Smith N., 2006, *Astrophys. J.*, **644**, 1151.

Smith R. A., Jones F. E., Chasmar R. P., 1968, *The Detection and Measurement of Infrared Radiation*, Oxford University Press, UK.

Smith W., 2007, *Modern Optical Engineering*, McGraw-Hill.

Snitzer E., 1961, *Phys. Rev. Lett.*, **7**, 444.

Sobottka S., Williams M., 1988, *IEEE Trans. Nucl. Sci.*, **35**, 348.

Srinivasan R., Anupama K., Saha S. K., Aditya R., 2008, *FPGA Based Design for CCD Camera Controller*, Presented at the 2nd National Workshop on Advanced Optoelectronic Materials & Devices.

Srinivasan R., Anupama K., Suneeta, Saha S. K., Aditya R., 2009, In *Emerging Trends in Electronic and Photonic Devices and Systems*, Eds., P. Chakrabarti, et al., MacMillan Publishers Ind. Ltd., New Delhi, 173.

Stefan J., 1879, *Sitzungsberichte der Mathematisch-Naturwissenschaftlichen Classe der kaiserlichen Akad. der Wissenschaften*, **79**, 391.

Stetson S. B., Reynolds D. B., Stapelbroek M. G., Stermer R. L., 1986, *SPIE*, **686**, 48.

Steward E. G., 1983, *Fourier Optics and Introduction*, John Wiley & Sons Inc., New York.

Stipčević M., Skenderović H., Gracin D., 2010, *Opt. Express*, **18**, 17448.

Streetman B. G., Banerjee S., 2000, *Solid State Electronic Devices*, Prentice Hall, New Jersey.

Strehl K., 1902, *Zeitschrift für Instrumentenkunde*, **22**, 213.

Strömgren B., 1956, *Vistas Astron.*, **2**, 1336.

Surya A., Saha S. K., 2014, *J. Opt.*, Springer.

Surya A., Saha S. K., Labeyrie A., 2014, *Mon. Not. R. Astron. Soc.*, **443**, 852.

Sze S. M., Ng K. K., 2007, *Physics of Semiconductor*, John Wiley & Sons, New Jersey.

Talmi Y., 1994, Intensified array detectors, in *Charge Transfer Devices in Spectroscopy*, Eds., J. V. Sweedler, et al., VCH Publishers, New York.

Tatarski V. I., 1961, *Wave Propagation in Turbulent Medium*, Dover, New York.

Tatulli E., et al., 2007, *Astron. Astrophys.*, **464**, 29.

Taylor R. E., 2005, Radioactive dating, in *Handbook of Archaelogical Sciences*, Eds., D. R. Brothwell and A.M. Pollard, John Wiley & Sons, New York, 21.

Teich M. C., 1968, *IEEE*, **56**, 37.

Teich M. C., 1977, Nonlinear heterodyne detection, in *Optical and Infrared Detectors*, Ed., R. J. Keyes, Springer-Verlag, Berlin.

Teich M. C., Keyes R. J., Kingston R. H., 1966, *Appl. Phys. Lett.*, **9**, 357.

Teich M. C., Wolga G. J., 1968, *Phys. Rev.*, **171**, 809.

Tewari S. P., 2014, Private communication.

Thiébaut E., Abe L., Blazit A., Dubois J.-P., Foy R., Tallon M., Vakili F., 2003, *SPIE*, **4841**, 1527.

Thilker D., et al. 2009, *Nature*, **457**, 990.

Thomson J. J., 1897, *The Electrician*, **39**, 104.

Timothy J. G., 1983, *Publ. Astron. Soc. Pac.*, **95**, 810.

Timothy J. G., 1985, *Opt. Eng.*, **24**, 1066.

Timothy J. G., 1986, *SPIE*, **687**, 109.

Timothy J. G., Bybee R. L., 1975, *Rev. Sci. Instrum.*, **46**, 1615.

Tissot J. L., Robert P., Durand A., Tinnes S., Bercier E., Crastes A., 2013, *Defence Sci. J.*, **63**, 545.

Tohyama S., Miyoshi M., Kurashina S., Ito N., Sasaki T., Ajisawa A., Tanaka Y., Kawahara A., Iida K., Oda N., 2006, *Opt. Eng.*, **45**, 014001.

Townes C. H., Bester M., Danchi W., Hale D., Monnier J., Lipman E., Everett A., Tuthill P., Johnson M., Walters D., 1998, *SPIE*, **3350**, 908.

Townes C. H., Schawlow A., 1958, *Phys. Rev.*, **112**, 1940.

Trakalo M., Webb P. P., Poirier P., McIntyre R. J., 1987, *Appl. Opt.*, **26**, 3594.

Tremsin A. S., Pearson J. F., Lees J. E., Fraser G. W., 1996, *Nucl. Instrum. Methods Phys. Res. A.*, **368**, 719.

Tulloch S., 2005, ING scientific detectors for astronomy 2005, *Astrophys. Space Sci. Lib.*, **336**, 303.

Tyndall J., 1870, *Total Reflexion, Notes about Light.*

Tyson R. K., 2000, 'Introduction' in *Adaptive Optics Engineering Handbook*, Ed., R. K. Tyson, Dekkar, New York, 1.

Van der Ziel A., 1968, *Solid State Physical Electronics*, Prentice-Hall INC, N.J.

van Vliet C., 1999, in *Noise in Physical Systems and 1/f Fluctuations*, Ed., C. Surya, Bentham Press, London, 3.

Vineeth C., Pant T. K., Devasia C. V., Sridharan R., 2007, *Geophys. Res. Lett.*, **34**, L14101.

Vineeth C., Pant T. K., Maria A., Geetha R., Devasia C. V., 2005, *Geophys. Res. Lett.*, **32**, L19101.

Vuillermet M., 2013, *Defence Sci. J.*, **63**, 550.

Walsh J. W. T., 1958, *Photometry*, Dover, New York.

Wang S. Y., Lin S. D., Wu W., Lee C. P., 2001, *Appl. Phys. Lett.*, **78**, 1023.

Weigelt G. P., et al., 2007, *Astron. Astrophys.*, **464**, 87.

Weigelt G. P., et al., 2012, *Astron. Astrophys.*, **541**, L9.

Weik M. H., 1997, *Fiber Optics Standard Dictionary*, Chapman & Hall, New York.

Widenhorn R., Dunlap J. C., Bodegom E., 2010, *IEEE Trans.*, **57**, 581.

Wien W. C., 1898, *Ann. der Phys.*, **301**, 1.

Willett C. S., 1974, *Introduction to Gas Lasers*, Oxford, Pergamon.

Wizinowich P., et al., 2000, *Pub. Astron. Soc. Pac.*, **112**, 315.

Wolf S., 2003, *Comput. Phys. Comm.*, **150**, 99.

Wolfe W. L., 1998, *Introduction to Radiometry*, SPIE Optical Press, Washington.

Worden S. P., Lynds C. R., Harvey J. W., 1976, *J. Opt. Soc. Am.*, **66**, 1243.

Wyszecki G., Stiles W. S., 1982, *Color Science: Concepts and Methods, Quantitative Data and Formulae*, Wiley, New York.

Yacobi B. G., 2003, *Semiconductor Materials: An Introduction to Basic Principles*, Springer, New York.

Yariv A., 1982, *An Introduction to Theory and Applications of Quantum Electronics*, John Wiley.

Young T., 1802, *Phil. Trans. R. Soc. Lond.*, **92**, 12.

Zhang H., et al., 2010, *Mon. Not. R. Astron. Soc.*, **401**, 1839.

Zworykin V. K., 1936, *L'Onde Électrique*, **15**, 265.

Index